Evolvability

Vienna Series in Theoretical Biology

Gerd B. Müller, editor-in-chief

Thomas Pradeu and Katrin Schäfer, associate editors

The Evolution of Cognition, edited by Cecilia Heyes and Ludwig Huber, 2000

Origination of Organismal Form, edited by Gerd B. Müller and Stuart A. Newman, 2003

Environment, Development, and Evolution, edited by Brian K. Hall, Roy D. Pearson, and Gerd B. Müller, 2004

Evolution of Communication Systems, edited by D. Kimbrough Oller and Ulrike Griebel, 2004

Modularity: Understanding the Development and Evolution of Natural Complex Systems, edited by Werner Callebaut and Diego Rasskin-Gutman, 2005

Compositional Evolution: The Impact of Sex, Symbiosis, and Modularity on the Gradualist Framework of Evolution, by Richard A. Watson, 2006

Biological Emergences: Evolution by Natural Experiment, by Robert G. B. Reid, 2007

Modeling Biology: Structure, Behaviors, Evolution, edited by Manfred D. Laubichler and Gerd B. Müller, 2007

Evolution of Communicative Flexibility, edited by Kimbrough D. Oller and Ulrike Griebel, 2008

Functions in Biological and Artificial Worlds, edited by Ulrich Krohs and Peter Kroes, 2009

Cognitive Biology, edited by Luca Tommasi, Mary A. Peterson and Lynn Nadel, 2009

Innovation in Cultural Systems, edited by Michael J. O'Brien and Stephen J. Shennan, 2009

The Major Transitions in Evolution Revisited, edited by Brett Calcott and Kim Sterelny, 2011

Transformations of Lamarckism, edited by Snait B. Gissis and Eva Jablonka, 2011

Convergent Evolution: Limited Forms Most Beautiful, by George McGhee, 2011

From Groups to Individuals, edited by Frédéric Bouchard and Philippe Huneman, 2013

Developing Scaffolds in Evolution, Culture, and Cognition, edited by Linnda R. Caporael, James Griesemer, and William C. Wimsatt, 2013

Multicellularity: Origins and Evolution, edited by Karl J. Niklas and Stuart A. Newman, 2016

Vivarium: Experimental, Quantitative, and Theoretical Biology at Vienna's Biologische Versuchsanstalt, edited by Gerd B. Müller, 2017

Landscapes of Collectivity in the Life Sciences, edited by Snait B. Gissis, Ehud Lamm, and Ayelet Shavit, 2017

Rethinking Human Evolution, edited by Jeffrey H. Schwartz, 2018

Convergent Evolution in Stone-Tool Technology, edited by Michael J. O'Brien, Briggs Buchanan, and Metin I. Erin, 2018

Evolutionary Causation: Biological and Philosophical Reflections, edited by Tobias Uller and Kevin N. Lala, 2019

Convergent Evolution on Earth: Lessons for the Search for Extraterrestrial Life, by George McGhee, 2019

Contingency and Convergence: Toward a Cosmic Biology of Body and Mind, by Russell Powell, 2020

How Molecular Forces and Rotating Planets Create Life, Jan Spitzer, 2021

Rethinking Cancer: A New Understanding for the Post-Genomics Era, edited by Bernhard Strauss, Marta Bertolaso, Ingemar Ernberg, and Mina J. Bissell, 2021

Levels of Organization in the Biological Sciences, edited by Daniel S. Brooks, James DiFrisco, and William C. Wimsatt, 2021

The Convergent Evolution of Agriculture in Humans and Insects, edited by Ted R. Schultz, Richard Gawne, and Peter N. Peregrine, 2022

Evolvability: A Unifying Concept in Evolutionary Biology?, edited by Thomas F. Hansen, David Houle, Mihaela Pavličev, and Christophe Pélabon, 2023

Evolvability

A Unifying Concept in Evolutionary Biology?

Edited by Thomas F. Hansen, David Houle, Mihaela Pavličev, and Christophe Pélabon

The MIT Press
Cambridge, Massachusetts
London, England

The MIT Press would like to thank the anonymous peer reviewers who provided comments on drafts of this book. The generous work of academic experts is essential for establishing the authority and quality of our publications. We acknowledge with gratitude the contributions of these otherwise uncredited readers.

This book was set in Times New Roman by Westchester Publishing Services. Printed and bound in the United States of America.

Library of Congress Cataloging-in-Publication Data

Names: Hansen, Thomas F., editor.
Title: Evolvability : a unifying concept in evolutionary biology? / edited by Thomas F. Hansen, David Houle, Mihaela Pavličev, and Christophe Pélabon.
Description: Cambridge, Massachusetts : The MIT Press, [2023] | Series: Vienna series in theoretical biology | Includes bibliographical references and index.
Identifiers: LCCN 2022038288 (print) | LCCN 2022038289 (ebook) | ISBN 9780262545624 (paperback) | ISBN 9780262374705 (epub) | ISBN 9780262374699 (pdf)
Subjects: LCSH: Evolution (Biology)—Philosophy.
Classification: LCC QH360.5 .E99 2023 (print) | LCC QH360.5 (ebook) | DDC 576.801—dc23/eng/20220920
LC record available at https://lccn.loc.gov/2022038288
LC ebook record available at https://lccn.loc.gov/2022038289

Contents

Series Foreword

Biology is a leading science in this century. As in all other sciences, progress in biology depends on the interrelations between empirical research, theory building, modeling, and societal context. Whereas molecular and experimental biology have evolved dramatically in recent years, generating a flood of highly detailed data, the integration of these results into useful theoretical frameworks has lagged behind. Driven largely by pragmatic and technical considerations, research in biology continues to be less guided by theory than seems indicated. By promoting the formulation and discussion of new theoretical concepts in the biosciences, this series intends to help fill important gaps in our understanding of some of the major open questions of biology, such as the origin and organization of organismal form, the relationship between development and evolution, and the biological bases of cognition and mind. Theoretical biology has important roots in the experimental tradition of early-twentieth-century Vienna. Paul Weiss and Ludwig von Bertalanffy were among the first to use the term *theoretical biology* in its modern sense. In their understanding the subject was not limited to mathematical formalization, as is often the case today, but extended to the conceptual foundations of biology. It is this commitment to a comprehensive and cross-disciplinary integration of theoretical concepts that the Vienna Series intends to emphasize. Today, theoretical biology has genetic, developmental, and evolutionary components, the central connective themes in modern biology, but it also includes relevant aspects of computational or systems biology and extends to the naturalistic philosophy of sciences. The Vienna Series grew out of theory-oriented workshops organized by the KLI, an international institute for the advanced study of natural complex systems. The KLI fosters research projects, workshops, book projects, and the journal *Biological Theory*, all devoted to aspects of theoretical biology, with an emphasis on—but not restriction to—integrating the developmental, evolutionary, and cognitive sciences. The series editors welcome suggestions for book projects in these domains.

Gerd B. Müller, Thomas Pradeu, Katrin Schäfer

1 Introduction: Evolvability

Thomas F. Hansen, Christophe Pélabon,
David Houle, and Mihaela Pavličev

The concept of evolvability emerged in the 1990s in association with new research fields with ambitions to extend or challenge mainstream evolutionary theory. This chapter briefly outlines debates and controversies surrounding the evolvability concept, and explains the motivation for this edited volume on the subject.

1.1 Motivation for This Book

In the early 1990s a novel concept emerged in evolutionary biology that soon became fashionable as a banner for a range of new research fields that challenged or expanded on neo-Darwinian orthodoxy. This concept was *evolvability,* the ability to evolve. Despite a variety of definitions and uses, the common denominator was a focus on the dispositions or preconditions for evolution by natural selection to occur. Many uses of the concept were associated with a structural, as opposed to a functional, perspective on evolution (e.g., Gould 2002; Amundson 2005; Wagner 2014), and it found popularity among those who sought to expand evolutionary theory in new directions and among critics of mainstream neo-Darwinism. The term also found use among more conventional evolutionary biologists as a label for new ways of measuring genetic and mutational variation (e.g., Houle 1992) and among those that wanted a less controversial term for constraints on evolution (see Brigandt 2015). Consequently, evolvability is, or is becoming, part of the theoretical foundation of fields as diverse as evolutionary developmental biology (evo-devo), evolutionary quantitative genetics (EQG), paleobiology, and artificial life. It also plays an increasing role in studies of macroevolution, in systems biology, and in digital and experimental evolution.

After 30 years of expanding research under the evolvability banner, it is time to take stock of what we have learned and to assess the influence and novelty of the concept itself. There is no shortage of reviews, perspectives, and assessments of evolvability (e.g., Alberch 1991; Wagner and Altenberg 1996; Love 2003; Nehaniv 2003; Hansen and Houle 2004; Schlichting and Murren 2004; Hansen 2006, 2016; Sniegowski and Murphy 2006; Hendrikse et al. 2007; Sterelny 2007, 2011; Pigliucci 2008; Brookfield 2009; Wagner 2010; Pavličev and Wagner 2012; Brown 2014; Kopp and Matuszewki 2014; Brigandt 2015; Vasas et al. 2015; Minelli 2017; Payne and Wagner 2019; Hansen and Pélabon 2021; Porto 2021; Watson 2021; Love et al. 2022; Nuño de la Rosa and Villegas 2022; Riederer et al. 2022), and the concept, if not always the term, has been instrumental in some influential books with structural perspectives on evolution (Kauffman 1993; Dennett 1995; Maynard Smith and Szathmáry 1995; Dawkins 1996; Raff 1996; Gerhart and Kirschner 1997; Gould 2002; A. Wagner 2005; G. P. Wagner 2014). Nevertheless, there is not yet any broad overview of the different lines of research that fall under the banner, and no comprehensive attempt has been made at synthesizing the knowledge gained in the different fields.

During the academic year 2019/2020, two of us (TFH and CP) organized a work group on "Evolvability: A new and unifying concept in evolutionary biology?" at the Centre of Advanced Study (CAS) in Oslo, to which we invited researchers with backgrounds in evo-devo, EQG, systems biology, paleobiology, and macroevolution, as well as philosophers and historians of biology. This book is an outgrowth of this project, and most of the contributors visited and participated in the discussions of the work group. Unfortunately, the project was cut short by the COVID pandemic in 2020, diminishing discussions on the macroevolutionary parts in particular. The resulting book is thus somewhat weighted toward development, population genetics, and microevolutionary perspectives, but we still see it as substantially synthetic across disciplines, and we hope it can stimulate further investigations and discussions of evolvability.

1.2 Debates and Controversies

Despite occasional earlier usage and related concepts (Sansom 2009; Brigandt 2015; Nuño de la Rosa 2017; Crother and Murray 2019), the concept of evolvability, or more precisely, *the evolution of evolvability,* arguably appeared in an essay by Dawkins (1988). This essay was part of an edited volume on artificial life and an outcome of a Santa Fe workshop on this topic. In his essay, Dawkins built on his "biomorph" artificial-life simulations and sought to characterize the fundamental preconditions for evolution to occur ("replication" and "embryology") and to ask how these preconditions themselves could evolve. He immediately sensed the radical nature of this question and started his essay with an apologetic assurance that he remained "a dyed-in-the-wool, radical neo-Darwinian." Why would such a question about the evolution of evolvability be thought a challenge to neo-Darwinism? The answer may no longer be obvious, but likely reflects a scarcity of such questions in the evolutionary biology of the time. Although similar questions had been asked on the edges of the main paradigm (e.g., Riedl 1978; Conrad 1983; Wagner 1986), the existence, or at least production, of genetic variation for natural selection to operate had the status of an axiomatic premise in textbooks and authoritative accounts of evolutionary theory. Asking new questions can happen within a paradigm, however, and need not be controversial. Okasha (2021) sees the ability to *endogenize* and explain previously external assumptions as a striking feature of evolutionary biology and a contributer to its success and generality. The endogenization of evolvability, in the form of studying the origin and structuring of genetic variability, is a paradigmatic example. The nonradical way of doing this can be seen in the books by Dennett (1995) and Maynard Smith and Szathmáry (1995), which both asked new and far-reaching questions about the origin and evolution of structures and systems that facilitate evolution, but situated these questions within and in support of the neo-Darwinian framework. In contrast, Kauffman (1993) placed his work on evolvable systems in opposition to mainstream evolutionary theory.

We can recognize four positions about the novelty of evolvability: (1) Evolvability is a radically new concept that transcends orthodox evolutionary theory. (2) Evolvability is a novel concept associated with new questions and research strategies, but these have been accommodated (endogenized) by standard evolutionary theory. (3) Evolvability is a new term with new sociological roles, but conceptually a continuation of older concepts, such as developmental constraint (e.g., Brigandt 2015). (4) There is nothing new but a pretentious name.

Reflecting perceptions of the status of the fields themselves, researchers in evo-devo may tend toward positions 1, 2, or 3, while evolutionary quantitative geneticists may tend toward positions 2, 3, or 4. The most extreme positions 1 and 4 are on display in the debates about the extended synthesis, in which evolvability is sometimes packaged with such concepts as niche construction, epigenetic inheritance, and mutation- or plasticity-driven evolution, and then either reified as overturning the old order (e.g., Laland et al. 2014) or dismissed as nothing new (e.g., Wray et al. 2014; Charlesworth et al. 2017).

Philosophers and historians of biology debate the historical and conceptual relationships among notions of evolvability in the different research fields. While some want a single unified concept (e.g., Sterelny 2007; Brown 2014), others see clusters of unrelated use (Love 2003; Pigliucci 2008; Nuño de la Rosa 2017). At first glance, the evolvability concepts in evo-devo and EQG appear different, with the former focused on the ability of the individual organism, or the genotype-phenotype map, to produce new potentially adaptive variants, and the latter focused on standing genetic variation and the ability of the population to respond to selection. There is crosstalk, however, and in EQG there has been increasing focus on how the genotype-phenotype map, in terms of pleiotropy, epistasis, and norms of reaction, acts to structure both mutational and standing variation (Hansen and Pélabon 2021). There is also a small, but persistent, stream of research aiming to connect evo-devo to population genetics and thus to population variation (e.g., Wagner and Altenberg 1996; Hansen 2006). In this endeavor, evolvability is a common focus, and if not unified, it may be unifying.

Unification is harder to perceive across scales of evolution. Perhaps reflecting the lack of a well-developed theory of macroevolution, there is no obvious connection between notions of microevolutionary evolvability describing genetic or mutational variability in quantitative terms and qualitative notions of macroevolutionary evolvability in terms of innovation and transitions among body plans. In his book on innovation, Andreas Wagner (2011) explicitly separated qualitative macroevolutionary "innovability" from evolvability, or from what he perceived as more quantitative, overly varied and "muddled" uses of evolvability. Innovation has both qualitative and quantitative aspects, however, and Günter Wagner (2014) distinguished between two types of innovation, with one being the qualitative origin of new character identities and the other being the more quantitative origin of new variational modalities.

Another point of contention regards the evolution of evolvability. Although initial concerns to demonstrate that evolvability could evolve at all now seem naive, there are more subtle matters to settle in terms of whether and when adaptations for evolvability can evolve, and the levels of selection that may cause such evolution. From the macroevolutionary point of view, one may ask whether changes in evolvability emerge gradually or if they require special events, such as transitions to states that permit qualitatively new evolutionary possibilities. The latter may involve the evolution of new character identities, body plans, or inheritance systems. Interest in the evolution of evolvability also leads to empirical questions about differences in evolvability among clades and traits, and to whether this may generate differences in diversity or disparity.

Finally, there are debates about the proper definition, characterization, and measurement of evolvability. In evo-devo, there are debates about the relationship of evolvability to modularity and integration, and in EQG there are debates about the scaling and parameterization of univariate and multivariate genetic variation that best predict evolvability.

1.3 The Chapters

Figure 1.1 provides a roadmap to the chapters in terms of where they connect to the disciplinary and conceptual landscape of evolvability research. The chapters may be read in any order depending on the interests of the reader.

1.3.1 Historical Aspects

This volume opens with a contribution from Nuño de la Rosa (chapter 2), in which she first reviews the bibliometric study of evolvability research presented in Nuño de la Rosa (2017). Indeed, the six evolvability "research fronts" identified as co-citation clusters in Nuño de la Rosa (2017) were a motivation and guide for our CAS working group. Although the six clusters, *Complex networks, Molecular evolution, Quantitative genetics, Population genetics, Marcroevolution,* and *Evo-devo,* may not correspond entirely to our figure 1.1, they illustrate how the evolvability concept is being used in different disciplines and trace some of the connections between them. Villegas et al. (chapter 3) update this analysis by adding citations between 2014 and 2021. They identify *Network analysis, Evo-devo, Quantitative genetics,* and *Molecular evolution* as the four main clusters of contemporary evolvability citation. Going back to chapter 2, Nuño de la Rosa then moves from a quantitative to a qualitative approach. She presents preliminary results from a set of interviews with researchers on their attitudes toward evolvability, conducted as a part of the CAS project. These interviews put on display a range of different opinions about the novelty and utility of the concept, as well as on its historical origin(s). Nuño de la Rosa then finishes her chapter with a philosophical discussion about the progress of science, even suggesting a new role for the term "evolvability" in explaining internal theoretical mechanisms that facilitate particular directions of research.

1.3.2 Conceptual Framework

Moving from history to philosophy, Villegas et al. (chapter 3) ask what conceptual roles are played by evolvability in the different research fronts. These include setting a research agenda and characterizing the phenomena to be studied. Evolvability also plays a role in explaining, predicting, or controlling other phenomena (evolvability as explanans), and as a target for explanation, prediction, or control (evolvability as explanandum). Villegas et al. then discuss the connections between these roles and the unity of the evolvability concept itself.

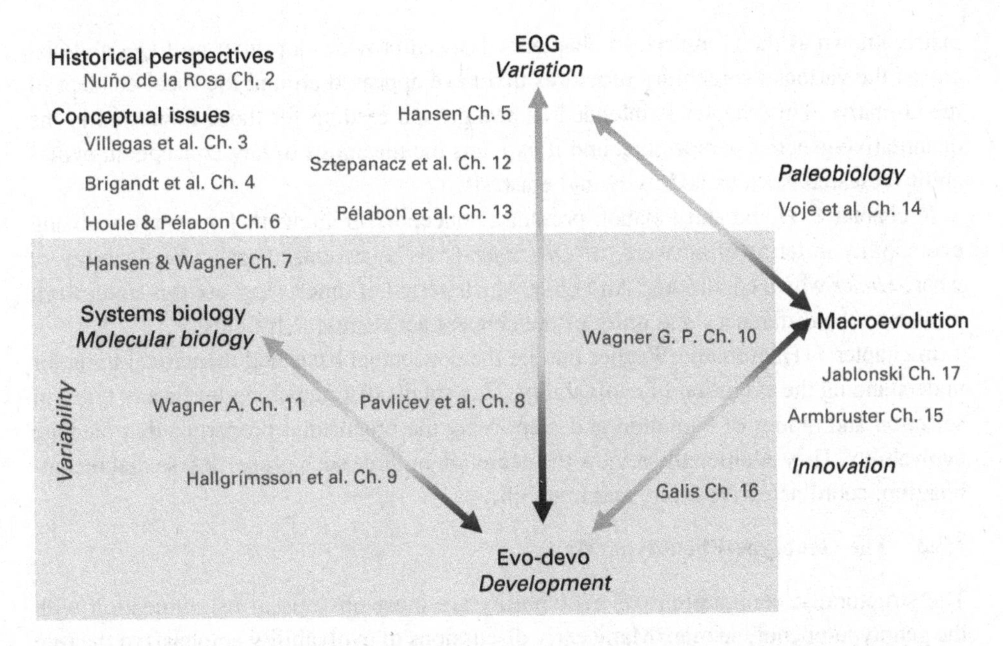

Figure 1.1
Roadmap for evolvability research: The chapters of this book are mapped out in relation to the research areas that we, the editors, see them as most connected to. We have indicated connections between research areas with arrows and placed some of the chapters in relation to these. The shaded area labeled "Variability" indicates focus on the disposition for variation (e.g., mutability), as opposed to variation as a disposition for evolution.

One of the novel aspects of evolvability is its dispositional nature (Wagner and Altenberg 1996; Love 2003; Nuño de la Rosa and Villegas 2022). Evolvability heralded a focus on dispositions or potentials in evolutionary research. In a key contribution, Günter Wagner and Lee Altenberg (1996) emphasized the distinction between variation and variability, with the latter as the potential for the former, and pointed out that these need to be treated and studied as different phenomena. Naturally, they linked evolvability with variability as joint dispositional concepts, and in most fields evolvability is operationalized as a disposition for variation or innovation. In chapter 4, Brigandt et al. discuss the nature of dispositional concepts in general and then investigate manifestations in the different areas of evolvability research. They discuss the utility of evolvability as disposition and the implications this has for unification across fields.

In many ways, quantitative genetics is the odd field out in evolvability research. In EQG, the concept of evolvability has a distinct origin as a name for a particular way of scaling genetic variation and as a replacement for the (dubious) use of heritability as a measure of evolutionary potential (Houle 1992; Hansen et al. 2011). Hence, in EQG, evolvability is mainly linked with realized nondispositional variation and less with variability. Properly constructed, however, it is still a dispositional concept, but now as a disposition for a population to respond to selection, which, at least in the short term, depends on (genetic) variation. Whether in name or not, evolvability is a major research topic in EQG. Hansen and Pélabon (2021) have argued that the key event establishing EQG as a research field was Lande and Arnold's (1983) conceptual separation of selection and evolvability in the form of the selection gradient and the additive genetic variance matrix known as the G-matrix. In chapter 5, Hansen provides a primer on EQG and discusses the various evolvability measures that have appeared around the focal concept of the G-matrix. This chapter is intended as background reading for those interested in the quantitative-genetics perspective, and it explains the meanings of key concepts in evolvability research, such as additivity and epistasis.

In chapter 6, Houle and Pélabon present a conceptual framework for operationalizing evolvability in terms of answers to "*Of-Under-Over*" questions. That is: Evolvability *of* what? *Under* which conditions? And *Over* which period of time? They use this framework to discuss the differences and unity of the concept across research fields.

In chapter 7, Hansen and Wagner outline the conceptual basis and theoretical tools for understanding the evolution of evolvability. This entails distinguishing levels and types of selection and modes of evolution and identifying the organismal properties that mediate evolvability. They additionally review the evolution of three such properties: sexual recombination, coordinated variation, and mutability.

1.3.3 The Genotype-Phenotype Map

The structuralist connotations of evolvability are most obvious in its connection with the genotype-phenotype map. Many early discussions of evolvability emphasized the role of the genotype-phenotype map in structuring variation (e.g., Dawkins 1988; Alberch 1991; Wagner and Altenberg 1996), and major determinants of evolvability, such as continuity, modularity, robustness, and epistasis, can be seen as properties of the genotype-phenotype map. In chapter 8, Pavličev et al. discuss the relationship between evolvability and the genotype-phenotype map from a theoretical perspective, with particular emphasis on pleiotropy (modularity) and epistasis (context dependency). They delineate different conceptualizations of the genotype-phenotype map and argue for a research program based on mechanistic modeling and integration of the genotype-phenotype map into evolutionary models.

In chapter 9, Hallgrímsson et al. discuss the genotype-phenotype map and its implications for evolvability from an empirical perspective grounded in developmental biology. They assess the empirical basis for dimensionality and integration of the map, and they analyze how discontinuous variation can arise from metastability in seemingly robust developmental systems.

In chapter 10, Günter P. Wagner presents a new model for a genotype-phenotype map based on basic principles of trait growth and uses it to show how developmental interdependencies may affect quantitative measures of evolvability. He further explores the consequences of changes in evolvability on the evolution of body shape and illustrates how evolvability may evolve as a by-product of selection on developmental processes.

Robustness toward genetic and environmental perturbations is a central aspect of the genotype-phenotype map with a complicated relationship to evolvability. At first sight, it seems that robustness would reduce evolvability by repressing the potential for variation. Robustness also provides opportunities for systems drift and accumulation of hidden variation, however, and may thus enhance the potential for evolutionary exploration and innovation (e.g., A. Wagner 2005, 2008). In chapter 11, Andreas Wagner explores these issues and reviews some of his extensive work on the relationship between robustness and evolvability.

1.3.4 Microevolutionary Perspectives

Microevolutionary perspectives on evolvability are mostly associated with quantitative genetics approaches or with experimental evolution. As mentioned in section 1.3.2, EQG is characterized by quantitative measures of evolvability designed to predict quantitative responses to defined selection pressures (Hansen, chapter 5; Houle and Pélabon, chapter 6). The question of what determines short-term evolvability in this sense is taken up in several chapters. In chapter 12, Sztepanacz et al. discuss the influence of mating systems and population structures on evolvability. They particularly address the effects of inbreeding on evolvability, and they discuss how evolvability relates to sexual selection. In chapter 13, Pélabon et al. provide further discussion of how ecological factors influence evolvability and its evolution. They discuss the influence of selection, gene flow and population size on the evolution of evolvability and take up the important question of how organismal plasticity and stress may interact with evolvability.

How far microevolutionary processes can be extrapolated to explain macroevolution has been a recurrent question in evolutionary biology for at least 80 years (Futuyma 2015). Accordingly, it has been suggested that patterns of short-term evolvability could limit or bias evolution at longer timescales (e.g., Schluter 1996; Hansen and Houle 2004). On one hand, estimates of standing genetic variation discussed in chapters 5, 6, 12, and 13 indicate that evolvability is usually sufficient to allow large changes on timescales above a few tens or hundreds of generations at least, and would not seriously constrain evolution on macroevolutionary timescales. On the other hand, there is accumulating evidence that quantitative measures of evolvability correlate with divergence among populations and species on fairly long timescales. This conundrum is reviewed in chapter 14 by Voje et al., who go on to discuss ideas for how these observations can be reconciled. They argue that progress would require developing models that link microevolutionary theory with the temporal dynamics of the fitness landscape.

1.3.5 Macroevolutionary Perspectives

While Voje et al. (chapter 14) consider macroevolution on moderately long timescales, they do so in terms of quantitative changes that can be directly related to microevolutionary measures of selection and evolvability. In contrast, many macroevolutionary phenomena involve qualitative changes that are not easily captured by or related to quantitative measures. The approaches of evo-devo, for example, often invoke developmental, physiological, or gene-regulatory mechanisms to understand the basis or potential for larger morphological changes.

In chapter 15, Armbruster discusses the developmental and physiological basis for repeated evolutionary patterns in the evolution of flowering plants. In doing so, he employs classical concepts in developmental evolution, such as allometry, heterochrony and exaptation, as well as newer evo-devo concepts such as co-option and modularity. He illustrates how these concepts can be used to understand the potential for evolution and discusses how to separate the roles of selectability and evolvability in macroevolution.

In chapter 16, Galis discusses constraints on early development and their consequences for the evolvability of animal body plans. She considers two hypotheses to explain the strongly conserved phase in early development known as the phylotypic stage. One hypothesis explains the phylotypic stage as consequence of pleiotropic constraints due to strong inductive interactions during this developmental stage, and the other explains it as a consequence of the genetic robustness of the gene networks that underlie organogenesis. Galis favors the former hypothesis.

A common characterization of macroevolution is evolution above the species level. In chapter 17, Jablonski takes up the effects of evolvability on speciation and extinction, and evaluates its role as an explanatory factor in diversification at the clade level. The general methodological challenge is to distinguish the effects on macroevolution of intrinsic factors, such as evolvability, from extrinsic factors, such as changes in the biotic or abiotic environment. This requires both comparative and paleobiological approaches, and new methods for their integration.

1.3.6 A New and Unifying Concept?

In the concluding chapter 18, Houle et al. summarize what we have learned about the question initially asked: Is evolvability a new and unifying concept in evolutionary biology? In this chapter, we consider both the points of views expressed in the other chapters of this book, as well as our own judgment and reflections after nearly 3 years of discussion and interaction with the participants of the CAS evolvability project.

1.4 Missing Pieces

Although these chapters cover a wide range of topics and offer different opinions and perspectives, there are some noticeable biases and omissions. A full treatment of evolvability research could have included discussions of major transitions, innovation, experimental evolution, regulatory evolution, niche construction, nongenetic evolution, and the origin of life. Separate chapters on mutation, molecular and biochemical evolution, modularity, and character identity would also have been natural, even if these topics are discussed in some of the chapters we do have. From the philosophical and historical points of view, we could have included treatments of pre-1990 research on evolutionary potential, arguments for a unified evolvability concept, and discussions of the extended evolutionary synthesis. Finally, we note the absence of perspectives from computer science, artificial life, and digital evolution.

Nevertheless, we hope that this book can be a useful source for contemporary debates and research on the potential for evolution.

Acknowledgments

Our main debt is to the Centre for Advanced Study (CAS) in Oslo for hosting and financing the work group that lead to this book. The staff at CAS provided excellent support both during and after the workshop. We thank all the researchers who participated in or visited the work group, and those who participated in the online journal-club discussions after the premature pandemic-related closing of the CAS project. You have all influenced this book, whether you contributed chapters or not. We also thank Gerd Müller and Anne-Marie Bono for encouragement and support during the preparation of this book.

References

Alberch, P. 1991. From genes to phenotype—dynamic-systems and evolvability. *Genetica* 84: 5–11.

Amundson, R. 2005. *The Changing Role of the Embryo in Evolutionary Thought: Roots of Evo-Devo.* Cambridge Studies in Philosophy and Biology. Cambridge: Cambridge University Press.

Brigandt, I. 2015. From developmental constraints to evolvability: How concepts figure in explanation and disciplinary identity. In *Conceptual Change in Biology: Scientific and Philosophical Perspectives on Evolution and Development,* edited by A. Love, 305–325. Dordrecht: Springer.

Brookfield, J. F. Y. 2009. Evolution and evolvability: Celebration Darwin 200. *Biology Letters* 5: 44–46.

Brown, R. L. 2014. What evolvability really is. *British Journal for the Philosophy of Science* 65: 549–572.

Charlesworth, D., N. H. Barton, and B. Charlesworth. 2017. The sources of adaptive variation. *Proceedings of the Royal Society B* 284: 20162864.

Conrad, M. 1983. *Adaptability: The Significance of Variability from Molecule to Ecosystem.* New York: Plenum Press.

Crother, B. I., and C. M. Murray. 2019. Early usage and meaning of evolvability. *Ecology and Evolution* 9: 3784–3793.

Dawkins, R. 1988. The evolution of evolvability. In *Artificial Life: The Proceedings of an Interdisciplinary Workshop on the Synthesis and Simulation of Living Systems,* edited by C. Langton, 202–220. Boston: Addison-Wesley.

Dawkins, R. 1996. *Climbing Mount Improbable*. New York: Viking.

Dennett, D. C. 1995. *Darwin's Dangerous Idea: Evolution and the Meanings of Life.* New York: Simon & Schuster.

Futuyma, D. J. 2015. Can modern evolutionary theory explain macroevolution? In *Macroevolution: Explanation, Interpretation and Evidence,* edited by E. Serrelli, and N. Gontier, 29–86. Dordrecht: Springer.

Gerhart, J., and M. Kirschner. 1997. *Cells, Embryos and Evolution: Towards a Cellular and Developmental Understanding of Phenotypic Variation and Evolutionary Adaptability.* Hoboken, NJ: Wiley.

Gould, S. J. 2002. *The Structure of Evolutionary Theory*. Cambridge, MA: Belknap.

Hansen, T. F. 2006. The evolution of genetic architecture. *AREES* 37: 123–157.

Hansen, T. F. 2016. Evolvability. In *The Encyclopedia of Evolutionary Biology,* Vol. 2, edited by R. M. Kliman, 83–89. Cambridge, MA: Academic Press.

Hansen, T. F., and D. Houle. 2004. Evolvability, stabilizing selection, and the problem of stasis. In *Phenotypic Integration: Studying the Ecology and Evolution of Complex Phenotypes,* edited by M. Pigliucci and K. Preston, 130–150. Oxford: Oxford University Press.

Hansen, T. F., and C. Pélabon. 2021. Evolvability: A quantitative-genetics perspective. *AREES* 52: 153–175.

Hansen, T. F., C. Pélabon, and D. Houle. 2011. Heritability is not evolvability. *Evolutionary Biology* 38: 258–277.

Hendrikse, J. L., T. E. Parsons, and B. Hallgrímsson. 2007. Evolvability as the proper focus of evolutionary developmental biology. *Evolution and Development* 9: 393–401.

Houle, D. 1992. Comparing evolvability and variability of quantitative traits. *Genetics* 130: 195–204.

Kauffman, S. A. 1993. *The Origins of Order: Self-Organization and Selection in Evolution.* Oxford: Oxford University Press.

Kopp, M., and S. Matuszewski. 2014. Rapid evolution of quantitative traits: Theoretical perspectives. *Evolutionary Applications* 7: 169–191.

Laland, K., T. Uller, M. Feldman, et al. 2014. Does evolutionary theory need a rethink? Yes, urgently. *Nature* 514: 161–164.

Lande, R., and S. J. Arnold. 1983. The measurement of selection on correlated characters. *Evolution* 37: 1210–1226.

Love, A. C. 2003. Evolvability, dispositions, and intrinsicality. *Philosophy of Science* 70: 1015–1027.

Love, A. C., M. Grabowski, D. Houle, et al. 2022. Evolvability in the fossil record. *Paleobiology* 48: 186–209.

Maynard Smith, J., and E. Szathmáry. 1995. *The Major Transitions in Evolution.* Oxford: Freeman.

Minelli, A. 2017. Evolvability and its evolvability. In *Challenging the Modern Synthesis: Adaptation, Development, and Inheritance,* edited by P. Huneman and D. Walsh, 211–238. Oxford: Oxford University Press.

Nehaniv, C. 2003. Editorial: Evolvability. *BioSystems* 69: 77–81.

Nuño de la Rosa, L. 2017. Computing the extended synthesis: Mapping the dynamics and conceptual structure of the evolvability research front. *Journal of Experimental Zoology B* 328: 395–411.

Nuño de la Rosa, L., and C. Villegas. 2022. Chances and propensities in evo-devo. *British Journal for the Philosophy of Science* 73: 509–533.

Okasha, S. 2021. The strategy of endogenization in evolutionary biology. *Synthese* 198: 3413–3435.

Pavličev, M., and G. P. Wagner. 2012. Coming to grips with evolvability. *Evolution: Education and Outreach* 5: 231–244.

Payne, J. L., and A. Wagner. 2019. The causes of evolvability and their evolution. *Nature Reviews Genetics* 20: 24–38.

Pigliucci, M. 2008. Opinion—Is evolvability evolvable? *Nature Reviews Genetics* 9: 75–82.

Porto, A. 2021. Variational approaches to evolvability: Short- and long-term perspectives. In *Evolutionary Developmental Biology: A Reference Guide,* edited by L. Nuño de la Rosa and G. Müller, 1112–1124. Dordrecht: Springer.

Raff, R. A. 1996. *The Shape of Life: Genes, Development, and the Evolution of Animal Form.* Chicago: University of Chicago Press.

Riederer, J. M., S. Tiso, T. J. van Eldijk, and F. J. Weissing. 2022. Capturing the facets of evolvability in a mechanistic framework. *TREE* 37: 430–439.

Riedl, R. 1978. *Order in Living Organisms: A Systems Analysis of Evolution.* Hoboken, NJ: Wiley.

Sansom, R. 2009. Evolvability. In *The Oxford Handbook of Philosophy of Biology,* edited by M. Ruse. Oxford: Oxford University Press.

Schlichting, C. D., and C. J. Murren. 2004. Evolvability and the raw materials for adaptation. In *Plant Adaptation: Molecular Genetics and Ecology,* edited by Q. C. B. Cronk, J. Whitton, R. H. Ree, and I. E. P. Taylor, 18–29. Ottawa: NRC Research Press.

Schluter, D. 1996. Adaptive radiation along genetic lines of least resistance. *Evolution* 50: 1766–1774.

Sniegowski, P. D., and H. A. Murphy. 2006. Evolvability. *Current Biology* 16: 831–834.

Sterelny, K. 2007. What is evolvability? In *Philosophy of Biology,* edited by M. Matthen and C. Stephens, 177–192. Amsterdam: Elsevier.

Sterelny, K. 2011. Evolvability reconsidered. In *The Major Transitions in Evolution Revisited,* edited by K. Sterelny and B. Calcott, 83–100. Cambridge, MA: MIT Press.

Vasas, V., C. Fernando, A. Szliagyi, I. Zachar, M. Santos, and E. Szathmáry. 2015. Primordial evolvability: Impasses and challenges. *Journal of Theoretical Biology* 381: 29–38.

Wagner, A. 2005. *Robustness and Evolvability in Living Systems.* Princeton, NJ: Princeton University Press.

Wagner, A. 2008. Robustness and evolvability: A paradox resolved. *Proceedings of the Royal Society B* 275: 91–100.

Wagner, A. 2011. *The Origins of Evolutionary Innovations: A Theory of Transformative Change in Living Systems.* Oxford: Oxford University Press.

Wagner, G. P. 1986. The systems approach: An interface between development and population genetic aspects of evolution. In *Patterns and Processes in the History of Life,* edited by D. M. Raup and D. Jablonski, 149–165. Dordrecht: Springer.

Wagner, G. P. 2010. Evolvability: The missing piece of the Neo-Darwinian synthesis. In *Evolution since Darwin: The First 150 Years,* edited by M. A. Bell, D. J. Futuyma, W. F. Eanes, and J. S. Levinton, 197–213. Sunderland, MA: Sinauer.

Wagner, G. P. 2014. *Homology, Genes, and Evolutionary Innovation.* Princeton, NJ: Princeton University Press.

Wagner, G. P., and L. Altenberg. 1996. Complex adaptations and the evolution of evolvability. *Evolution* 50: 967–976.

Watson, R. A. 2021. Evolvability. In *Evolutionary Developmental Biology: A Reference Guide,* edited by L. Nuño de la Rosa and G. Müller, 134–148. Dordrecht: Springer.

Wray, G. A., H. E. Hoekstra, D. J. Futuyma, et al. 2014. Does evolutionary theory need a rethink? No, all is well. *Nature* 514: 161–164.

2 A History of Evolvability: Reconstructing and Explaining the Origination of a Research Agenda

Laura Nuño de la Rosa

This is one of the main reasons why most practicing scientists don't end up working on things that are fundamental: Because they don't know the history of things.
—Russ Lande

This chapter addresses the origination of evolvability research with the aim of contributing more generally to the reconstruction and explanation of the recent history of evolutionary biology. I combine co-citation analysis and first-person reconstructions of the history of the field obtained from a series of interviews with evolutionary biologists who were and/or are currently active in evolvability studies. After a preliminary methodological reflection, I present a reconstruction of the multiple origins of evolvability research. In the last section of the chapter, I make use of cultural evolution theory to discuss two kinds of explanations that might account for this pattern: "Selectionist" explanations highlight aspects of the methodological and intellectual landscape that promoted the acceptability and diffusion of the evolvability perspective; "evolvability" explanations address the role of internal, theoretical developments involved in the origination and diversification of evolvability research. Although selectionist explanations have been largely explored, internal factors accounting for the evolvability of scientific concepts and theories remain relatively neglected. I argue that explaining the recent history of evolvability research from this perspective provides promising insights to our understanding of science dynamics.

2.1 Introduction

Although the idea of evolvability, as for almost every topic in evolutionary biology, can be traced back to Darwin, evolvability as a research program only emerged in the late 1990s. Compared to the concern with constraints, which remained steady throughout the 1990s and has fallen gradually over the past decade, the interest in evolvability underwent an exponential growth in the first half of the 2000s and has continued to increase after 2005 at a slower pace (see figure 2.1). Since the late 990s, evolvability research has increased and diversified in several domains of biology, being considered by some as one of the main "expanders" of the Evolutionary Synthesis (Pigliucci 2009).

This chapter addresses the historical origins of evolvability research with the aim of contributing more generally to the reconstruction and explanation of the recent history of evolutionary biology. Even more broadly, I aspire to shed some light on one of the most fascinating questions in the history of science, namely, "[b]y what processes do intellectual

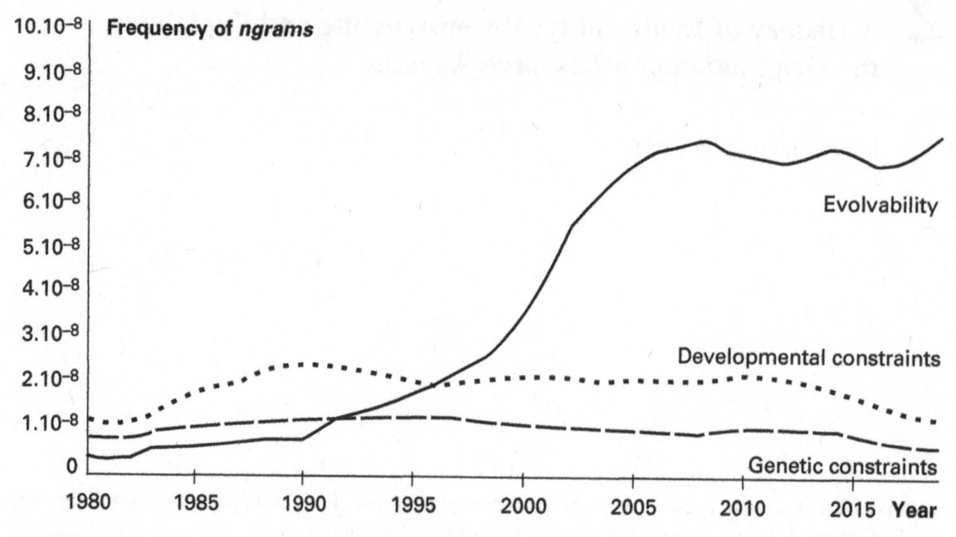

Figure 2.1
Graph charting the frequency of use of the phrases (*ngrams*) "evolvability," "genetic constraints," and "developmental constraints" in Google Books written in English between 1980 and 2019. The *y* axis corresponds to the number of times an ngram (e.g., evolvability) is used as compared to the total number of words (or two-word phrases) in all books published each year (*x* axis). This is the reason that the numbers in the *y* axis are so small, but what matters is the comparison between the frequencies of each phrase. Adapted from the graph generated by Google Books Ngram Viewer.

innovations originate, spread, and establish themselves within a scientific tradition?" (Toulmin 1967, 460).

Over the past few years, different bibliometric methods, including temporal reconstructions of citation landscapes and co-word analyses, have been employed for studying the dynamics of scientific fields (Chavalarias and Cointet 2013). Co-citation networks, where nodes correspond to cited references and links to co-citation relationships (Small and Griffith 1974), are particularly useful for analyzing and mapping the conceptual and dynamical structure of research agendas. In a previous work, I used co-citation analysis to study the interdisciplinary structure of evolvability research (Nuño de la Rosa 2017; see also Villegas et al., chapter 3).[1] I concluded that the evolvability research program shows a complex conceptual structure organized in six partly overlapping disciplinary clusters: complex network analysis, molecular evolution, quantitative genetics, population genetics, evo-devo, and macroevolutionary studies (see figure 2.2).

Although recent studies have looked at the early meanings and usages of the evolvability concept (Crother and Murray 2019), the origination of evolvability as a research program organized around a set of coordinated problem agendas (Brigandt and Love 2012) remains largely unexplored. To inquire into the generation of the clusters and the resulting interdisciplinary structure of evolvability research identified in my previous work, here I compare evolvability co-citation networks from different time periods (see figure 2.3 in section 2.3). Moreover, with the aim of interrogating the biographical, societal, and theo-

1. References to chapter numbers in the text are to chapters in this volume.

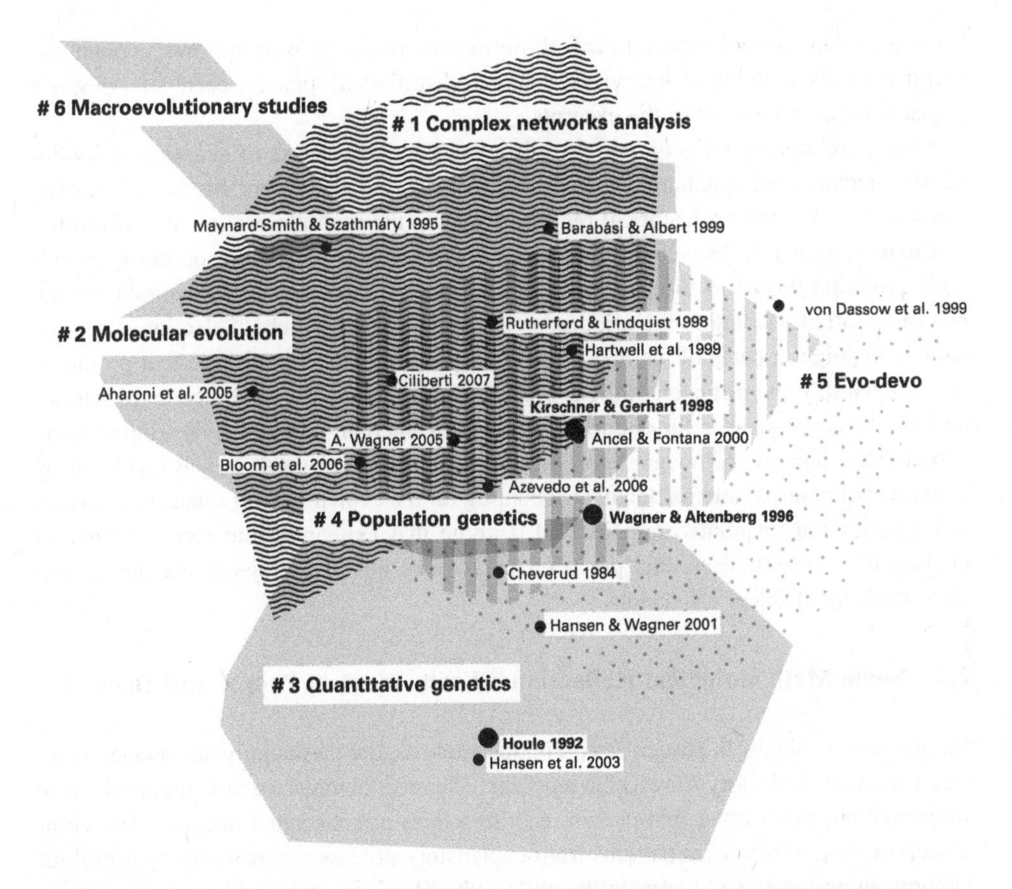

Figure 2.2
Evolvability co-citation network. Modified from Nuño de la Rosa (2017), Figure 3. Nodes represent the most co-cited publications within the network. Co-citation links have been removed. The clusters group publications linked together by a higher number of co-citation links. Bibliographic records were gathered from the Web of Science, based on a topic search for papers on "evolvability" or "evolutionary adaptability" published in English between 1970 and 2014. The data file, containing 1,039 full records and cited references, was imported to CiteScape 3.9.R7 to map a single network. Check the Methods section in Nuño de la Rosa (2017) for more details.

retical processes generating this pattern, I combine quantitative bibliometric analysis with a series of in-depth interviews with evolutionary biologists who were and/or are currently active in evolvability studies. In alphabetical order, this chapter includes content from 25 interviews with Scott Armbruster, Steve Arnold, Richard Dawkins, Frietson Galis, Benedikt Hallgrímsson, Thomas Hansen, David Houle, Gene Hunt, Johannes Jaeger, Russ Lande, Lee Hsiang Liow, Michael Lynch, Joanna Masel, Mihaela Pavličev, Joshua Payne, Massimo Pigliucci, Christophe Pélabon, Arthur Porto, Peter Schuster, Arlin Stoltzfus, Jacqueline Sztepanacz, Masahito Tsuboi, Kjetil Voje, Andreas Wagner, and Günter P. Wagner. Most interviewees were participants of the project on evolvability that took place between 2019 and 2020 at the Centre for Advanced Study (CAS) in Oslo. Some of them (Arnold, Dawkins, Jaeger, Lande, Masel, Pigliucci, Schuster, Stoltzfus, and A. Wagner) were not directly involved in the CAS project but were interviewed either because of their historical role in the origination of evolvability research in different disciplinary areas or because of their current work on evolvability-related issues. In what follows, content and literal excerpts from these interviews will be identified by placing each interviewee's surname in parentheses after the excerpts.

After a preliminary reflection on the advantages and drawbacks of quantitative bibliometric methods and qualitative interviews for studying the recent history of science (section 2.2), I present a historical reconstruction of the multiple origins of evolvability research (section 2.3). In section 2.4, I make use of some conceptual resources from cultural evolution theory to discuss two kinds of explanations that might account for the historical pattern leading to the origination of evolvability research: "Selectionist" explanations highlight aspects of the methodological and intellectual landscape that promoted the acceptability and diffusion of the evolvability perspective; "evolvability" explanations address the consequences for empirical research of internal, theoretical developments in evolutionary biology. Although selectionist explanations have been largely explored by cultural evolutionists, internal factors accounting for the evolvability of scientific concepts and theories remain relatively neglected. I argue that explaining the recent history of evolvability research from this perspective provides novel, promising insights to our understanding of science dynamics.

2.2 Some Methodological Reflections: From Ideas to People and Back

Bibliometric methods, in particular co-citation analysis, are particularly advantageous for reconstructing the history of research agendas. The most obvious merit is that they avoid subjective biases affecting first-person reconstructions of conceptual lineages. The virtuous decoupling of historical patterns from explanatory processes is particularly useful for circumventing theoretical biases in the philosophy of science, where historical narratives are often shaped to illustrate one's preferred approach (Chavalarias et al. 2022). Importantly, bibliometric methods also allow us to separate intellectual descent from conceptual convergence, or the emergence of the same research themes in different disciplinary environments. Co-citation networks are particularly useful in this regard, insofar as they include not only publications on our topic of interest but also the references cited in these publications, which permits reconstruction of the disciplinary background they belong to. Co-citation analysis is also a good sieve for separating intellectual traditions from mere historical precedence. The history of evolvability research abounds in cases of predecessors who did not have an influence in subsequent studies. Andreas Wagner refers to a paper on protein evolution (Lipman and Wilbur 1991) that preceded Peter Schuster and colleagues' work on neutral networks, but that he was unaware of when he started working on evolvability-related issues (A. Wagner). Frietson Galis's thinking on the evolvability of body plans was highly influenced by Rudy Raff (1996), although she later discovered that it was Klaus Sander (1983) who first characterized the phylotypic stage more than a decade earlier. Some of the precedent works that were neglected by a scientific community are often written by the same authors who are later acknowledged as the founders of the field. Günter Wagner's work is a good example. Although his article in *Evolution* (G. P. Wagner and Altenberg 1996) is widely perceived as a founding work on evolvability, he identifies a series of papers starting in 1981 (G. Wagner 1981) as his first work on the evolution of evolvability. As we will see in this chapter, many more earlier works can be identified as precedents of evolvability-related ideas. However, when following the history of a scientific field rather than a specific idea, the question is not whether there are historical precedents of this idea, but rather when it became a well-identified research agenda: "Very often, the topics arrive early, but the question is whether they precipitate sustained research effort or not" (G. Wagner).

Finally, the dynamics of ideas are relatively independent of their individual carriers, and co-citation analysis allows us to detect wide patterns of usage of methodological and conceptual resources that are independent of the explicit purpose for which they were originally conceived. An illustrative example is the mutation (M-) matrix, a quantitative genetics parameter that describes the effects of new mutations on genetic variances and covariances (Jones et al. 2007). Although introduced in the 1980s in the field of quantitative genetics, the M-matrix is not even recognized by its users as a measure of the potential to evolve: "turning a science into ideology only works by simplifying things, not seeing things that go beyond your framework [. . .] you are doing it yourself even if you don't want to" (G. Wagner). Another example is the separation between selection and the evolutionary response to selection in the multivariate version of the breeder's equation introduced by Russ Lande and Steve Arnold in their seminal paper (Lande and Arnold 1983). This separation is interpreted in several interviews as a key step in evolvability research, but Lande and Arnold themselves highlight different aspects of their contribution to the topic (see section 2.4.2).

Notwithstanding their advantages, bibliometric methods also have serious flaws. First, many works are cited but not read. This leads to miscitations and propagation of "mutated memes," as Conner and Lande (2014) warn about many references to the botanist Raissa Berg. Berg (1960) is one of the earliest proponents of a modularity hypothesis for the differential evolvability of floral and vegetative traits. She argued that floral traits were subject to stabilizing selection due to specialized pollination, leading to decoupling of the floral and vegetative traits into separate "correlation pleiades." This claim is often mistakenly interpreted, according to Conner and Lande, as entailing that flowers are tightly integrated organs (but see Armbruster et al. 2014). Second, some papers are retrospectively magnified. There are plenty of "courtesy citations" to works perceived by scientific communities as important ancestors of a given idea, even when these works did not have a real influence. For instance, Andreas Wagner refers to Alberch (1991) as the "go-to" citation on early work on genotype-phenotype maps, a paper that he himself cites, even though he does not think it offers a clear treatment of the topic (A. Wagner). Moreover, even when citations are not mere courtesy, it is hard to know what their real influence was. For instance, in the introduction to their paper, Lande and Arnold (1983) present their measure of selection as a kind of response to Gould and Lewontin's (1979) criticism of adaptationism. However, when asked about the influence of this criticism, Arnold denies that it played any role: "Gould and Lewontin 1979 didn't acknowledge the fact that measuring selection is a way to tackle the adaptationist paradigm. It's basically a rhetorical paper that enjoys and generates conversation, but not research" (Arnold). A further phenomenon that is hard to detect from mere citation patterns are rediscoveries. For instance, although modularity was a well-known topic in the zoological literature, it only became a topic of interest in plant evolution after the rediscovery of Berg's work (Lande 1979, 1980). Lande's citations of Berg's works led Jim Cheverud to read Berg, and they caused Scott

Armbruster to write an article testing her ideas (Armbruster et al. 1999), which in turn brought Berg's thinking to the attention of more botanists (Cheverud, Armbruster).

A second group of limitations of bibliometric methods has to do with how scientific knowledge is transmitted. Citations do not reflect the transmission of exact copies of ideas but interpretations of them. This is a well-known shortcoming of meme-like models of cultural evolution (Sterelny 2006): The transmission of information is not about copying but about inferential reconstruction, and reconstructions are highly biased by a multiplicity of psychological and social factors. Therefore, cultural transmission, including the transmission of scientific ideas, can be highly inaccurate. Nonetheless, many factors have been shown to enhance the reliability of cultural transmission, from group size (Henrich and Boyd 2002) to epistemic scaffolds, such as well-designed specialized vocabularies (Sterelny 2006). In science dynamics, both the increase in size of scientific communities and the epistemic scaffolding that accompanies the institutionalization of disciplines promote the reliability of transmission of scientific ideas. However, when scientific fields are nascent, the small size of the disciplinary community and the lack of epistemic scaffolds make bibliometric methods poor tools for reconstructing the origination of a field. This situation can linger for decades in some fields. This is the case, according to Joanna Masel, for theoretical population genetics, which still lacks mentors and an acceptable textbook that would serve the purpose of training people in a common background (Masel).

Finally, just as fossils are imprints of ecological interactions, very often citations are just traces of true intellectual exchanges. Many anecdotes from our interviews illustrate this point, such as that of the day when Steve Arnold, one of the founders of evolutionary quantitative genetics, met Pere Alberch, one of the originators of evolutionary developmental biology:

> I remember being at a party [. . .] when Pere Alberch had come to give a seminar in Chicago. We both had a few drinks, and he started attacking evolutionary quantitative genetics. So I started beating him about this, and he got very heated. My feeling was that he didn't understand my field, whereas I understood his field and its limitations rather more vividly than he did. So we were in each other's faces, battling for our approaches. *I think this captures the essence more than the subtleties of who is referencing whom.* (Arnold, my emphasis)

Asking the very actors who were involved in the origination of our field of interest seems like a natural choice to interpret and complement the information obtained by citation analysis. Although the list of interviewed evolutionary biologists referred to above is by no means exhaustive, I believe it is a fair representation of the individual "carriers" of the evolvability program. Interviewees include the authors of some of the seminal works of evolvability research identified by co-citation analysis, as well as biologists from a variety of disciplinary approaches (paleontology, evolutionary genetics, evo-devo, molecular evolution, evolutionary computing, and systems biology) and taxonomic specialties (molecular, plant, and animal evolution). Nonetheless, two important biases in the choice of people need to be acknowledged. First, quantitative geneticists are clearly overrepresented compared to biologists with other backgrounds. Second, senior and male scholars have a disproportionately higher representation compared to female and young biologists. The first bias is inherited from the composition of the members of the CAS project and is partly compensated for by interviews with biologists who visited the CAS and a few who were external to the project. The second bias is attributable to the historiographical nature of the interviews project. Although partially mitigated by including younger and women researchers when considering the totality of topics covered in the interviews, the historiographical aims of this chapter impede the correction of this bias.

2.3 The Multiple Origins of Evolvability Research

Phylogenetic approaches to the history of science attempt to reconstruct patterns of conceptual descent with modification that are assumed to be independent of mechanisms explaining these patterns (Lennox 2001). Patterns of conceptual descent include not only genealogical intellectual relationships but also the tempo of conceptual change, which, just like organic evolution, shows periods of stasis and change.

A preliminary comparison of co-citation networks of evolvability works published in different time periods delivers a good first approximation to the origination of evolvability research (see figure 2.3). Concerning the tempo, a clear explosion of interest in evolvability occurred in the early 2000s, as shown by the burst in citations in the 5 years separating the two networks (see also figure 2.1). Regarding the pattern of conceptual descent, the origins of evolvability studies seem as heterogeneous as the agenda itself, although interdisciplinary exchanges appear to have been crucial in its expansion and diversification. Figure 2.3a, comprising references cited from 1990 to 2000, shows a clear dominance of quantitative genetics research on evolvability. The 1992 article by David Houle occupies the center of the network, connected to a cloud of works on heritability in quantitative genetics (Mousseau and Roff 1987; Falconer 1989; Messina 1993) on the left, and two closely linked papers (G. Wagner and Altenberg 1996; Kirschner and Gerhart 1998) on the right. In figure 2.3b, covering the period between 2000 and 2005, the developmental evolutionary approach to evolvability, represented by G. Wagner and Altenberg 1996 and Kirschner and Gerhart 1998, has moved to the center, while quantitative genetics publications on evolvability are displaced to the right, indicating that they no longer represent the central conceptual framework in evolvability research. In turn, there is an explosion of connections to new disciplinary fields, including computational evolution (Kauffman 1993; Altenberg 1994) and neutral networks (Schuster et al. 1994; Huynen et al. 1996; van Nimwegen et al. 1999; Ancel and Fontana 2000) on the left, and canalization (G. Wagner et al. 1997; Rutherford and Lindquist 1998; Rutherford 2000), molecular evolution and evo-devo (von Dassow and Munro 1999; von Dassow et al. 2000) on the right.

Historical roots differ, depending on the disciplinary context. Evolvability means different things (Pigliucci 2008; Brigandt et al., chapter 4) and plays different roles (Villegas et al., chapter 3) in evolutionary biology. This plurality reflects the historical reconstructions of the origination of each disciplinary approach to evolvability, as well as the perceptions of how evolvability studies relate to classical work in evolutionary biology. In turn, this heterogeneity in evolvability research reflects the heterogeneity of biology as a discipline. Steve Arnold points out this issue when asked about the different views of evolution endorsed by quantitative geneticists and evolutionary molecular biologists: "It's probably asking too much that the generalizations coming out of evolutionary quantitative genetics are going to find a receptive ear in molecular evolution circles [. . .] we are literally talking different languages, and literally thinking about different empirical systems" (Arnold).

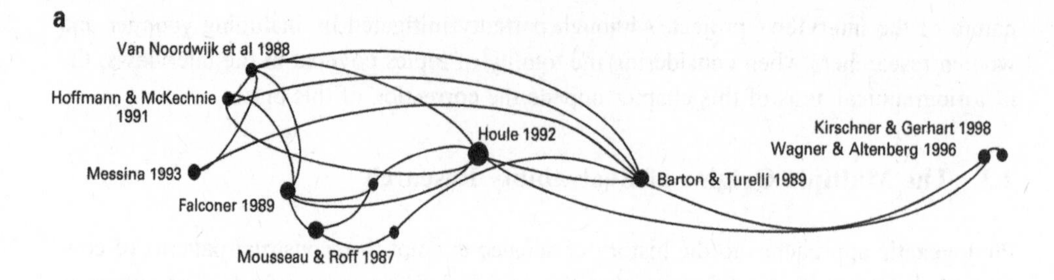

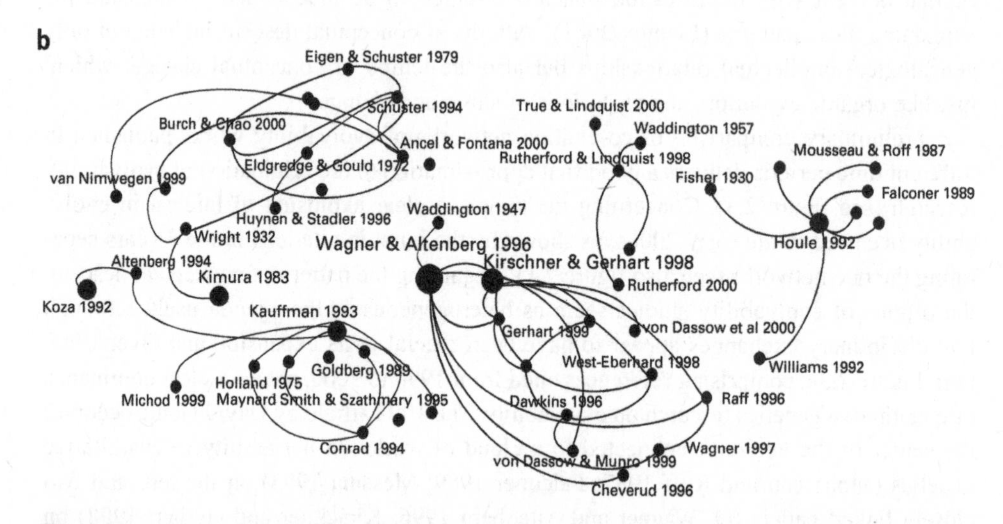

Figure 2.3
Co-citation networks on evolvability. Nodes correspond to cited references and edges to co-citation links. Bibliographic records were gathered from the Web of Science based on a search for publications containing "evolvability" in their title, keywords, or abstract, and published in English between 1990 and 2005. Full records and cited references were downloaded. This datafile was imported into VOSviewer, version 1.6.17 to build and visualize the co-citation networks. Figure 2.3a corresponds to works published between 1990 and 2000 with more than five citations, and cited references. Figure 2.3b corresponds to publications between 2000 and 2005 with more than seven citations, and cited references. Only the main nodes and a few representative co-citation links are shown. Navigable networks can be found in the book's online resources.

In considering evolvability as the ability of populations to respond to selection, evolutionary geneticists tend to see a clear continuity between themes discussed in classical population genetics and what is now called evolvability. In her interview, theoretical population geneticist Joanna Masel recognizes some of the classical work of the founders of population genetics (e.g., Fisher's on genetic linkage or the red queen hypothesis on the evolution of sex) as works on "the challenges to rapid adaptation." Although these works emphasized sex rather than other concepts currently related to evolvability (e.g., robustness or modularity), they were still works on "the limits to the rate of adaptation," which Masel interprets as limits to evolvability. According to Stoltzfus, this early identification of evolvability with rapid adaptation was influenced by the belief that the age of the earth might be much younger than the current estimate of 4.5 billion years. This was "one reason why Haldane and Fisher believed that evolution would take place based on abundant standing variation: it's faster" (Stoltzfus).

From the perspective of a quantitative geneticist such as David Houle, research on evolvability concerns the more general question of "what are the conditions under which evolution is possible?" In this view, "the intellectual tradition of what is now evolvability" also includes classical works in population genetics:

> When I wrote the 1992 paper I thought these were things that everybody was already thinking about. I would trace the ancestry of that back to Fisher and Wright at least [. . .]. What are the conditions under which evolution is possible? Wright's shifting balance theory, for example, is in that sense about evolvability: he had this idea about the combination of internal states, population states and selective environment that would enable new things to come about in evolution. Fisher similarly worked on that [. . .]. For example, Fisher microscope models: how to approach an optimum? This is kind of about evolvability. (Houle)

From this perspective, "far from being something that was ignored," evolvability is "something that wasn't named evolvability in the past" (Houle). In particular, Houle links the origins of his work on evolvability to his early interest in heterosis, or the problem of the advantage of being a heterozygote. He interprets this interest as "a component of a larger problem," namely, the maintenance of genetic variation: "Implicitly (or explicitly, depending on how you want to see it), that connects to the issue of what we now call evolvability, because without genetic variation, you don't have any ability to evolve at all." Houle recalls that at the time he started his PhD, there was an important controversy over the maintenance of variation in life-history traits. According to the famous table in Falconer's textbook in quantitative genetics (showing up in our network in the 1989 edition; see figure 2.3), life-history traits were less heritable than morphological traits and therefore were regarded as less responsive to selection. In thinking of fly wings, Houle noticed that wings did not vary much and that the concepts of heredity and evolvability were not the same thing. This led him to introduce a new mean-standardized measure of evolvability, defined as the ability of populations to respond to selection (Houle 1992). Why he decided to call this ability "evolvability" did not rely on him being aware of other uses of the term. He did not know about Dawkins's paper (1988), although he recalls that Stuart Kauffman visited his university when he was a graduate student and he might have heard the word (Houle).

Houle's definition and measure of evolvability has become the central reference for evolutionary quantitative geneticists (see figure 2.3b). In the view of the interviewees in this field, the ability of populations to respond to selection is not a new topic of research but instead a new measure of such a capacity: "a way of scaling genetic variance, and not necessarily [. . .] a thing in itself" (Sztepanacz). In Christophe Pélabon's view, the interest in evolvability was already present in the work on heritability, insofar as heritable traits were those considered to respond to selection. According to him, the novelty of evolvability does not come from the idea of evolvability itself but rather from the recognition "that traits can differ in their evolvability." Pélabon quotes Mousseau and Roff (1987), an important node in our network (see figure 2.3a), as one of the first works to deal with this issue. Although Mousseau and Roff discussed heritability and therefore used a poor method of standardization, Pélabon thinks that the question that traits differed in their ability to respond to selection was already present in that paper. Furthermore, if one includes the maintenance of variation in the definition of evolvability, previous works on such phenomena as mutational meltdown (Gabriel and Lynch 1993), or the inability of populations to maintain themselves due to the accumulation of deleterious mutations, might also be considered as precedents of evolvability research (Lynch).

Paleontologists (Hunt, Liow) and quantitative geneticists working at the intersection of micro and macroevolution (Porto, Tsuboi, Voje) were all interested in constraints early in their careers and see the latter as the flip side of evolvability. The idea that species and clades differ in their intrinsic ability to evolve is also regarded as an old idea in macroevolutionary studies (see Jablonski, chapter 17). Nonetheless, some of the publications identified as seminal works of this approach are perceived by some evolutionary geneticists as endorsing a very different view of evolution. In this regard, the key reference showing up in our network is Eldredge and Gould (1972) on punctuated equilibrium (see figure 2.3b). Researchers working on macroevolutionary scales see evolvability as a core component of the punctuated-equilibrium hypothesis, "in that an increase in evolvability during speciation allows new species to appear, while species that have emerged seem to have rather low evolvability, since they are not changing at all" (Voje). Two other recurrent works in our interviews with paleontologists are Vermeij (1973) on the role of dimensionality for the evolutionary "versatility" of body plans, and Lloyd and Gould (1993) on species selection on variability.

Younger evolutionary geneticists working on macroevolution tend to perceive "a natural continuity" between classical evolutionary genetics and current studies on evolvability. Masahito Tsuboi associates evolvability with the interest in "how genetic variation arises in the first place," a concern that has not been much addressed in evolutionary genetics but was present early on (Tsuboi). In contrast, biologists endorsing a mechanistic understanding of evolvability do not recognize clear precedents in evolutionary genetics. This is true for Massimo Pigliucci, who does not see any intellectual ancestor of evolvability research in classical population genetics. Molecular evolutionist Arlin Stoltzfus argues that mutational biases in the introduction of variation were neglected in classical evolutionary genetics. However, the received view of evolution as selection shifting gene frequencies from abundant variation in gene pools, started to collapse with the irruption of comparative genetics in the 1960s. In this regard, Stoltzfus identifies an important mechanistic shift in the 1980s, when biologists "began to think about evolution like mutationists, and to treat evolution as a Markov chain of mutation-fixation events." This new way of thinking and modeling evolution led to the emergence of molecular evolution, an approach that "was absolutely not predicted" by standard theory (Stoltzfus). This "shift to mutationism" is interpreted as an instance of what H. Allen Orr (2002) has called the "curious disconnect" between the mathematical models and the verbal theory of evolutionary genetics (Stoltzfus; see also Stoltzfus 2012, 2017).

In the field of population genetics, Mihaela Pavličev has recently argued that there were some parallel attempts in the 1980s to integrate physiological and developmental mechanisms with evolutionary theory (see Pavličev 2016 and references therein). However, these early attempts did not leave a mark in population genetics and faded from view in the 1990s.

Thomas Hansen describes the field of quantitative genetics when he moved to the University of Oregon in 1992 as "a field in expansion," where there was a lot of enthusiasm, "because people had started to do field studies of selection, and there were expectations around the ability to study the effect of genetic constraints." However, he also "perceived it as narrow, in the sense that where variation came from was not problematized, and mutations were not studied that much." Moreover, studies of mutations "were mostly estimating mutational variances, which is descriptive work. There wasn't much theory around how the properties of mutations are generated, or what the consequences of them are" (Hansen).

The neutral networks approach to evolvability is represented in the co-citation network by several papers from the mid-1990s (Schuster et al. 1994; Huynen et al. 1996; Ancel and Fontana 2000; Burch and Chao 2000), in which the ability to evolve is modeled as the accessibility of phenotypes in mutational neighborhoods (see figure 2.3b). Interviewees working in this field (Schuster, Payne, A. Wagner) agree on referring to Sewall Wright's landscapes as originating a new way of looking at evolution, but they identify in the "molecular evolution" a turning point in the understanding of the genotype-phenotype map. According to Peter Schuster, one of the founders of this approach, classical molecular biology perished in the late 1970s and was progressively replaced by a much more complex view of the genotype-phenotype relationship. Schuster refers to his work and that of his collaborators as being instrumental in introducing the genotype-phenotype mapping as an intermediate bridge not included in Wright's model (Schuster). Younger scholars in computational evolution, such as Josh Payne, agree with this view:

> The seeds of evolvability research were planted by Sewall Wright and John Maynard Smith. I think that concepts like genotype spaces and adaptive landscapes that are so central to evolvability research have been around for some time, but I don't think that they were really thinking about evolvability the way that we think evolvability now: the ability of mutation to bring phenotypic variation [. . .] that's part of what they were thinking of, but it wasn't what was driving the research. (Payne)

This reconstruction is clearly reflected in the network (figure 2.3b, bottom-left), where Wright (1932) and Maynard Smith and Szathmáry (1995) are important nodes that connect publications endorsing a neutral networks approach to evolvability. In turn, this cluster is peripherally connected to a paper by Eigen and Schuster (1979) representing one of the oldest subclusters of the network, namely, studies on evolvability in the field of the origins of life. In this context, the capacity to evolve is seen as a key condition for the successful origination of life as we know it. In the late 1970s, debates on the conditions for autocatalysis and self-enhancement for life origination were followed by debates on the requirements for biotic entities to undergo Darwinian evolution. In this context, Eigen and Schuster's work on the error threshold (Eigen 1971; Eigen and Schuster 1979) can be read as work on evolvability, namely, what the limits of mutation rates are that make Darwinian evolution possible (Schuster).

In the fields of evolutionary systems biology and evolutionary developmental biology, evolvability is seen as dependent on the internal properties of developmental systems. In the co-citation network, the field of evo-devo is represented by works on the developmental determinants of evolvability (Kirschner and Gerhart 1998; West-Eberhard 1998), and the role of developmental modularity (Raff 1996; von Dassow and Munro 1999; von Dassow et al. 2000) and integration (Cheverud 1996) (see figure 2.3b). In her interview, Frietson Galis agrees that evolvability issues were "definitely" present before the 1990s, but she recognizes a different set of ancestors of evolvability research. She refers to William Bateson on structural variation, Waddington and Schmalhausen on developmental plasticity, stabilizing selection, and genetic assimilation, Vermeij (1973) on the versatility of body plans, and Raff (1996) on the effect of developmental modularity on their conservation and variability. Johannes Jaeger associates his interest in developmental evolvability with his frustration with reductionist views of development and evolution, and he recalls Kauffman (1993) and Goodwin (1994) as his major inspirations for an alternative, structuralist approach to evolvability.

Evolutionary biologists working at the intersection of evolutionary genetics and developmental evolution identify the "organismal perspective" advanced by Rupert Riedl as a major influence on their thinking on evolvability:

> the core deficit [of population genetics] was the complete elimination of the theoretical importance of the organism, basically screening off everything that has to do with the organism so collapsing into one single parameter, fitness, on the one hand, and leaving it open at the lower level, at the genome, that is, genome and fitness. (G. Wagner)

Riedl's book *Order in Living Organisms: A System Theory of Evolution* (1978) does not show up in our network, but recurrently turns up in the interviews as a major influence for "a small group of people that were either connected to or inspired by Riedl's ideas" (G. Wagner). In this context, Wagner interprets the work that he, together with a few other people (including Jim Cheverud), were doing in the 1980s as an attempt to build a "conceptual infrastructure" or a "theoretical framework" that covered the gap between population genetics and organismal biology (G. Wagner). Günter Wagner, Andreas Wagner, and Mihaela Pavličev had Riedl as a professor during their graduate studies at the University of Vienna, and they all recall his lectures as being both unclear and inspiring. In his interview, Cheverud recalls having been highly influenced by Riedl's ideas. Cheverud read Riedl's book before writing his paper on morphological integration (Cheverud 1982), and the publication of that paper inaugurated his relationship with G. Wagner, with whom he has collaborated ever since. In this theoretical context, the link between evolvability and *internal selection* is regarded as distinctively crucial. In Pavličev's view, the original idea that modularity enhances evolvability (as formulated in G. Wagner and Altenberg 1996 but dating back to Olson and Miller 1958 and Riedl 1978) "included function" into the definition of evolvability (Pavličev). Modularity was not only conceived in terms of dimensionality, or the number of traits affected by mutations, but also referred to how mutations in functional modules were more likely to be selected than those breaking the function. According to Pavličev, "this aspect has been dropped in later usage of modularity, treating only the variational part, the reduced dimensionality that can be exposed to any kind of external selection." Yet, in contrast to external selection, internal selection is predictable, as it is dependent on the organismal structure: "there are unconditionally deleterious and likely advantageous or neutral directions." Therefore, "evolvability should include internal selection. Without it, we are measuring variation or variability, essentially selectability" (Pavličev). Frietson Galis's work on the role of negative pleiotropic effects in constraining the evolvability of body plans (starting in Galis 1999) aligns with this perspective.

Finally, several interviewees from different disciplinary areas locate the novelty of evolvability research in issues related not to evolvability itself, but to its evolution. For instance, when Richard Dawkins first wrote about the evolution of evolvability (Dawkins 1988), he saw it "as a heresy," an exception to his "emphasis on microevolutionary pressures." Although he now thinks that he "was wrong about it being heretical," Dawkins related the heresy to what he then interpreted as group selection when it was actually "clade selection" (Williams 1992). In his view, the heterodoxy of evolvability had nothing to do with the role of constraints in evolution:

I am perfectly happy with the idea that in what I called biomorph land [. . .] some corridors are harder to go down than others [. . .]. So if you think there is a controversy between internalist and externalist thinking, then that is my concession to internalism, but I never thought of it as a concession, because I think that it was obvious. (Dawkins)

Joanna Masel also emphasizes lineage selection as a recent expansion of the classical frame of evolutionary theory that was developed in connection with the evolution of evolvability in the early 2000s, a topic to which she contributed (Masel and Bergman 2003).

The conceptual roots of evolvability research cannot be attributed to a common intellectual descent. Instead, parallel roots lead to what is still not regarded as a single concept. Nonetheless, interdisciplinary exchanges did play an important role in the explosion and diversification of evolvability studies in the 2000s. There were many precedents in attempting to set up this interdisciplinary research agenda. The 1989 Dahlem Conference included a group discussion on the evolution of evolvability (Arnold et al. 1989). The discussion was coordinated by Steve Arnold and included people as diverse as Pere Alberch, Vilmos Csányi, Richard Dawkins, Sharon B. Emerson, Bernd Fritzsch, Tim J. Horder, John Maynard Smith, Matthias J. Starck, Elisabeth Vrba, Günter Wagner, and David Wake. The group report included almost every topic that later has been discussed in the evolvability literature, from the role of developmental constraints to levels of selection. However, the ideas discussed did not precipitate an alternative research front until the 2000s. A series of books published in the mid-1990s (Maynard Smith and Szathmáry 1995; Dawkins 1996; Raff 1996) seem to have played an important role in this regard. They all show up in the co-citation network (see figure 2.3b) and are acknowledged by at least one of our interviewees (Hansen) to have been influential in his thinking on evolvability.

More local, interdisciplinary interactions concern relationships between paleontology and quantitative genetics, between computer science and molecular evolution, and between theoretical chemistry and neutral network models. Jim Cheverud (1982, 1988) describes the novelty of his early work on morphological integration as the result of bridging phenotypic studies on integration from paleontology (Olson and Miller 1958) with agricultural studies. When asked about his major influences at that time, Cheverud refers to authors from different specialties, including Berg, Gould, Riedl, and Lande and Arnold, with whom he interacted in the early 1980s in the Chicago area. Another area of disciplinary overlap, reflected in the collaborations among Peter Schuster, Günter Wagner, Peter Stadler, and Andreas Wagner, was that between theoretical chemistry and molecular evolution. In turn, the emergence of evolutionary approaches in engineering and computer science acted as the enabling factor for evolvability becoming a research agenda. In the field of evolutionary engineering, G. Wagner points to Ingo Rechenberg's (1973) book as one of the earliest attempts to solve the evolvability problem. Rechenberg was a German aircraft engineer who developed an evolutionary method based on random changes for solving complex optimization problems. This method included "a feedback loop that optimizes the evolvability of that device." According to Wagner, the founders of evolutionary genetic programming reached the same conclusion:

you can successfully use random change as a way to improve things if and only if the variational process is tuned appropriately to solve these problems. So evolvability is not a trivial state of any replicating and varying system but needs to be built into the system in order [for evolution] to be possible. (G. Wagner)

In contrast to evolutionary engineering, computer science is significatively represented in the co-citation network (figure 2.3b) as an influential cluster of publications on genetic programming (Holland 1975; Goldberg 1989; Koza 1992) constituting the intellectual base for computational studies on evolvability (Kauffman 1993; Altenberg 1994). This ascendency is manifest in Richard Dawkins's seminal work (Dawkins 1988). In his interview, Dawkins admits that he was highly influenced by his attendance of an Artificial Intelligence workshop in Los Alamos in 1987, where he met Chris Langton, Stuart Kauffman, and Craig Reynolds.

2.4 Explaining the Origination of Evolvability Research

My analysis of the origination pattern of evolvability research shows that there were historical precedents of almost all relevant components of the evolvability research agenda, but they did not precipitate into such an agenda until the 2000s. This pattern of "conceptual lag" is not unique to evolvability research. Other approaches, such as eco-evo-devo (Love 2015) were also drafted in the late 1970s–early 1980s but did not crystallize as research agendas until 20 years later. Why did these new perspectives have to wait two decades to be pursued as core research programs in evolutionary biology? Which factors determine scientists' choices among available intellectual variants? (Toulmin 1967).

To answer this question, I will apply conceptual tools from cultural evolution theory as they have recently been applied to epistemic evolution and in particular to science dynamics (Richerson et al. 2013; Mesoudi et al. 2013; chapters 2–4 in Love and Wimsatt 2019; Fadda 2021). Cultural evolution theory applies models and metaphors drawn from evolutionary biology to explain the evolution of culture, including the history of science. Nonetheless, my use of these conceptual tools for explaining the origin and diversification of evolvability research will be intentionally metaphorical. I do not consider evolvability as a cultural replicator, nor do I endorse a population approach to the differential reproduction of cultural variants (Richerson and Boyd 2005). There are substantial differences between theoretical and biological variation (Thagard 1980) that I will not discuss here. From my perspective, the main advantage of evolutionary philosophies of science is that they allow us to offer integrated accounts of traditionally opposed perspectives of science, namely, internalist narratives based on the rationality of scientific progress and externalist reconstructions of the social norms governing scientific communities (Fadda 2021). In what follows, I embrace evolutionary explanations as loose analogies that help organize the many factors at play in the origination of evolvability research into two broad, nonmutually exclusive, explanatory kinds, namely, (1) "selectionist" explanations, and (2) "evolvability" explanations. While the former have been largely explored by evolutionary epistemologists, I will advance a novel internalist approach for the origination of evolvability research that might be generalizable to other episodes in the history of science.

2.4.1 Selection Criteria and the Dynamics of the Academic Landscape

Evolutionary epistemologists have mainly looked at science as a selection process (Hull 2001). Selection criteria concern the epistemic standards of what scientists find acceptable, and changes in these standards often depend on changes in the intellectual landscape that foster the acceptability of new theoretical perspectives. Interviewees appeal to several factors transforming the intellectual landscape for the acceptability of the evolvability perspective and moving it from theoretical debates to empirical studies. These factors include the incorporation of new and simpler model systems, such as prions, RNA molecules, or minimal genetic networks (Masel); the development of new molecular methods for engineering proteins in the late 1990s and early 2000s (A. Wagner); the discovery of genetic similarities between regulatory genes in the 1990s and the establishment of phylogenetic methods in the 1980s (G. Wagner); and the development of computational technologies since the early 1990s (A. Wagner). All these methodological innovations "helped forge an experimental, empirical paradigm that was meeting a field that was ready to move in that direction" (G. Wagner).

Changes in epistemic standards concern how conceptual innovations meet novel technological and conceptual niches, but dissemination of scientific ideas also depends on social criteria that bias their selection by individual scientists. In cultural evolution, "context biases" refer to sociological factors, such as the status or prestige of individuals, or the frequency of ideas in a given community, that play a role in the dissemination of ideas (Fadda 2021). For instance, prestige biases appeal to the disposition of individuals to instantiate the practices of successful individuals. A clear example of the role of academic status and social prestige of scientists in the acceptance and dissemination of evolvability ideas is that of Günter Wagner. As mentioned above, before the publication of his article with Altenberg in 1996, Wagner had written a series of papers on evolvability starting in 1981. However, as he himself recognizes, the fact that these papers were published "in an obscure place and by an obscure person" might explain why the basic idea did not make an impression in the field (G. Wagner). In contrast, several interviewees refer to Wagner moving to Yale in 1991 (Galis), gaining a MacArthur fellowship a year later (Cheverud) or publishing that paper in the well respected journal *Evolution* as playing a key role in the reception of evolvability ideas.

"Conformist biases" refer to the tendency to adopt the most common practice in a given population (Boyd and Richerson 1985). The use of heritability instead of evolvability as a measure of the potential to respond to selection is a good example of "intellectual inertia" in evolutionary quantitative genetics (Houle, Hansen). As a structured set of methodological and conceptual practices, conformist biases align with the kind of sociological factors that explain the ideological consistency of scientific communities during normal periods of science (Kuhn 1970). In turn, when a field is in a state of crisis, conformist biases are less likely to be followed, while new concepts and paradigms are more prone to proliferate (Thagard 1980). It might be argued that the proliferation of evolvability studies in the 2000s was enhanced by the critical interrogation of the foundations of the Modern Synthesis that started to grow on several disciplinary grounds in the late 1980s (see section 2.3).

Dialectical styles play an important role in the construction of scientific consensus. In his celebrated book on the origins of population genetics, Provine (1974, 25) argued that if Mendelians and biometricians had worked with, instead of against each other, the mathematical synthesis between Mendelian inheritance and natural selection attained by population genetics might have occurred 15 years earlier. Can the same be said about population geneticists and evo-devoists, or between micro- and macroevolutionists in the 1970s and early 1980s? Many of our interviewees agree that the clash between microevolutionists and paleontologists after the publication of the punctuated equilibrium hypothesis (Eldredge and Gould 1972) made integration a difficult enterprise. Mihaela Pavličev also points to dialectical styles as one of the factors accounting for the disappearance of mechanistic approaches in population genetics after the attempts made in the 1980s: "there were very powerful, present, people in the field that were probably very strongly advocating for their own approach to the questions" (Pavličev).

Dialectical styles do not characterize all members of opposite sides. For instance, Jim Cheverud's collaborators (Pavličev, Porto, and G. Wagner) agree that his nonconfrontational style facilitated synthesis. Nonetheless, the relationship between paleontologists and evolutionary biologists is no longer regarded as contentious as it was in the 1970s: "the relationship is a little more either benign neglect in terms of evolutionary theory and among paleontologists to a little bit more of people cooperative and interested on both sides" (Hunt); "I think that, more and more, graduate students come to know something about the fossil record as well as quantitative approaches in biology" (Liow). Regarding the conflict between organismal and statistical approaches to evolution, many young researchers are simply unaware of the existence of such a conflict (Pavličev). Massimo Pigliucci resorts to Plank's principle of generational replacement (Kuhn 1970, 151) to account for the resistance of evolutionary biologists to study evolvability, which he regards as a core component of the Extended Synthesis:

> Most of the people that are resistant to the Extended Evolutionary Synthesis are what I would consider at this point the old guard: Michael Lynch [. . .] Doug Futuyma, Jerry Coyne [. . .]. All of these people are still among the major critics of the Extended Synthesis and they're all on their way out in terms of their influence on the field and in terms of their careers. New generations come in, and now it's easy for the new students to talk about plasticity, niche construction, and evolvability. It's kind of like a second nature because they grew up with that literature and they don't see it as problematic. (Pigliucci)

Another populational factor in the spread of scientific ideas concerns the institutional "maturation" of ideas. When asked about the explosion of interest in evolvability in the mid-1990s, Pigliucci argues that scientific ideas need some time to mature, reaching a threshold when suddenly, enough researchers start working on new research topics.

> It may have a snowball effect when these students start to work on the topic and later start an academic career, request funding for that sort of stuff, and eventually they themselves are called by granting agencies to adjudicate grants, so they tend to fund that kind of research. So it takes about 20 years for that kind of development to occur. (Pigliucci)

The publication of reviews and popular science books on evolvability-related issues (see section 2.3) might have triggered this snowball effect in younger generations of evolutionary biologists.

2.4.2 From Conceptual Constraints to Evolvability of Theoretical Components

Together with selectionist explanations, evolution of culture theorists have pointed to the importance of constraints internal to practices, behaviors, or ideas accounting for the evolution of cultural variants. "Content biases" concern epistemic preferences of scientists based on what are perceived as theoretical virtues of scientific ideas (Fadda 2021). As opposed to context biases, they comprise *intrinsic* properties, such as simplicity or generality, that make some cultural items more prone to be copied than others. Independently of the academic status of their carriers, the success of some papers boosting research in evolvability notably depended on the clarity with which ideas that had been previously advanced were formulated.

This was the case of G. Wagner and Altenberg's 1996 paper or that of A. Wagner (2008) on the relationship between robustness and evolvability, a link already hinted at by previous works (e.g., Schuster et al. 1994) (Payne). In quantitative genetics, Houle and Hansen also had to publish the same ideas in more digestible ways (Hansen et al. 2011) to propagate the mean-standardized measure of evolvability in their scientific community (Hansen). In contrast, other concepts and tools, such as mutation matrices, have met greater resistance to being incorporated into the methodological repertoire of evolutionary biologists because of the intrinsic difficulties associated with their estimation. Massimo Pigliucci believes that, compared to plasticity, evolvability itself is a more difficult concept, and he speculates this might explain why it took longer to become a research agenda: "evolvability is less easy to grasp at a conceptual level, it has been explored for less time than plasticity, there are different types and levels of evolvability, and [it] is far more difficult to study empirically, especially in macroevolutionary-leaning aspects of the evolvability question." Notwithstanding its complexities, Pigliucci argues that evolvability has other internal, theoretical virtues related to *generality* that might explain its late but resounding success as a fundamental concept in evolutionary biology:

> I think evolvability plays a particularly interesting role, partly because it is such a high-level concept that can be applied widely, while none of the other concepts actually work the same way. You don't talk about phenotypic plasticity of a clade, or epigenetic inheritance between species. Niche construction gets closer because it can actually expand to different levels, but evolvability is really such a broad concept that it can expand on everything, from within population variation to major transitions: that's pretty much the entire span of evolutionary biology. So that's one reason why I think it is a fundamental concept. (Pigliucci)

In what might be interpreted as a population approach to "cultural evolvability," Mesoudi et al. (2004) cite a study showing that more heterogeneous teams of researchers make more discoveries than do teams composed of scientists with similar disciplinary backgrounds. They interpret this increase in scientific productivity as resulting from the ability of heterogeneous scientific communities to generate more variation in research outputs on which selection can act. In our case, interactions of researchers from different disciplinary backgrounds seem to have been instrumental in the diversification of evolvability research, as described in section 2.3. As I have argued elsewhere (Nuño de la Rosa 2017), overlapping disciplinary areas in evolvability research act as "trading zones" (Galison 1999) of concepts and methodologies that are later translated and operationalized in different disciplinary specialties.

Together with the structure of scientific communities, internal determinants of science development play an important role in the proliferation and diversification of research agendas. Some cultural evolution theorists have explored how links between social practices might confer different cultural evolvabilities on these practices (Sterelny 2006). Some practices (e.g., those related to social interaction) tend to be strongly associated, whereas others (e.g., technological and craft skills) tend to evolve in a more modular way, insofar as they can be adopted without influencing one another. Philosophers of science have also paid attention to the role of integration between scientific concepts and resulting patterns of conceptual covariation (Brigandt 2013; Love 2015). The ability of the evolvability conceptual framework to connect related concepts in evolutionary biology seems to have been crucial in the dissemination of evolvability ideas. For instance, David Houle believes that the accidental use of the same word in different disciplinary contexts (e.g., computer science and quantitative genetics) had a great effect in this regard. Johannes Jaeger comes to a similar conclusion: "it's good to have a term like that because, even if it means different things to different people, it focuses people on certain types of questions." In each disciplinary field, evolvability is also seen as a "meta-concept," insofar as it connects many related concepts into a unified research agenda. Benedikt Hallgrímsson regards it this way for the field of evo-devo:

> Evolvability as a concept is sort of a meta-evo-devo concept, because it needs to refer to robustness, integration, modularity, constraints, and all these core concepts of evolutionary developmental biology; and it makes sense to use it because there are some times when you want to refer to the collection of all those things, so you talk about evolvability. Or you want to refer to the connections of all those things and how they relate to the nondevelopmental determinants of evolvability, such as population genetics concepts. (Hallgrímsson)

Connections among concepts determine how they are transmitted through time and across disciplinary contexts. Just like traits, concepts not only travel in clusters, but during their journey, they also individuate in different theoretical contexts:

> You can take every concept in modern biology [. . .] and trace it back. What you find is that the concept exists in different contexts, I guess changed and reinterpreted by the context it has every time. [. . .] People that conceptualize new things are often not the best people to define it in a concrete way, because when things are first articulated, they are fuzzy and poorly connected to other concepts. Then we refine them, and we interpret them as time goes on. (Hallgrímsson)

The emergence of evolvability research in the field of quantitative genetics offers an exemplary case study to investigate the individuation of a scientific concept in a given disciplinary background. In his interview, Thomas Hansen identifies "two events or theoretical developments in evolutionary biology that set the stage for the study of evolvability as something separate from selection" (see also Hansen and Pélabon 2021). The first event had to do with the separation of *selection from inheritance* in Fisher's Fundamental Theorem of 1930 (see Frank 2012). In Darwin's original formulation, selection and inheritance were deeply entangled. The Modern Synthesis inherited this view, and the implication of Fisher's theorem passed unnoticed until Price reinterpreted it in the early 1970s, making the separation between selection and transmission explicit: "What this [separation] allows is that we can theoretically and empirically study selection without bothering about inheritance, which is something that we really couldn't do before" (Hansen).

The second, and independent, event was the separation between *selection and constraints* in quantitative genetics introduced by Lande and Arnold (1983) when they wrote the response to selection as the G-matrix multiplied by the selection gradient, operationalized as multiple regression: "This provided two tools to study both variation and selection in the field. So people could go out and estimate selection gradients in the field without worrying about genetics at all. And at the same time, you can study the G-matrix independently of the selection, which typically happened in the lab, and later also in actual populations" (Hansen). Given that in Houle's (1992) definition of evolvability, the G-matrix is what determines evolvability in the short term, the separation between the G-matrix and the selection gradient was a fundamental step in the autonomization of evolvability as a research agenda. According to Hansen, it was this separation that "really facilitated the study of evolvability as a separate entity."

As already mentioned, neither Lande nor Arnold interpret their article in these terms. When asked about the novelty of their contribution, they both highlight the extension of the breeder's equation to multiple characters and the measure of selection as multiple regression. Neither of them refers to the implications of this measure for the separation between selection and constraints, a distinction that, according to Lande, he had previously introduced (Lande 1979, 1982). This is again an example of theoretical developments that are not consciously intended by their creators: "In developing a way to measure selection, they just happen to set things up in a way that made it possible to study evolvability" (Hansen). Indeed, in using variance standardization of traits, Lande and Arnold reintroduced the correlation between the selection gradient and the G-matrix, leading to a paradoxical situation: "Since everybody copied their approach, we got a situation where the theory was conceptually correct but was empirically implemented in a way that had all the old problems. This is what David [Houle] and the rest of us have been trying to straighten out with the mean-standardizations" (Hansen).

From this perspective, the mean-standardized measure of evolvability is not "just one of several measures" (Lande) of a theory that was already at work. Instead, it results from a theoretical reinterpretation in the evolvability framework that takes into account the centrality of scale in evolutionary biology. This argument needs to be understood within the more general framework relating meaning and measurement in biological theory (Houle et al. 2011; Houle and Pélabon, chapter 6). From this perspective, the life of scientific concepts can be understood as the result of a transition from verbal models to measurement. Hansen and Houle (2008) identify the lack of such a theory as one of the reasons that evolvability has received relatively little attention until recently. Recent work on evolvability has been precisely "about creating this theoretical context for it to be meaningful and operational" (Hansen). For instance, the move from constraints to evolvability in the work on *Dalechampia* (Hansen et al. 2003a,b) depended on the operationalization of ideas that had been previously formulated in a vague manner. An example is *conditional evolvability,* which was used "as a method for quantifying evolutionary modularity, which was a very important concept in the 1990s but [. . .] was rather vague" (Hansen).

The modularity of theoretical components can be applied to further distinctions accounting for the autonomous development of evolvability as a research agenda. In particular, the *separation between the G-matrix and the M-matrix* in quantitative genetics can be interpreted as a third event in theoretical decoupling that allowed evolvability to be studied independently of selection. In classical evolutionary theory, the mutational input was treated as a fixed parameter, while evolvability research is precisely interested in understanding how mutational effects and mutation rates evolve:

> I think the connection between evo-devo and evolutionary quantitative genetics came through that: because we need the genotype-phenotype [GP] map to study mutational effects. [. . .]. If we think about the GP map as an abstract mathematical function, as you move around in this landscape, mutational effects will change. Basically, a change in the effect of mutations is the same as a change in the effect of the difference between two alleles that segregate in a population. This is determined by the GP map and therefore we need to mathematicize the GP map and put it into the theoretical population genetics framework. (Hansen)

In this context, a further conceptual distinction that has contributed to the independent study of the variational determinants of evolvability has been that between *biological and statistical epistasis*. While biological epistasis refers to the dependency of the phenotypic effects of mutations on the genetic context, statistical epistasis refers to the statistical deviation from additivity in a given population. This conceptual distinction arose gradually in the early 1990s, starting with Cheverud and Routman (1996). Since then, Hansen, G. Wagner, and collaborators (e.g., Hansen et al. 2006) have worked on elucidating the effects of epistasis on the selection response. In this regard, the individuation of the epistasis concept is again regarded as resulting from it becoming a meaningful concept that can be appropriately measured and integrated with population genetics theory:

> In classical population genetics theory [epistasis] is largely something to average over, because the main concern was the changing gene frequency or the change in the mean of a character under natural selection, and there the influence of epistasis is small. [. . .] However, if you think that variational properties of organisms change in evolution, you want to know how they do it, and thus there has to be fundamentally a question of how epistatic effects and other forms of context-dependency contribute to and are involved in evolutionary change at the population genetic level. [. . .] How to properly define epistasis is the problem of how to define a quantitative concept, which is measurement theory. (G. Wagner)

I have argued that the modularization of theoretical components accounts for the autonomous development of evolvability as a research agenda, at least in quantitative genetics. But why evolvability instead of constraints? As mentioned in section 2.1, the ascendancy of evolvability in the 1990s coincides with the decline, or at least the stasis, of constraints (see figure 2.1). However, Ingo Brigandt (2015) has argued that we should not interpret this pattern as a replacement but rather as a transformation of the concern with constraints into a research program on evolvability. From this perspective, evolvability was instrumental in overcoming the vigorous debate between constraints and selection that predated evolutionary biology in the 1980s and 1990s. By emphasizing the positive side of constraints, from the prevention to the facilitation of change, evolvability played an important role in dissolving this dichotomy (Love 2015). Our interviewees agree that constraints and evolvability are two sides of the same coin and that research on constraints can be easily translated into research on evolvability. However, they perceive evolvability as a more productive framework: "Even if one can literally map almost everything from evolvability back into that framework, it's been, I think, just a better way of thinking about it" (Hunt). Younger scholars working on microevolutionary studies of macroevolutionary variation agree: Although evolutionary geneticists working on macroevolutionary timescales were convinced that phenotypic variation would eventually be found in every dimension, "the idea that there are dimensions in which evolution proceeds faster was very attractive" (Porto). Therefore, the theoretical decoupling of evolvability from selection was not a mere autonomization: It required the integration of evolvability into a common theoretical context where selection was a key explanatory component. Studies on the evolution of evolvability played a crucial role in this integration of evolvability in evolutionary theory, insofar as they opened up the possibility of explaining evolvability as a result of selection (Okasha 2018).

The origination and diversification of an interdisciplinary research agenda such as evolvability is a complex episode in science dynamics. Only a simultaneous consideration of explanatory approaches can do full justice to the multidimensional phenomena involved in cultural evolution (Love and Wimsatt 2019). An evaluation of the relative weight of the external social and internal theoretical factors involved in the origination of evolvability research is beyond the scope of this chapter, but I believe the major factors in play have been at least outlined here. In particular, I have argued that the individuation of scientific concepts in different theoretical contexts, as well as the modularization of theory components, played an important role in enabling the independent development and diversification of evolvability research. I hope to have persuaded the reader that biological studies on evolvability have promising metatheoretical consequences for the understanding of science itself.

Acknowledgments

Interviews were conducted between 2019 and 2021 in the framework of the project "Evolvability: a new and unifying concept in evolutionary biology," funded by the Centre for Advanced Study (CAS) at the Norwegian Academy of Science and Letters. I am greatly indebted to CAS for its generous funding, which together with a grant from the Spanish Ministry of Economy and Competitiveness (project PGC2018-099423-B-I00) made this research possible. Thomas Hansen and Christophe Pélabon have been enthusiastic supporters of the interviews project, which received additional financial support from CAS (project UCM-ART.83 246-2020). I am greatly indebted to Silvia Basanta and very specially to Cristina Villegas for their support in preparing and transcribing the interviews. I thank all participants of the CAS project for stimulating discussions on the history of evolvability research. My greatest acknowledgment goes to the interviewees, who gave selflessly of their time to answer my questions. They all have revised the content of their interview transcript and approved its use for academic purposes. This approval does not entail endorsement of any of the claims here defended or reproduced from others.

References

Alberch, P. 1991. From genes to phenotype: Dynamical systems and evolvability. *Genetica* 84: 5–11.

Altenberg, L. 1994. The Evolution of Evolvability in Genetic Programming. *Advances in Genetic Programming* 3: 47–74.

Ancel, L. W., and W. Fontana. 2000. Plasticity, evolvability, and modularity in RNA. *Journal of Experimental Zoology B* 288: 242–283.

Armbruster, W. S., V. S. Di Stilio, J. D. Tuxill, T. C. Flores, and J. L.V. Runk. 1999. Covariance and decoupling of floral and vegetative traits in nine neotropical plants: A re-evaluation of Berg's correlation-pleiades concept. *American Journal of Botany* 86: 39–55.

Armbruster, W. S., C. Pélabon, G. H. Bolstad, and T. F. Hansen. 2014. Integrated phenotypes: Understanding trait covariation in plants and animals. *Philosophical Transactions of the Royal Society B* 369: 20130245.

Arnold, S. J., P. Alberch, V. Csányi, R. Dawkins, S. B. Emerson, B. Fritzch, T. J. Horder, J. Maynard Smith, M. J. Starck, E. S. Vrba, G. P. Wagner, and D. B. Wake. 1989. How do complex organisms evolve? In *Complex Organismal Functions: Integration and Evolution in Vertebrates,* edited by D. B. Wake and G. Roth, 403–433. Chichester, UK: Wiley.

Berg, R. L. 1960. The ecological significance of correlation pleiades. *Evolution* 14: 171–180.

Boyd, R., and P. J. Richerson. 1985. *Culture and the Evolutionary Process.* London: University of Chicago Press.

Brigandt, I. 2013. Integration in biology: Philosophical perspectives on the dynamics of interdisciplinarity. *Studies in History and Philosophy of Biological and Biomedical Sciences* 44: 461–465

Brigandt, I. 2015. From developmental constraint to evolvability: How concepts figure in explanation and disciplinary identity. *Boston Studies in the Philosophy and History of Science* 307: 305–325.

Brigandt, I., and A. C. Love. 2012. "Conceptualizing evolutionary novelty: Moving beyond definitional debates." *Journal of Experimental Zoology B* 318: 417–427.

Burch, C. L., and L. Chao. 2000. Evolvability of an RNA virus is determined by its mutational neighbourhood. *Nature* 406: 625–628.

Chavalarias, D., and J. P. Cointet. 2013. Phylomemetic patterns in science evolution—The rise and fall of scientific fields. *PLOS ONE* 8: e54847.

Chavalarias, D., P. Huneman, and T. Racovski. 2022. Using phylomemies to investigate the dynamics of science. In *The Dynamics of Science: Computational Frontiers in History and Philosophy of Science,* edited by G. Ramsey and A. De Block. Pittsburgh: Pittsburgh University Press.

Cheverud, J. M. 1982. Phenotypic, genetic, and environmental morphological integration in the cranium. *Evolution* 36: 499–516.

Cheverud, J. M. 1988. A comparison of genetic and phenotypic correlations, *Evolution* 42: 958–968.

Cheverud, J. M. 1996. Developmental integration and the evolution of pleiotropy. *American Zoologist* 36: 44–50.

Cheverud, J. M., and E. J. Routman. 1996. Epistasis as a source of increased additive genetic variance at population bottlenecks. *Evolution* 50: 1042–1051.

Conner, J. K., and R. Lande. 2014. Raissa L. Berg's contributions to the study of phenotypic integration, with a professional biographical sketch. *Philosophical Transactions of the Royal Society B* 369: 20130250.

Crother, B., and C. Murray. 2019. Early usage and meaning of evolvability. *Ecology and Evolution* 9: 3784–3793.

Dawkins, R. 1988. The evolution of evolvability. In *Artificial Life: The Proceedings of an Interdiciplinary Workshop on the Synthesis and Simulation of Living Systems,* edited by C. Langton, 201–220. Santa Fe, NM: Addison Wesley.

Dawkins, R. 1996. *Climbing Mount Improbable.* London: Norton.

Eigen, M. 1971. Self-organization of matter and the evolution of biological macromolecules. *Die Naturwissenschaften* 58: 465–523.

Eigen, M., and P. Schuster. 1979. *The Hypercycle. A Principle of Natural Self-Organization.* Berlin: Springer Verlag.

Eldredge, N., and S. J. Gould. 1972. Punctuated equilibria: An alternative to phyletic gradualism. In *Models in Paleobiology,* edited by T. Schopf, 82–115. San Francisco: Freeman, Cooper & Co.

Fadda, A. 2021. Population thinking in epistemic evolution: Bridging cultural evolution and the philosophy of science. *Journal for General Philosophy of Science* 52: 351–369.

Falconer, D. S. 1989 [1960]. *Introduction to Quantitative Genetics.* New York: Longman Scientific and Technical.

Frank, S. A. 2012. Wright's adaptive landscape versus Fisher's fundamental theorem. In *The Adaptive Landscape in Evolutionary Biology,* edited by E. I. Svensson and R. Calsbeek, 41–57. Oxford: Oxford University Press.

Gabriel, W., and M. Lynch. 1993. Muller's ratchet and mutational meltdowns. *Evolution* 47: 1744–1757.

Galis, F. 1999. Why do almost all mammals have seven cervical vertebrae? Developmental constraints, Hox genes, and cancer. *Journal of Experimental Zoology B* 285: 19–26.

Galison, P. 1999. Trading zone: Coordinating action and belief. *The Science Studies Reader,* 137–160. New York: Routledge.

Goldberg, D. E. 1989. *Genetic Algorithms in Search, Optimization, and Machine Learning.* Addison-Wesley.

Goodwin, B. 1994. *How the Leopard Changed Its Spots.* London: Weidenfeld & Nicolson.

Gould, S. J., and R. C. Lewontin. 1979. The spandrels of San Marco and the Panglossian paradigm: A critique of the adaptationist programme. *Proceedings of the Royal Society B* 205: 581–598.

Hansen, T. F, and D. Houle. 2008. Measuring and comparing evolvability and constraint in multivariate characters. *Journal of Evolutionary Biology* 21: 1201–1219.

Hansen, T. F., and C. Pélabon. 2021. Evolvability: A quantitative-genetics perspective. *AREES* 52: 153–175.

Hansen, T. F., C. Pélabon, W. S. Armbruster, and M. L. Carlson. 2003a. Evolvability and genetic constraint in *Dalechampia* blossoms: Components of variance and measures of evolvability. *Journal of Evolutionary Biology* 16:754–766.

Hansen, T. F., W. Armbruster, M. Carlson, and C. Pélabon. 2003b. Evolvability and genetic constraints in *Dalechampia* blossoms: Genetic correlations and conditional evolvability. *Journal of Experimental Zoology B* 296: 23–39.

Hansen, T. F., J. M. Alvarez-Castro, A. J. R. Carter, J. Hermisson, and G. P. Wagner. 2006. Evolution of genetic architecture under directional selection. *Evolution* 60: 1523–1536.

Henrich, J., and R. Boyd. 2002. On modelling cognition and culture: Why cultural evolution does not require replication of representations. *Culture and Cognition* 2: 87–112.

Holland, J. H. 1975. *Adaptation in Natural and Artificial Systems.* Cambridge, MA: MIT Press.

Houle, D. 1992. Comparing evolvability and variability of quantitative traits. *Genetics* 130: 195–204.

Houle, D., C. Pélabon, G. P. Wagner, and T. F. Hansen. 2011. Measurement and meaning in biology. *Quarterly Review of Biology* 86: 1–32.

Hull, D. L. 2001. *Science and Selection: Essays on Biological Evolution and the Philosophy of Science.* Cambridge: Cambridge University Press.

Huynen, M. A., P. F. Stadler, and W. Fontana. 1996. Smoothness within ruggedness: The role of neutrality in adaptation. *PNAS* 93: 397–401.

Jones, A. G., S. J. Arnold, and R. Bürger. 2007. The mutation matrix and the evolution of evolvability. *Evolution* 61: 727–745.

Kauffman, S. A. 1993. *The Origins of Order: Self-organization and Selection in Evolution.* Oxford: Oxford University Press.

Kirschner, M. W., and J. C. Gerhart. 1998. Evolvability. *PNAS* 95: 8420–8427.

Koza, J. R. 1992. *Genetic Programming: On the Programming of Computers by Means of Natural Selection.* Cambridge, MA: MIT Press.

Kuhn, T. 1970 (2nd ed.) *The Structure of Scientific Revolutions.* Chicago: University of Chicago Press.

Lande, R. 1979. Quantitative genetic analysis of multivariate evolution, applied to brain-body size allometry. *Evolution* 33: 402–416.

Lande, R. 1980. The genetic covariance between characters maintained by pleiotropic mutations. *Genetics* 94: 203–215.

Lande, R. 1982. A quantitative genetic theory of life history evolution. *Ecology* 63: 607–615.

Lande, R., and S. J. Arnold. 1983. The measurement of selection on correlated characters. *Evolution* 37: 1210–1226.

Lennox, J. G. 2001. History and philosophy of science: A phylogenetic approach. *História, Ciências, Saúde-Manguinhos* 8: 655–669.

Lipman, D. J., and W. L. Wilbur, 1991. Modelling neutral and selective evolution of protein folding. *Proceedings of the Royal Society* 245: 7–11.

Lloyd, E. A., and S. J. Gould, 1993. Species selection on variability. *PNAS* 90: 595–599.

Love, A. C. (editor) 2015. *Conceptual Change in Biology: Scientific and Philosophical Perspectives on Evolution and Development.* Dordrecht: Springer.

Love, A. C., and W. C. Wimsatt (editors). 2019. *Beyond the Meme: Development and Structure in Cultural Evolution.* Minneapolis: University of Minnesota Press.

Masel J., and A. Bergman. 2003. The evolution of the evolvability properties of the yeast prion [PSI+]. *Evolution* 57: 1498–1512.

Maynard Smith, J., and E. Szathmáry. 1995. *The Major Transitions in Evolution.* Oxford: Oxford University Press.

Mesoudi, A., A. Whiten, and K. N. Laland. 2004. Perspective: Is human cultural evolution Darwinian? Evidence reviewed from the perspective of the *Origin of Species. Evolution* 58: 1–11.

Mesoudi, A., K. N. Laland, and R. Boyd. 2013. The cultural evolution of technology and science. In *Cultural Evolution,* edited by P. J. Richerson and M. H. Christiansen, 193–216. Cambridge, MA: MIT Press.

Messina, F. J. 1993. Heritability and 'evolvability' of fitness components in *Callosobruchus maculatus. Heredity* 71: 623–629

Mousseau, T. A., and D. A. Roff. 1987. Natural selection and the heritability of fitness components. *Heredity* 59: 181–197.

Nuño de la Rosa, L. 2017. Computing the extended synthesis: Mapping the dynamics and conceptual structure of the evolvability research front. *Journal of Experimental Zoology B* 328: 395–411.

Okasha, S. 2018. The strategy of endogenization in evolutionary biology. *Synthese* 198: 3413–3435

Olson, E., and R. Mi̇ller. 1958. *Morphological Integration.* Chicago: University of Chicago Press.

Orr, H. A. 2002. The population genetics of adaptation: The adaptation of DNA sequences. *Evolution* 56: 1317–1330.

Pavličev, M. 2016. Pleiotropy and its evolution: Connecting evo-devo and population genetics. In *Evolutionary Developmental Biology: A Reference Guide,* edited by L. Nuño de la Rosa and G. B. Müller, 1087–1096. Cham, Switzerland: Springer.

Pigliucci, M. 2008. Is evolvability evolvable? *Nature Reviews Genetics* 9 (1): 75–82.

Pigliucci, M. 2009. An extended synthesis for evolutionary biology. *Annals of the New York Academy of Sciences* 1168: 218–228.

Provine, W. B. 1974. *The Origins of Theoretical Population Genetics.* Chicago: University of Chicago Press.

Raff, R. 1996. *The Shape of Life: Genes, Development, and the Evolution of Animal Shape.* Chicago: University of Chicago Press.

Rechenberg, I. 1973. *Evolutionsstrategie.* Stuttgart: Friedrich Frommann Verlag.

Richerson, P. J., and R. Boyd 2005: *Not by Genes Alone: How Culture Transformed Human Evolution.* Chicago: University of Chicago Press.

Richerson, P. J., M. H. Christiansen, A. Mesoudi, et al. 2013. The cultural evolution of technology and science. In *Cultural Evolution: Society, Technology, Language and Religion,* edited by M. H. Christiansen and P. J. Richerson, 193–216. Cambridge, MA: MIT Press.

Riedl, R. 1978. *Order in Living Organisms: A Systems Analysis of Evolution.* Wiley.

Rutherford, S. L. 2000. From genotype to phenotype: Buffering mechanisms and the storage of genetic information. *BioEssays* 22: 1095–1105.

Rutherford, S. L., and S. Lindquist. 1998. Hsp90 as a capacitor for morphological evolution. *Nature* 396 (6709): 336–342.

Sander K. 1983. The evolution of patterning mechanisms: Gleanings from insect embryogenesis and spermatogenesis. In *Development and Evolution,* edited by B. C. Goodwin, N. Holder, and C. C. Wylie, 137–159. Cambridge: Cambridge University Press.

Schuster, P., W. Fontana, P. F. Stadler, and I. L. Hofacker. 1994. From sequences to shapes and back: A case study in RNA secondary structures. *Proceedings of the Royal Society B* 255: 279–284.

Small, H., and B. C. Griffith. 1974. The Structure of Scientific Literatures I: Identifying and Graphing Specialties 4(1). *Science Studies:* 17–40.

Sterelny, K. 2006. The evolution and evolvability of culture. *Mind & Language* 21: 137–165.

Stoltzfus, A. 2012. Constructive neutral evolution: Exploring evolutionary theory's curious disconnect. *Biology Direct* 7: 1–13.

Stoltzfus, A. 2017. Why we don't want another "Synthesis." *Biology Direct* 12: 1–12.

Thagard, P. 1980. Against evolutionary epistemology. *PSA: Proceedings of the Biennial Meeting of the Philosophy of Science Association* 1: 187–196.

Toulmin, S. 1967. The evolutionary development of science. *American Scientist* 55: 456–471.

Van Nimwegen, E., J. P. Crutchfield, and M. Huynen. 1999. Neutral evolution of mutational robustness. *PNAS* 96: 9716–9720.

Vermeij, G. J. 1973. Adaptation, versatility, and evolution. *Systematic Biology* 22: 466–477

Von Dassow, G., and E. Munro. 1999. Modularity in animal development and evolution: Elements of a conceptual framework for EvoDevo. *Journal of Experimental Zoology B* 285: 307–325.

Von Dassow, G., E. Meir, E. M. Munro, and G. M. Odell. 2000. The segment polarity network is a robust developmental module. *Nature* 406: 188–192.

Wagner, A. 2008. Robustness and evolvability: A paradox resolved. *Proceedings of the Royal Society B* 275: 91–100.

Wagner, G. P. 1981. Feedback selection and the evolution of modifiers. *Acta Biotheoretica* 30: 79–102.

Wagner, G. P., and L. Altenberg. 1996. Perspective: Complex Adaptations and the Evolution of Evolvability. *Evolution* 50: 967–976.

Wagner, G. P., G. Booth, and H. Bagheri-Chaichian. 1997. A population genetic theory of canalization. *Evolution* 51: 329–347.

West-Eberhard, M. J. 1998. Evolution in the light of developmental and cell biology, and vice versa. *Proceedings of the National Academy of Sciences of the United States of America* 95 (15): 8417–8419.

Williams, G. C. 1992. *Natural Selection: Domains, Levels, and Challenges.* Oxford: Oxford University Press.

Wright, S. 1932. The roles of mutation, inbreeding, crossbreeding and selection in evolution. *Proceedings of the Sixth International Congress on Genetics* 1: 356–366.

3 Conceptual Roles of Evolvability across Evolutionary Biology: Between Diversity and Unification

Cristina Villegas, Alan C. Love, Laura Nuño de la Rosa,
Ingo Brigandt, and Günter P. Wagner

Biologists and philosophers have noted the diversity of interpretations of evolvability in contemporary evolutionary research. Different clusters of research identified by co-citation patterns or shared methodological orientation sometimes concentrate on distinct conceptions of evolvability. We examine five different activities where the notion of evolvability plays conceptual roles in evolutionary biological investigation: setting a research agenda, characterization, explanation, prediction, and control. Our analysis of representative examples demonstrates how different conceptual roles of evolvability are quasi-independent and yet exhibit important relationships across scientific activities. It also provides resources to detail two distinct strategies for how evolvability can help synthesize disparate areas of research and thereby potentially serve as a unifying concept in evolutionary biology.

3.1 Introduction

Evolvability is a property of living systems that refers broadly to their capacity, ability, or potential to evolve. However, the property is conceptualized in different ways when used by biologists (see also Brigandt et al., chapter 4).[1] For example, some researchers attribute evolvability to populations and construe it in terms of the ability to respond to selection (Houle 1992), whereas others attribute evolvability to organisms and understand it as the capacity to generate heritable phenotypic variation (G. Wagner and Altenberg 1996; Kirschner and Gerhart 1998). Some biologists and philosophers have noted the diverse interpretations of evolvability found in contemporary evolutionary research, leading Brown (2014) to describe the evolvability literature as conceptually confusing.

One philosophical response to this situation is to identify a central or core meaning for the concept of evolvability. Differences in conceptualization are then understood as variations on this primary or basic meaning, such as "the joint causal influence of . . . internal features [of populations] on the outcomes of evolution" (Brown 2014, 549). However, it quickly becomes difficult to specify what counts as an internal feature of a population (Love 2003). Similar difficulties arise when attempts are made to identify the essence of a scientific concept (e.g., "gene"; Griffiths and Stotz 2013). Another response is to argue

1. References to chapter numbers in the text are to chapters in this volume.

that diverse interpretations correspond to distinct phenomena (Pigliucci 2008a), although this response raises the question of why the same term "evolvability" is used.

A third response is to analyze what these different conceptualizations accomplish in scientific reasoning. It assumes that the variation in conceptualization is there for a reason and plays some functional role. Understanding these functional roles is potentially relevant to ongoing empirical inquiry, because, once understood, they can be more actively marshaled to perform scientific tasks. Complementary possibilities for functional roles include (a) tracking distinct methodological approaches to a phenomenon of interest, (b) representing distinct scientific aims (either in or across disciplines), and (c) locating different commitments about the significance of a concept in a set of theoretical assumptions (e.g., is it central or peripheral to a particular explanation?) or with respect to its range of application (e.g., is the concept intended to apply only under particular circumstances or be fully general?). The present analysis adopts the orientation of this third response and is motivated by empirical evidence that points toward these possible functional roles as being operative across evolvability research.

A recent, large-scale citation analysis demonstrates that there are several co-citation clusters of research that concentrate on distinct conceptions of evolvability, either from a specific disciplinary or shared methodological orientation (Nuño de la Rosa 2017). These clusters map onto six broad disciplinary approaches: evo-devo, complex network analysis, molecular evolution, population genetics, quantitative genetics, and macroevolutionary studies. However, the clusters overlap and do not cleanly separate along disciplinary lines. This overlap is suggestive of links across different fields of evolutionary inquiry that might correspond to different functional roles. These links could help synthesize associated theoretical commitments among conceptions of evolvability and their evidential underpinnings into a more general perspective on evolutionary processes.

This chapter takes as its starting point the different interpretations of evolvability and these intriguing patterns of usage in and across research clusters. We leverage this diversity to address the question of what *conceptual roles* evolvability plays across evolutionary biology. In particular, we identify and examine multiple scientific activities in which the concept of evolvability plays a role in evolutionary biological investigation: setting a research agenda, characterization, explanation, prediction, and control. Our primary goal is to better grasp how the notion of evolvability is functioning in the investigative practices of evolutionary biologists. The existence of different possible conceptual roles provides a rationale for why we might expect to find distinct interpretations of a central concept, a pattern that can be observed for many central concepts in biology, such as "species" (Hey 2001) or "gene" (Griffiths and Stotz 2013), as well as in other sciences, such as chemistry or physics (e.g., "hardness" in materials science; Wilson 2006). Additionally, an understanding of differences in conceptual roles for distinct activities could yield resources to bridge different investigative approaches and thereby provide routes to synthesize findings about evolvability across disciplinary boundaries.

Our analysis of how biologists use the concept of evolvability to fulfill distinct roles in their various activities also can foster an understanding of its success or failure in accomplishing investigative work (see also Nuño de la Rosa, chapter 2). Once we better understand that different conceptions can exhibit distinct roles in various scientific activities and how they do so, then we are positioned to ask whether a particular conception *can or should* play a specific role in inquiry. This type of question can be elaborated to scrutinize how these distinct roles are related to one another and whether (and to what degree) these relations facilitate the successful investigation of evolvability. This approach takes on special significance, because one or more of the roles that evolvability plays might serve to unify disparate areas of research in evolutionary biology.

We commence our analysis by distinguishing five different activities relevant to evolutionary biology where evolvability plays a role: setting a research agenda, characterization, explanation, prediction, and control (section 3.2). Next, we turn to questions about how different activities can be related to one another or are jointly operative in evolvability research (section 3.3). Finally, we argue that focusing on the role of evolvability in the activity of setting a research agenda could be strategic for unifying fields of study, such as evo-devo, complex network analysis, quantitative genetics, and macroevolutionary studies in contemporary evolutionary research (section 3.4).

3.2 Conceptual Roles for Evolvability

A concept plays a role as a *tool* when it is used to accomplish a particular end in the context of a scientific activity, or as a *target* when it represents a particular aim for an activity of scientific inquiry. Concepts can play the same role in different activities, and different roles can operate in the same activity. In what follows, we range over research programs and disciplinary approaches to illuminate different instances of roles of evolvability in various scientific activities.

3.2.1 Setting a Research Agenda

Setting a research agenda functions to guide ongoing investigative efforts and motivate future research. One positive effect of the concept of evolvability emerging and increasing in prominence in the 1990s was to encourage investigation into the scope and generation of phenotypic variation independently of its selective value (Nuño de la Rosa, chapter 2). Hendrikse et al. (2007) illustrate this vision of evolvability as establishing a research agenda, singling it out as a central problem in biology and the primary problem of evo-devo (see also Minelli 2010). Although this framing may foster the neglect of other crucial questions in evo-devo (Müller 2021), it shows how an approach can set its own research agenda around evolvability questions that were not answered or even articulated by either developmental biology or traditional evolutionary biology.

A research agenda not only highlights phenomena in need of investigation but also has an internal architecture that gives direction to scientific investigation and coordinates efforts across research groups (Love 2008, 2013). Such a *problem structure* consists of systematic relations between the individual component questions that make up the agenda. For example, the concept of evolutionary novelty functions to set an agenda, directing and coordinating attempts to account for the origin of characters (Brigandt and Love 2012). The problem structure of such a research agenda indicates how different explanatory contributions are to be synthesized (Love 2021), such as how modifications in lower-level traits (e.g., gene regulatory mechanisms) yield changes in higher-level traits (e.g., cellular interactions and tissue formation). Similarly, the research agenda of evolvability has its own problem structure. The proposal by Hendrikse et al. (2007) suggests some of the relevant problem structure for an evolvability research agenda. They articulate two related domains: "(i) Bias in the direction of variation generated" and "(ii) Modulation of the amount of variation generated" (Hendrikse et al. 2007, 396). From their evo-devo perspective, it is crucial to understand the interrelated developmental-genetic basis of both domains. Additional structure includes how investigations of phenomena contributing to evolvability—such as modularity, heterochrony, morphological integration, and canalization—can be coordinated.

Concepts that play roles in agenda setting can provide concrete guidance for a specific approach or field, such as evolvability research in evo-devo (Hendrikse et al. 2007). However, an agenda-setting concept also can be a tool for mapping out a landscape of research that is relevant to multiple biological fields. The landscape of such a research agenda can be described as a "trading zone" (Galison 1999): an interdisciplinary area of collaboration in which members of different scientific communities exchange concepts, methods, and results that are then translated into the specific language of these different communities. Thus, evolvability need not only be a central problem for evo-devo; it may well function to set a research agenda across evolutionary biology, with a problem structure capable of coordinating interdisciplinary research and even uniting efforts from various fields (Brigandt 2015b; Nuño de la Rosa 2017; see figure 3.1). For example, theoretical insights about the relationship between modularity and evolvability have been shared across disciplines, even though the notion of modularity is defined differently in terms of topological connections (computational evolution), developmental interactions (evo-devo), or constrained pleiotropic effects (quantitative genetics). The concept of evolvability's role as a *tool* for setting a research agenda that coordinates interdisciplinary research makes it a natural candidate to consider when exploring how evolvability might serve as a unifying concept (see section 3.4).

3.2.2 Characterization

In a research agenda, it is crucial for scientists to adequately characterize the phenomenon of evolvability. Often this characterization involves finding one or more working definitions. The way a phenomenon is characterized or defined "sets the frame" of an inquiry—it tells researchers what to attend to and what needs to be predicted, explained, or controlled (Colaço 2018). Characterizing involves distinguishing artifacts from genuine results or one phenomenon from another; identifying normal precipitating, inhibiting, and modulating conditions; and detailing the amount of variation possible for a phenomenon to exhibit (Craver and Darden 2013). In most of these situations, evolvability plays a role as the *target* of characterization by representing what the phenomenon is, the conditions that permit its manifestation, or how it differs from other biological phenomena. For example, characterizing evolvability as "the ability of a population to respond to directional selection" (Hansen and Houle 2008) helps distinguish a capacity for phenotypic change from the strength and direction of selection. An elaborated conception from quantitative genetics in terms of additive genetic variance provides a specific characterization of the causal basis of evolvability that details how it and directional selection operate as separate factors that result in phenotypic change (Hansen 2006; Hansen, chapter 5).

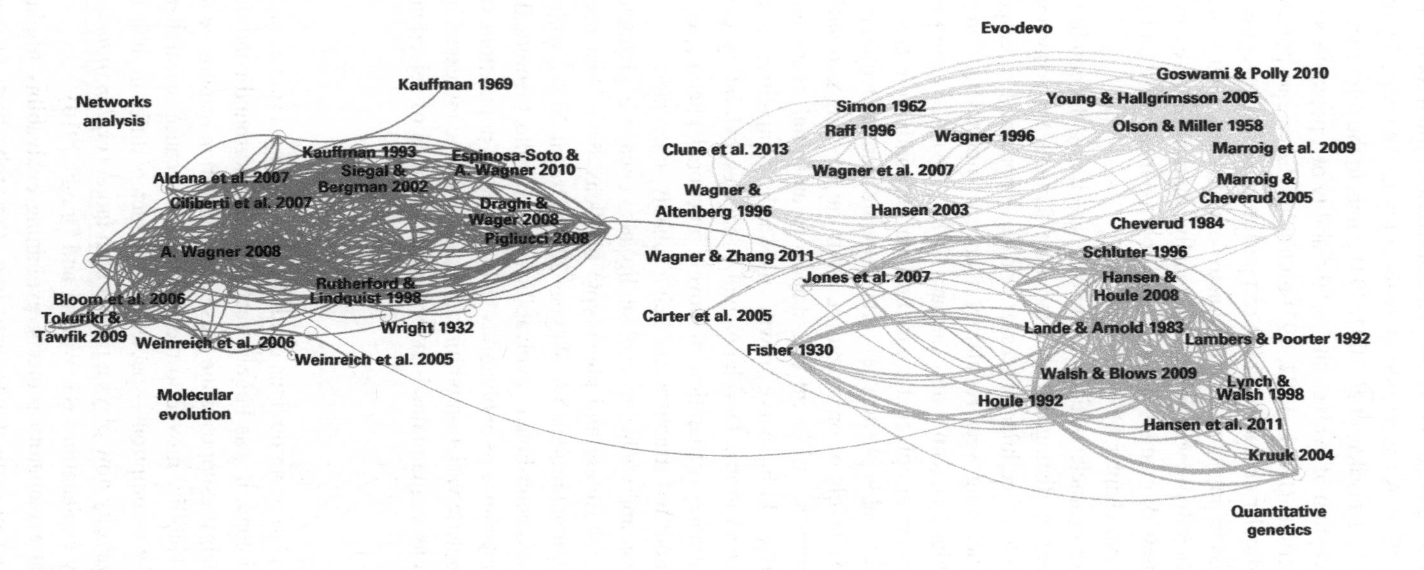

Figure 3.1
Co-citation evolvability network. Nodes correspond to the most-cited references in recent evolvability literature and links to co-citation links (i.e., to the frequency of being cited together). The network shows both a clustering of references by domains and an interdisciplinary landscape where many works are cited across disciplines. Full bibliographic records were gathered from the Institute for Scientific Information's Web of Science based on a search for publications containing "evolvability" in their title, keywords, or abstract, published in English between 2010 and 2021 (1,469 papers and 60,057 cited items). Software used: VOSviewer (version 1.6.17). A navigable network can be found in the book's online resources: https://mitpress.mit.edu/9780262545624/evolvability/.

Several different characterizations of evolvability are present in the scientific literature (Pigliucci 2008a; Nuño de la Rosa 2017). These characterizations often focus on different features in need of investigation. A conception of evolvability as "the capacity of a developmental system to evolve" (Hendrikse et al. 2007, 394) points to the relevance of properties of development; a conception of evolvability as "the ability of a population to respond to natural or artificial selection" (Houle 1992, 195) highlights the role of population and variational structure. A particular characterization can perform useful conceptual work by implying that research needs to pay specific attention to some feature, such as the generation of novel or adaptive phenotypic variation. Additionally, characterizing evolvability in a detailed theoretical framework enables its quantification for measuring and predictive purposes (Houle and Pélabon, chapter 6).

Evolvability as a target of characterization can involve specifying the conditions under which it can be precipitated or distinguishing alternative features that contribute to its occurrence. Different aspects of cellular processes and developmental mechanisms can contribute to evolvability, including weak regulatory linkage, compartmentation (modularity), and exploratory behavior (Kirschner and Gerhart 1998, 2005). For example, exploratory behavior—the generation of epigenetic variation that responds to interactions with other components to produce viable phenotypes—can generate many potential phenotypic states and operate on many levels of organization, from the growing and shrinking of microtubules in a cell (permitting different cell shapes) to an initial overabundance of axons and synaptic connections during nerve growth followed by competitive axon pruning (resulting in functional innervation). Developmental processes exhibiting exploratory behavior permit the evolutionary generation of novel, functional phenotypes, such as muscles of a limb with a modified structure still being reliably innervated.

Theoretical and simulation approaches to evolvability also engage in characterization. The evolutionary roles of robustness and phenotypic plasticity have been investigated theoretically using computational models (A. Wagner 2005; Draghi 2019), which illuminates how these properties can contribute to evolvability. For example, theoretical analyses of the manifestation or maintenance of evolvability in hypothetical populations can ascertain whether the range of genetic variation within populations can be increased by phenotypic plasticity or if plasticity is maintained under repeated rounds of selection (Draghi and Whitlock 2012).

3.2.3 Explanation

Although evolvability seems to be an obvious candidate for playing a role as an explanatory tool in evolutionary biology, it was largely treated as a background condition in the past, because classical models presupposed the presence of variation responsive to natural selection. Research on evolvability moved to the foreground because researchers recognized that it was a nontrivial assumption—species and characters differ in their ability to respond to selection. Evolvability now plays a role as both target (*explanandum*) and tool (*explanans*) in evolutionary explanations (G. Wagner and Draghi 2010).

As the *target* of different explanatory projects, explaining evolvability might refer to identifying general properties of evolvable systems (e.g., their robustness or modularity) or unraveling the causal basis of the differential capacities of traits to evolve (e.g., additive genetic variance in quantitative genetics or developmental properties in evo-devo; Hallgrímsson et al., chapter 9). For instance, pleiotropic relationships between floral and vegetative pigments account for the evolvability of floral color, resulting in diversification (Armbruster 2002, and chapter 15). Alternatively, the goal can be to understand evolvability as the result of evolutionary principles, such as direct selection for a group-level adaptation, the accumulation of neutral changes in complex genomes, or indirect selection acting on phenotypic traits or their underlying developmental architecture (Hansen 2011; Hansen and Wagner, chapter 7).

Evolvability is used as a *tool* to explain a wide range of evolutionary phenomena, ranging from the plausibility of life (Vasas et al. 2012), the evolution of complexity (G. Wagner and Altenberg 1996), and metazoan diversification (Gerhart and Kirschner 2007) to specific evolutionary pathways. Evolvability is an explanatory tool for a variety of evolutionary trajectories in specific traits, including body shape (Bergmann et al. 2020), the stability of wing shape compared to the lability of life history traits in *Drosophila* (Houle et al. 2017), or differences between vegetative and floral traits (Hansen et al. 2007).

These different situations can be partially understood in terms of distinct meanings for scientific explanation. Evolvability explanations found in quantitative genetics conform to the covering-law model of scientific explanation (Hempel 1965), where phenomena (e.g., the evolution of a quantitative trait) are explained by subsuming them under law-like generalizations (e.g., the Lande equation; Lande 1979). In contrast, evolvability explanations in evo-devo approaches involve mechanistic reasoning, where explaining a phenomenon means breaking it down into interacting parts that are organized to produce, underlie, or maintain it (Sterelny 2011; Craver and Darden 2013). Other evolvability explanations need not refer to specific mechanisms but can capture the space of possible and plausible changes or behaviors that arise from diverse causal processes (Brigandt 2015a; Austin and Nuño de la Rosa 2021; Nuño de la Rosa and Villegas 2022). For instance, robustness can facilitate evolvability by means of the accumulation of hidden variation, but this can be achieved by different mechanisms. Similarly, using evolvability as an explanatory tool is relevant to both actual changes a trait underwent and changes it could potentially undergo (e.g., in response to various selection differentials or mutation rates; see Brigandt et al., chapter 4).

This variety of legitimate conceptions of explanation helps account for the existence of different approaches to explaining evolvability (target) or using evolvability to explain other evolutionary phenomena (tool). Scientific theories, concepts, and models are only explanatory in a context-dependent fashion (Woodward 2014). Explanations of evolvability take different forms, depending on the investigative approach used and the type of question addressed, often in a discipline-dependent manner, just like scientific explanations in other domains.

3.2.4 Prediction

In many situations, evolutionary biologists aim to predict the evolutionary trajectory of a biological system. Prediction involves inferences from models, theories, and empirical knowledge about a phenomenon to some unobserved empirical fact. In some philosophical models of scientific explanation, an explanation and a prediction have the same logical structure, but the ability to quantitatively predict need not yield mechanistic explanations—explanation and prediction are often decoupled (Scriven 1959). This decoupling demands a separate treatment of prediction as a scientific activity where evolvability can play a role.

Inferring unobserved facts fulfills at least two distinct aims in scientific practice. First, predicting specific outcomes can serve as a basis for guiding future action, such as intervening in a phenomenon to achieve different goals (e.g., designing artificial selection experiments or making policy recommendations with respect to environmental problems). Second, predictions are associated with the testability of hypotheses and models (e.g., Popper 2002 [1963]). A good scientific model is expected to make specific predictions that are empirically testable. Failed predictions point to difficulties with the model that require revision. Hypotheses and models about evolvability are often tested by comparing experimental results with specific predictions.

Evolvability can play a role as the *target* of predictions. Sometimes, rather than directly measuring evolvability, evolutionary biologists infer the evolutionary capacity of systems from prior knowledge. For example, robustness measures are a good proxy (and therefore predictive) of the evolutionary potential (i.e., evolvability) of the RNA virus $\phi6$ under thermal stress in experimental studies (Ogbunugafor et al. 2009). In addition, scientists may want to predict changes in evolvability when some conditions of the system vary. For example, epistatic models of the genotype-phenotype map predict changes in evolvability on the basis of the type of directional epistatic interactions, whether positive or negative (Carter et al. 2005; Hansen 2011).

Evolvability is also an important *tool* for prediction as a part of a well-developed theoretical model that allows for precise measurement (see Houle and Pélabon, chapter 6). There are at least two domains where evolvability measures fit this criterion: quantitative genetics and the evolution of neutral networks. In quantitative genetics, a trait's evolvability is a measure of the capacity of that trait to change its phenotypic value in response to directional selection in a population. This measure enables researchers to predict mean phenotypic change of a trait under specific selective pressures, such as wing shape divergence among *Drosophila* species under directional selection (Hansen and Houle 2008). Crucially, evolvability can be a tool in predictions that test evolutionary hypotheses, such as whether there are differences in the evolutionary potential of life history traits and morphology (Price and Schluter 1991; Houle 1992). Similarly, it can play this role in guiding future action, such as the degree of resiliency and adaptability to drastic ecological changes in conservation biology (Gienapp et al. 2017) or the evolutionary dynamics of drug resistance in the context of medical research (e.g., Polster et al. 2016).

Neutral network approaches also have sufficiently developed theoretical models to facilitate evolvability playing a role in prediction. In this orientation, evolvability is a measure of the ability of a system to produce heritable phenotypic variation (A. Wagner 2008). From this measure, one can predict the ratio of neutral evolutionary change, provided one explicitly models the structure of the genotypic space and mutation rates. Examples include making predictions about gene regulatory circuit evolution (Payne et al. 2014) and could be extended to more complex systems, such as microbiome ecological interactions relevant to the development of medical treatments (Widder et al. 2016).

Sometimes predictions do not refer to future events but to unobserved past ones and are distinguished as *retrodictions.* Retrodictions are important for reconstructing the evolutionary past, especially in macroevolutionary studies of evolvability, and they can be an indicator of the predictive potential of a model (see Jablonski, chapter 17). The quantitative genetics sense of evolvability can play a role in macroevolutionary retrodictions when phenotypic matrices are used as a proxy for genotypic matrices (Hunt 2007). Their use facilitates evaluating specific theoretical models of evolution using fossil record data (Love et al. 2022). However, the extrapolation of these measures to macroevolutionary retrodictions is contested, because the parameters measured in extant populations over geological time spans can be unstable. For example, patterns of body size evolution in the fossil record diverge depending on different timescales used in analyses (Uyeda et al. 2011). Yet there is growing evidence that evolvability can predict patterns of macroevolution at surprisingly long timescales, such as standing genetic variation in a population of *Drosophila melanogaster* being strongly correlated with phenotypic divergence across 40 million years of evolution in Drosophilidae (Houle et al. 2017; Voje et al., chapter 14).

3.2.5 Control

That evolvability can be used to make predictions suggests it can play a role in the activity of *control,* which involves scientific practices that use explicit modifications or experimental interventions to change target systems. These interventions—whether an amino acid substitution in a protein or an adjustment to a computer algorithm—can contribute to a better understanding of natural systems or yield novel artifacts, features, or processes. The former can be seen for studies of evolvability in which aspects of its causal basis are manipulated, thus being the *target* of control, either experimentally or in simulations. In simulations, evolvability can be controlled through the manipulation of a G-matrix (genetic variance-covariance matrix) under the same selection gradient conditions (see also Hansen, chapter 5; Pélabon et al., chapter 13), which can result in different kinds of evolutionary divergence (Jones et al. 2018). Similarly, perturbations of the connectivity of molecular networks in computational models facilitate the identification of network topology changes that confer increased evolvability on some genotypes (Ancel and Fontana 2000). The latter can involve limiting evolvability through genetically engineering pesticide resistance in crops or enhancing it by facilitating the spread of genetic variation in a population for purposes of conservation (Campbell et al. 2017). Additionally, when attempting to increase yield-related characteristics of wheat (Nadolska-Orczyk et al. 2017), the correlated change in traits connected by a pleiotropic genetic architecture can be subject to control during breeding.

The use of directed evolution in protein engineering is another locus for the activity of controlling evolvability (Bloom et al. 2006). This research manipulates the capacity of evolvability through iterated selection processes to achieve proteins with specific properties (Bornscheuer et al. 2019). For example, enzymes used in industrial applications are subject to temperatures that often exceed (both in intensity and duration) those found in natural biological systems. Manipulating evolvability to create more thermostable enzymes via directed evolution permits more efficient and widespread use of such enzymes in these applications (Rigoldi et al. 2018). This involves theoretical tools familiar to evolutionary biologists to identify trade-offs (e.g., fitness landscapes), such as those between stability and solubility due to stabilizing mutations on the protein surface that increase hydrophobicity (Broom et al. 2017).

Evolvability can also be the target of control in conservation biology. One strategy for species preservation is to maintain the adaptive potential of populations for evolution through breeding protocols, such as the strategic selection of founders with a particular genetic architecture that can decrease the intensity of inbreeding depression (Allendorf et al. 2010). The overall effect is to maintain or increase levels of additive genetic variance. The manipulation of these forms of variation relevant to evolvability, rather than just variation per se (some of which might be neutral rather than adaptive), can lead to more effective conservation efforts and may avoid unintended outcomes (Campbell et al. 2017).

3.3 Interrelationships between Conceptual Roles

Thus far we have treated each of the scientific activities in which evolvability plays a conceptual role independently (see summary in table 3.1). However, different conceptual roles are often present simultaneously across activities and, more importantly, bear significant relationships to one another. For example, in certain contexts, a predictive model can be considered to explain the phenomena it predicts (see section 3.2.4). Sometimes predictions may refer to already observed data that can be fit into a particular model or theory for explanatory purposes. Thus, predictive accuracy can be a measure of explanatory power, such as in quantitative genetics, where the ability of the additive variance to predict the response to selection is taken as evidence of additive variation explaining short-term evolvability. Similarly, failure of prediction also can guide the search for a better explanation. Problems with additive variance predicting the evolvability of a population over longer periods might indicate that mechanistic accounts of changes in the structure of genotype-phenotype maps are needed to complement statistical descriptions that typically figure in evolutionary genetics (Hansen 2006; Hansen, chapter 5: Sztepanacz et al., chapter 12; Voje et al., chapter 14), or that the alignment of the genetic variance-covariance matrix with the direction of selection cannot be extrapolated to the longer term, given the populational context dependency of G-matrices. However, knowledge about the nature of this alignment and its stability on shorter timescales could be used to augment evolvability by manipulating processes of directed evolution, such as for selective breeding in conservation efforts.

Controlling evolvability (see section 3.2.5) through directed evolution increases understanding of what kinds of properties promote the ability to evolve and therefore can have an impact on characterization and explanation. For example, studies in protein engineering have demonstrated that the evolvability of proteins is facilitated by thermodynamic stability that engenders mutational robustness (Bloom et al. 2006; Tokuriki and Tawfik 2009). Even though the activity of control emphasizes what can be created or made experimentally (e.g., an enzyme with specific catalytic properties), researchers engaged in this

Table 3.1
Examples of the concept of evolvability being used as a tool or target in different scientific activities

	Setting a research agenda	Characterization	Explanation	Prediction	Control
Evolvability as a tool	Maps out the structure of collaborative research		Explains the evolution of complexity	Predicts evolutionary trends	
Evolvability as a target		Additive genetic variance characterizes evolvability	Exploratory behavior explains evolvability	Robustness measures are a proxy for evolvability	Creating more thermostable enzymes implies controlling their evolvability

manipulation are also concerned with prediction. Synthetic biologists aim to predict so that they can control how biological artifacts will behave outside laboratory conditions or in unforeseen environments. Success in the manipulation of evolvability correlates with advances in prediction. Similarly, the manipulation of specific genetic aspects of a developing organism can lead to a more precise account of what evolvability is (i.e., its characterization), as well as to a better explanation of its causal basis (see section 3.2.3).

Additionally, there are cases in which characterizing evolvability more precisely increases the capacity of researchers to control it. A richer characterization of evolvability also provides a clearer conception of what is in need of explanation (i.e., evolvability as *explanandum*). It therefore has the potential to yield better resources for using evolvability to explain patterns of trait origination or distribution in a lineage (i.e., evolvability as *explanans*). Different characterizations of evolvability can lead to different preferred explanations. If we characterize evolvability as the robustness of a trait as represented by a neutral network that confers a greater capacity for exploring phenotypic space, we may explain it in terms of the evolution of resistance to genetic perturbations (A. Wagner 2008). Different characterizations of evolvability also help shape an investigative agenda, providing structure to the research questions that evolutionary biologists ask (e.g., "can we predict how a trait will evolve under the manipulation of a particular genetic variable, which was determined to be a key contributor to modularity?").

Despite the existence of many connections among different conceptual roles of evolvability, these connections are not deductively necessary. A good prediction does not necessarily yield a good explanation and vice versa. First, one can make successful predictions without adequate explanations. Modularity might be a good predictor of evolvability, but this does not mean that modularity necessarily explains evolvability under all circumstances (see Hallgrímsson et al., chapter 9; Houle and Pélabon, chapter 6). Speciation rates might be indicators of phenotypic evolvability under some circumstances (Rabosky et al. 2013), but they do not necessarily account for why evolvability might be linked to lineage diversification. Predictions of an outcome based on a quantitative model often fall short of a mechanistic explanation that would capture relevant components that causally generate the outcome. Second, one can explain without prediction. Evolutionary biology can provide good explanations of past evolutionary events but is unable to offer good predictions of the evolutionary future due to unpredictability entailed by historical contingency (Scriven 1959; Beatty 1995; Blount et al. 2018). Manipulating a protein to increase evolvability does not automatically translate into more robust or precise predictions about population-level responses. Conversely, the ability to predict a trend under certain circumstances may not afford increased capacity to manipulate current conditions. The characterization of different contributors to evolvability (e.g., distinguishing modularity and phenotypic plasticity), whether through theoretical modeling or experimentation, does not immediately yield an explanation for how they make this contribution. Setting a research agenda shapes what counts as an explanation, organizing the lines of inquiry necessary to formulate an adequate account of evolvability, but it does not select from among the candidate explanatory factors or determine how they combine to provide an appropriate explanation. And having a good candidate explanation for evolvability with strong empirical and theoretical support does not mean the task of characterization is finished. Further exploration of the properties of developing organisms and aspects of population structure has the potential to reveal hitherto unknown dimensions of what evolvability is.

We label these complex relationships among conceptual roles *quasi-independence.* The independence of roles in different scientific activities makes it possible for progress to occur differentially across the diverse landscape of research into evolvability. Biologists can advance in understanding evolvability with respect to prediction but not necessarily with respect to explanation. Similarly, advances may occur in one approach to explaining evolvability but not in others. However, the roles are not fully independent but only quasi-independent; different lines of research can sometimes exhibit correlative progress. Advances in our characterization of evolvability can be linked to advances in our abilities to predict or manipulate it. Critically, quasi-independence makes it possible for different disciplinary approaches to favor or emphasize one or more roles or scientific activities over others. Evo-devo has focused on characterizing and explaining evolvability, largely leaving aside predictive and control aspects. Quasi-independence also implies that roles and their associated conceptions in different activities are not in direct competition and therefore can coexist in evolutionary biological research. Advances concerning explanation do not require a trade-off in progress with respect to prediction.

One final corollary of quasi-independence is that no role is necessarily more fundamental than another. Consequently, no single scientific activity is expected to predominate. If we achieve an adequate explanation of evolvability, it would not preempt evolvability's distinctive conceptual roles in prediction, characterization, or control. This is because what it means to explain a phenomenon varies across biological fields and research questions, and because the aims and means of prediction, characterization, or control cannot be reduced to those of explanation. The quasi-independence among roles and activities also suggests that no characteristic orientation (e.g., mechanistic explanation as an approach's central aim) or preferred conception of evolvability is primary.

3.4 Implications for Unification

Evolvability is present in most, if not all, branches of evolutionary biology even if it appears in a scientific activity under the guise of different conceptions or roles. Thus, the extent to which there is unificatory potential for evolvability partially rests upon the extent to which evolutionary biology is a unified discipline. Although a potential synthesis across evolutionary approaches is being discussed in some contexts (e.g., Pigliucci and Müller 2010), the fields that compose evolutionary biology are diverse in their goals and methodologies. The evolvability concept also reflects this situation, with specific combinations of conceptions and roles falling along natural divisions among disciplinary or methodological approaches (Nuño de la Rosa 2017; see figure 3.1).

Another issue to keep in view is that there are different kinds of unification. A classical view of science identifies its progress with theory unification across domains. This identification resonates with the idea that evolvability already has or should be a "unified" notion in the sense of having one primary meaning and one preeminent explanatory role (e.g., Sterelny 2007; Brown 2014), all in the context of a consensus account of evolutionary theory. However, given the diversity of theoretical contexts across evolutionary disciplines and the heterogeneity of activities within those disciplines, exploring the unificatory potential of conceptual roles across diverse activities—rather than aiming to reduce all of them to one fundamental meaning or role (such as explanation)—seems more promising

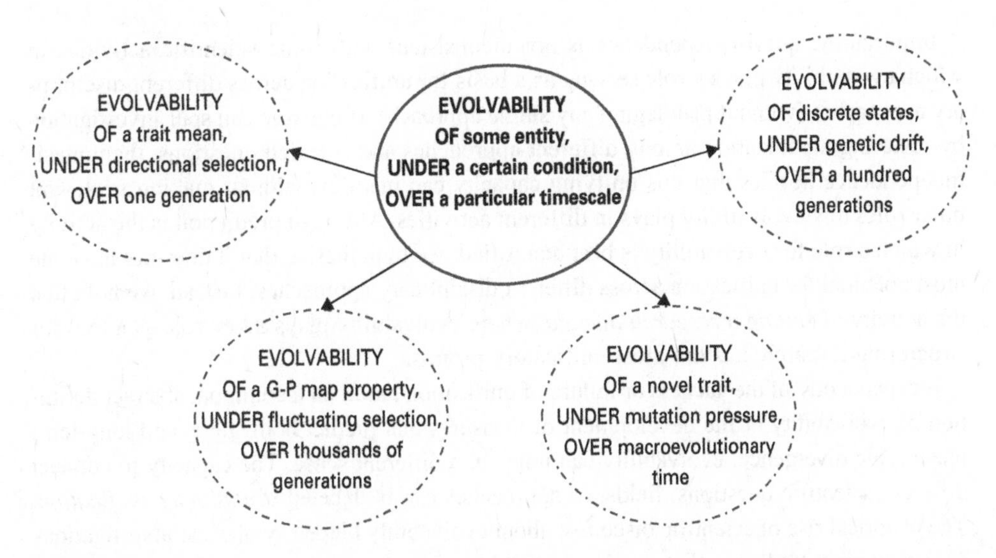

Figure 3.2
Definitional unification. Understood as a measurable disposition, this schema is intended to unify different conceptions of evolvability (see Houle and Pélabon, chapter 6). Note that the different possibilities displayed in the schema are not exhaustive, nor do they correspond to definitions of evolvability in disciplines. Instead, they are illustrations of parameterizations that would result in a measurable sense of evolvability.

for understanding the relationships among these different approaches in evolutionary biology (Brigandt and Love 2012).

One possibility is having a unified definition of evolvability in a particular scientific role (*definitional unification*). Among the activities in which evolvability plays a role, prediction stands out for its degree of theoretical development, especially in quantitative genetics. This follows from a precise mathematical characterization that can be obtained from measures in artificial and natural populations. For example, evolvability measures have been used to predict wing shape divergence in *Drosophila* species (Hansen and Houle 2008). A unified notion of evolvability that facilitates prediction lies behind Houle and Pélabon's (chapter 6) proposal, who argue that all conceptions of evolvability assume it is a dispositional property and develop a framework for undertaking meaningful measurements of this disposition in empirical cases. Such a framework involves identifying the relevant features that evolve (i.e., evolvability *of*) and the applicable conditions *under* and timescales *over* which evolution takes place. This framework shows one way in which evolvability could serve to unify evolutionary research: it yields an abstract scheme that encompasses many concrete definitions found in the literature (see figure 3.2). Different conceptions simply focus on different, concrete "of-under-over" aspects: evolvability *of* a quantitative trait *under* directional selection *over* multiple generations versus evolvability *of* new phenotypic variants *under* mutation in a certain developmental architecture *over* millions of years. Although this framework offers a good strategy for measuring evolvability because it is characterized in a number of different disciplinary contexts, it does not provide a framework for linking together different research questions (e.g., how explanations of short-term evolvability connect with explanations of long-term evolvability) or scientific activities (e.g., how the short-term prediction on the basis of the additive variance is related to the mechanistic explanation of evolvability).

Importantly, quasi-independence is not inconsistent with some scientific activities in which evolvability plays a role serving as a basis for unification across different disciplinary approaches (without privileging any single approach). If one role can spur investigation by fostering organization among different approaches and research questions, then quasi-independence implies that this unifying capacity can manifest without making irrelevant other roles that evolvability plays in different activities. Although prediction is the activity in which a role for evolvability is best quantified, we hypothesize that it does not have the most potential for unification across different disciplinary approaches. Instead, we hold that the activity of *setting a research agenda,* where evolvability plays a key role as a tool for structuring research, has the most unificatory promise.

Independently of the success or failure of unification based on a common, abstract definition of evolvability or the development of measures that predict both short- and long-term phenotypic divergence, evolvability can unify in a different sense. The capacity to connect different scientific questions, fields, or approaches can be labeled *disciplinary unification.* The historical rise of scientific discourse about evolvability helped synthesize investigations across research traditions that previously had been largely unrelated (see Nuño de la Rosa, chapter 2). This synthesizing role corresponds to the activity of setting a research agenda. A concept that sets a research agenda not only motivates further scientific efforts but also structures ongoing research and coordinates disciplinary contributions (Love 2021). This is because the agenda represented by the concept consists of many component questions, which are related in systematic ways (see section 3.2.1), such as the amount of phenotypic variation that can be generated and biases in the direction of variation (Hendrikse et al. 2007).

Figure 3.3 offers one illuminating (if incomplete) perspective on the structure of the problem agenda associated with evolvability. Although there are other ways of articulating the landscape of evolvability research (see figure 6.1 in Houle and Pélabon, chapter 6), any account will have the agenda-setting benefit of mapping out some connections among fields and approaches. Figure 3.3 captures disciplines, phenomena, and clusters of questions in evolvability-related research. Thus the figure makes it possible to display several patterns of existing research, such as evo-devo inquiry into developmental phenomena relevant to phenotypic variability (e.g., the modularity of the genotype-phenotype map). At the same time, the figure not only includes phenotypic variability, phenotypic variation, and actual evolutionary change as phenomena directly germane to evolvability research, it also depicts how they are related to one another and other phenomena. Variation, including the covariation among different characters, can be measured in actual populations. Together with additional factors, such as selection, theoretical models can then predict the resulting microevolutionary change. However, if one wants to understand what leads to and accounts for patterns of phenotypic variation, further issues need to be investigated. In addition to the impact of population processes on variability (e.g., mating systems, population size), the potential for phenotypic variability must be dissected in terms of the structure of the genotype-phenotype map (e.g., modularity, robustness), all of which can be enriched by an investigation of epistatic patterns and the underlying developmental architecture. Moreover, if one wants to understand how these microevolutionary changes lead to macroevolutionary patterns, further investigation into long-term changes in these factors is required. By foreshadowing how these different contributions from evolvability

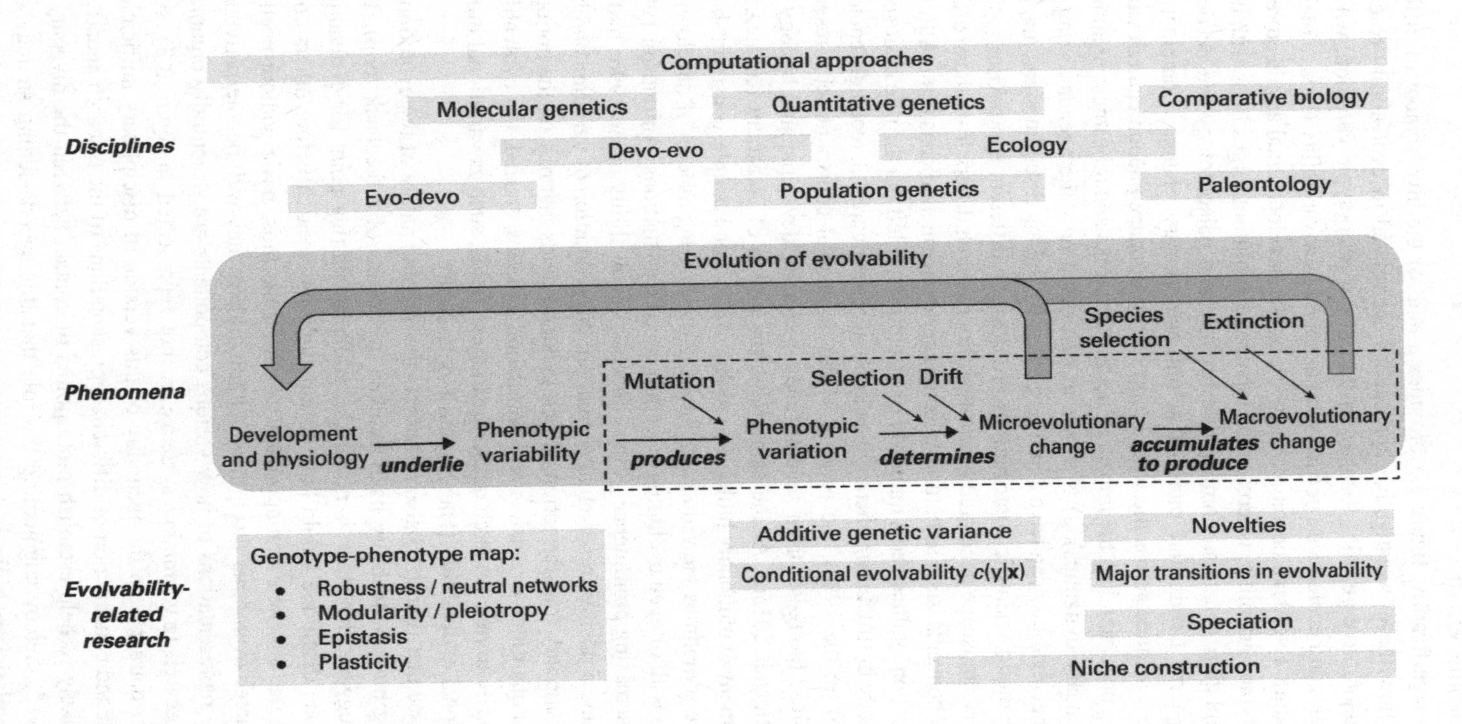

Figure 3.3
Evolvability as a multidisciplinary research agenda. Different disciplines representing different approaches to evolvability are keyed to phenomena on different timescales that bear specific relationships to one another. These disciplines and phenomena also exhibit correspondence with questions found in evolvability research, such as the genotype-phenotype map, conditional evolvability, or major transitions in evolvability. The dashed rectangle represents the classical picture of evolutionary research before the evolvability research agenda developed.

research can be connected, the concept of evolvability sets a research agenda that coordinates various scientific efforts.

Perhaps most significantly, figure 3.3 provides a map of the involvement of different disciplines and how these investigative approaches are related in evolvability research. For example, quantitative genetics focuses on the role of phenotypic variation, whereas population genetics and ecology are needed to understand how variability leads to realized evolutionary change. Evo-devo encompasses work on the developmental architecture that underlies phenotypic variability, whereas developmental evolution forges links with quantitative and population genetics. Comparative biology and paleontology are needed to investigate long-term trends and rates of actual evolutionary change; computational approaches have relevance across the whole loop of evolutionary phenomena relevant to evolvability (e.g., understanding the impact of neutral networks and robustness or simulating microevolutionary dynamics). This shows in rich detail how the agenda-setting role of the concept of evolvability can have a unifying effect by linking disciplines and mapping out connections among scientific contributions provided by different approaches.

The disciplinary landscape generated by evolvability research differs from those offered by other classical concepts, such as natural selection or developmental constraint. Although the notion of selection synthesized a manifold of disciplines, it left out the role of development and physiology in structuring phenotypic variability (see figure 3.3, rectangle enclosed by a dashed line). In the 1980s, the concept of developmental constraint set a research agenda that involved biologists from several fields, including developmental biology and paleontology (Brigandt 2015b). At the same time, the notion of constraint had negative connotations for many evolutionary biologists (Arthur 2015). For example, constraint-based explanations were sometimes viewed as emphasizing the limiting aspects of development and as competing with or even excluding selection-based explanations (Amundson 1994). In contrast, the increasing prominence of the concept of evolvability in the 1990s had the advantage of setting a positive research agenda about the generation of variation and how it conditions evolutionary transformation, which included fields that rely on the notion of selection, such as quantitative genetics and population genetics. Indeed, the evolvability agenda has given rise to new research questions in these fields, such as variational modularity and conditional evolvability (Hansen and Houle 2008).

It is crucial to see that figure 3.3 also represents the ongoing nature of the evolutionary process, including how evolvability itself evolves (by means of a feedback loop). This ongoing nature suggests that there is no single, preferred starting point for evolvability research, where one discipline would have to conclude its research before others could initiate or contribute, or where one approach would be the most basic without needing explanatory resources from other disciplines. Although there may well be alternative and equally legitimate representations of how various components are structurally organized in evolvability research (e.g., mating systems are not represented in figure 3.3), such alternatives do not detract from the fecundity of this version of disciplinary unification. In fact, the pursuit and construction of different representations of the research agenda of evolvability are likely to help establish more points of contact between the diversity of approaches involved, thereby augmenting the role that this agenda-setting strategy can play in unifying evolutionary biological investigations around the concept of evolvability.

3.5 Conclusion

We began our analysis with the observation that variation in the conceptualization of evolvability in research likely plays some functional role in the reasoning endeavors of scientists. After examining the roles of evolvability in five different scientific activities (section 3.2), we interrogated how its different roles and conceptions can be related to one another or are jointly operative in diverse ways because of their quasi-independence (section 3.3). In closing, we described two candidate strategies for addressing whether evolvability might play a unifying role in evolutionary biology: definitional unification and disciplinary unification (section 3.4). Although we argued that the latter appears to harbor a more encompassing basis for unification across many fields and approaches in evolutionary biology (based on its role of setting a research agenda), the value of our analysis stands independently of this claim. Explicit scrutiny of the conceptual roles that evolvability plays in contemporary evolutionary biology helps show how a rich and variegated space of possibilities can be utilized by researchers to facilitate fruitful interdisciplinary lines of investigation and thereby yield a deeper understanding of evolvability.

Acknowledgments

We thank all participants of the "Evolvability: A New and Unifying Concept in Evolutionary Biology?" project, and especially Thomas Hansen and Christophe Pélabon, for stimulating discussions and comments on a draft of this chapter. We also acknowledge the discussion among participants of the "Evolvability across Biological Disciplines" session at the 2021 meeting of the International Society for the History, Philosophy and Social Studies of Biology. All coauthors are greatly indebted to the Centre for Advanced Study (CAS) at the Norwegian Academy of Science and Letters for its generous funding, which made this research project possible. Cristina Villegas and Laura Nuño de la Rosa were also supported by the Spanish Ministry of Economy and Competitiveness (project PGC2018-099423-B-I00). Ingo Brigandt's work was additionally supported by the Social Sciences and Humanities Research Council of Canada (Insight Grant 435-2016-0500). Alan Love was supported by the John M. Dolan Professorship and the Winton Chair in the Liberal Arts. Günter Wagner and Alan Love are supported, in part, by the John Templeton Foundation (grant #61329).

References

Allendorf, F. W., P. A. Hohenlohe, and G. Luikart. 2010. Genomics and the future of conservation genetics. *Nature Reviews Genetics* 11: 697–709.

Amundson, R. 1994. Two concepts of constraint: Adaptationism and the challenge from developmental biology. *Philosophy of Science* 61: 556–578.

Ancel, L. W., and W. Fontana. 2000. Plasticity, evolvability, and modularity in RNA. *Journal of Experimental Zoology B* 288: 242–283.

Armbruster, W. S. 2002. Can indirect selection and genetic context contribute to trait diversification? A transition-probability study of blossom-colour evolution in two genera. *Journal of Evolutionary Biology* 15: 468–486.

Arthur, W. 2015. Internal factors in evolution: The morphogenetic tree, developmental bias, and some thoughts on the conceptual structure of evo-devo. In *Conceptual Change in Biology: Scientific and Philosophical Perspectives on Evolution and Development*, edited by A. C. Love, 343–363. Dordrecht: Springer.

Austin, C. J., and L. Nuño de la Rosa. 2021. Dispositional properties in evo-devo. In *Evolutionary Developmental Biology: A Reference Guide,* edited by L. Nuño de la Rosa and G. Müller, 469–481. Cham, Switzerland: Springer.

Beatty, J. 1995. The evolutionary contingency thesis. *Concepts, Theories, and Rationality in the Biological Sciences* 45: 81.

Bergmann, P. J., G. Morinaga, E. S. Freitas, D. J. Irschick, G. P. Wagner, and C. D. Siler. 2020. Locomotion and palaeoclimate explain the re-evolution of quadrupedal body form in *Brachymeles* lizards. *Proceedings of the Royal Society B* 287: 20201994.

Bloom, J. D., S. T. Labthavikul, C. R. Otey, and F. H. Arnold. 2006. Protein stability promotes evolvability. *PNAS* 103: 5869–5874.

Blount, Z. D., R. E. Lenski, and J. B. Losos. 2018. Contingency and determinism in evolution: Replaying life's tape. *Science* 362: eaam5979.

Bornscheuer, U. T., B. Hauer, K. E. Jaeger, and U. Schwaneberg. 2019. Directed evolution empowered redesign of natural proteins for the sustainable production of chemicals and pharmaceuticals. *Angewandte Chemie International Edition* 58: 36–40.

Brigandt, I. 2015a. Evolutionary developmental biology and the limits of philosophical accounts of mechanistic explanation. In *Explanation in Biology: An Enquiry into the Diversity of Explanatory Patterns in the Life Sciences,* edited by P.-A. Braillard and C. Malaterre, 135–173. Dordrecht: Springer.

Brigandt, I. 2015b. From developmental constraint to evolvability: How concepts figure in explanation and disciplinary identity. In *Conceptual Change in Biology: Scientific and Philosophical Perspectives on Evolution and Development,* edited by A. C. Love, 305–325. Dordrecht: Springer.

Brigandt, I., and A. C. Love. 2012. Conceptualizing evolutionary novelty: Moving beyond definitional debates. *Journal of Experimental Zoology B* 318: 417–427.

Broom, A., Z. Jacobi, K. Trainor, and E. M. Meiering. 2017. Computational tools help improve protein stability but with a solubility tradeoff. *Journal of Biological Chemistry* 292: 14349–14361.

Brown, R. L. 2014. What evolvability really is. *British Journal for the Philosophy of Science* 65: 549–572.

Campbell, C. S., C. E. Adams, C. W. Bean, and K. J. Parsons. 2017. Conservation evo-devo: Preserving biodiversity by understanding its origins. *TREE* 32: 746–759.

Carter, A. J. R., J. Hermisson, and T. F. Hansen. 2005. The role of epistatic gene interactions in the response to selection and the evolution of evolvability. *Theoretical Population Biology* 68: 179–196.

Colaço, D. 2018. Rip it up and start again: The rejection of a characterization of a phenomenon. *Studies in History and Philosophy of Science* 72: 32–40.

Craver, C. F., and L. Darden. 2013. *In Search of Mechanisms: Discoveries across the Life Sciences.* Chicago: University of Chicago Press.

Draghi, J. 2019. Phenotypic variability can promote the evolution of adaptive plasticity by reducing the stringency of natural selection. *Journal of Evolutionary Biology* 32: 1274–1289.

Draghi, J. A., and M. C. Whitlock. 2012. Phenotypic plasticity facilitates mutational variance, genetic variance, and evolvability along the major axis of environmental variation. *Evolution* 66: 2891–2902.

Galison, P. 1999. Trading zone: Coordinating action and belief. In *The Science Studies Reader,* edited by B. Mario, 137–160. New York: Routledge.

Gerhart, J. C., and M. W. Kirschner. 2007. The theory of facilitated variation. *PNAS* 104: 8582–8589.

Gienapp, P., S. Fior, F. Guillaume, J. R. Lasky, V. L. Sork, and K. Csilléry. 2017. Genomic quantitative genetics to study evolution in the wild. *TREE* 32: 897–908.

Griffiths, P., and K. Stotz. 2013. *Genetics and Philosophy: An Introduction.* Cambridge: Cambridge University Press.

Hansen, T. F. 2006. The evolution of genetic architecture. *AREES* 37: 123–157.

Hansen, T. F. 2011. Epigenetics: Adaptation or contingency? In *Epigenetics: Linking Genotype and Phenotype in Development and Evolution,* edited by H. Benedikt and K. H. Brian, 357–376. Oakland, CA: University of California Press.

Hansen, T. F., and D. Houle. 2008. Measuring and comparing evolvability and constraint in multivariate characters. *Journal of Evolutionary Biology* 21: 1201–1219.

Hansen, T. F., C. Pélabon, and W. S. Armbruster. 2007. Comparing variational properties of homologous floral and vegetative characters in *Dalechampia scandens:* Testing the Berg hypothesis. *Evolutionary Biology* 34: 86–98.

Hansen, T. F., C. Pélabon, and D. Houle. 2011. Heritability is not evolvability. *Evolutionary Biology* 38: 258.

Hempel, C. G. 1965. *Aspects of Scientific Explanation and Other Essays in the Philosophy of Science.* New York: Free Press.

Hendrikse, J. L., T. E. Parsons, and B. Hallgrímsson. 2007. Evolvability as the proper focus of evolutionary developmental biology. *Evolution & Development* 9: 393–401.

Hey, J. 2001. *Genes, Categories, and Species: The Evolutionary and Cognitive Cause of the Species Problem.* Oxford: Oxford University Press.

Houle, D. 1992. Comparing evolvability and variability of quantitative traits. *Genetics* 130: 195–204.

Houle, D., G. H. Bolstad, K. Van der Linde, and T. F. Hansen. 2017. Mutation predicts 40 million years of fly wing evolution. *Nature* 548: 447–450.

Hunt, G. 2007. Evolutionary divergence in directions of high phenotypic variance in the ostracode genus *Poseidonamicus. Evolution* 61: 1560–1576.

Jones, A. G., R. Bürger, and S. J. Arnold. 2018. The G-matrix simulator family: Software for research and teaching. *Journal of Heredity* 109: 825–829.

Kirschner, M. W., and J. C. Gerhart. 1998. Evolvability. *PNAS* 95: 8420–8427.

Kirschner, M. W., and J. C. Gerhart. 2005. *The Plausibility of Life: Resolving Darwin's Dilemma.* New Haven, CT: Yale University Press.

Kitcher, P. 1984. Species. *Philosophy of Science* 51: 308–333.

Lande, R. 1979. Quantitative genetic analysis of multivariate evolution, applied to brain:body size allometry. *Evolution* 33: 402–416.

Love, A. C. 2003. Evolvability, dispositions, and intrinsicality. *Philosophy of Science* 70: 1015–1027.

Love, A. C. 2008. Explaining evolutionary innovations and novelties: Criteria of explanatory adequacy and epistemological prerequisites. *Philosophy of Science* 75: 874–886.

Love, A. C. 2013. Interdisciplinary lessons for the teaching of biology from the practice of evo-devo. *Science & Education* 22: 255–278.

Love, A. C. 2021. Interdisciplinarity in evo-devo. In *Evolutionary Developmental Biology: A Reference Guide,* edited by L. Nuño de la Rosa and G. Müller, 407–423. Cham, Switzerland: Springer.

Love, A. C., M. Grabowski, D. Houle, et al. 2022. Evolvability in the fossil record. *Paleobiology* 48: 186–209.

Minelli, A. 2010. "Evolutionary developmental biology does not offer a significant challenge to the neo-Darwinian paradigm." In *Contemporary Debates in the Philosophy of Biology,* edited by F. Ayala and R. Arp, 213–226. Malden, MA: Wiley-Blackwell.

Müller, G. B. 2021. Evo-devo's contributions to the extended evolutionary synthesis. In *Evolutionary Developmental Biology: A Reference Guide,* edited by L. Nuño de la Rosa and G. Müller, 1127–1138. Cham, Switzerland: Springer.

Nadolska-Orczyk, A., I. K. Rajchel, W. Orczyk, and S. Gasparis. 2017. Major genes determining yield-related traits in wheat and barley. *Theoretical and Applied Genetics* 130: 1081–1098.

Nuño de la Rosa, L. 2017. Computing the extended synthesis: Mapping the dynamics and conceptual structure of the evolvability research front. *Journal of Experimental Zoology B* 328: 395–411.

Nuño de la Rosa, L., and C. Villegas. 2022. Chances and propensities in evo-devo. *The British Journal for the Philosophy of Science* 73: 509–533.

Ogbunugafor, C. B., R. C. McBride, and P. E. Turner. 2009. Predicting virus evolution: The relationship between genetic robustness and evolvability of thermotolerance. *Cold Spring Harbor Symposia on Quantitative Biology* 74: 109–118.

Payne, J. L., J. H. Moore, and A. Wagner. 2014. Robustness, evolvability, and the logic of genetic regulation. *Artificial Life* 20: 111–126.

Pigliucci, M. 2008a. Is evolvability evolvable? *Nature Reviews Genetics* 9: 75–82.

Pigliucci, M. 2008b. What, if anything, is an evolutionary novelty? *Philosophy of Science* 75: 887–898.

Pigliucci, M., and G. Müller, eds. 2010. *Evolution, the Extended Synthesis.* Cambridge, MA: MIT Press.

Polster, R., C. J. Petropoulos, S. Bonhoeffer, and F. Guillaume. 2016. Epistasis and pleiotropy affect the modularity of the genotype-phenotype map of cross-resistance in HIV-1. *Molecular Biology and Evolution* 33: 3213–3225.

Popper, K. 2002 [1963]. *Conjectures and Refutations: The Growth of Scientific Knowledge.* New York: Routledge.

Price, T., and D. Schluter. 1991. On the low heritability of life-history traits. *Evolution* 45: 853–861.

Rabosky, D. L., F. Santini, J. Eastman, S. A. Smith, B. Sidlauskas, J. Chang, and M. E. Alfaro. 2013. Rates of speciation and morphological evolution are correlated across the largest vertebrate radiation. *Nature Communications* 4: 1958.

Rigoldi, F., S. Donini, A. Redaelli, E. Parisini, and A. Gautieri. 2018. Engineering of thermostable enzymes for industrial applications. *APL Bioengineering* 2: 011501.

Scriven, M. 1959. Explanation and prediction in evolutionary theory. *Science* 130: 477–482.

Sterelny, K. 2007. What is evolvability? In *Philosophy of Biology*, edited by M. Matthen and C. Stephens, 163–178. Amsterdam: Elsevier.

Sterelny, K. 2011. Evolvability reconsidered. In *The Major Transitions in Evolution Revisited*, edited by B. Calcott and K. Sterelny, 83–100. Cambridge, MA: MIT Press.

Tokuriki, N., and D. S. Tawfik. 2009. Stability effects of mutations and protein evolvability. *Current Opinion in Structural Biology* 19: 596–604.

Uyeda, J. C., T. F. Hansen, S. J. Arnold, and J. Pienaar. 2011. The million-year wait for macroevolutionary bursts. *PNAS* 108: 15908–15913.

Vasas, V., C. Fernando, M. Santos, S. Kauffman, and E. Szathmáry. 2012. Evolution before genes. *Biology Direct* 7: 1.

Wagner, A. 2005. *Robustness and Evolvability in Living Systems*. Princeton, NJ: Princeton University Press.

Wagner, A. 2008. Robustness and evolvability: A paradox resolved. *Proceedings of the Royal Society B* 275: 91–100.

Wagner, G. P., and L. Altenberg. 1996. Complex adaptations and the evolution of evolvability. *Evolution* 50: 967–976.

Wagner, G. P., and J. Draghi. 2010. Evolution of evolvability. In *Evolution: The Extended Synthesis*, edited by G. Müller and M. Pigliucci, 379–399. Cambridge, MA: MIT Press.

Widder, S., R. J. Allen, T. Pfeiffer, et al. 2016. Challenges in microbial ecology: Building predictive understanding of community function and dynamics. *The ISME Journal* 10: 2557–2568.

Wilson, M. 2006. *Wandering Significance: An Essay on Conceptual Behavior*. Oxford: Oxford University Press.

Woodward, J. 2014. Scientific explanation. In *The Stanford Encyclopedia of Philosophy*, edited by E. N. Zalta. http://plato.stanford.edu/entries/scientific-explanation.

4 Evolvability as a Disposition: Philosophical Distinctions, Scientific Implications

Ingo Brigandt, Cristina Villegas, Alan C. Love, and Laura Nuño de la Rosa

A disposition or dispositional property is the capacity, ability, or potential to exhibit some outcome. Evolvability refers to a disposition to evolve. This chapter discusses why the dispositional nature of evolvability matters—why philosophical distinctions about dispositions can have scientific implications. To that end, we build a conceptual toolkit with vocabulary from prior philosophical analyses using a different disposition (protein foldability) and then apply this toolkit to address several methodological questions related to evolvability. What entities are the bearers of evolvability? What features causally contribute to the disposition of evolvability? How does evolvability manifest? The various possible answers to these questions available from philosophical distinctions suggest implications for why the concept of evolvability as a disposition is useful in evolutionary research. These include (1) securing scientific virtues (e.g., explanatory depth and generalization, prediction or retrodiction, and control or manipulation) and (2) fostering interdisciplinary collaboration through the coordination of definitional diversity and different types of inquiry. Together these implications facilitate concentration on a variety of research questions at different levels of organization and on distinct timescales, all of which should be expected for a complex dispositional property such as evolvability.

4.1 Introduction

A disposition or dispositional property is the capacity, ability, or potential to exhibit some outcome. Since something can possess an ability without currently displaying it, dispositions are attributed when it makes sense to distinguish between something having a property and manifesting that property. For example, the fragility of a window is its disposition to break under certain conditions (e.g., on impact from a rock). The window is fragile even if it never breaks, and we can make claims about its degree and kind of fragility despite its manifestation not being directly observed. This kind of property contrasts with nondispositional or so-called categorical properties of the window, such as size or shape, for which a distinction between the property and its manifestation is irrelevant. Dispositional properties are common in biology (Hüttemann and Kaiser 2018), including the foldability of proteins or the differentiability or pluripotency of cells. The most recognizable example in evolutionary biology is fitness, where the general capacity to survive and reproduce is distinct from actual reproductive success (Mills and Beatty 1979). There also is an abundance of dispositional notions in the field of evolutionary developmental biology (evo-devo), such as forms of phenotypic plasticity, including the capacity to exhibit different morphological traits in seasonal polyphenisms (Austin 2017; Austin and Nuño de la Rosa 2021).

Evolvability—the core notion of this volume—is a disposition. Scientific definitions make this dispositional character plain by construing evolvability as the "*ability* to respond to selection" (Houle 1992, 195), the "*capacity* to generate heritable phenotypic variation" (Kirschner and Gerhart 1998, 8420), or the "*ability* to produce adaptive variants" (G. Wagner and Altenberg 1996, 970, emphasis added in all quotes). Evolvability can but need not manifest in a higher rate of evolution, because two populations with an identical ability to evolve may come to exhibit different evolutionary outcomes due to chance—just as in the case of fitness as a capacity to reproduce—or due to the two populations being exposed to different environmental conditions with different regimes of natural selection. Twenty-seven years ago, G. Wagner and Altenberg (1996) noted that *evolvability* and *variability* are dispositions and should be distinguished from actual evolutionary outcomes and observable variation (see also Hansen 2006). This distinction encourages making the dispositional property of evolvability a target for scientific investigation, especially how this capacity has changed through the history of life (i.e., the evolution of evolvability). Indeed, the increasing prominence of the notion of evolvability marks an important change in recent scientific theorizing (Brigandt 2015; Nuño de la Rosa 2017; and chapter 2[1]), and its dispositional character may be part of the reason that the concept of evolvability nowadays enjoys widespread use in a variety of biological fields investigating evolution.

One task of this chapter is to argue that the dispositional nature of evolvability matters—to explain why philosophical distinctions about dispositions can have scientific implications—and to show what is distinctive about evolvability compared to other dispositions in evolutionary biology. However, as the limited array of evolvability definitions described above makes plain, different conceptions of evolvability are used by different disciplines, from quantitative genetics and evo-devo to evolutionary systems biology and paleontology. We therefore address what kind of disposition evolvability is in these different contexts, and how these conceptions and disciplinary approaches to such a complex disposition are related. To this end, we first build a philosophical toolkit related to dispositions, before deploying it both to make sense of the plurality of evolvability definitions and to draw implications for its scientific significance and explanatory value as a dispositional concept. Importantly, our analysis does not provide definitive claims about specific cases of evolvability but rather generates resources for more nuanced interpretations relevant to further empirical and theoretical evaluations.

4.2 Vocabulary: Building a Toolkit for Evolvability

To analyze evolvability as a disposition, it is useful to build a conceptual toolkit based on vocabulary from prior philosophical discussions. Although there are different, often non-overlapping bodies of literature in philosophy that reflect on dispositions (e.g., metaphysics, philosophy of science, and philosophy of probability), it is possible to identify some core ideas that are helpful (see Choi and Fara 2021 for an overview). To do so, we focus on a different disposition—protein foldability—so that the basic distinctions are laid out in advance of their application to evolvability. Although protein foldability provides sufficient biological details to illustrate the value of most of the dispositional vocabulary (see box 4.1),

1. References to chapter numbers in the text are to chapters in this volume.

Box 4.1
Philosophical vocabulary for dispositions

Background conditions: The various circumstances present when a disposition is manifested, may or may not be relevant (e.g., temperature while glass shatters).

Bearer: The entity that has a disposition or exhibits a dispositional property. This entity can be an individual or an aggregate and can be considered as either a token or a type (i.e., an individual or a class of similar individuals; a single population of entities or a class of similar populations of entities).

Causal basis: The underlying properties determining that the capacity applies (e.g., amino acid sequence in a foldable polypeptide).

Disposition or Dispositional property: The capacity itself (e.g., fragility, foldability, or evolvability).

Explanatory depth: The aim of an account of a disposition that demonstrates how the causal basis contributes to the process of disposition manifestation, why a particular manifestation is regularly achieved and relatively stable, or what effects might arise from various possible stimulus conditions.

Intrinsic/Extrinsic: Whether the causal basis that determines the disposition is fully internal to the disposition's bearer (e.g., its parts and their interactions) or whether features external to the bearer (e.g., environmental components) also contribute to the capacity obtaining.

Manifestation: The actual display of the disposition (e.g., the glass breaking or the protein folding correctly), which might vary in rate (slower or faster), intensity, or number (e.g., different correct conformations of an intrinsically disordered protein).

Probabilistic/Deterministic: Whether a disposition manifests always (deterministically) or only some of the time (probabilistically, as a propensity) under the appropriate stimulus conditions.

Single-track/Multitrack: Whether a disposition always has a single outcome (manifestation) or whether there are several distinct outcomes, depending on the stimulus condition.

Stimulus (Triggering) conditions: The circumstances leading to the disposition's manifestation (e.g., a solid object hitting glass).

Token/Type: Dispositions can be ascribed to single entities (tokens) or to classes/kinds of entities (types).

there are limitations to the example as a template that we note in section 4.3 before turning to evolvability.

Foldability is the disposition of a protein to have its linearly arranged structural components—amino acids—rearranged ("folded") into higher-order, 3-dimensional configurations to accomplish cellular functions. Biologists attribute this disposition to proteins because proteins adopt one or more 3-dimensional (tertiary) structures or conformations that are typically necessary for them to be functional. (These conformations exhibit variation that can be represented with probability distributions.) The process of achieving a 3-dimensional structure is sometimes described as spontaneous (Campbell and Reece 2002). This appeal to spontaneity denotes something happening as soon as certain conditions arise. Just as glass shatters on projectile impact (fragility), a protein has a disposition to fold (foldability) into one or more conformations with functional properties under appropriate *stimulus conditions* in its chemical environment. The display of a tertiary structure and corresponding function is a *manifestation* of the disposition. The capacity of foldability *explains* the resulting patterns of folded-and-functional proteins (the manifestation). Such a dispositional property can be ascribed to an individual macromolecule (a *token*) or to a class of the same polypeptides (a *type*). However, predicting how this folding occurs and understanding mechanistically why a specific functional outcome is achieved have been refractory questions. Some of the empirical details making these questions difficult to solve serve to further build a toolkit for talking about dispositions.

Initially, protein folding was understood to result from physical properties of the component amino acid residues in a polypeptide (e.g., hydrophobic residues avoid interaction with surrounding water by segregating to internal regions). These components and their interactions were thought to constitute the *causal basis* of foldability: "the native conformation is determined by the totality of interatomic interactions and hence by the amino acid sequence" (Anfinsen 1973, 223). Some of the strongest evidence for this idea came from experiments on the denaturation and refolding of ribonucleases in vitro. These experiments suggested that the causal basis was *intrinsic* to the linear polypeptide; that is, the polypeptide's primary structure contains all the information required for achieving a 3-dimensional, functional conformation. This causal basis provides increased *explanatory depth* (Weslake 2010) to our understanding of protein folding, accounting for both how the process of folding occurs and why one or more functional conformations are regularly achieved and relatively stable.

However, the proteins investigated in these experiments did not fold as rapidly as in their cellular context (a discrepancy in the *rate* of expected manifestation). This suggested that there were *background conditions* (Hüttemann and Kaiser 2018) to be considered for protein folding, such as particular environmental factors that enhanced folding speed. For example, Anfinsen's group found that folding was faster when the in vitro solution contained an enzyme from the endoplasmic reticulum. Some proteins do not fold functionally (i.e., the disposition fails to manifest) or do not fold functionally as commonly under in vitro conditions. Such background conditions can therefore be relevant by modulating the disposition's manifestation (e.g., its rate), even when they are not stimulus conditions that initiate a particular manifestation in the first place.

Another conceptual issue arises from the frequency of successful manifestation: whether a disposition is *deterministic* (always manifesting under a stimulus condition) or *probabilistic* (manifesting under a stimulus condition with some quantitative frequency). The existence of the unfolded protein response in cells (Hetz 2012), where only some proportion of a protein species folds functionally, points toward foldability as a probabilistic disposition or *propensity.* A further complication relates to the assumption that proteins have a single native conformation. Intrinsically disordered proteins highlight that the sequence-structure-function relationship for foldability is not universal; many proteins do not form stable and static 3-dimensional configurations but instead assume many structural conformations over time under different stimulus conditions (Tompa 2010). The same disposition can yield multiple outcomes (i.e., a *multitrack* disposition), and these outcomes can be quantified with a probability distribution over the possible manifestations. Most dispositions are probabilistic in nature, manifesting only with a certain probability, but the multitrack foldability of intrinsically disordered proteins reminds us that there can be several probabilities of manifesting in different ways.

These additional issues with the manifestation of foldability prompted detailed studies of the background conditions, which revealed the importance of distinct chaperone proteins that guide protein folding during de novo synthesis, quality control, and the response to stress in the crowded environments of cells. Sometimes chaperones provide a sequestered domain and at other times actively facilitate folding. Even when mutations are introduced that lead to altered amino acid components in a polypeptide, functional folding can be induced by the overproduction of chaperones (Maisnier-Patin et al. 2005). Depending on how foldability is modeled, this role of chaperones can be understood differently, either as appropriate background or stimulus conditions or as *extrinsic* components of (or contributors to) the causal basis of the disposition (Hüttemann and Love 2011). At a minimum, the causal basis of foldability is more complex and relationally intertwined than previously thought (see section 4.3.2).

4.3 Methodological Questions: Putting the Toolkit to Use

This philosophical toolkit for studying dispositional properties orients us to several methodological questions that frame research on evolvability. For example, at what rates and on what timescales does evolvability manifest? Are the contributors to the causal basis of evolvability only intrinsic to the bearer of evolvability? What is being measured when studying evolvability empirically (causal basis, disposition, manifestation, or something else)? Additionally, further questions not directly addressed in the example of protein foldability must be explored, such as the difference between an individual entity versus an aggregate of entities having a disposition, the quantitative measure and comparison of dispositions, the distinctive behavior of dispositions at different timescales, or the evolutionary transformation of dispositions. We now turn to these and allied questions with special attention to the variety of definitions of evolvability, including why different definitions need not be in conflict and how different construals of evolvability are related.

4.3.1 What Are the Bearers of the Dispositional Property of Evolvability?

In the case of complex dispositions, it is useful to clarify what the *bearer* of a disposition is. For evolvability, the bearer is whatever entity is evolvable and thus possesses the capacity to evolve. Once we have specified a particular bearer of evolvability, we are better positioned to examine the constituents of its causal basis (section 4.3.2). In addition to common cases where an *individual* object possesses a disposition, we need to consider that an *aggregate* of individual objects also can have a dispositional property (which also matters for protein foldability; Invernizzi et al. 2012). Both options appear in scientific definitions of evolvability. When evolvability is interpreted as a population-level capacity, it is an aggregate of individual organisms that bear the disposition: "the ability of a *population* to respond to selection"" (Houle 1992, 195, emphasis added; see also Flatt 2005; G. Wagner 2014). In contrast, the bearer of evolvability appears to be an individual object if the disposition is construed as a property of an organism: "evolvability is an *organism's* capacity to generate heritable phenotypic variation" (Kirschner and Gerhart 1998, 8420, emphasis added; see also Yang 2001). For accounts that focus on the evolvability of a species or a higher taxon (Hopkins 2011; McGuire and Davis 2014; see also Jablonski, chapter 17), both options are utilized: Sometimes a lineage is conceptualized as an individual (an entity persisting as the same thing across time even while its constituent parts change; Rieppel 2007), and sometimes as an aggregate of organisms that compose an evolvable taxon.

How should we interpret these divergent viewpoints about what possesses evolvability? One possibility is to adopt a more abstract definition, such as by understanding evolvability as the "ability of a *biological system* to produce phenotypic variation" (Payne and Wagner 2019, 24, emphasis added). A "biological system" could be an individual (an organism is a biological system) or an aggregate of individuals, including a whole population of conspecific organisms or an interacting microbial system consisting of individuals from various species. However, it is practically necessary for many scientific investigations to have an account that more concretely specifies the bearer of evolvability. In population and quantitative genetics, it is typically populations of individuals (or their traits) that are assumed to be the units of evolutionary change: Populations (as aggregates) possess evolvability in the sense of having the ability to respond to selection. In evo-devo, the focus is on how the configuration of development for organisms biases phenotypic evolution. In such a context, evolvability is conceptualized as a property of organisms in the sense of the ability to generate functional phenotypic variants. These different definitions of evolvability need not be in conflict: They can refer to different aspects of the evolutionary process or be useful tools that have appropriate uses in the context of different approaches.

At the same time, the notion that an organism would be evolvable appears misguided; only a population or lineage of individuals evolves, not an organism. This issue can be resolved by making use of the toolkit distinction between types and tokens. One token organism cannot evolve, but a type or kind of organism exemplified in a population of individuals can undergo evolutionary change. This type or kind is representative, sometimes only implicitly, of a population or aggregated unit. Thus, the above definitions that refer to an organism's evolvability can be read as referring to a *type* of organism, such as organisms from a particular species or exhibiting shared constellations of traits. Likewise, when a developmental system is seen as the bearer of evolvability—"evolvability, the capacity of a *developmental system* to evolve" (Hendrikse et al. 2007, 394, emphasis added)—it is not a token developmental system of one individual organism, but the type of developmental system that is evolvable (Nuño de la Rosa and Villegas 2022). An analogous bearer of evolvability is the genotype-phenotype map: "evolvability as a property of the *genotype-phenotype map* (the genetic system) and not as a population property" (Hansen 2006, 129, emphasis added). As noted above, another prominent bearer of evolvability is a *trait,* given that phenotypic traits (Roseman et al. 2010; Opedal et al. 2017) or characters (G. Wagner and Altenberg 1996) evolve. Traits on quite different levels of organization can exhibit evolvability, from molecular traits (e.g., protein domains and transcription factor binding sites) and physiological traits (e.g., metabolic rates and immune responses), to complex morphological traits (e.g., insect wings and vertebrate limbs) and life-history traits (e.g., size at maturity and clutch size).

Although different interpretations concentrate on a variety of relevant bearers, it is possible to navigate among these interpretations. The same researcher may offer a generic or abstract characterization of evolvability as well as a specific or concrete characterization of evolvability, and then use them for distinct purposes. For instance, from the perspective of quantitative genetics, Houle (1992) construes evolvability generally as "the ability of a *population* to respond to natural or artificial selection" (195, emphasis added), while also providing a quantitative measure for the evolvability of a specific *phenotypic trait* (in terms of the trait's coefficient of additive genetic variation). The latter permits a comparison of the abilities of different specific traits in the same population to respond to selection; a generic definition only stating that evolvability is a capacity of a population would not accomplish this. A specific definition focusing on traits also facilitates comparisons of the same trait in different populations. Although different bearers often yield different evolvability dispositions, which are relevant for different purposes or in different scientific fields, in section 4.4 we also point to interrelations among these different construals of evolvability.

4.3.2 What Is the Causal Basis of Evolvability?

Once the bearer of evolvability is defined (whether it be traits, organisms, populations, or taxa), a subsequent concern is to understand the disposition's causal basis: What makes a particular bearer evolvable? For example, G. Wagner and Altenberg (1996, 967) focus on the evolvability of characters and argue that what determines this is the structure of the genotype-phenotype map. This structure governs the way in which random genetic mutation translates into nonrandom, structured, and possibly adaptive phenotypic variation for characters (Nuño de la Rosa and Villegas 2022). Many accounts point to the genotype-phenotype map and similar features of development underlying the variational properties of phenotypes as the disposition's *causal basis:* "evolvability . . . is largely *a function of* the developmental system's ability to generate variation" (Hendrikse et al. 2007, 394, emphasis added).

As noted (section 4.3.1), the genotype-phenotype map can itself be the *bearer* of evolvability. This bearer is operative in models of how the genotype-phenotype map as an entity evolves or can evolve, including discussions of whether a particular genotype-phenotype map is an adaptation (Pavličev and Hansen 2011; see also Hansen and Wagner, chapter 7). This point is a reminder of the value of the philosophical toolkit. What is a *bearer* in one analysis of evolvability can be a *causal basis* in another analysis. It is crucial to be explicit about whether a feature, such as the genotype-phenotype map, is considered the disposition's bearer (the entity possessing the ability to evolve) or the causal basis (of some other bearer of evolvability, e.g., a character). Keeping this in view can facilitate the identification of connections across different interpretations as well as possible ambiguities in how researchers are using evolvability. For example, Hansen (2006) claims that he "follow[s] Wagner & Altenberg (1996) in defining evolvability as *a property of* the genotype-phenotype map (the genetic system) and not as a population property" (129, emphasis added). Although this claim appears to assert that the genotype-phenotype map (and not a population) is the disposition's bearer, Hansen also states that the genotype-phenotype map "*determines* the variational properties of the phenotype" (123, emphasis added). The latter suggests that the genotype-phenotype map is the causal basis, while the phenotype is the bearer (which aligns with our earlier reconstruction of G. Wagner and Altenberg 1996). More generally, equivocal statements like "evolvability is a property of X" could be intended to mean either "evolvability is a capacity of X" (which refers to the *bearer* X) or "evolvability is a function of X" (which refers to the causal basis X; see the quote earlier in this section from Hendrikse et al. 2007). In addition to clarifying potential ambiguities, the distinction we are making here serves as a reminder of distinct research tasks: The genotype-phenotype map needs to be investigated both as a bearer of and the causal basis for evolvability.

Focusing on the genotype-phenotype map as the causal basis of evolvability tends to accent features *internal* to organisms: "The evolvability of an organism is its *intrinsic* capacity for evolutionary change. . . . It is a function of the range of phenotypic variation the genetic and developmental architecture of the organism can generate" (Yang 2001, 59). These internal properties can then be examined in more detail, such as how modularity is conducive to evolvability (G. Wagner and Altenberg 1996; Kirschner and Gerhart 1998; Pavličev and Hansen 2011; Pavličev et al., chapter 8; Hallgrímsson et al., chapter 9; Jablonski, chapter 17). Additionally, we can ask whether evolvability is always an intrinsic capacity. Our formulation of the toolkit using protein foldability raised the possibility that biological dispositions can have an *extrinsic* causal basis in the sense that some features external to the bearer are relevant to the disposition being present (Hüttemann and Kaiser 2018; Love 2003). A protein's foldability is not always due to its internal amino acid sequence but can depend on the presence of chaperones.[2] If so, such external factors are not just stimulus conditions that can trigger the disposition's manifestation (e.g., directional selection in the case of evolvability) or mere background conditions: They would need to be included in models of the disposition's nature (see box 4.1).

In the case of evolvability, the distinction between internal and external is ultimately a matter of theoretical perspective: A methodological choice should be made about whether to investigate some factors external to an evolvable entity because they impact evolvability or to relegate them to the background conditions under which evolvability takes place. Consider several scenarios where evolvability may not be an intrinsic disposition but also can be seen as contingent on external factors. If the focus is on a phenotypic trait as the bearer of evolvability, then it may be the case that extrinsic features compose at least part of the causal basis of this trait's evolvability, such as pleiotropic relations to other traits or the frequency of the trait in the whole population. Regarding another bearer, a taxon, there are a variety of considerations that point to an extrinsic basis, including interactions with entities in the environment that have an impact on evolvability (Love 2003; Sterelny 2011; Jablonski, chapter 17). From a paleontological perspective, species and higher taxa can possess evolvability also in terms of having lower extinction rates and the ability to undergo adaptive radiation and diversify. A taxon's extinction rate can depend on its geographic range (Jablonski 1987), but this is not primarily an intrinsic, internal property of the organisms making up a taxon, because it can include landscape topography or ecological diversity across the range (Love 2003). In cases of ecosystem engineering, a taxon's ability to undergo radiation and diversification is dependent on transformed ecological conditions, where the resulting ecological feedback can yield self-propagating radiations (Erwin 2012). A classic example is the radiation of mammals due to empty niches left by the extinction of non-avian dinosaurs (Alroy 1999). Such abiotic conditions as well as ecological interactions with other taxa would causally contribute to the basis of a taxon's evolvability relationally or extrinsically.

Methodologies in evo-devo make it natural to view evolvability as a disposition with an intrinsic basis, where evolvability resides within an organism whose developmental architecture yields its ability to generate phenotypic variation. At the same time, some evo-devo and eco-devo phenomena relevant to the generation of novelty are difficult to fit into an intrinsic causal basis framework. For example, symbiosis has contributed to the origin of multicellularity as well as to multiple origins of herbivory (Gilbert 2020). In

2. Other dispositional properties exhibit a similar context dependence. For example, the "stemness" (a dispositional property) of a cell can depend on the cell being situated in an appropriate stem cell niche (Laplane and Solary 2019).

niche construction, organisms not only encounter a given environment to which they adapt, but their physiological and behavioral activities also modify the environment in an adaptive fashion, resulting in feedback loops and organism-environment coevolution (Scott-Phillips et al. 2014; Clark et al. 2020). Examples include the evolution of lactose intolerance in humans influencing the rate and direction of evolution (Scott-Phillips et al. 2014) and horned dung beetle larvae manipulating the surrounding dung, which influences morphological development, sexual dimorphism, and life history (Schwab et al. 2016). In both symbiosis and niche construction, ecological interactions between the organism and some of the biotic or abiotic entities in its environment have undergone adaptive evolution and increased the evolvability of organisms in the lineage. Such ongoing evolutionary interactions transform features internal as well as external to organisms, exhibiting an iterative dynamic between "internalization" of environmental factors and "externalization" of an organism's structures (Laubichler and Renn 2015).

Organism-environment interactions, theoretically captured as G×E (gene-environment interaction), likewise matter for quantitative genetics (Hansen, chapter 5; Houle and Pélabon, chapter 6). As additive genetic variance determines the response to selection, one common quantitative measure of a phenotype's evolvability is $I_A = V_A/\bar{X}^2$, where V_A is the phenotype's additive genetic variance (the proportion of phenotypic variance due to additive genetic effects), and $\bar{X}$ is the phenotype's mean value in the population (Hansen et al. 2003). A trait's V_A in a specific population—with a fixed genetic composition—can change when the environment changes (just as the trait's heritability is environment dependent; Rice 2012). Due to G×E, a phenotype's value can change together with the environment, in turn changing the phenotypic variance across the population. Similarly, the phenotype's mean value $\bar{X}$ can be subject to change when the environment changes—without the population and its genetic composition changing. Therefore, the evolvability $I_A = V_A/\bar{X}^2$ for a trait in a given population can be different if this population is in two different environments (see also Pélabon et al., chapter 13). Thus, evolvability as construed in quantitative genetics might not be solely an intrinsic property of a population but might also be dependent on the population's environment; so the disposition's causal basis is partly extrinsic.

An overarching lesson that emerges from these considerations of why or when evolvability might have an extrinsic, relational causal basis is that these judgments depend on the specific definition of evolvability being used. If on some definition the manifestation is not just phenotypic variation, but also stipulates that the phenotypic variation must be *adaptive,* then the external context of individual organisms or populations of organisms has to be considered, because a trait's adaptiveness is relative to the environment. In a similar vein, the relevant causal basis also varies depending on the bearer of evolvability. For example, population size is an intrinsic factor for the ability of a *population* to respond to selection, but population size is obviously external to an *organism's morphological structure,* which forms the core of evo-devo models of evolvability. If the bearer possessing evolvability is not an organism with its developmental structure, but instead is understood as a larger biological system (which contains organism-environment interactions), then this system exhibiting evolvability would also be the complete causal basis, rendering the disposition intrinsic (Nuño de la Rosa and Villegas 2022).

Because of these different possible methodological choices and explanatory strategies, it may be preferable to talk in terms of *causal contributors to evolvability* rather than one unique causal basis. This explicitly acknowledges the diversity of individual causal factors involved: from the internal properties of individuals (e.g., the genotype-phenotype map and developmental architecture) to population-level features (e.g., genetic variance), from aspects of the abiotic or biotic environment of the individuals composing a population to organism-environment interactions. Which of these causal contributors are actively investigated with respect to evolvability is a *methodological choice*. Some biologists may consider environmental features as a causal background condition and focus on the genotype-phenotype maps of different taxa as the primary driver of evolvability; others will take some environmental factors (e.g., selection regimes) as stimulus conditions for the manifestation of evolvability. Still others may incorporate organism-environment reciprocal interactions in their explanatory frameworks in cases where niche construction or ecosystem engineering yield evolvability. In some situations, examining specific causal contributors (while relegating others to background conditions) may be a standard methodological preference adopted by some research groups; in other cases, an adjudication of what contributors must be included to yield an adequate explanation of a specific instance of evolvability may be required (Baedke et al. 2020).

4.3.3 How Does Evolvability Manifest?

Given this plurality of bearers and causal contributors, what characterizes evolvability as a distinct property? Dispositions are defined by their manifestation—and sometimes by their stimulus conditions, too—regardless of the variety of possible causal mechanisms that bring about this manifestation. For example, foldability is individualized as a disposition by a folding behavior under certain conditions (such as being in a solvent), despite the many distinct amino acid compositions that can be the causal basis of foldability. Therefore, even if evolvability can be instantiated by many diverse causal contributors in different contexts, it is a single dispositional property because it is defined by a manifestation: evolving in a certain fashion. Importantly, many if not all living systems are trivially evolvable or have been in their evolutionary history. However, when foldability is invoked, the aim is to specify a tendency of certain amino acid sequences to fold in a specific way as compared to other sequences. The same is true for evolvability: Evolvability is invoked to specify that some living systems (or configurations thereof) tend to evolve more readily than others (e.g., to diverge more or to evolve at a higher rate). These comparisons might include vertical contrasts with the ancestors of those systems or horizontal contrasts with other lineages (see also Jablonski, chapter 17). This is a restricted, comparative sense of evolvability that is distinct from the less informative and quasi-universal capacity to evolve (Love 2003).

Thus, to make sense of a comparative notion of evolvability in a specific context, it is crucial to specify the kind of evolutionary manifestation in view. Different definitions of evolvability focus on different possible manifestations. For example, if evolvability is defined as the ability of a population to respond to selection (Houle 1992), then this property will be manifested in a particular pattern of population-level phenotypic change across a certain number of generations (when under a selective regime), where that selection regime can be considered as the stimulus condition. If evolvability is understood as the ability to generate *new* phenotypic variants (A. Wagner 2005) or *novel* forms (Klingenberg 2005), then it will be expressed in the form of novelties or morphological disparity. In these cases, mutations are typically considered to be the stimulus conditions. If evolvability is conceived as the capacity to provide *adaptive* phenotypic variation (Payne and Wagner 2019), it will manifest in the alignment of new variants with the direction of selection, which contrasts with definitions focused on the production of heritable phenotypic variation without reference to its adaptive value. The meaning of "new," "adaptive," or "diverse" needs to be specified in each context for these manifestations of evolvability to be defined precisely.[3] For example, a character can be considered new if it is not homologous to an ancestral character or provides novel potential for future variation (Brigandt and Love 2012).

It is not only the kind of evolutionary change (adaptive, new, diverse, etc.) that must be specified; in addition, an appropriate timescale needs to be chosen. The complexities mentioned so far should make it obvious that there is no unique, privileged timescale for evolvability. Instead, each different approach will require a timescale that is relevant to its methodological perspective and research question, while ensuring that the timescale used to measure the stimulus condition (e.g., the period during which an average selection differential applies) matches the timescale for the ensuing manifestation of a population's or taxon's evolvability (Houle and Pélabon, chapter 6). Theoretical models and field studies in quantitative genetics will typically choose shorter timescales than paleontological investigations will. An important question in this area is the extent to which results from microevolutionary timescales can be projected to mesoevolutionary or macroevolutionary scales; much of the information available for current populations, such as the genetic basis of variation or selective pressures, is inaccessible for evolution on longer timescales (see also Jablonski, chapter 17). Limitations to such projectability justify the need to use more than one timescale in evolvability studies, especially given that different approaches also might focus on a different kind of evolvability (e.g., a population's evolvability in terms of changes in a phenotypic trait as opposed to a taxon's evolvability in terms of differential phylogenetic branching and extinction rates).

Once a suitable type of and timescale for the manifestation of the disposition to evolve are specified, it becomes more meaningful to talk about biological systems being more or less evolvable and, importantly, some systems being more evolvable than others.[4] In this sense, evolvability is a graded disposition, manifesting with different strengths in different systems, which in turn can show a capacity to evolve *more often* or at a *higher rate* than in other systems. To make such claims in a precise manner, however, there must be a way to measure evolvability (Houle and Pélabon, chapter 6). The dispositions of particular *tokens* are not directly measurable simply because, by definition, capacities as such are not observable properties. However, one can measure a disposition's strength by other means. One straightforward way to do this is by measuring rates of manifestation in aggregates or *types*. Mutation rates, rates of short-term response to selection, or long-term morphological disparity may serve as a proxy for the evolvability of some biological

3. This point reveals an analogy to protein foldability as a disposition. Rather than understanding foldability as the generation of any tertiary structure, one might require that foldability involves the production of a specific functional protein.

4. Our conceptual distinctions align with how Houle and Pélabon (chapter 6) make evolvability and its measurement context specific. In their "evolvability of . . . under . . . over . . ." framework, the evolvability "of" designates what we call the bearer, the evolvability "under" (e.g., directional selection) specifies the stimulus condition, and the evolvability "over" (e.g., a few generations) is equivalent to our timescale.

systems. For example, Landry et al. (2007) measured mutation rates in gene networks and suggested that these rates can be a good estimate of the evolvability of gene expression under no selection. However, since evolvability is not always manifested, this approach has limitations. A population can be very evolvable and yet fail to evolve, because it has not experienced the appropriate stimulus conditions or encountered interfering factors, just like a potentially fit phenotype can fail to spread and reach fixation in a population due to genetic drift. A different way to measure evolvability indirectly is by measuring one or more of its corresponding causal contributors. This approach offers a basis for comparative claims in terms of the same properties constituting a causal basis for evolvability in different biological systems. For example, one trait of a population might have a higher evolvability *than* another trait in the same population (because it has more underlying additive genetic variation), or one taxon can be more evolvable with respect to a trait *than* another taxon is, because its developmental architecture permits the generation of more morphological variation in a specific direction.

This analysis brings us to a major distinction in our philosophical toolkit: deterministic dispositions versus probabilistic dispositions (propensities). Although some dispositions always exhibit their characteristic manifestation when the stimulus conditions obtain, there are cases in which the manifestation will occur with a certain probability. Probabilities are generally preferred for measuring propensities in science (Cartwright 1989). An atom's radioactive decay is probabilistic, just like biological fitness is when understood as the capacity of an organism to survive and reproduce. Evolvability is likewise a probabilistic disposition (Brown 2014; Nuño de la Rosa and Villegas 2022), because there is rarely a single effect that will necessarily follow whenever it manifests. Instead, changes in selective pressures or genetic alterations, as triggers of evolvability, can bring about a range of different results, resulting in a probabilistic pattern of possible evolutionary trajectories.

A related idea from our toolkit is *multitrack* dispositions. Consider evolvability in an RNA model with a genotype-phenotype map that relates primary RNA sequences to their folded secondary structure. In this type of model, evolvability can be measured as the probability of a point mutation yielding one among several possible novel secondary structures (Nuño de la Rosa and Villegas 2022). Each of those secondary structures is a different possibility—the secondary structures are different *tracks* that can be followed in the manifestation of evolvability. The same holds for the evolvability of a phenotypic trait. Due to the architecture of the genotype-phenotype map, different phenotypes will have different probabilities of arising from one phenotypic starting point, which can be captured by a *probability distribution* across outcome phenotypes, that is, by multiple tracks (Stadler et al. 2001). For the purpose of theoretical modeling, researchers generally have to assign probabilistic estimates, including for unlikely outcomes if the aim is to theoretically investigate a variety of potential outcomes. But the primary reason that evolvability is probabilistic and multitrack derives from evolutionary events being subject to stochasticity. Whether a new phenotype will occur or how common it will be in a population is, at least in part, a matter of chance with respect to a variety of processes: sexual reproduction, population composition, mutational events, developmental noise, and environmental fluctuation. This stochasticity is especially pertinent for the manifestation of evolvability on longer timescales, where accidental events also can result in the extinction of whole taxa (Jablonski, chapter 17). This issue highlights the role of *contingency* and chance in evolution (Blount et al. 2018).

Philosophers have typically assumed that the stimulus condition (triggering condition) and manifestation of a disposition are one-time *events*. In the case of glass having the disposition to break, the stimulus condition of being hit by an object and the manifestation of breaking are events with a short and well-demarcated duration. Only recently have some philosophers come to appreciate that, at least in the case of biological dispositions, the stimulus conditions or manifestation may be an *ongoing process* (Hüttemann and Kaiser 2018). This conceptual point is relevant for evolvability. The stimulus conditions for evolvability may be the occurrence of natural selection or some other ecological conditions, such as extinction events or factors producing random drift. Even if discrete changes in selective pressures prompt changes in the manifestation of evolvability for traits, neither selection nor the response of biological systems to it are time-point events; instead, they are best understood as ecological conditions or processes extended over longer durations of time. As a result, evolvability becomes a moving target. Not only are the stimulus conditions and manifestation of the disposition processes without a fixed termination point, but the biological system itself also undergoes modifications as evolvability manifests, which may entail a change in the system's evolvability (e.g., through modifications of one or more causal contributors). In this case, the original disposition has undergone modification before its manifestation could terminate. The evolutionary alteration of gene regulatory networks might reduce pleiotropy and thereby increase evolvability: The activation of a system's evolvability can result in increases to its longer-term evolvability.

4.4 Implications: Why Is the Conceptualization of Evolvability as a Disposition Useful?

Having surveyed how the philosophical toolkit related to dispositions applies to different aspects of evolvability, a natural question is how the dispositional character of evolvability matters for ongoing biological inquiry. For example, various efforts are dedicated to quantitatively measuring and representing evolvability (e.g., using G-matrices and M-matrices, or genotype networks) to make accurate evolutionary *predictions* (Hansen, chapter 5; Houle and Pélabon, chapter 6). Since evolvability is not merely a property of one specific population, it is desirable to establish *generalizations* across populations and taxa. In addition to the scientific aims of predicting some outcome under natural circumstances or generalizing conclusions from isolated investigations, evolvability is also useful for the purpose of *control* of or intervention in natural systems, including breeding and conservation efforts or for protein engineering (Villegas et al., chapter 3).

Knowing about evolvability permits one to manipulate natural systems because evolvability is a *causal* capacity. In general, dispositions are scientifically important because they embody causal potency—dispositions bring about their manifestations (Austin 2017). This points to their *explanatory* role because a disposition explains the occurrence of its characteristic manifestation. In addition to documenting phenotypic variation, a primary aim of evolutionary biology is to understand the very *ability* to produce variation, exhibit evolutionary transformation, and yield phylogenetic diversification, precisely because evolvability explains these evolutionary outcomes of interest. The same is true for other dispositions in evolutionary biology (Austin and Nuño de la Rosa 2021). Phenotypic plasticity (as the capacity to generate several phenotypes) explains the possibility of various phenotypes resulting from one genotype, as well as why a particular phenotype resulted from environmental circumstances. Developmental robustness explains why a functional phenotype was maintained despite environmental or genetic perturbations during ontogeny. Likewise, phenotypic integration as the covariability of characters accounts for how a few genetic mutations can generate changes in many traits, often in a coordinated and functional fashion (see also Hallgrímsson et al., chapter 9). And modularity is the basis for the capacity of different organismal characters to vary quasi-independently of one another, which is instrumental to explaining the possibility of ongoing adaptation. However, evolvability may be distinctive as a disposition. For example, plasticity and organismal robustness to environmental perturbation primarily pertain to short-term effects, whereas evolvability also captures long-term evolutionary potential, thereby accounting for change and innovation across longer timescales.

One of the most important features of evolvability as a disposition is that it does not just explain one characteristic manifestation or the *actual* evolutionary outcome observed (see also Sterelny 1996; Brown 2014). Instead, the disposition of evolvability has *explanatory depth,* because a biological system's evolvability explains a whole *range of possible manifestations* under different potential stimulus conditions. This can be seen in quantitative genetics, where evolvability is understood as the ability of a population to respond to selection through phenotypic change. The stimulus condition of primary interest is the strength of selection acting on traits, and the manifestation is a change in the population's (mean) phenotypes. In the case of several phenotypic traits, evolvability can be captured by the additive genetic variances and covariances of traits (Hansen 2016; Hansen, chapter 5; Pélabon et al., chapter 13). This quantitative measure permits a prediction of how these phenotypic traits would change in each *possible* scenario if a particular strength of selection were to act on these traits. This general approach not only tracks one particular pattern of within-population phenotypic change to infer the selection pressures that obtained, but it also aims to predict the phenotypic change that would result from any of many different possible selection pressures. This type of prediction is useful for anticipating the effectiveness of a potential breeding strategy or ecological intervention (among other things). As a consequence of a population's evolvability (including trait covariances and different amounts of additive genetic variances for different traits), the phenotypic response may well deviate from the direction of selection (Hansen 2016). Unlike simply documenting actual phenotypic variation, this type of study of a disposition facilitates an understanding of how a population would respond to various possible situations.

In a similar vein, paleontologists may try to understand a range of possible evolutionary trajectories in addition to explaining a particular trajectory of morphological change seen in the fossil record. Would a larger or different range of outcomes have been possible from an ancestral starting point? Would a particular evolutionary outcome known from the fossil record be likely to reoccur if evolution restarted from this ancestral point? Evolvability—as the capacity of taxa to generate specific patterns of morphological transformation and diversification—can account for the possibility or difficulty of obtaining alternative outcomes (Love et al. 2022). Paleontological research also can employ computer simulations to investigate these types of scenarios. For example, although extinction events have a destructive effect in the short run, one may explore whether it is possible (and under which conditions it is possible) for extinctions to accelerate evolution *in the long run.* Prior extinction events might have selected for taxa that can rapidly occupy vacated niches;

these taxa have a high evolvability with respect to future extinction events as a possible stimulus condition of evolvability (Lehman and Miikkulainen 2015). Studies of evolutionary novelty (including in evo-devo) may have one particular morphological transition in view, such as the origin of the nervous system. Considered from the vantage point of evolvability, this morphological transition can be investigated more broadly with respect to the range of possible evolutionary outcomes under different stimulus conditions in light of assumed developmental contributors to the causal basis.

A recurring research task for evolutionary biologists is to understand *how evolvability itself evolves* (G. Wagner and Altenberg 1996; Draghi and Wagner 2008; Hansen and Wagner, chapter 7). Yet the question of the evolution of capacities has been absent from the classical philosophical literature on dispositions. Here we encounter a seeming puzzle, given the dispositional nature of evolvability. Selection can act on actual variation but not on the potential for variation: "the basic problem with the evolution of evolvability is that selection cannot act on potentials or *abilities*—only on results." (Watson 2021, 143) Once again, the philosophical toolkit suggests a way to get traction on this puzzle (see box 4.1). Selection acts on the *manifestations* of the evolvability disposition (resulting in evolutionary change), but this does not amount to selecting for the disposition itself. Importantly, selection can also act on the *causal basis* of a biological disposition. One relevant causal contributor to evolvability is the genetic architecture of organisms (Hansen 2006) and the structure of organismal development (Kirschner and Gerhart 2005). Genetic architecture and developmental mechanisms are concrete traits on which natural selection can act. As a result, even though selection does not act directly on a disposition, the evolution of the disposition to evolve can be understood in terms of selection on one or more contributors to evolvability's causal basis. Modifications of development yielding a phenotype-genotype map with an altered modular or covariational structure among traits can result in substantial evolutionary changes in evolvability (see also Hallgrímsson et al., chapter 9; Hansen and Wagner, chapter 7). This issue is another important place where the dispositional nature of evolvability matters: it serves to clarify distinct ways in which it could evolve by pointing to its relevant causal basis or manifestation.

We noted above (section 4.1) the presence of various definitions of evolvability. Importantly, there need not be a single evolvability disposition; instead, different scientific fields can focus legitimately on different concrete dispositions. For example, the evolvability of a population in terms of changes in phenotypic traits can be a different disposition from the evolvability of a clade in terms of taxonomic diversification. Our toolkit helps clarify that there are different kinds of evolvability and that the corresponding dispositions can be subject to distinct investigative projects. Most importantly, in the face of definitional diversity, the philosophical toolkit illustrates that different definitions can be legitimate and need not be in competition. For instance, the bearer of a disposition can be an individual or an aggregate. Correspondingly, the capacity to generate novel morphological variation may be a property of an organism's developmental system, but evolvability as the ability to respond to selection is a property of a population or aggregate of individuals.

We also distinguished between the disposition, its manifestation, and the disposition's causal basis. One reason these distinctions matter is that the definition of evolvability (e.g., simply in terms of generating variation, or more specifically, in terms of adaptive variation) has implications for the corresponding manifestation and causal basis. Once the choice of a particular definition with its bearer of evolvability is explicitly recognized, it increases clarity about the relevant manifestation and causal basis. An additional reason that these distinctions are useful is that different scientific activities can target different aspects of a disposition (see also Villegas et al., chapter 3). *Predicting* the outcome of evolvability concerns its manifestation, whereas *explaining* evolvability is about understanding the disposition's causal basis—an investigative context where developmental mechanisms are relevant, even if these do not suffice for the endeavor of making predictions for specific populations.

Although different definitions of evolvability may not be reducible to one another, and different fields may focus on different kinds of evolvability, there still can be systematic relations between them that are the subject of investigation. For example, individual organisms (as the bearers according to some definitions) have the disposition to generate adaptive variation, which in turn forms the causal basis for the evolvability of populations (the focus of other definitions). One can likewise reveal connections between different approaches to investigating evolvability. To the extent that the same kind of evolvability requires investigation from different perspectives because it is a complex phenomenon, studies of evolvability in evolutionary biology demand *interdisciplinary collaboration.* One biological field may address specific aspects of a disposition, such as paleontology elucidating phylogenetic patterns of diversification (the manifestation of taxon evolvability) or attempt to understand the ecological and biogeographical conditions that (as part of the causal basis) enabled high rates of diversification and reduced rates of extinction. One field might concentrate on the manifestation (or the disposition), while another is needed to uncover aspects of the causal basis. A case in point is when quantitative genetics measures realized phenotypic and genetic variation, whereas evo-devo seeks to characterize the developmental architecture of organisms that constitutes key aspects of the causal basis of evolvability. Insofar as different approaches investigate different aspects of evolvability, there are connections among such approaches that structure interdisciplinary collaboration (see Villegas et al., chapter 3). Because evolvability occurs on multiple timescales and is exhibited by entities on different levels of organization, contributions from different biological fields are needed to dissect its multifaceted complexity. And given the diversity of causal contributors to evolvability—from the internal constitution of organisms to organism-environment interactions and various ecological conditions—an interdisciplinary strategy seems most appropriate for understanding it. Keeping in view that evolvability is a disposition and utilizing the resources of the philosophical toolkit described in this chapter are critical to accomplishing this task.

Acknowledgments

We thank all participants of the "Evolvability: A New and Unifying Concept in Evolutionary Biology?" CAS project, and especially Thomas Hansen, Christophe Pélabon, David Houle, and Benedikt Hallgrímsson for comments on a draft of this chapter. We also acknowledge the discussion among participants of the "Evolvability across biological disciplines" session at the 2021 meeting of the International Society for the History, Philosophy and Social Studies of Biology. All coauthors are greatly indebted to the Centre for Advanced Study (CAS) at the Norwegian Academy of Science and Letters for its generous funding, which made this research project possible. Cristina Villegas and Laura Nuño de la Rosa were also supported by the Spanish Ministry of Economy and Competitiveness (project PGC2018-099423-

B-I00). Ingo Brigandt's work was additionally supported by the Social Sciences and Humanities Research Council of Canada (Insight Grant 435-2016-0500). Alan Love was supported in part by the John Templeton Foundation (grant #61329), the John M. Dolan Professorship, and the Winton Chair in the Liberal Arts.

References

Alroy, J. 1999. The fossil record of North American mammals: Evidence for a Paleocene evolutionary radiation. *Systematic Biology* 48: 107–118.

Anfinsen, C. B. 1973. Principles that govern the folding of protein chains. *Science* 181: 223–230.

Austin, C. J. 2017. Evo-devo: A science of dispositions. *European Journal for Philosophy of Science* 7: 373–389.

Austin, C. J., and L. Nuño de la Rosa. 2021. Dispositional properties in evo-devo. In *Evolutionary Developmental Biology: A Reference Guide,* edited by L. Nuño de la Rosa and G. Müller, 469–481. Cham, Switzerland: Springer.

Baedke, J., A. Fábregas-Tejeda, and F. Vergara-Silva. 2020. Does the extended evolutionary synthesis entail extended explanatory power? *Biology & Philosophy* 35: 20.

Blount, Z. D., R. E. Lenski, and J. B. Losos. 2018. Contingency and determinism in evolution: Replaying life's tape. *Science* 362: eaam5979.

Brigandt, I. 2015. From developmental constraint to evolvability: How concepts figure in explanation and disciplinary identity. In *Conceptual Change in Biology: Scientific and Philosophical Perspectives on Evolution and Development,* edited by A. C. Love, 305–325. Dordrecht: Springer.

Brigandt, I., and A. C. Love. 2012. Conceptualizing evolutionary novelty: Moving beyond definitional debates. *Journal of Experimental Zoology B* 318: 417–427.

Brown, R. L. 2014. What evolvability really is. *British Journal for the Philosophy of Science* 65: 549–572.

Campbell, N. A., and J. B. Reece. 2002. *Biology,* 6th edition. San Francisco: Benjamin Cummings.

Cartwright, N. 1989. *Nature's Capacities and Their Measurement.* Oxford: Oxford University Press.

Choi, S., and M. Fara. 2021. Dispositions. In *The Stanford Encyclopedia of Philosophy,* edited by E. N. Zalta. https://plato.stanford.edu/entries/dispositions/.

Clark, A. D., D. Deffner, K. Laland, J. Odling-Smee, and J. Endler. 2020. Niche construction affects the variability and strength of natural selection. *American Naturalist* 195: 16–30.

Draghi, J., and G. P. Wagner. 2008. Evolution of evolvability in a developmental model. *Evolution* 62: 301–315.

Erwin, D. H. 2012. Novelties that change carrying capacity. *Journal of Experimental Zoology B* 318: 460–465.

Flatt, T. 2005. The evolutionary genetics of canalization. *Quarterly Review of Biology* 80: 287–316.

Gilbert, S. F. 2020. Developmental symbiosis facilitates the multiple origins of herbivory. *Evolution & Development* 22: 154–164.

Hansen, T. F. 2006. The evolution of genetic architecture. *AREES* 37: 123–157.

Hansen, T. F. 2016. Evolvability, quantitative genetics of. In *Encyclopedia of Evolutionary Biology,* edited by R. M. Kliman, 83–89. Oxford: Academic Press.

Hansen, T. F., C. Pélabon, W. S. Armbruster, and M. L. Carlson. 2003. Evolvability and genetic constraint in *Dalechampia* blossoms: Components of variance and measures of evolvability. *Journal of Evolutionary Biology* 16: 754–766.

Hendrikse, J. L., T. E. Parsons, and B. Hallgrímsson. 2007. Evolvability as the proper focus of evolutionary developmental biology. *Evolution & Development* 9: 393–401.

Hetz, C. 2012. The unfolded protein response: Controlling cell fate decisions under ER stress and beyond. *Nature Reviews Molecular Cell Biology* 13: 89–102.

Hopkins, M. J. 2011. How species longevity, intraspecific morphological variation, and geographic range size are related: A comparison using late Cambrian trilobites. *Evolution* 65: 3253–3273.

Houle, D. 1992. Comparing evolvability and variability of quantitative traits. *Genetics* 130: 195–204.

Hüttemann, A., and M. I. Kaiser. 2018. Potentiality in biology. In *Handbook of Potentiality,* edited by K. Engelhard and M. Quante, 401–428. Dordrecht: Springer.

Hüttemann, A., and A. C. Love. 2011. Aspects of reductive explanation in biological science: Intrinsicality, fundamentality, and temporality. *British Journal for the Philosophy of Science* 62: 519–549.

Invernizzi, G., E. Papaleo, R. Sabate, and S. Ventura. 2012. Protein aggregation: Mechanisms and functional consequences. *International Journal of Biochemistry & Cell Biology* 44: 1541–1554.

Jablonski, D. 1987. Heritability at the species level: Analysis of geographic ranges of Cretaceous mollusks. *Science* 238: 360.

Kirschner, M. W., and J. C. Gerhart. 1998. Evolvability. *PNAS* 95: 8420–8427.

Kirschner, M. W., and J. C. Gerhart. 2005. *The Plausibility of Life: Resolving Darwin's Dilemma*. New Haven, CT: Yale University Press.

Klingenberg, C. P. 2005. Developmental constraints, modules, and evolvability. In *Variation: A Central Concept in Biology*, edited by B. Hallgrímsson and B. K. Hall, 219–247. Burlington, MA: Elsevier.

Landry, C. R., B. Lemos, S. A. Rifkin, W. J. Dickinson, and D. L. Hartl. 2007. Genetic properties influencing the evolvability of gene expression. *Science* 317: 118.

Laplane, L., and E. Solary. 2019. Towards a classification of stem cells. *eLife* 8: e46563.

Laubichler, M. D., and J. Renn. 2015. Extended evolution: A conceptual framework for integrating regulatory networks and niche construction. *Journal of Experimental Zoology B* 324: 565–577.

Lehman, J., and R. Miikkulainen. 2015. Extinction events can accelerate evolution. *PLOS ONE* 10: e0132886.

Love, A. C. 2003. Evolvability, dispositions, and intrinsicality. *Philosophy of Science* 70: 1015–1027.

Love, A. C., M. Grabowski, D. Houle, et al. 2022. Evolvability in the fossil record. *Paleobiology* 48: 186–209.

Maisnier-Patin, S., J. R. Roth, Å. Fredriksson, T. Nyström, O. G. Berg, and D. I. Andersson. 2005. Genomic buffering mitigates the effects of deleterious mutations in bacteria. *Nature Genetics* 37: 1376–1379.

McGuire, J. L., and E. B. Davis. 2014. Conservation paleobiogeography: The past, present and future of species distributions. *Ecography* 37: 1092–1094.

Mills, S. K., and J. H. Beatty. 1979. The propensity interpretation of fitness. *Philosophy of Science* 46: 263–286.

Nuño de la Rosa, L. 2017. Computing the extended synthesis: Mapping the dynamics and conceptual structure of the evolvability research front. *Journal of Experimental Zoology B* 328: 395–411.

Nuño de la Rosa, L., and C. Villegas. 2022. Chances and propensities in evo-devo. *British Journal for the Philosophy of Science* 73: 509–533.

Opedal, Ø. H., G. H. Bolstad, T. F. Hansen, W. S. Armbruster, and C. Pélabon. 2017. The evolvability of herkogamy: Quantifying the evolutionary potential of a composite trait. *Evolution* 71: 1572–1586.

Pavličev, M., and T. F. Hansen. 2011. Genotype-phenotype maps maximizing evolvability: Modularity revisited. *Evolutionary Biology* 38: 371–389.

Payne, J. L., and A. Wagner. 2019. The causes of evolvability and their evolution. *Nature Reviews Genetics* 20: 24–38.

Rice, S. H. 2012. The place of development in mathematical evolutionary theory. *Journal of Experimental Zoology B* 318: 480–488.

Rieppel, O. 2007. Species: Kinds of individuals or individuals of a kind. *Cladistics* 23: 373–384.

Roseman, C. C., K. E. Willmore, J. Rogers, C. Hildebolt, B. E. Sadler, J. T. Richtsmeier, and J. M. Cheverud. 2010. Genetic and environmental contributions to variation in baboon cranial morphology. *American Journal of Physical Anthropology* 143: 1–12.

Schwab, D. B., H. E. Riggs, I. L. G. Newton, and A. P. Moczek. 2016. Developmental and ecological benefits of the maternally transmitted microbiota in a dung beetle. *American Naturalist* 188: 679–692.

Scott-Phillips, T. C., K. N. Laland, D. M. Shuker, T. E. Dickins, and S. A. West. 2014. The niche construction perspective: A critical appraisal. *Evolution* 68: 1231–1243.

Stadler, B. M. R., P. F. Stadler, G. P. Wagner, and W. Fontana. 2001. The topology of the possible: Formal spaces underlying pattern of evolutionary change. *Journal of Theoretical Biology* 213: 241–274.

Sterelny, K. 1996. Explanatory pluralism in evolutionary biology. *Biology and Philosophy* 11: 193–214.

Sterelny, K. 2011. Evolvability reconsidered. In *The Major Transitions in Evolution Revisited*, edited by B. Calcott and K. Sterelny, 83–100. Cambridge, MA: MIT Press.

Tompa, P. 2010. *Structure and Function of Intrinsically Disordered Proteins*. Boca Raton, FL: CRC Press.

Wagner, A. 2005. *Robustness and Evolvability in Living Systems*. Princeton, NJ: Princeton University Press.

Wagner, G. P. 2014. *Homology, Genes, and Evolutionary Innovation*. Princeton, NJ: Princeton University Press.

Wagner, G. P., and L. Altenberg. 1996. Complex adaptations and the evolution of evolvability. *Evolution* 50: 967–976.

Watson, R. A. 2021. Evolvability. In *Evolutionary Developmental Biology: A Reference Guide*, edited by L. Nuño de la Rosa and G. Müller, 133–148. Cham, Switzerland: Springer.

Weslake, B. 2010. Explanatory depth. *Philosophy of Science* 77: 273–294.

Yang, A. S. 2001. Modularity, evolvability, and adaptive radiations: A comparison of the hemi- and holometabolous insects. *Evolution & Development* 3: 59–72.

5 Variation, Inheritance, and Evolution: A Primer on Evolutionary Quantitative Genetics

Thomas F. Hansen

Evolutionary quantitative genetics (EQG) emerged as a research paradigm in the 1980s based on operational tools for studying variation, inheritance, and selection in field and lab studies. In this chapter, I review the conceptual foundations of EQG as well as newer developments, with particular emphasis on the representation of evolvability and constraints.

5.1 Introduction

Quantitative genetics grew out of the biometry initiated by Francis Galton in the late 1800s. Galton wanted a quantitative science of inheritance and variation, and he was the first to systematically collect and analyze data on variation and inheritance in human populations (Bulmer 2003). He provided the first quantitative evidence for resemblance of offspring to their parents, and thereby for a central part of Darwin's theory of evolution by natural selection. Unfortunately, Galton made mistakes in the interpretation of his findings, which may have contributed to the long delay in the acceptance of natural selection on quantitative variation as the main force of evolution. Galton observed that the offspring of extreme (e.g., selected) parents tended to be more similar than their parents to the population average and took this to be a force working against selection. In Galton's model, selection could only make transient changes that would regress back toward the starting point. This led him and others to the view that selection on "biometric" variation was impotent and irrelevant to macroevolution, which instead must happen through discrete qualitative "mutations" that allow permanent change in the species (Provine 1971).

These misunderstandings were corrected when it was realized that biometric variation was based on discrete Mendelian genes. Even if offspring deviate less than their parents from the population mean on average (the heritability is less than unity), they are still different from the mean, and in the Mendelian model of inheritance, this difference is permanent and will not regress. This is captured in the breeder's equation: $\Delta\bar{z} = h^2 S$, where the change in the mean of a trait, z, from one generation to the next is equal to the heritability, h^2, multiplied with the selection differential, S, the average difference between selected parents and the population mean. By the time Fisher wrote his foundational paper on quantitative genetics in 1918, it was becoming clear that quantitative variation was compatible with a genetic architecture consisting of many Mendelian factors of small effect, and that selection on such variation could produce nonregressive open-ended evolution.

Importantly, it was understood that recombination between Mendelian factors could produce phenotypes that were far outside the observed range of variation, and that selection could thus do more than picking out the most extreme "lineage", as was commonly thought at the turn of the century (Beatty 2016). Fisher's model also revealed that only a part of the genetic variation, the so-called additive component, is reliably inherited during a response to selection

These insights are the foundation of quantitative genetics, but they did not earn the field a central place in the modern synthesis. Instead, the reduction of biometric variation to Mendelian genetics seems to have been taken as a license to focus on one- or two-locus allele-frequency dynamics, while leaving implicit the evolution of complex phenotypes. Quantitative genetics survived and developed in the applied breeding sciences, but for the next 50 years, it was less important to the development and application of evolutionary theory.

This situation changed in the 1980s with the emergence of a new evolutionary quantitative genetics (EQG), which soon became a major driver of empirical research in evolutionary biology. The emergence of EQG happened in two steps. The first was a set of theoretical papers by Russ Lande in the late 1970s (e.g., Lande 1976a,b, 1979, 1980). In these papers, Lande derived equations describing the evolution and maintenance of variation in quantitative traits under selection, drift, and mutation. He hypothesized that genetic variation in polygenic traits was stable and able to support open-ended evolution on macroevolutionary time scales. The second step was the development of operational measures of selection and constraints that provided a framework for empirical studies of the microevolutionary process. In particular, Lande and Arnold (1983) presented the selection-gradient approach for studying selection with multiple regression techniques. This allowed simple and effective ways of separating direct and indirect selection, and studying modes and causes of selection (e.g., Arnold 1983; Arnold and Wade 1984a,b; Wade and Kalisz 1990).

The focal point of EQG is the *Lande equation,* $\Delta\bar{\mathbf{z}} = \mathbf{G}\boldsymbol{\beta}$, which expresses the response to selection of a multivariate (vector) trait, $\mathbf{z}$, as a product of the G-matrix, $\mathbf{G}$, describing additive genetic variation in the trait vector and the selection gradient, $\boldsymbol{\beta}$, describing directional selection on the traits. This provides a neat separation between evolvability on one side and selection on the other. Each of these can be studied statistically in isolation and then be brought together for interpretation in a common theoretical framework. This heuristic generated two interlinked empirical research programs: One centered on field studies of selection based on the selection gradient, and the other centered on lab and field studies of genetic constraints and evolvability based on the G-matrix.

In this chapter, I first explain the statistical model of the genotype-phenotype map initiated by Fisher in 1918 before I present the basic theory of EQG with emphasis on the operationalization of selection and evolvability. I focus on basic principles, concepts, and measurement, and I do not review the many empirical findings of EQG research over the past 40 years. Although a comprehensive overview of these findings is not available, some elements can be found in Charmantier et al. (2014) and Hansen and Pélabon (2021).

5.2 The Statistical-Genetics Model of Variation and Inheritance

At the core of quantitative genetics sits a statistical model of the genotype-phenotype map that was introduced by Fisher (1918). To understand this model as it stands today (e.g., Lynch and Walsh 1998), imagine that we were asked to predict the stature of a random individual, which we will take to be Francis Galton. If we have no knowledge about Galton, except perhaps that he is a male, the best we can do is to guess that he is at the (male) average of the population. If we denote stature with z, and the population average is $\bar{z} = 170\ cm$, our initial guess for Galton's stature is

$$z = \bar{z} = 170\ cm.$$

Now, say we learn that Galton was a carrier of a specific allele at the locus *ZBTB38*, which codes for a transcription factor that is involved in the regulation of the thyroxin hydroxylase gene. This locus was found to have one of the largest effects on human stature in the Genome-Wide-Association study (GWAS) of Gudbjartsson et al. (2008). Carriers of the specific allele are on average about $0.5\ cm$ taller than the population at large. With this information, we predict that the stature of Galton is

$$z = \bar{z} + \alpha_{zbtb38} = 170\ cm + 0.5\ cm = 170.5\ cm.$$

The entity α is the average deviation (excess) of the allele. Such deviations can be defined for all the alleles that segregate in the population, and if we knew the average deviations of all the alleles in Galton's genome, we could sum them together to get a better prediction of his stature. The real Galton was tall, so say that the sum is a positive $10\ cm$; then our prediction is

$$z = \bar{z} + \sum_i \alpha_i = 170\ cm + 10\ cm = 180\ cm,$$

where the sum is over all alleles in the given genome (i.e., twice the number of loci for a diploid organism). This sum, $A = \Sigma_i \alpha_i$ is known as the breeding value of the individual, because with some assumptions, we can predict the average trait of offspring from the average of the breeding values of their parents. The breeding value of Galton is thus $A = 10\ cm$. For simplicity, I will also refer to A and the α as additive effects.

Knowing the additive effect of his genes gives us a better prediction of Galton's height but is still inaccurate, because we have no biological basis for our assumption that the average deviations can be added together. For example, imagine that Galton had not one, but two copies of the tall allele on the *ZBTB38* locus. Then this should add $1\ cm$ to his predicted height, but if it was sufficient with only one copy for the biological effect, this prediction would be wrong. We can account for this possibility by adding a dominance deviation, defined as the average deviance of the phenotype of the carriers of the two alleles from the additive prediction based on the two alleles. If it did not matter whether there were one or two tall *ZBTB38* alleles, their dominance deviation would be $\delta = -0.5\ cm$. We can calculate such dominance deviations for the allele pairs at all Galton's loci, and if we add them together we get a dominance effect $D = \Sigma_i \delta_i$. If we pretend that $D = -5\ cm$ for Galton, we get

$$z = \bar{z} + A + D = 170\ cm + 10\ cm - 5\ cm = 175\ cm,$$

as a better prediction of his stature. Likewise, there is no biological reason for adding effects across loci, and for each pair of alleles at different loci, call them i and j, we can define an additive-by-additive epistatic deviation $\alpha\alpha_{ij}$ as the average deviation of their carriers from the additive prediction. The sum of these effects over all pairs of alleles is $AA = \Sigma_i \Sigma_j \alpha\alpha_{ij}$. Adding this term will account for pairwise deviations from the additive prediction but not higher-order interactions. It may be that individuals who carry three specific alleles will

A_1	B_1
A_2	B_2

$$z = \bar{z} + \alpha_{A1} + \alpha_{A2} + \alpha_{B1} + \alpha_{B2} \qquad (A)$$
$$+ \delta_{A1A2} + \delta_{B1B2} \qquad (D)$$
$$+ \alpha\alpha_{A1B1} + \alpha\alpha_{A1B2} + \alpha\alpha_{A2B1} + \alpha\alpha_{A2B2} \qquad (AA)$$
$$+ \alpha\delta_{A1A2B1} + \alpha\delta_{A1A2B2} + \alpha\delta_{A1B1B2} + \alpha\delta_{A2B1B2} \qquad (AD)$$
$$+ \delta\delta_{A1A2B1B2} \qquad (DD)$$
$$z = \bar{z} + A + D + AA + AD + DD$$

Figure 5.1
The statistical-genetics model for a trait determined by two loci: The genetic value of an individual with alleles A_1 and A_2 at one locus and B_1 and B_2 at the other locus is decomposed into all possible component effects. The four first-order additive effects are given on the first line, the two dominance effects on the second line, the four pairwise *AA* epistatic effects on the third line, the four third-order *AD* epistatic effects on the fourth line, and the single fourth-order *DD* epistatic effect is given on the fifth line.

deviate on average from the prediction based on the additive, dominance, and pairwise epistatic effects involving those three alleles. This deviation can then be added to make an even better prediction. Such average deviations from lower-order predictions can be calculated for any set of alleles, and if we could calculate and add together the effects of all possible subsets of alleles from Galton's genome, we would get a perfect prediction of his total genotypic value, *G*. In figure 5.1 this is illustrated for a trait influenced by two loci. In addition to the *AA* effect, we recognize *AD* effects due to interactions among a pair of alleles at one locus and a single allele at another locus and *DD* effects due to interactions among all four alleles at two loci. With more loci, there are also *AAA* effects due to deviations of triplets of alleles at different loci and higher-order effects.

Adding all these effects together yields the total epistatic effect, *I*. If Galton's total epistatic value is $I = AA + AD + DD + AAA +$ (higher-order effects) $= 3\ cm$, we get

$$z = \bar{z} + A + D + I = 170\ cm + 10\ cm - 5\ cm + 3\ cm = 178\ cm,$$

and Galton's total genetic value is $G = A + D + I = 8\ cm$. We predict that his genes would make him 8 *cm* taller than the population average, but this is still not a perfect prediction, because it leaves out effects of the environment. Just as each individual has a unique set of alleles, each individual experiences a unique set of environmental influences in terms of different diets, diseases, amounts of stress, and so forth. For each specific environmental influence, one can define an average deviation from the population mean of individuals experiencing this factor, and these deviations can then be summed up to an environmental value, *E*, which can be added to the genetic value. Belonging to the gentry, Galton likely experienced an environment favorable to growth, so we assign him a positive score, say, $E = 6\ cm$. With this we arrive at the prediction

$$z = \bar{z} + G + E = 170\ cm + 8\ cm + 6\ cm = 184\ cm,$$

which should be close to Galton's true height, but just as genes may interact in their effects, some combinations of genes and environments may have effects that deviate from the sum of their separate effects. For example, an individual experiencing a calcium deficit in his diet may be disproportionally affected if he also carries alleles that weakens his calcium uptake. Such deviations are called genotype-by-environment, $G \times E$, interactions, and they can be summed and added to the genotypic and environmental values. If Galton's $G \times E$ value is –1 *cm*, we get

$$z = \bar{z} + G + E + G \times E = 170\ cm + 8\ cm + 6\ cm - 1\ cm = 183\ cm,$$

which would be a perfect prediction of Galton's stature, provided we have included and perfectly measured all genetic and environmental effects.

It is not even remotely possible to measure all relevant effects, but this is not the point. The model is useful because it can be combined with (Mendelian) rules of inheritance and segregation of alleles to understand trait variation and inheritance. The population variance in the phenotype, V_P, can be decomposed into contributions from each of the components discussed above by assuming that they are statistically independent of one another:

$$V_P = V_G + V_E + V_{G \times E} = V_A + V_D + V_I + V_E + V_{G \times E},$$

and each of these can be decomposed further. For example, $V_I = V_{AA} + V_{AD} + V_{DD} + V_{AAA}$ + higher-order effects. Also, by assuming Hardy-Weinberg and linkage equilibrium, which is equivalent to assuming that all alleles segregate independently, we can decompose each variance component into contributions from each average deviation. For example, the additive genetic variance is the sum of the variances of all the average deviations of alleles, $V_A = \Sigma_i \mathrm{Var}[\alpha_i]$, where $\mathrm{Var}[\alpha_i]$ is the population variance in the average deviation of single alleles. The assumption that all alleles segregate independently may seem severe, but covariances among alleles and genotypes can be added to the decomposition or incorporated into the definition of the variance components. The additive variance is usually understood as including these covariances, and the sum of the variances of the average deviations is technically known as the "genic" variance.

Even though all the statistical variance components contribute to differences among individuals, they are not inherited in the same way. Figure 5.2 illustrates the inheritance of the different components with a two-locus model. Our individual from figure 5.1 is mated to another individual, and an offspring is produced by picking one allele from each locus from each parent for the offspring's genotype. Comparing the genotypes of the mother and her offspring reveals that only a minority of the statistical components determining the mother's genetic value are to be found in her offspring. The two share half of their additive deviations, but only one out of the many dominance and epistatic deviations. From the perspective of the offspring, all its additive deviations are inherited from its two parents, but only half of its *AA* epistatic deviations and none of the deviations involving dominance in any form. The reason for the latter is that an individual can never inherit two alleles at one locus from the same parent. Its dominance deviations are therefore different from those of its parents. An exception occurs if the two parents are related. If they are, they may carry some alleles that are identical by descent, and then the offspring may end up with the same two allele copies as a parent at some loci (see Lynch and Walsh 1998 for details).

On the population level, these observations imply that all the additive genetic deviations in a population have been inherited (in a rearranged form) from the parental population, but none of the dominance deviations and only fractions of epistatic deviations are inherited. The noninherited genetic deviations have been generated de novo by recombination. This has consequences for evolution by natural selection. Selection acts similarly on all

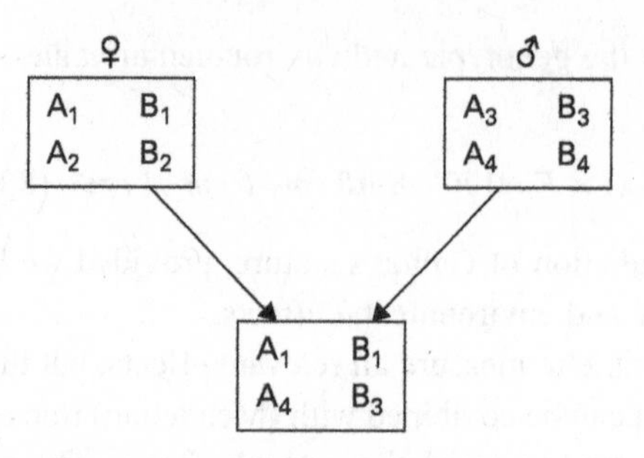

$$z_{mother} = \bar{z} + \alpha_{A1} + \alpha_{A2} + \alpha_{B1} + \alpha_{B2} + \delta_{A1A2} + \delta_{B1B2}$$
$$+ \alpha\alpha_{A1B1} + \alpha\alpha_{A1B2} + \alpha\alpha_{A2B1} + \alpha\alpha_{A2B2} + \alpha\delta_{A1A2B1} + \alpha\delta_{A1A2B2} + \alpha\delta_{A1B1B2} + \alpha\delta_{A2B1B2} + \delta\delta_{A1A2B1B2}$$

$$z_{child} = \bar{z} + \alpha_{A1} + \alpha_{A4} + \alpha_{B1} + \alpha_{B3} + \delta_{A1A4} + \delta_{B1B3}$$
$$+ \alpha\alpha_{A1B1} + \alpha\alpha_{A1B3} + \alpha\alpha_{A4B1} + \alpha\alpha_{A4B3} + \alpha\delta_{A1A4B1} + \alpha\delta_{A1A4B3} + \alpha\delta_{A1B1B3} + \alpha\delta_{A4B1B3} + \delta\delta_{A1A4B1B3}$$

Figure 5.2
Inheritance of a two-locus trait: Two individuals mate to produce an offspring, which inherits the alleles A_1 and B_1 from the mother, and the alleles A_4 and B_3 from the father. The full genotypic values of the mother and the offspring are spelled out. The dashed arrows indicate the shared effects. These are two additive effects, α_{A1} and α_{B1}, and one epistatic effect, $\alpha\alpha_{A1B1}$. This shows that half the additive effects and one quarter of the *AA* epistatic effects are shared between a parent and the offspring. Also note that all the additive effects of the offspring are inherited from the two parents, while none of the dominance effects and only fractions of the epistatic effects are inherited.

components, but only selection on additive variation is fully transmitted to the next generation. Although some epistatic deviations are also inherited, and indeed contribute to the selection response, this response is transient in the sense that it is broken down by recombination.

For these reasons, it is the additive component of variance that determines the short-term evolvability of a population, which explains the focus on the additive variance in EQG. In particular, it explains the central role of the additive genetic variance matrix, **G**, which summarizes the additive genetic variances and covariances of a set of traits. For a vector, $\mathbf{z}=\{z_1, z_2, z_3\}^T$, of three traits, the G-matrix is

$$\mathbf{G} = \begin{bmatrix} V_{A1} & C_{A12} & C_{A13} \\ C_{A12} & V_{A2} & C_{A23} \\ C_{A13} & C_{A23} & V_{A3} \end{bmatrix},$$

where V_{Ai} is the additive genetic variance of trait *i*, and C_{Aij} is the additive genetic covariance of trait *i* and *j*. Variance matrices describe variances of vectors, and the phenotypic variance matrix, denoted **P**, can be decomposed into additive genetic, dominance, epistatic, and environmental components just as described for the variance of a univariate trait. The inheritance also works in the same way. A covariance can be thought of as the amount of variance that is shared between two variables, and just as for variances, it is the additive part that is stably inherited.

The statistical model also provides the tools for estimation of the different variance components. By comparing the genetic makeup of relatives, as was done for mother and offspring in figure 5.2, it is easy to compute what they share. In the case of parent and offspring, we saw that they share half the additive variance, a quarter of the additive-by-additive variance, and none of the dominance variance. Similar computations can be done for other types of relatives. Fullsibs also share half of the additive variance and a quarter of the additive-by-additive variance, but additionally, they share a quarter of the dominance variance. Excess similarity (i.e., covariance) of fullsibs compared to parent with offspring can thus be used to estimate dominance variance. Halfsibs are particularly well suited to estimate additive variance, because they share a quarter of the additive variance but no dominance and only one sixteenth of the additive-by-additive variance. More generally, known or estimated pedigrees can be used to find the shared components of any pair of individuals in a population, and this can be used in a mixed-model framework to estimate the different components of variance as well as other aspects of genetic architecture (Lynch and Walsh 1998; Wilson et al. 2010; Jensen et al. 2014; Morrissey et al. 2014).

5.3 Selection in EQG

Natural selection occurs when individuals in a population have different properties, and these properties affect the ability of the individuals to survive and reproduce. If the properties are heritable in the sense that offspring are (statistically) similar to their parents, then we have evolution by natural selection. Evolution by natural selection can thus be seen as a two-step process with a first selection step and a second transmission step, in which the properties of the selected parents are transmitted to their offspring. Although transmission can be complicated, the mathematics of selection is simple.

An episode of selection can be described mathematically as a mapping from a set of individuals before selection to another set of individuals after selection (Price 1970; Arnold and Wade 1984a; Kerr and Godfrey-Smith 2009). Since the goal is to understand the effects of selection on some property, the individuals are classified into types that share the property of interest (e.g., Otsuka 2019). The types may be speckled and melanic moths, as in the classical textbook example of selection due to industrial pollution (Majerus 1998); all individuals that share a particular allele or genotype, as in classical population genetics; or all individuals that share a particular value of a quantitative trait, say a stature of 183 *cm*. The key point is that each type is assigned a fitness, a number that represents its contribution to the set after selection. In biological terms, the fitness is the ability of the type to survive and reproduce, and in mathematical terms, it is a random variable with a type-dependent distribution across individuals (Hansen 2017). In practice, we consider only the expected value of the fitness for a type (as in the propensity interpretation of fitness). Formally, the fitness of a type, z, is defined as the ratio of the amount of the type after selection, N_z', and the amount before selection, N_z. This yields the relation

$$N_z' = W(z)N_z,$$

where $W(z)$ is the absolute fitness of type z. Usually, the interest is not in the absolute amount of the type, but in its frequency relative to other types. If p_z is the frequency (fraction of total amount) of type z before selection, then some algebra yields its frequency after selection:

$$p_z' = w(z)p_z,$$

where $w(z) = W(z)/\bar{W}$, with $\bar{W}$ the mean fitness of the population. The $w(z)$ is the relative fitness of the type. Hence, absolute fitnesses describe changes in amount of types, and relative fitnesses describe changes in the frequency of types. From this equation, we can also calculate that the change in the mean trait value due to selection alone (i.e., the selection differential S) equals the covariance between the trait value and relative fitness:

$$S = Cov[w(z), z].$$

Price (1970) added transmission to this equation:

$$\Delta\bar{z} = Cov[w(z), z] + \overline{w(z)(z'(z) - z)},$$

where $z'(z)$ is the expected trait value of offspring of parents with trait value z, and the average in the transmission term is taken over the population distribution of z. The whole transmission term can be interpreted as the average difference in trait value between offspring and their parents. It is weighted with fitness to account for some types contributing more offspring than others.

The breeder's equation can be derived from the Price equation by assuming that the parent-offspring regression is linear with slope equal to the heritability, h^2. Then it follows from simple geometry that the average difference between offspring and their parents is $(h^2 - 1)S$, and we get

$$\Delta\bar{z} = S + (h^2 - 1)S = h^2 S.$$

Although the breeder's equation has been the standard representation of the response to selection in quantitative genetics, the foundation of EQG was based on a conceptually important rearrangement of the equation. I refer to this rearrangement as the Lande equation, and in the univariate case, it goes as follows:

$$\Delta\bar{z} = h^2 S = \frac{V_A}{V_P} S = V_A \frac{S}{V_P} = V_A \beta,$$

where $\beta = S/V_P$ is the selection gradient (i.e., the linear regression slope of relative fitness on the trait). Given this and defining heritability as $h^2 = V_A/V_P$, the Lande and the breeder's equations are mathematically equivalent. They are not conceptually equivalent, however, because they make different assumptions about what entities go together as quasi-independent units. In the case of the breeder's equation, these entities are the heritability and the selection differential; and in the case of the Lande equation, they are the additive variance and the selection gradient. Choices of which conceptual entities to measure and use in models constrain our thinking and interpretation of results, and the Lande equation paved the way for the study of evolvability separate from selection in a way that is not apparent with the breeder's equation (see section 5.8).

An instructive alternative derivation of the Lande equation is to assume that the breeding value, A, is the only component of the genotype that is transferred from parents to offspring, and that the breeding value of the offspring is equal in expectation to the breeding value of the parents. Then the evolutionary change in the trait equals the evolutionary change in the breeding value, and we get

$$\Delta\bar{z} = \Delta\bar{A} = Cov[w(z), A] = V_A \frac{Cov[w(z), A]}{V_A} = V_A \beta_A,$$

where β_A is the selection gradient on the breeding value of the trait. Hence, we see that the Lande equation also follows from the assumptions that only the breeding value is transmitted and that the phenotypic selection gradient, β, equals the (additive) genetic selection gradient, β_A. The latter is not a trivial assumption, because selection on phenotype and genotype may differ (e.g., Morrissey et al. 2010; Reid and Sardell 2011).

Lande (1979) defined the selection gradient theoretically in terms of the derivative (gradient) of relative fitness with respect to the trait (vector), but Lande and Arnold (1983) proposed to study it as a statistical regression of relative fitness on the trait (vector). The regression approach provides a simple method for studying the trait-fitness relationship in natural populations and is one reason for the success of EQG as an empirical research program. All one needs to use this are measures of traits and realized fitnesses of individuals, and then the coefficients from a multiple regression of relative fitness on the traits estimate the elements of the vector selection gradient

$$\boldsymbol{\beta} = \begin{bmatrix} \beta_1 \\ \beta_2 \\ \beta_3 \end{bmatrix},$$

shown here for three traits. Here, β_i, the partial selection gradient on trait *i*, yields the linear selection on trait *i* controlling for indirect selection due to correlations with other (included) traits. This technique can be used to distinguish direct and indirect selection, and it lends itself to experimental and observational studies in which hypothesized causes of selection can be tested by comparing selection gradients in their presence and absence. In this respect, the Lande-Arnold framework and the selection gradient are superior to the breeder's equation and the selection differential, which does not easily separate direct and indirect selection.

A key step in applying the framework is to find a useful measure of fitness. Most studies use proxy variables called "fitness components" that can be assumed to represent fitness over an episode of selection when other factors are kept constant. The trick is to find a fitness component that captures the causal mechanism under investigation. In a study of sexual selection on a mating display, one may, for example, measure the number of matings an individual obtains during a mating season. This will not capture all selection on the trait (e.g., through survival), but it may reasonably capture the influence of sexual selection.

Selection gradients are tools to describe fitness landscapes. The selection gradient points in the direction of steepest ascent in the fitness landscape, and its length is a measure of how steep the landscape is in this direction. The response to selection may not follow the selection gradient, however. As illustrated in figure 5.3, the response is generally deflected toward directions of higher evolvability (Lande 1979; Arnold 1994; Schluter 1996; Hansen and Houle 2008). Information about the curvature of the fitness landscape can also be obtained through second-order selection gradients, which can be estimated by including quadratic and interaction terms in the regression. These are collected in the so-called gamma matrix (e.g., Blows 2007), here illustrated for three traits:

$$\boldsymbol{\Gamma} = \begin{bmatrix} \gamma_{11} & \gamma_{12} & \gamma_{13} \\ \gamma_{12} & \gamma_{22} & \gamma_{23} \\ \gamma_{13} & \gamma_{23} & \gamma_{33} \end{bmatrix},$$

where the diagonal γ_{ii} terms are the second derivatives with respect to traits. Negative values indicate stabilizing selection on the trait, and positive values indicate disruptive

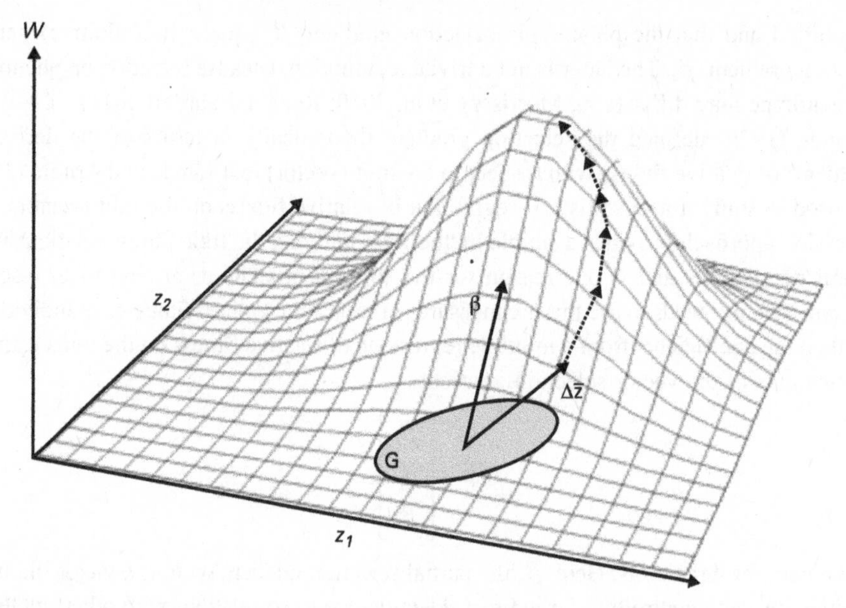

Figure 5.3
Evolution on a fitness landscape: The surface plots fitness against two traits. The ellipse **G** represents the shape of the G-matrix. The selection gradient **β** points from the population mean in the direction of the steepest ascent in the fitness landscape. The response to selection, $\Delta\bar{z}$, is deflected from the selection gradient toward directions of higher evolvability. With a single-peaked landscape, subsequent responses (dotted lines) will curve toward the peak and eventually bring the population mean to peak fitness. In every generation, mean fitness will increase with $e(\boldsymbol{\beta})|\boldsymbol{\beta}|^2$, where $e(\boldsymbol{\beta})$ is evolvability in the direction of the selection gradient and $|\boldsymbol{\beta}|$ is the magnitude of the selection gradient.

selection. Be aware that the quadratic regression coefficients need to be multiplied by 2 to yield the diagonal γ_{ii} (Stinchcombe et al. 2008). The off-diagonal γ_{ij} terms describe correlational selection (i.e., how the selection gradient on one trait changes when another trait is being changed).

5.4 Evolvability and Constraints in EQG

Although the tools for empirical study of selection were one reason for success of EQG, the other reason was the possibility for studying genetic constraints with the G-matrix (e.g., Lande 1979; Cheverud 1984; Arnold 1992). The G-matrix determines how much the selection response is deflected from the selection gradient, and how easy it is to evolve in different directions in morphospace. With three traits, the Lande equation yields the following expression for the response to selection in trait z_1:

$$\Delta\bar{z}_1 = V_{A1}\beta_1 + C_{A12}\beta_2 + C_{A13}\beta_3.$$

The first term is due to direct selection on the trait itself. The other terms show that selection on z_2 and z_3 (i.e., β_2 and β_3) causes a correlated response in our focal trait if there is additive genetic covariance between the traits.

The G-matrix describes genetic constraint (or evolvability) in two ways. The first is in terms of how much additive genetic variance there is in the different traits and trait combinations,

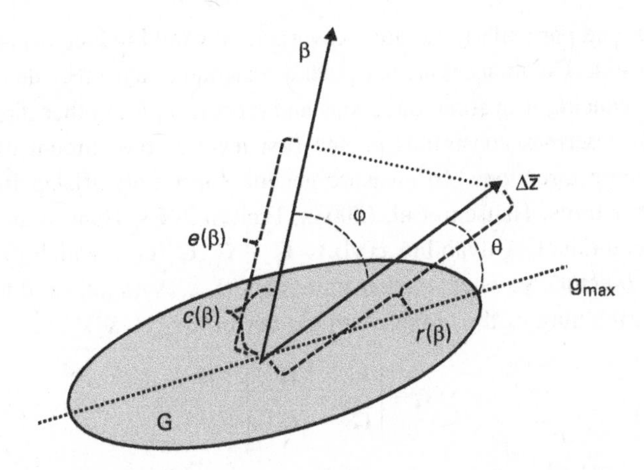

Figure 5.4
The geometry of evolvability and constraint measures. The figure shows the response, $\Delta\bar{\mathbf{z}}$, to a selection gradient, $\boldsymbol{\beta}$, as determined by an additive genetic variance matrix $\mathbf{G}$, with variation distributed as indicated by the ellipse. The major axis of variation in $\mathbf{G}$ is the vector $\mathbf{g}_{max}$, giving the direction of maximum evolvability. Schluter (1996) termed this vector the *line of least genetic resistance* and proposed to use the angle, θ, between this and the response as a measure of (lack of) constraint. The *evolvability*, $e(\boldsymbol{\beta})=\boldsymbol{\beta}^T\mathbf{G}\boldsymbol{\beta}$, in the direction of the selection gradient is defined as the length of the projection of the response on the gradient. It measures how far the population mean is shifted in the direction of selection. This value is less than the length of the response vector, which is the *respondability*, $r(\boldsymbol{\beta})=\sqrt{\boldsymbol{\beta}^T\mathbf{G}^2\boldsymbol{\beta}}$. The ratio between the evolvability and the respondability equals the cosine of the angle φ between the response and the selection gradient, and it measures how close the response tracks the direction of selection. Marroig et al. (2009) termed this the *flexibility*. The *conditional evolvability* in the direction of the selection gradient, $c(\boldsymbol{\beta})=(\boldsymbol{\beta}^T\mathbf{G}^{-1}\boldsymbol{\beta})^{-1}$, measures the evolvability if the response is constrained to follow the gradient by selection against deviations, as on an adaptive ridge. The ratio between the conditional and the unconditional evolvability is the *autonomy*: $a(\boldsymbol{\beta})=c(\boldsymbol{\beta})/e(\boldsymbol{\beta})$. It measures the fraction of genetic variance in direction $\boldsymbol{\beta}$ that is free from constraints. All equations are derived in Hansen and Houle (2008) and require standardizing $\boldsymbol{\beta}$ to unit length. Most of these statistics can be computed with G. Bolstad's R-package "evolvability" (Bolstad et al. 2014).

and hence their capability to respond to selection. The second is that it shows how selection on other traits can interfere with (or enhance) evolution on a focal trait through genetic covariances.

Figure 5.4 illustrates some statistics for describing the potential for evolution conveyed by a G-matrix. First, because evolvability may differ in different directions in phenotype space, Hansen and Houle (2008) proposed a measure of evolvability in a vector direction $\mathbf{z}$ as $e(\mathbf{z})=\mathbf{z}^T\mathbf{G}\mathbf{z}/|\mathbf{z}|^2$. This statistic has some desirable properties for a multivariate measure of evolvability: (1) It is the additive genetic variance in the direction of the vector, and thus, (2) it reduces to the additive genetic variance for a univariate trait. (3) If measured along a selection gradient, it is the length of the projection of the response (predicted from the Lande equation) on this gradient. It thus measures the ability to respond in the direction of selection and accordingly, (4) it is proportional to the increase in mean fitness due to selection. But note that $e(\mathbf{z})$ is not a measure of the expected magnitude of the selection response. It is a measure of the ability to respond in the direction of selection and would perhaps have been better named adaptability rather than evolvability. On the other hand, it is also (5) proportional to the expected change (variance) in direction $\mathbf{z}$ due to genetic drift.

The evolvability averaged over all directions in morphospace is $\bar{e}=\mathrm{Tr}[\mathbf{G}]/d$, where Tr is the trace operator, and d is the dimensionality (rank) of the G-matrix. One interesting aspect of this average is that it is unaffected by covariances. Intuitively, one might think that covariances, and particularly negative covariances, would reduce overall evolvability, but this is not the case. Covariances are not general constraints. What they do is to redistribute evolvability by reducing it in some directions and increasing it in other directions.

The $e(\mathbf{z})$ thus describes constraints on the first level as the amount of transmissible variation in specific directions. To measure genetic constraints arising from interfering selection on other traits, Hansen et al. (2003a; Hansen 2003; Hansen and Houle 2008) proposed the conditional evolvability $c(\mathbf{y}|\mathbf{x}) = \mathbf{G}_y - \mathbf{G}_{yx}\mathbf{G}_x^{-1}\mathbf{G}_{xy}$, which gives the evolvability of a trait (vector), $\mathbf{y}$, when another trait (vector), $\mathbf{x}$, is not allowed to change. This is based on a partitioning of the G-matrix of the vector $\mathbf{z} = \{\mathbf{y}^T, \mathbf{x}^T\}^T$ as

$$\mathbf{G} = \begin{bmatrix} \mathbf{G}_y & \mathbf{G}_{yx} \\ \mathbf{G}_{xy} & \mathbf{G}_x \end{bmatrix},$$

where $\mathbf{G}_y$ and $\mathbf{G}_x$ are the variance matrices of the vectors $\mathbf{y}$ and $\mathbf{x}$, respectively, and $\mathbf{G}_{yx} = \mathbf{G}_{xy}^T$ is their covariance matrix. The conditional evolvability equals the component of additive variance in a trait vector that is independent of another (defined) trait vector. If there is directional selection on a trait, $\mathbf{y}$, that is correlated with other traits, $\mathbf{x}$, under stabilizing selection, the initial response to selection would be determined by $e(\mathbf{y})$, but indirect selection on the constraining traits would shift them from their optima, inducing counteracting selection that would decrease the response of the focal trait. The response in the focal trait would decay toward a rate given by $c(\mathbf{y}|\mathbf{x})$. If the stabilizing selection is strong, this rate is approached in a handful of generations (Hansen et al. 2019). Conditioning on all orthogonal directions yields the statistics $c(\mathbf{z}) = (\mathbf{z}^T\mathbf{G}^{-1}\mathbf{z}/|\mathbf{z}|^2)^{-1}$, illustrated in figure 5.4, which measures how easy it is to move along an adaptive ridge in direction $\mathbf{z}$. The $c(\mathbf{z})$ is inversely proportional to how much fitness must increase along the ridge to drive a given change.

The conditional evolvability, and derived measures such as autonomy, can be seen as operational measures of modularity, because they quantify the ability of parts to evolve independently of one another. The concept of modularity has long antecedents (e.g., Olson and Miller 1958; Berg 1959), but it became a focus of research in evodevo, where it is treated almost as a synonym of evolvability. The idea is that the evolvability of complex integrated organisms requires organization into functional parts (modules), such as limbs or organs, that can evolve quasi-independently of other parts (Riedl 1977; Cheverud 1984; Wagner and Altenberg 1996; Wagner 2014). Consequently, there appeared many studies of modularity and integration through patterns of trait correlation in G- or P-matrices (reviewed in Melo et al. 2016). This has motivated a variety of methods and statistics for testing and comparing constraints (e.g., Lande 1979; Schluter 1996; Hansen et al. 2003a, 2019; Hansen and Houle 2008; Hohenlohe and Arnold 2008; Marquez 2008; Agrawal and Stinchcombe 2009; Hine et al. 2009; Kirkpatrick 2009; Mitteroecker 2009; Hansen and Voje 2011; Chevin 2013; Bolstad et al. 2014; Grabowski and Porto 2017; Sztepanacz and Houle 2019; Cheng and Houle 2020).

5.5 The Role of Mutation

Studying constraints with the G-matrix is predicated on $\mathbf{G}$ being somewhat constant on relevant time scales. Lande (1976b, 1980) put forward the hypothesis that a constant $\mathbf{G}$ conveying high evolvability could be maintained in a balance between mutation and stabilizing selection. This hypothesis is part of the foundation of EQG and is implicit in much EQG research, but it soon came under debate (e.g., Turelli 1984), and it became clear that its validity depends on poorly known empirical information about patterns of selection, genetic architecture, and mutational input. This debate motivated empirical research, including direct attempts to estimate changes and differences in **G** among different species (Arnold et al. 2008; McGlothlin et al. 2018), and attempts to estimate the rate of mutational input to quantitative traits (Houle and Kondrashov 2006; Halligan and Keightley 2009).

The mutation matrix **M** is defined as the amount of new additive genetic variance that arises in a trait vector in each generation by mutation. Under the Gaussian assumptions favored by Lande, the equilibrium **G** under symmetric stabilizing selection with the mean at the optimum is

$$\hat{\mathbf{G}} = \sqrt{n}\,\boldsymbol{\Gamma}^{-\frac{1}{2}}\left(\boldsymbol{\Gamma}^{\frac{1}{2}}\mathbf{M}\boldsymbol{\Gamma}^{\frac{1}{2}}\right)^{\frac{1}{2}}\boldsymbol{\Gamma}^{-\frac{1}{2}},$$

where $\boldsymbol{\Gamma}$ is the matrix of second-order selection gradients; n is the (effective) number of loci affecting the trait vector; and the $\boldsymbol{\Gamma}$-matrix is for the genotype, and not the phenotype, and thus potentially dependent on the environmental variance. This equation is derived from equation 28.37c in Walsh and Lynch (2018) based on Lande (1980). In addition to the "Gaussian" assumption of normal distribution of genetic effects at each locus, it assumes weak selection, additivity, equivalent loci, and linkage equilibrium. The equation shows that although **G** tends to increase with input of mutational variance and decrease with the strength of stabilizing selection, the relationship is nonlinear with no simple proportionality with either. This can be seen by calculating the evolvability along eigenvectors, **v**, of $\boldsymbol{\Gamma}$ as $e(\mathbf{v}) = \sqrt{ne_m(\mathbf{v})/\gamma}$, where γ is strength of stabilizing selection (magnitude of corresponding eigenvalue), and $e_m(\mathbf{v}) = \mathbf{v}^T\mathbf{M}\mathbf{v}/|\mathbf{v}|^2$ is the mutational variance in the direction of **v**. Note that the equilibrium depends on genetic architecture beyond the M-matrix in the form of the number of loci affecting the traits. Beyond the Gaussian, weak-selection and additivity assumptions, the equilibrium also depends on details of genetic architecture, such as size, bias, and pleiotropy of mutational effects, as well as on population structure and the exact mode of selection (Walsh and Lynch 2018). Epistasis also affects the maintenance of both additive and nonadditive genetic variance (Hermisson et al. 2003).

Hence, it cannot be assumed that **M** and **G** are equivalent in their effects on evolvability, as is often implicit in verbal discussions. The processes maintaining variation must be considered in addition to the input of variation. Nevertheless, the amount and pattern of mutational input is essential to understanding both short-term evolvability and long-term constraints. Subsequent empirical research has vindicated Lande's hypothesis that mutational input is sufficient to maintain high evolvability in the face of selection. This vindication has come both through direct estimates of **M** and other mutational parameters, and through a growing realization that many quantitative traits are influenced by a very large number of locations in the genome. The latter not only generates a large mutational target size, it also makes the Gaussian assumptions more likely.

Houle (1998) proposed that mutational target size was a major determinant of levels of genetic variation. Traits and trait categories with the potential to be affected by many genomic changes are likely to experience more mutation and to harbor more genetic variation, and this factor may explain more of the "variation in variation" than differences in strengths of selection. This hypothesis explains why life-history traits and fitness tend to have more additive genetic variation than morphological traits despite presumably stronger selection on the former. Life-history traits, such as survival and fertility, have large mutational target sizes, because they are influenced by most aspects of morphology, physiology, and behavior, and thus "inherit" their mutational input. Consequently, the mutational input to life-history traits can be so large as to be able to replenish standing levels of variation in mere tens of generations (Houle 1998; Halligan and Keightley 2009).

5.6 Genetic Drift in EQG

Adaptive genetic evolution is driven by (direct) natural selection, but many evolutionary changes are not adaptive. We have seen that indirect selection and correlated responses are important, and mutation, recombination, migration, and genetic drift also cause change. Genetic drift refers to changes due to random sampling of alleles in finite populations. For a polygenic trait (vector) with a linear genotype-phenotype map, the change in the mean from generation to generation due to drift is normally distributed with mean zero and a variance matrix equal to $\mathbf{G}/N_e$, where N_e is the effective population size, an inverse measure of the strength of genetic drift (Lande 1976a, 1979). Even if there is no trend to the changes, the trait mean will undergo random fluctuations that shifts it away from the ancestral state. The variance of these fluctuations in a trait, $\mathbf{z}$, will increase linearly with time (in generations) and be proportional to the evolvability $e(\mathbf{z})$ in the direction of the trait vector.

This prediction assumes that the G-matrix stays constant, but another effect of genetic drift is to reduce the genetic variance and thus the G-matrix each generation with an expected factor of $1 - 1/2N_e$. Given an input $\mathbf{M}$ of mutational variance, the change in $\mathbf{G}$ in a "neutral" model (i.e., due to mutation and drift alone) is then $\Delta\mathbf{G} = (1 - 1/2N_e)\mathbf{G} + \mathbf{M}$, and from this, we infer a neutral equilibrium of $\hat{\mathbf{G}} = 2N_e\mathbf{M}$. Hence, if there is no selection, we expect more genetic variation in larger populations, and we expect the G-matrix to be proportional to the M-matrix. Lynch (1990) noted that this result implies that the effective population size will drop out of the variance of trait change from generation to generation, which will be equal to $2\mathbf{M}$. Hence, we expect neutral evolution to be equally fast in small and large populations. This is the quantitative-genetics analog of Kimura's molecular clock, in which the substitution rate depends only on the mutation rate and not on the population size. It implies that the expected variance of change in a trait vector, $\mathbf{z}$, in a neutral model is equal to $2e_m(\mathbf{z})$ per generation with $e_m(\mathbf{z})$, defined in section 5.5, being a mutational evolvability.

Although the drift and neutral models are unrealistic as standalone hypotheses for the evolution of quantitative traits, they yield insights into the process of evolution. Contrary to intuition, the neutral model predicts high rates of evolution on macroevolutionary time scales. Lynch (1990) used estimates of mutational variance to argue that almost all observed changes in the fossil record were too small to be compatible with the neutral model, and of course also with sustained directional selection. Hence, we reach the inescapable conclusion that quantitative traits must be under some form of stabilizing selection on long time scales.

In the presence of selection, genetic drift is not likely to be important on the phenotypic level beyond a few generations (the molecular level is another matter; see Lynch 2007).

One instance of this concerns the role of stochastic peak shifts in evolution. We have seen that selection will bring a population to the nearest adaptive peak, but it need not be the highest peak. The deflections of the evolutionary path caused by the G-matrix may influence which peak is reached but have no partiality for higher peaks. Hence, selection may often act as a constraint that keeps populations from finding better adaptations, and stochastic peak shifts have a potentially important role in allowing populations to explore the broader adaptive landscape. Lande (1985), however, showed that the expected waiting time to attain a successful peak shift scales with the drop in fitness when crossing the valley to the power of N_e, which means that only small populations can cross only shallow valleys in a reasonable time, or to put it bluntly: Significant peak shifts do not occur by drift. Thus populations will typically find themselves trapped on local peaks, and larger evolutionary changes must involve some form of change in the adaptive landscape.

5.7 The Environment in EQG

One attraction of quantitative genetics is the incorporation of environmental sources of individual differences. Despite frequent appeals to high heritability, most variation in quantitative traits tend to be environmental in origin. Hansen and Pélabon (2021) found that the median heritability from 2,536 estimates from wild populations was 31%, meaning that more than 2/3 of the variance is not additive genetic. Even for 1-dimensional morphological measurements, which tend to have the highest heritabilities, only a median 40% of the variance was additive genetic. For life-history traits, which are more susceptible to environmental influences, the median heritability was merely 18%. Some of the nonadditive variance may still be genetic, but environmental sources of variation are clearly important.

The environmental variance can also be broken down into different sources. One important component in many organisms is the effect of parents on their offspring. In mammals, where maternal care and provision are all-important, the maternal component of variance can be substantial (Mousseau and Fox 1998). Parental effects must be considered when estimating genetic variance components and can have consequences for evolution by selection (e.g., Lynch 1987; Kirkpatrick and Lande 1989; Arnold 1994; Wilson et al. 2005; Day and Bonduriansky 2011).

It is convenient to distinguish between micro- and macroenvironmental effects. The former are the effects of many small or unknown sources that must be treated stochastically, as is done with the environmental variance components. Macroenvironmental effects, however, are the result of variation in influential and measurable environmental variables that are somewhat stable for the individual in question. These are often described with reaction norms, which are functions that describe how the expected trait value of a genotype varies with an environmental variable (figure 5.5); for example, body size in relation to food availability. Genetic variation in the slope or shape of reaction norms, that is, genotype-by-environment interaction, sets up the potential for evolutionary changes in plasticity (Via and Lande 1985). Plasticity usually refers to reaction norms that are adapted to create favorable trait values in the particular environment encountered by the individual. The ability to express defense compounds or morphology in the presence of predators or herbivores are standard examples. Not all norms of reaction are adaptive, however; for

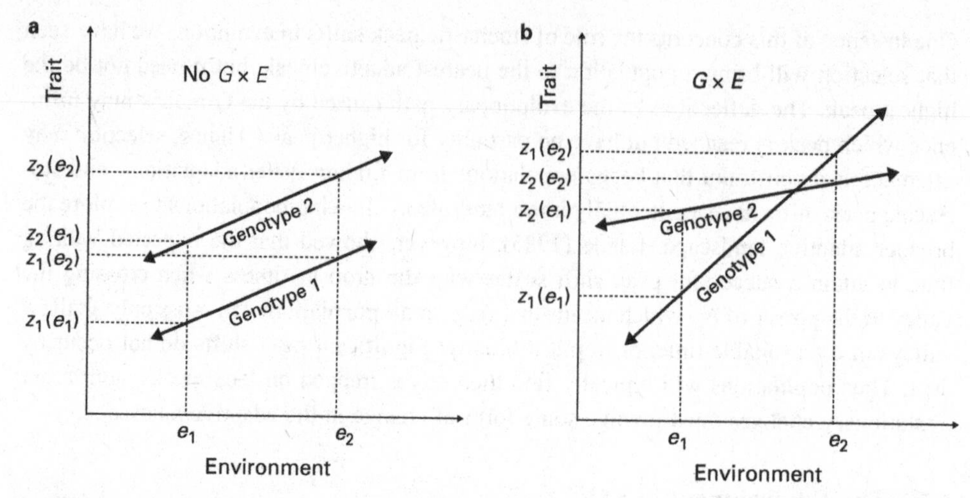

Figure 5.5
Reaction norms: The reaction norms of two genotypes are plotted, and the resulting trait values are given for two states of an environmental variable. Panel A shows a case with no $G\times E$ interaction. The reaction norms have the same shape, so that the phenotypic differences between genotype 1 and genotype 2 are the same in all environments. In panel B, the shapes of the reaction norms differ, so that the difference between the genotypes in environment e_1 is different from the difference in environment e_2, which yields $G\times E$ interaction. With $G\times E$ interaction, the shape of the reaction norm, and thus plasticity, can evolve.

example, growing to small size on limited food is more an outcome of necessary constraints than an adaptive plastic response.

Whether adaptive or not, plasticity and norms of reaction create substantial variation both within and between populations. In a review of reciprocal-transplant experiments, Stamp and Hadfield (2020) found that some 70% of local population differences could be ascribed to plastic responses and only 30% to genetic differences. Plasticity is not just an alternative to genetic adaptation, however, and there has been much discussion on how the two can interact in evolution. Depending on circumstances, plasticity can either facilitate or buffer genetic adaptation (e.g., Paenke et al. 2007).

Traits with lots of genetic variance also tend to have lots of environmental variance. One surprising consequence of this is that heritabilities do no not reflect amount of additive genetic variance (Houle 1992; Hansen et al. 2011; see figure 5.6). The most likely explanation for this puzzling finding is that traits that are generally decanalized or complex with many parts and developmental contingencies tend to be simultaneously sensitive to genetic and environmental perturbations. Cheverud (1988) conjectured that patterns of genetic and environmental correlation are also similar, and based on this, he controversially proposed that the phenotypic P-matrix could be used as substitute for the G-matrix. This idea is approaching a methodological principle in paleobiology and morphometrics, where P-matrices are routinely used in place of G-matrices to study evolvability and constraints (Love et al. 2022). Although we have seen that substituting **P** for **G** cannot be justified by unspecific appeals to "high" heritability, it is an open question when the environmental E-matrix is sufficiently similar to **G** to make **P** a more accurate estimator of the shape of **G** than the often imprecisely estimated **G** itself.

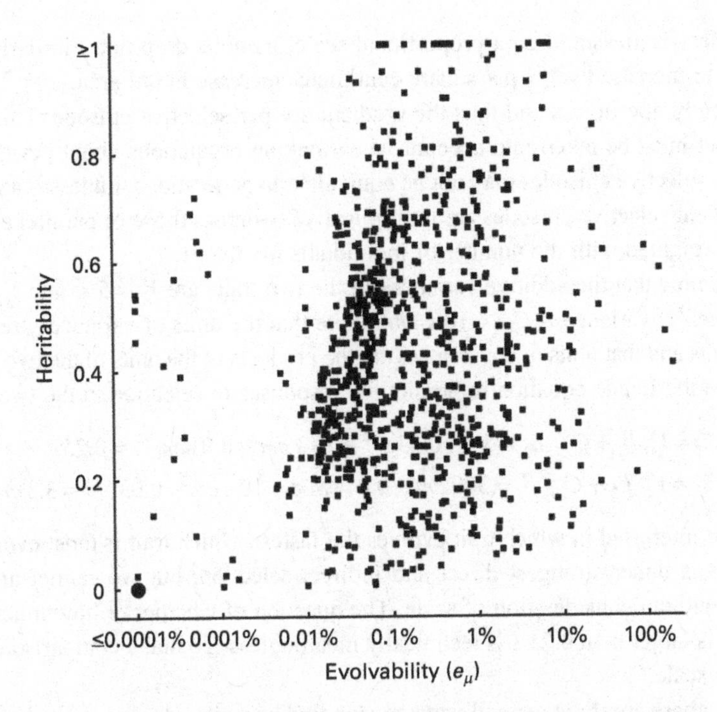

Figure 5.6
Effects of scaling: No relationship between heritability and mean-scaled additive variance (evolvability), as illustrated for 929 1-dimensional morphological measurements. The correlation is 0.04, which means that only a fraction of a percent of the variance in one variable is explained by the other. The big dot in the lower-left corner represents 24 nonpositive estimates. Data from Hansen and Pélabon (2021).

5.8 Issues of Scale and Measurement

In contrast to classical evolutionary genetics, which is concerned with categorical traits on nominal scales, quantitative genetics works with "quantitative" traits on continuous scales. This provides the opportunity for more accurate predictions and tests of the theory, and facilitates better measurement, statistics and quantitative interpretation. Unfortunately, this quantitative promise is often unfulfilled. Evolutionary biologists are accustomed to advanced statistics but are often unaccustomed with quantification. A tradition has developed that allows qualitative interpretations of quantitative results through significance tests and a general neglect of magnitudes, scales and meanings of estimated parameters (Houle et al. 2011; Morrissey 2016).

One manifestation of this tradition is a widespread neglect of units. To see how units work in the focal Lande equation, consider its application to two traits in an imaginary study of sexual selection in a peacock-like bird: display duration, z_1, measured in seconds, and tail area, z_2, measured in square centimeters. Assume a selection gradient is estimated as $\beta = \{0.03\ s^{-1}, 0.01\ cm^{-2}\}^T$. Note that units of the gradient are inverses of trait units. The partial selection gradient on display duration, $\beta_1 = 0.03\ s^{-1}$, says that if display duration is increased by one second, then fitness (i.e., the fitness component in question) is increased by 3% of its mean. This is because a selection-gradient analysis must use relative fitness, which is fitness measured in units of its own mean. The unit of β_1 is thus fitness per second, but as fitness is measured on a proportional scale, its units drop out. Similarly, fitness is predicted to increase by 1% per square centimeter increase in tail area.

Technically, the fitness and thus the gradient are per selective episode. This is usually implicit but must be taken into account when making predictions about per generational change, as selective episodes may not be equivalent to generations. Fitnesses and gradients of subsequent selective episodes are multiplicative, whereas those of parallel episodes are additive (weighted with the number of individuals involved).

Assume now that the additive variances of the two traits are $V_{A1} = 5\ s^2$ and $V_{A2} = 300\ cm^4$, and that their covariance is $C_{A12} = 10\ s\ cm^2$. Note that the units of variances are squares of the trait units and that units of covariances are the products of the units of the two traits. Now we can use the Lande equation to predict the responses to selection in the two traits:

$$\Delta\bar{z}_1 = V_{A1}\,\beta_1 + C_{A12}\beta_2 = 5\,s^2 \cdot 0.03\,s^{-1} + 10\,s\,cm^2 \cdot 0.01\,cm^{-2} = 0.25s,$$
$$\Delta\bar{z}_2 = V_{A2}\,\beta_2 + C_{A12}\beta_1 = 300\,cm^4 \cdot 0.01\,cm^{-2} + 10\,\mathrm{s}\,cm^2 \cdot 0.03\,s^{-1} = 3.3\,cm^2.$$

We may be interested in which trait evolves the fastest, which trait is most evolvable, and which trait is under strongest direct and indirect selection, but we cannot answer such questions without consideration of scale. The question of whether or how much a change of 3.3 cm^2 is larger than 0.25 s is technically meaningless. To make comparisons, we need a common scale.

In EQG, there are three general ways to standardize traits: (1) divide by the population phenotypic standard deviation, (2) divide by the population mean, and (3) use a log scale. I will refer to the first as variance standardization and to the second as mean standardization. Working on a log scale is in practice the same as mean standardization. Both are proportional scales. The difference is that log transformation expresses trait changes in proportion to the trait value, while mean standardization expresses it in proportion to the trait mean. Unless changes are large, numerical and interpretational differences between mean and log scaling are minute.

Variance standardization is most common. It converts additive genetic variances into heritabilities, expresses selection gradients in units of fitness change per standard deviation change in the trait, and gives selection responses and differentials in units of standard deviations. On the assumption that phenotypic standard deviations have the same meaning for different traits, this allows comparison and answers to the above questions. It is problematic, however, because the phenotypic standard deviation is itself intertwined with the entities it is used to scale; additive variances and covariances in particular This creates a correlation between the measure stick and the entities to be measured that can obscure interpretation.

As illustrated in figure 5.6, there is no relationship between variance-standardized additive variances (i.e., heritabilities) and mean-standardized additive variances. This is caused by the correlation between genetic and environmental variances, and probably also by the fact that epistatic variance components scale with powers of the additive variance (Hansen and Wagner 2001), which means that increasing additive variance tends to decrease the proportion of the genetic variance that is additive. So the choice of scale is not trivial, and specifically, heritabilities should not be used to measure evolvability (Hansen et al. 2011).

Houle (1992) was the first to demonstrate problems with variance standardization and to propose mean standardization as an alternative. He showed that life-history traits, which have low heritabilities, in fact tend to have high amounts of additive genetic variation, whereas morphological traits, which have larger heritabilities, in fact tend to have less additive variation than life-history traits. This finding overturned the hypothesis that stronger selection tends to remove genetic variation from life-history traits and paved the way for the hypothesis that levels of additive variation are more influenced by mutational input. Fitness itself often has high evolvability, but typically very low heritability (Hendry et al. 2018; Hansen and Pélabon 2021).

If we apply mean standardization to the univariate Lande equation, we get

$$\frac{\Delta\bar{z}}{\bar{z}} = \left(\frac{V_A}{\bar{z}^2}\right)(\beta\bar{z}) = e_\mu \beta_\mu,$$

where e_μ is the mean-scaled "evolvability," and β_μ is the mean-scaled selection gradient (Hansen et al. 2003b; Hereford et al. 2004). The mean-scaled selection gradient gives the proportional increase in fitness with a proportional increase in the trait. The mean-scaled evolvability can be interpreted as the predicted proportional increase in the trait mean per generation under unit selection ($\beta_\mu = 1$). Note that mean-scaled evolvabilities are often given as coefficients of additive variation, CV_A, the square root of e_μ, but this is not a good quantitative measure of evolvability, because the response to selection is proportional to trait variance and not to standard deviation.

If mean display duration is $\bar{z}_1 = 10\ s$ and mean tail area is $\bar{z}_2 = 200\ cm^2$, then mean standardizing our example yields

$$\frac{\Delta\bar{z}_1}{\bar{z}_1} = \left(\frac{V_{A1}}{\bar{z}_1^2}\right)(\beta_1\bar{z}_1) + \left(\frac{C_{A12}}{\bar{z}_1\bar{z}_2}\right)(\beta_2\bar{z}_2) = 0.05 \cdot 0.3 + 0.005 \cdot 2 = 0.025,$$

$$\frac{\Delta\bar{z}_2}{\bar{z}_2} = \left(\frac{V_{A2}}{\bar{z}_2^2}\right)(\beta_2\bar{z}_2) + \left(\frac{C_{A12}}{\bar{z}_1\bar{z}_2}\right)(\beta_1\bar{z}_1) = 0.0075 \cdot 2 + 0.005 \cdot 0.3 = 0.017,$$

which predicts that mean display duration will increase by 2.5% per generation, while mean tail area will increase by 1.7%. Hence, on a proportional scale, display duration would evolve faster. Tail area has a stronger effect on fitness, however, because $\beta_{\mu 2} = 2$ means that a 1% increase in tail area would increase fitness by 2% in comparison to $\beta_{\mu 1} = 0.3$, meaning that a 1% increase in display duration would increase fitness by 0.3%. The univariate evolvability of display duration, $e_\mu(z_1) = 0.05$, predicting a 5% increase per generation under unit selection, is much larger than for tail area, however, as $e_\mu(z_2) = 0.0075$ predicts a 0.75% increase in tail area per generation under unit selection. The strong selection on tail area also induces more indirect selection on display duration, contributing to its larger selection response. Although the two traits reinforce each other in this example, we can compute conditional evolvabilities to show that they do not contain much potential for constraining each other. The conditional evolvabilities are $c_\mu(z_1 | z_2) = 0.047$ and $c_\mu(z_2 | z_1) = 0.0070$, implying a mere 7% reduction in evolvability of each trait if the other was under stabilizing selection.

Note that the selection gradient β measures strength of selection in terms of effect of potential change on fitness, regardless of variation in the trait. The selection differential S, in contrast, depends on trait variation, and if there were a lot of environmental variation in display duration, which would be likely, its selection differential may well be larger than that of tail area. This would not change the response to selection, however, as the environmental variation would not be transmitted. It would manifest as a lower heritability of display duration. An issue with the breeder's equation is thus that its components, the selection differential and the heritability, tend to be negatively correlated. Counterintuitively, traits with high heritabilities or large selection differentials are not more likely to evolve fast.

Proportional scales provide an intuitive way to quantify evolvability. The number of generations it takes to increase a trait with a factor k is approximately $t_k = \ln(k)/e_\mu \beta_\mu$, and we can take the number of generations to double the trait under unit selection, $t_2 = \ln(2)/e_\mu$, as a convenient quantification of evolutionary potential. Applying this to estimates of evolvability taken from the recent review of Hansen and Pélabon (2021) shows that quantitative traits usually have large potentials for change on macroevolutionary time scales. Although the median evolvability for 1-dimensional morphological traits of $e_\mu \approx 0.1\%$ sounds tiny, it allows a trait to double in the geological eyeblink of $t_2 \approx 700$ generations under unit selection. The median evolvability of fitness is $e_\mu \approx 1.3\%$, and because fitness is under unit directional selection by definition, this number implies that selection would continuously double genetic fitness each 50 generations if no other forces were involved.

Thoughtless standardizations have caused much interpretational damage in EQG and related fields. The misapplication of heritability as a measure of evolutionary potential is the most obvious example, but variance standardization has also wreaked havoc on many studies of selection and rates of evolution. It is particularly dangerous in multivariate studies, in which some form of standardization is usually required to compare traits. In studies based on P-matrices, variance standardization will convert P-matrices to correlation matrices; all information about trait variation will be lost; and any inference about modularity, integration, and evolutionary potential will be dubious.

Mean standardization and log scales largely avoid the rubber-scale problems of variance standardization, but some traits are not meaningful or need careful interpretation on proportional scales (e.g., Hereford et al. 2004; Stinchcombe 2005; Hansen et al. 2011; Houle et al. 2011; Matsumura et al. 2012; Opedal et al. 2017; Pélabon et al. 2020). As much as possible, interpretation should involve original scales, biological information, and theoretical context. Even if a quarter-second change in display duration cannot be directly compared to a square-centimeter change in tail area, these figures may have biological meaning to a researcher familiar with the organism and relevant theory. Results should be back-transformed and evaluated on original scales. This is particularly important when arbitrary nonlinear transformations have been applied for statistical reasons.

5.9 Quantitative Genetics and the Genotype-Phenotype Map

The statistical representation of variation, inheritance, and selection is well suited to describe evolution with simple operational models. Its disadvantage is that underlying biological complexities are hidden, and sometimes this matters for understanding evolution and particularly for understanding constraints. The G- (and M-) matrix only captures linear and often temporary constraints. To understand evolutionary potential for larger changes, the underlying biological causality needs be considered. With the application of quantitative genetics to evolutionary questions, this became increasingly apparent, and the effects of underlying development, physiology and genetic architecture on quantitative genetic variation are increasingly taken into consideration.

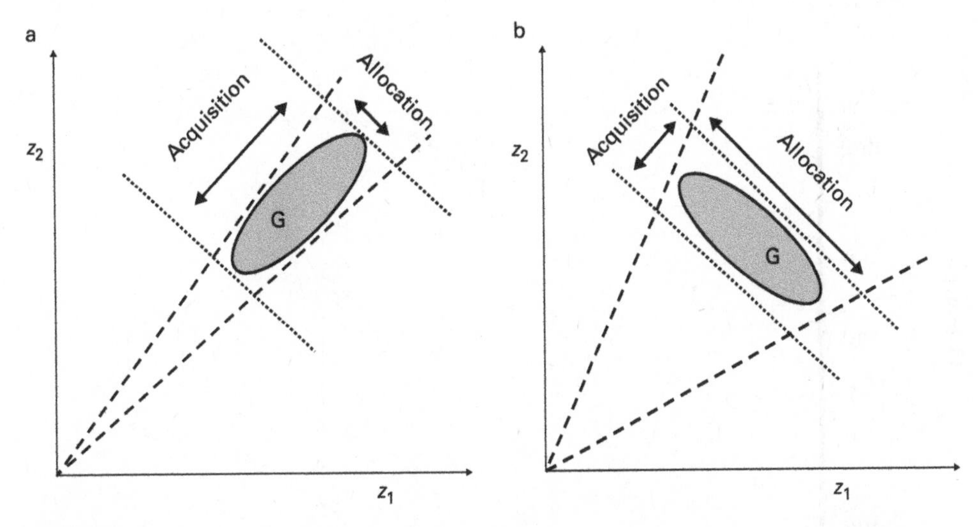

Figure 5.7
Effects of pleiotropy on the G-matrix. Two traits, z_1 and z_2, are determined by two genetically determined processes controlling how much resources are acquired and allocated to the growth of the traits. In (a) there is more variation in acquistion, causing positive covariance between the traits. In (b) there is more variation in the allocation, causing negative covariance between the traits.

One example, illustrated in figure 5.7, shows how two traits that are integrated on the biological level can still display all kinds of covariance structures. The traits are assumed to grow together on a shared resource that is then allocated between them, and their covariance is determined by how much variation there is in the shared resource and in the mechanism of allocation. This model has been used to illustrate problems with using the G- (or P-) matrix to study trade-offs or constraints between traits (e.g., Riska 1986; Houle 1991; Fry 1993). Even if there is a biological trade-off, covariances may still be positive or absent. More generally, it illustrates how a mixture of positive and negative pleiotropy can generate different patterns of covariance between traits.

Although pleiotropy (allelic differences that affect several traits) will usually be the main source of genetic covariance between traits, genetic covariance can also arise through linkage disequilibrium (covariance in occurrence) between alleles affecting the respective traits. Covariance through linkage disequilibrium can arise from genetic drift or nonrandom mating. It can also be built by correlational selection, but because it is broken down by recombination, it is unlikely to be substantial for polygenic traits with many unlinked sets of loci (e.g., Lande 1984). Gains from selection on variation due to linkage disequilibrium will be transient, but they may still be important if there are special mechanisms maintaining the disequilibrium. One example arises in the Fisher runaway model of sexual selection, when assortative mating between individuals with a particular trait and individuals with a mating preference for that trait creates linkage disequilibrium between alleles affecting the trait and alleles affecting the preference. Subsequent selection on the trait caused by the mating preference in the population will then also cause indirect selection on the preference genes. Hence, the preference indirectly selects itself, causing a potential for runaway evolution.

The main victim of the statistical representation of the genotype-phenotype map has been epistasis. There is a widespread misconception in quantitative genetics that epistasis

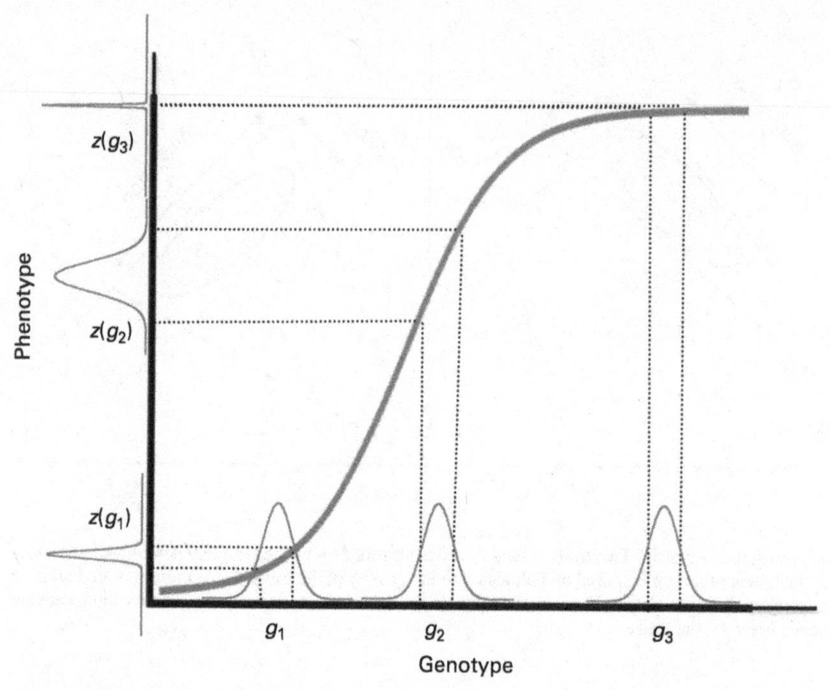

Figure 5.8
The genotype-phenotype map and the evolution of evolvability: Three different genetic backgrounds, g_1, g_2, and g_3, map the same amount of molecular variation to different levels of genetic variation in the phenotype. From g_1 to g_2 positive directional epistasis increases variation, and from g_2 to g_3 negative directional epistasis decreases variation. Note that there is negative directional epistasis in both directions from g_2. The phenotype is constrained from evolving beyond the range from $z(g_1)$ to $z(g_3)$. Based on Hansen (2015).

has only transient effects on evolution by selection and can be ignored. This conclusion is based on the finding that selection on epistatic variance only leads to changes in linkage disequilibrium and not to allele-frequency changes. Creating epistatic variance is not the only influence of epistasis, however. Biological epistasis also changes the biological (and thus statistical) effects of alleles, which may change the additive variance and thus the response to selection. Such changes are not transient and may be substantial (Hansen 2013). A key point is that permanent effects depend on systematic patterns of epistasis (Carter et al. 2005). Systematic positive epistasis in which allele substitutions with positive effects on the trait also tend to increase the effects of other allele substitutions on the trait will increase the additive variance and accelerate the response of a positively selected trait (figure 5.8). Systematic negative epistasis will have the opposite effects and can lead to canalization and evolutionary standstill (Hansen et al. 2006; Le Rouzic and Alvarez-Castro 2016). Nondirectional epistasis, in which positive and negative effects balance, will have no immediate effect on the response to selection. The statistical description of epistasis is blind to biological patterns in the interactions in the sense that the epistatic variance components are the same, regardless of presence or sign of directionality. EQG thus lacked operational tools to quantify relevant effects of epistasis, and some prominent researchers confused the absence of epistasis from quantitative-genetics theory with evidence against its importance. This gap is being ameliorated by consideration of explicit, theoretical or empirical genotype-phenotype maps (e.g., Cheverud and Routman 1995; Hansen and

Wagner 2001; Alvarez-Castro and Carlborg 2007; Le Rouzic 2014; Morrissey 2014, 2015; Hansen 2015; Alvarez-Castro 2016; Milocco and Salazar-Ciudad 2020), but empirical research on relevant patterns of epistasis is still scarce.

The emergence of genomics and molecular methods has provided many new insights into the genetic architecture of quantitative traits. Marker-based QTL and GWAS studies not only identify genes affecting traits but are also increasingly used to illuminate the general structure of the genotype-phenotype map. Here I only highlight one recent development: The increasing realization that many traits are influenced by segregating variation at thousands of locations in the genome (e.g., Boyle et al. 2017; Pitchers et al. 2019; Jakobson and Jarosz 2020). This conclusion contrasts with the picture provided by the first molecular analyses of quantitative traits in the 1990s, which often identified one or a few genes with large effects. These were mostly statistical artifacts, and even later studies finding dozens or hundreds of "significant" genes could usually not explain more than a fraction of the genetic variance. This became known as the problem of the "missing heritability." Subsequent studies with larger samples and better methods have traced the missing genetic variance to a huge number of segregating variants with very small effects. This "omnigenic" model (Boyle et al. 2017; Liu et al. 2019) implies that every gene in the genome has potential and often real effects on every quantitative trait. The likely reason is that a multitude of convoluted, indirect pathways exist for genes to affect traits, which again implies universal pleiotropy and a huge potential for the influence of epistatic interactions and for the generation of novel trait values by recombination (Hansen and Pélabon 2021). While there must be variation in the degree of directness of how genes influence a trait, it is debatable whether this model implies a distinction between core and peripheral genes as suggested by Liu et al. (2019).

5.10 Conclusion: Explaining Macroevolution?

The EQG research program was motivated by Lande's audacious hypothesis that simple evolutionary predictions based on a constant G-matrix could illuminate macroevolutionary change. Today there are vibrant research fields that use the EQG framework to understand short- and long-term evolution in the context of natural and sexual selection, life-history theory, behavior, and morphology (e.g., Boake 1994; Schluter 2000; Arnold et al. 2001; Pigliucci and Preston 2004; Roff 2007; Polly 2008; Svensson and Calsbeek 2012; Arnold 2014; Charmantier et al. 2014; Hendry 2017). The framework is frequently used to interpret long-term evolution in paleontological time series or comparative data. While the literal extrapolation of simple EQG models beyond a handful of generations may seem futile, as neither selection nor evolvability are likely to remain constant when the phenotype is changing, the EQG approach has still contributed many theoretical and empirical insights that must be part of the foundation for quantitative analysis of both micro- and macroevolution. These include:

1. Genetic architectures are highly polygenic and pleiotropic.
2. Mutational input to quantitative traits is high and explains variation in variation.
3. Short-term evolvability depends on additive variance.
4. Heritability is not evolvability.
5. Evolvability of low-dimensional traits is high from a macroevolutionary perspective.

6. Selection is strong, directional, and fluctuating on short time scales.
7. Selection is stabilizing over larger phenotypic and temporal ranges.
8. Indirect selection is important.
9. Neutral evolution is fast.
10. Peak shifts by drift do not occur.
11. Plasticity is important.
12. The evolution of evolvability (additive and mutational variance) depends on epistasis.
13. Microevolution is rapid and fluctuating but usually stationary.

An intuitive inference is that macroevolutionary dynamics should be determined by changes in the adaptive landscape and that genetic constraints are unimportant. At least the latter part of this inference is premature, however. There are indications that the structure of the G-matrix is related to rates of evolution and among-species variation even on million-year time scales (Voje et al., chapter 14),[1] and even if high evolvabilities allow rapid microevolution, we lack an understanding of how far traits can be extended without inducing pleiotropic or epistatic constraints.

Pleiotropic constraints arise when selection on a focal trait due to pleiotropy induces indirect selection on the genetic basis of other traits. Such changes are unlikely to be favorable and may cause counteracting selection to repair the damage. This effect is what is captured by conditional evolvability, but the measures I have discussed only quantify the effects of specific macroscopic traits, while we would like to know the effects from the entire genome. Universal pleiotropy predicts that selection on a trait would induce multitudes of minor changes in the genetic basis of essentially all other traits, and the consequences of this remains to be worked out.

Epistatic constraints arise when selection reduces genetic variance through negative directional epistasis. A genotype-phenotype map as shown in figure 5.8 would induce negative epistasis in both directions from the middle, and if the trait is selected toward the edges, it would become canalized and lose evolvability. Whether such structuring of genotype-phenotype maps is common is unknown, but it is a potential explanation for stationarity of microevolutionary change.

In conclusion, EQG provides a powerful toolbox for quantitative analysis of microevolution, but fundamental misconceptions of scale and the lack of a distinction between biological and statistical effects have hampered empirical and theoretical interpretation. When these problems are recognized, EQG should be a pillar for interpreting both micro- and macroevolution.

Acknowledgments

Thanks to the members of the evolvability group at the Centre for Advanced Study in Oslo and to members of the evolvability journal club for discussions and comments. David Houle and Christophe Pélabon made helpful comments on the manuscript.

1. References to chapter numbers in the text are to chapters in this volume.

References

Agrawal, A. F., and J. R. Stinchcombe. 2009. How much do genetic covariances alter the rate of adaptation? *Proceedings of the Royal Society B* 276: 1182–1191.

Alvarez-Castro, J. M. 2016. Mapping genetic architectures: A shift in evolutionary quantitative genetics. In *The Encyclopedia of Evolutionary Biology*, vol. 2, edited by R. M. Kliman, 127–135. Oxford: Academic Press.

Alvarez-Castro, J. M., and O. Carlborg. 2007. A unified model for functional and statistical epistasis and its application in quantitative trait loci analysis. *Genetics* 176: 1151–1167.

Arnold, S. J. 1983. Morphology, performance and fitness. *American Zoologist* 23: 347–361.

Arnold, S. J. 1992. Constraints on phenotypic evolution. *American Naturalist* 140: S85–S107.

Arnold, S. J. 1994. Multivariate inheritance and evolution: A review of concepts. In *Quantitative Genetic Studies of Behavioral Evolution*, edited by C. R. B. Boake, 17–48. Chicago: University of Chicago Press.

Arnold, S. J. 2014. Phenotypic evolution: The ongoing synthesis. *American Naturalist* 183: 729–746.

Arnold, S. J., and M. J. Wade. 1984a. On the measurement of natural and sexual selection: Theory. *Evolution* 38: 709–719.

Arnold, S. J., and M. J. Wade. 1984b. On the measurement of natural and sexual selection: Applications. *Evolution* 38: 720–734.

Arnold, S. J., M. E. Pfrender, and A. G. Jones. 2001. The adaptive landscape as a conceptual bridge between micro- and macroevolution. *Genetica* 112/113: 9–32.

Arnold, S. J., R. Bürger, P. A. Hohenlohe, B. C. Ajie, and A. G. Jones. 2008. Understanding the evolution and stability of the G-matrix. *Evolution* 62: 2451–2461.

Beatty J. 2016. The creativity of natural selection? Part I: Darwin, Darwinism, and the mutationists. *Journal of the History of Biology* 49: 659–684.

Berg, R. L. 1959. A general evolutionary principle underlying the origin of developmental homeostasis. *American Naturalist* 93: 103–105.

Blows, M. W. 2007. A tale of two matrices: Multivariate approaches in evolutionary biology. *Journal of Evolutionary Biology* 20: 1–8.

Boake, C. R. B. (Editor). 1994. *Quantitative Genetic Studies of Behavioral Evolution*. Chicago: University of Chicago Press.

Bolstad, G. H., T. F. Hansen, C. Pélabon, M. Falahati-Anbaran, R. Pérez-Barrales, and W. S. Armbruster. 2014. Genetic constraints predict evolutionary divergence in *Dalechampia* blossoms. *Philosophical Transactions Royal Society B* 369: 20130255.

Boyle, E. A., Y. I. Li, and J. K. Pritchard. 2017. An expanded view of complex traits: From polygenic to omnigenic. *Cell* 169: 1177–1186.

Bulmer M. 2003. *Francis Galton: Pioneer of Heredity and Biometry*. London: The Johns Hopkins University Press.

Carter, A. J. R., J. Hermisson, and T. F. Hansen. 2005. The role of epistatic gene interactions in the response to selection and the evolution of evolvability. *Theoretical Population Biology* 68: 179–196.

Charmantier, A., D. Garant, and L. E. B. Kruuk. (Editors). 2014. *Quantitative Genetics in the Wild*. Oxford: Oxford University Press.

Cheng, C., and D. Houle. 2020. Predicting multivariate responses of sexual dimorphism to direct and indirect selection. *American Naturalist* 196: 391–405.

Cheverud, J. M. 1984. Quantitative genetics and developmental constraints on evolution by selection. *Journal of Theoretical Biology* 110: 155–171.

Cheverud, J. M. 1988. A comparison of genetic and phenotypic correlations. *Evolution* 42: 958–968.

Cheverud, J. M., and E. J. Routman. 1995. Epistasis and its contribution to genetic variance components. *Genetics* 139: 1455–1461.

Chevin, L. M. 2013. Genetic constraints on adaptation to a changing environment. *Evolution* 67: 708–721.

Day, T., and R. Bonduriansky. 2011. A unified approach to the evolutionary consequences of genetic and nongenetic inheritance. *American Naturalist* 178: E18–E36.

Fisher, R. A. 1918. The correlation between relatives on the supposition of Mendelian inheritance. *Transactions of the Royal Society of Edinburgh* 3: 399–433.

Fry, J. D. 1993. The "general vigor" problem: Can antagonistic pleiotropy be detected when genetic covariances are positive? *Evolution* 47: 327–333.

Grabowski, M., and A. Porto. 2017. How many more? Sample size determination in studies of morphological integration and evolvability. *Methods in Ecology and Evolution* 8: 592–603.

Gudbjartsson, D. F., G. B. Walters, G. Thorleifson, et al. 2008. Many sequence variants affecting diversity of adult human height. *Nature Genetics* 40: 609–615.

Halligan, D. L., and P. D. Keightley. 2009. Spontaneous mutation accumulation studies in evolutionary genetics. *AREES* 40: 151–172.

Hansen, T. F. 2003. Is modularity necessary for evolvability? Remarks on the relationship between pleiotropy and evolvability. *Biosystems* 69: 83–94.

Hansen, T. F. 2013. Why epistasis is important for selection and adaptation. *Evolution* 67: 3501–3511.

Hansen, T. F. 2015. Measuring gene interactions. In *Epistasis: Methods and Protocols,* edited by J. H. Moore, and S. M. Williams, 115–143. New York: Humana Press.

Hansen, T. F. 2017. On the definition and measurement of fitness in finite populations. *Journal of Theoretical Biology* 419: 36–43.

Hansen, T. F., and D. Houle. 2008. Measuring and comparing evolvability and constraint in multivariate characters. *Journal of Evolutionary Biology* 21: 1201–1219.

Hansen, T. F., and C. Pélabon. 2021. Evolvability: A quantitative-genetics perspective. *AREES* 52: 153–175.

Hansen, T. F., and K. L. Voje. 2011. Deviation from the line of least resistance does not exclude genetic constraints: A comment on Berner et al. (2010). *Evolution* 65: 1821–1822.

Hansen, T. F., and G. P. Wagner. 2001. Modeling genetic architecture: A multilinear model of gene interaction. *Theoretical Population Biology* 59: 61–86.

Hansen, T. F., W. S. Armbruster, M. L. Carlson, and C. Pélabon. 2003a. Evolvability and genetic constraint in *Dalechampia* blossoms: Genetic correlations and conditional evolvability. *Journal of Experimental Zoology B* 296: 23–39.

Hansen, T. F., C. Pélabon, W. S. Armbruster, and M. L. Carlson. 2003b. Evolvability and genetic constraint in *Dalechampia* blossoms: Components of variance and measures of evolvability. *Journal of Evolutionary Biology* 16: 754–765.

Hansen, T. F., J. M. Alvarez-Castro, A. J. R. Carter, J. Hermisson, and G. P. Wagner. 2006. Evolution of genetic architecture under directional selection. *Evolution* 60: 1523–1536.

Hansen, T. F., C. Pélabon, and D. Houle. 2011. Heritability is not evolvability. *Evolutionary Biology* 38: 258–277.

Hansen, T. F., T. M. Solvin, and M. Pavličev. 2019. Predicting evolutionary potential: A numerical test of evolvability measures. *Evolution* 73: 689–703.

Hendry, A. P. 2017. *Eco-evolutionary Dynamics.* Princeton, NJ: Princeton University Press.

Hendry, A. P., D. J. Schoen, M. E. Wolak, and J. M. Reid. 2018. The contemporary evolution of fitness. *AREES* 49: 457–476.

Hereford, J., T. F. Hansen, and D. Houle. 2004. Comparing strengths of directional selection: How strong is strong? *Evolution*: 58: 2133–2143.

Hermisson, J., T. F. Hansen, and G. P. Wagner. 2003. Epistasis in polygenic traits and the evolution of genetic architecture under stabilizing selection. *American Naturalist* 161: 708–734.

Hine, E., S. F. Chenoweth, H. D. Rundle, and M. W. Blows. 2009. Characterizing the evolution of genetic variance using genetic covariance tensors. *Philosophical Transactions of the Royal Society B* 364: 1567–1578.

Hohenlohe, P. A., and S. A. Arnold. 2008. MiPoD: A hypothesis-testing framework for microevolutionary inference from patterns of divergence. *American Naturalist* 171: 366–385.

Houle, D. 1991. Genetic covariance of fitness correlates: What genetic correlations are made of and why it matters. *Evolution* 45: 630–648.

Houle, D. 1992. Comparing evolvability and variability of quantitative traits. *Genetics* 130: 195–204.

Houle, D. 1998. How should we explain variation in the genetic variance of traits? *Genetica* 102/103: 241–253.

Houle, D., and A. Kondrashov. 2006. Mutation. In *Evolutionary Genetics: Concepts and Case Studies,* edited by C. W. Fox and J. B. Wolf, 32–48. Oxford: Oxford University Press.

Houle D., C. Pélabon, G. P. Wagner, and T. F. Hansen. 2011. Measurement and meaning in biology. *Quarterly Review of Biology* 86: 3–34.

Jakobson, C. M., and D. F. Jarosz. 2020. What has a century of quantitative genetics taught us about nature's genetic toolkit? *Annual Review of Genetics* 54: 439–464.

Jensen, H., M. Szulkin, and J. Slate. 2014. Molecular quantitative genetics. In *Quantitative Genetics in the Wild,* edited by A. Charmantier, D. Garant, and L. E. B. Kruuk, 209–227. Oxford: Oxford University Press.

Kerr, B., and P. Godfrey-Smith. 2009. Generalization of the Price equation for evolutionary change. *Evolution* 63: 531–536.

Kirkpatrick, M. 2009. Patterns of quantitative genetic variation in multiple dimensions. *Genetica* 136: 271–284.

Kirkpatrick, M., and R. Lande. 1989. The evolution of maternal characters. *Evolution* 43: 485–503.

Lande, R. 1976a. Natural selection and random genetic drift in phenotypic evolution. *Evolution* 30: 314–334.

Lande, R. 1976b. The maintenance of genetic variability by mutation in a polygenic character with linked loci. *Genetical Research* 26: 221–235.

Lande, R. 1979. Quantitative genetic analysis of multivariate evolution, applied to brain:body size allometry. *Evolution* 33: 402–416.

Lande, R. 1980. The genetic covariance between characters maintained by pleiotropic mutations. *Genetics* 94: 203–215.

Lande, R. 1984. The genetic correlation between characters maintained by selection, linkage and inbreeding. *Genetics Research* 44: 309–320.

Lande, R. 1985. Expected time for random genetic drift of a population between stable phenotypic states. *PNAS* 82: 7641–7645.

Lande, R., and S. J. Arnold. 1983. The measurement of selection on correlated characters. *Evolution* 37: 1210–1226.

Le Rouzic, A. 2014. Estimating directional epistasis. *Frontiers in Genetics* 5: 198.

Le Rouzic, A., and J. M. Alvarez-Castro. 2016. Epistasis-induced evolutionary plateaus in selection responses. *American Naturalist* 188: E134–E150.

Liu, X., Y. I. Li, and J. K. Pritchard. 2019. Trans effects on gene expression can drive omnigenic inheritance. *Cell* 177: 1022–1034.

Love, A. C., M. Grabowski, D. Houle, et al. 2022. Evolvability in the fossil record. *Paleobiology* 48: 186–209.

Lynch, M. 1987. Evolution of intrafamilial interactions. *PNAS* 84: 8507–8511.

Lynch, M. 1990. The rate of morphological evolution in mammals from the standpoint of the neutral expectation. *American Naturalist* 136: 727–741.

Lynch, M. 2007. *The Origins of Genome Architecture*. Sunderland, MA: Sinauer.

Lynch, M., and B. Walsh. 1998. *Genetics and Analysis of Quantitative Characters*. Sunderland, MA: Sinauer.

Majerus, M. E. N. 1998. *Melanism: Evolution in Action*. Oxford: Oxford University Press.

Marquez, E. J. 2008. A statistical framework for testing modularity in multidimensional data. *Evolution* 62: 2688–2708.

Marroig, G., L. T. Shirai, A. Porto, F. B. de Oliveria, and V. De Conto. 2009. The evolution of modularity in the mammalian skull II: Evolutionary consequences. *Evolutionary Biology* 36: 136–148.

Matsumura, S., R. Arlinghaus, and U. Dieckmann. 2012. Standardizing selection strengths to study selection in the wild: A critical comparison and suggestions for the future. *BioScience* 62: 1039–1054.

McGlothlin, J. W., M. E. Kobiela, H. V. Wright, D. L. Mahler, J. J. Kolbe, J. B. Losos, and E. D. III Brodie. 2018. Adaptive radiation along a deeply conserved genetic line of least resistance in *Anolis* lizards. *Evolution Letters* 2–4: 310–322.

Melo, D., A. Porto, J. M. Cheverud, and G. Marroig. 2016. Modularity: Genes, development, and evolution. *AREES* 47: 463–486.

Milocco, L., and I. Salazar-Ciudad. 2020. Is evolution predictable? Quantifying the accuracy of quantitative genetics predictions for complex genotype-phenotype maps. *Evolution* 74: 230–244.

Mitteroecker, P. 2009. The developmental basis of variational modularity: Insights from quantitative genetics, morphometrics, and developmental biology. *Evolutionary Biology* 36: 377–385.

Morrissey, M. B. 2014. Selection and evolution of causally covarying traits. *Evolution* 68: 1748–1761.

Morrissey, M. B. 2015. Evolutionary quantitative genetics of non-linear developmental systems. *Evolution* 69: 2050–2066.

Morrissey, M. B. 2016. Meta-analysis of magnitudes, differences, and variation in evolutionary parameters. *Journal of Evolutionary Biology* 29: 1882–1904.

Morrissey, M. B., L. E. B. Kruuk, and A. J. Wilson. 2010. The danger of applying the breeder's equation in observational studies of natural populations. *Journal of Evolutionary Biology* 23: 2277–2288.

Morrissey, M. B., P. de Villemereull, B. Doligez, and O. Gimenez. 2014. Bayesian approaches to the quantitative genetic analysis of natural populations. In *Quantitative Genetics in the Wild*, edited by A. Charmantier, D. Garant, and L. E. B. Kruuk, 228–253. Oxford: Oxford University Press.

Moússeau, T. A., and C. W. Fox. (Editors). 1998. *Maternal Effects as Adaptations*. Oxford: Oxford University Press.

Olson, E. C., and R. L. Miller. 1958. *Morphological Integration*. Chicago: University of Chicago Press.

Opedal, Ø. H., G. H. Bolstad, T. F. Hansen, W. S. Armbruster, and C. Pélabon. 2017. The evolvability of herkogamy: Quantifying the evolutionary potential of a composite trait. *Evolution* 71: 1572–1586.

Otsuka, J. 2019. Ontology, causality, and methodology of evolutionary research programs. In *Evolutionary Causation: Biological and Philosophical Reflections,* edited by T. Uller and K. N. Laland, 247–264. The Vienna Series in Theoretical Biology. Cambridge, MA: MIT Press.

Paenke, I., B. Sendhoff, and T. J. Kawecki. 2007. Influence of plasticity and learning on evolution under directional selection. *American Naturalist* 170: E47–E58.

Pélabon, C., C. H. Hilde, S. Einum, and M. Gamelon. 2020. On the use of the coefficient of variation to quantify and compare trait variation. *Evolution Letters* 4: 180–188.

Pigliucci, M., and K. Preston. (Editors). 2004. *Phenotypic Integration: Studying the Ecology and Evolution of Complex Phenotypes.* Oxford: Oxford University Press.

Pitchers, W., J. Nye, E. J. Marquez, A. Kowlaski, I. Dworkin, and D. Houle. 2019. A multivariate genome-wide association study of wing shape in *Drosophila melanogaster. Genetics* 211: 1429–1447.

Polly, P. D. 2008. Developmental dynamics and G-matrices: Can morphometric spaces be used to model phenotypic evolution? *Evolutionary Biology* 35: 83–96.

Price, G. R. 1970. Selection and covariance. *Nature* 227: 520–521.

Provine, W. B. 1971. *The Origins of Theoretical Population Genetics.* Chicago: University of Chicago Press.

Reid, J. M., and R. J. Sardell. 2011. Indirect selection on female extra-pair reproduction? Comparing the additive genetic value of maternal half-sib extra-pair and within-pair offspring. *Philosophical Transactions of the Royal Society B* 279: 1700–1708.

Riedl, R. J. 1977. A systems-analytical approach to macro-evolutionary phenomena. *Quarterly Review of Biology* 52: 351–370.

Riska, B. 1986. Some models for development, growth, and morphometric correlation. *Evolution* 40: 1303–1311.

Roff, D. A. 2007. A centennial celebration for quantitative genetics. *Evolution* 61: 1017–1032.

Schluter, D. 1996. Adaptive radiation along genetic lines of least resistance. *Evolution* 50: 1766–1774.

Schluter, D. 2000. *The Ecology of Adaptive Radiation.* Oxford: Oxford University Press.

Stamp, M. A., and J. D. Hadfield. 2020. The relative importance of plasticity versus genetic differentiation in explaining between population differences; a meta-analysis. *Ecology Letters* 23: 1432–1441.

Stinchcombe, J. R. 2005. Measuring natural selection on proportional traits: Comparisons of three types of selection estimates for resistance and susceptibility to herbivore damage. *Evolutionary Ecology* 19: 363–373.

Stinchcombe, J. R., A. F. Agrawal, P. A. Hohenlohe, S. J. Arnold, and M. W. Blows. 2008. Estimating nonlinear selection gradients using quadratic regression coefficients: Double or nothing? *Evolution* 62: 2435–2440.

Svensson, E. I., and R. Calsbeek. (Editors). 2012. *The Adaptive Landscape in Evolutionary Biology.* Oxford: Oxford University Press.

Sztepanacz, J. L., and D. Houle. 2019. Cross-sex genetic covariances limit the evolvability of wing-shape within and among species of *Drosophila. Evolution* 73: 1617–1633.

Turelli, M. 1984. Heritable genetic variation via mutation-selection balance: Lerch's Zeta meets the abdominal bristle. *Theoretical Population Biology* 25: 138–193.

Via, S., and R. Lande. 1985. Genotype-environment interaction and the evolution of phenotypic plasticity. *Evolution* 39: 505–522.

Wade, M. J., and S. Kalisz. 1990. The causes of natural selection. *Evolution* 44: 1947–1955.

Wagner, G. P. 2014. *Homology, Genes, and Evolutionary Innovation.* Princeton, NJ: Princeton University Press.

Wagner, G. P., and L. Altenberg. 1996. Complex adaptations and the evolution of evolvability. *Evolution* 50: 967–976.

Walsh, B., and M. Lynch. 2018. *Evolution and Selection of Quantitative Traits.* Oxford: Oxford University Press.

Wilson, A. J., D. W. Coltman, J. M. Pemberton, A. D. J. Overall, K. A. Byrne, and L. E. B. Kruuk. 2005. Maternal genetic effects set the potential for evolution in a free-living vertebrate population. *Journal of Evolutionary Biology* 18: 405–414.

Wilson, A. J., D. Reale, M. N. Clements, et al. 2010. An ecologist's guide to the animal model. *Journal of Animal Ecology* 79: 13–26.

6 Measuring Evolvability

David Houle and Christophe Pélabon

The term evolvability is used in many different contexts in evolutionary biology, and this multiplicity has generated confusion. Here we examine the varied usages of evolvability from the perspective of conceptual measurement theory, the consideration of what attributes should be measured to quantify a concept. We argue that there is a shared conception of evolvability as a disposition, the propensity to evolve should the conditions promoting evolution occur. Even with that shared conception, identifying the properties that confer evolvability requires making explicit which entity's evolvability is of interest, the specific stimulus that would cause evolution, and the timescale over which evolution happens. We refer to these descriptors as *Of, Under,* and *Over,* respectively. Once *Of, Under,* and *Over* are clear, attributes of the properties that confer evolvability can be identified and then measured. Focusing on both the commonalities and the differences of evolvability concepts should help us develop the theoretical understanding that is a precursor to appropriate measurement.

6.1 Introduction

Evolvability is the disposition of a population to evolve (G. Wagner and Altenberg 1996; Love 2003; Hansen 2006; Brigandt et al., chapter 4).[1] The word is, however, used to refer to many somewhat different dispositions (Pigliucci 2008; Minelli 2017; Nuño de la Rosa 2017), and there is an even larger menagerie of attributes that are hypothesized to cause or shape evolvability. These attributes include mutation, pleiotropy, robustness, modularity, variability, and variation, among many others. On its face, this multiplicity of usages and attributes related to the concept suggests that evolvability is not a unifying concept.

The life cycle of concepts in science progresses from intuitive assertions through precise verbal models to increasingly rigorous mathematical models. In parallel, scientists identify relevant attributes in nature and learn to quantify them with increasing accuracy. With a mature understanding of evolvability, we could aspire to have the equivalent of a thermometer, call it an evometer, which would tell us how much disposition to evolve a population has. Such understanding implies that we would have a complementary theoretical appreciation for what that disposition actually is, and what it predicts about evolution.

1. References to chapter numbers in the text are to chapters in this volume.

From our current state of knowledge, the idea of an evometer seems plainly ridiculous. It is less ridiculous, however, if we remember that not so many centuries ago, there was no such thing as a thermometer either (McGee 1988; Chang 2004). Without a thermometer, an assertion that yesterday was hotter than today could not be verified without a component of storytelling, memory, or appeal to authority. And what does it mean to be hotter, anyway?

We offer the example of temperature and heat to make two points. First, having an evometer would imply knowing both what we mean by evolvability, and what role it plays in evolution. Chang (2004, 8) repeats the truism that "the scientific study of heat began with the invention of the thermometer," although physicists might substitute the term "energy" for "heat." Hand (2004, 2) quotes Lord Kelvin, who did so much to advance our understanding of the nature of energy on this point: "when you can measure what you are speaking about and express it in numbers you know something about it; but when you cannot measure it . . . your knowledge of it is of a meagre and unsatisfactory kind; it may be the beginning of knowledge, but you have scarcely, in your thoughts advanced to the stage of science."

Second, just because we cannot now conceive of an evometer to match each proposed concept of evolvability is no reason to discard that concept. If discussions of heat had been deemed too unscientific to be worthy of attention by Galileo and the others who rediscovered the thermoscope around 1600 (McGee 1988), we might have neither a well-developed theory of thermodynamics nor the ability to measure temperature today.

Many contributions to the literature on evolvability are at this pre-measurement level. For example, Hendrikse et al. (2007, 394) argue that evolvability, "the capacity of a developmental system to evolve" is the "central question" of evolutionary developmental biology. The authors focus on the ability of a developmental system to "generate" variation as the key to evolvability and separate this ability into bias in the direction of phenotypic effects and the amount of phenotypic variation generated. They further associate four other concepts with the study of this ability: integration, modularity, constraint, and canalization. This clearly argued essay has been influential and widely cited, and we agree that development shapes evolvability through features such as these. And yet this paper makes no reference to quantification whatsoever. We are only at the "beginning of knowledge" when it comes to some aspects of evolvability.

In this chapter, we consider evolvability and the attributes that may shape it from the perspective of measurement (Hand 2004). In particular, we focus on the principles that should guide our choice of attributes to be measured when measuring evolvability rather than on the details of how to quantify particular attributes. Houle et al. (2011) called this conceptual measurement theory.

We argue that the great complexities that immediately confront anyone wishing to measure the disposition to evolve can be surmounted by focusing on two important distinctions that are rarely made explicit. The first is that evolvability is caused by different attributes depending on the evolving entity, the evolutionary forces acting, and the timescale considered. The second is that many of the attributes that shape evolvability, including those featured in Hendrikse et al. (2007), have complex and nonmonotonic relationships with evolvability. Although we may be able to measure attributes such as modularity, they only measure evolvability under carefully delineated assumptions.

6.2 The Language of Dispositions

All conceptions of evolvability share a focus on the evolutionary future, that is, on predicting the possible state of a population in the future. This focus defines evolvability as a dispositional concept (Love 2003). We adopt much of the philosophical vocabulary concerning dispositions that Brigandt et al. (chapter 4) review. A disposition is the potential for some entity to manifest a particular change, when subjected to some set of stimulus conditions under appropriate background conditions. In the case of evolvability, the evolving entity must be a population of organisms or some set of populations. Most importantly for measurement, a disposition must be at least partly due to an intrinsic property of the bearers of the disposition. The bearer of a disposition may be distinct from the entity that manifests the change. Only populations can evolve, but the bearers of the properties hypothesized to generate evolvability can be either a population as a whole or a typical member of the population, such as an organism, a genome, or their components, a gene or an organ, which is termed a *type.* The individual bearers of the disposition are referred to as *tokens,* which may differ in their properties from the type.

This philosophical vocabulary only deals with qualitative outcomes in response to a discrete stimulus, such as the shattering of a fragile object when struck. The intrinsic properties that enable the disposition are also generally treated as discrete. An object is considered either fragile or not, even though fragile objects differ in their fragility. Bearers, however, can possess the relevant intrinsic property to different degrees. We refer to quantifiable intrinsic properties as *attributes.* Similarly, we seek to quantify the *stimulus strength* and the *manifestation strength.* The manifestation strength depends on both the amount of the attribute present and the stimulus strength.

To put this vocabulary into use, consider first a well-known dispositional property, solubility. The stimulus condition is placing a solid in a solvent, and the manifestation is solvation, passing into solution. What makes solubility a disposition is that a solid (e.g., sugar) will not pass into solution until exposed to a stimulus (e.g., water), whether that exposure happens today or next year. The attributes that determine solubility include the intermolecular interactions of the solid and solvent. These properties are responsible for differences in solubility, stimulus strength, and manifestation strength. For example, salt and sugar are both soluble in water. If we hold the stimulus strength constant at 1 liter of 20°C water, we can dissolve 2,000 g/l of sugar, but only 360 g/l of salt. Sugar is thus 5.5 times more soluble than salt in water.

With respect to evolvability, it is relatively straightforward to determine the manifestation strength (how much evolution has occurred). It is much more challenging to determine how much of that change is due to the disposition to evolve, and how much to stimulus strength (see Armbruster, chapter 15, and Jablonski, chapter 17, for further discussion in the context of macroevolution).

6.3 The Evolvability Multiverse

To apply the language of dispositions to evolvability, we need to recognize the diversity of usages of the concept (Pigliucci 2008; Nuño de la Rosa 2017). Nuño de La Rosa's

(2017) comprehensive review of the foundational literature suggests that the different conceptions of evolvability do share important common elements, such as the recognition that evolution consists of changes in the complement of genotypes in populations. Despite those commonalities, important differences preclude reducing evolvability to a single measurable disposition.

Nuño de La Rosa (2017) noted four "conceptual tensions" that characterize the differences among evolvability concepts: (1) Should research on evolvability focus on variability or variation? (2) Are we interested in all evolutionary changes, or just those of adaptive significance? (3) Do we want to focus on evolutionary innovations or novelties rather than other evolutionary changes? (4) Which organismal characteristics should evolvability be applied to and over what timescales? These questions partly echo the distinction that Pigliucci (2008) made between variation-, variability-, and innovation-based evolvability concepts. In Pigliucci's view, variation-based measures apply at short timescales, variability-based measures at intermediate timescales, and innovation-based ones at long timescales.

Part of the diversity in evolvability concepts arises because aspects of evolvability can be measured at four different biological levels: genetics, the genotype-phenotype (GP) map, variability, and variation (figure 6.1). The genetic level consists of processes that alter genotypes, including mutation, recombination, and segregation. The GP map level incorporates the tangled web of processes through which genotypes shape phenotypes. It includes development and physiology at cellular, tissue, organ, and whole-body scales. Variability is the disposition for a genetic change filtered through the GP map to generate phenotypic change. Variation refers to the differences in a particular population that result from the variability and the history of that population. This scheme explicitly distinguishes variation and variability, and it implicitly acknowledges the range of timescales by separating elements of evolvability that tend to evolve at a low rate (e.g., mutation, and the GP map) from variation, which can change rapidly (Pélabon et al., chapter 13). Some evolution is driven by stimuli that arise above the population level, such as extinction (Jablonski 2008). We did not try to represent those drivers in figure 6.1, which would require at least one additional level to represent species-level properties, such as species ranges (Grantham 2007; Villegas et al., chapter 3).

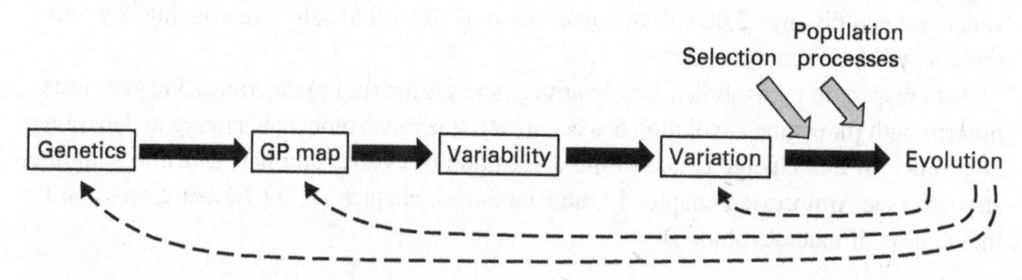

Figure 6.1
Flow of biological processes that shape evolution. Aspects of evolvability can be conceptualized at each of the levels shown in boxes: genetics, GP map, variability, and variation. Gray arrows represent processes that cause evolution through their effects on variation, notably natural selection and population processes, including drift and gene flow. The dashed lines represent the effects that evolution may have on variation, genetic processes (e.g., mutation and recombination), and the GP map.

To see how we can apply the language of dispositions to this scheme, let's consider mutation rate, the canonical evolvability attribute at the genetic level. The bearer of mutation rate is an individual organism, including especially its genome and internal physiological state. Mutation rate can be measured on the gametes produced by a single individual (see, e.g., Wang et al. 2012), although many estimates are made on the descendants of a set of individuals (e.g., Haag-Liautard et al. 2007). Such estimates of mutation rate refer to an individual as the representative of a type, specifically a typical genotype, itself a product of past population-level processes. Thus, the individual and its genome contain the relevant information necessary to study mutation rate. Although we might choose to measure mutation rate in only part of the genome, such as one gene, the actual rate is determined both by the properties of that gene and by the genomic and organismal milieu in which the gene resides. At the next level, the GP map specified by the entire genetic and developmental system of the individual determines whether a genomic variant has a phenotypic effect as well as the magnitude and direction of any effect. Mutation and the GP map together determine the variability of offspring potentially produced by each genotype. Ultimately, the mutation rate affects the rate of evolution of a population through its effect on the amount and nature of variation produced by the population, the evolving entity.

6.4 Of, Under, Over

Measurement is the process by which we assign numbers to the attributes of entities so that the mathematical relationships among the numbers capture the empirical relationships between the original entities (Krantz et al. 1971; Hand 2004). The theoretical context for measurement tells us what inferences we would like to make. In the case of evolvability, the entities are populations of organisms, or higher-level aggregations of populations, and the theoretical context is the hypothesis that intrinsic properties of individuals or populations give them the ability to evolve. Evolution is not a unitary process, and therefore the notion that "ability to evolve" is a single unitary disposition cannot be correct. The forces that cause evolution in one trait may differ from those affecting another trait; the attributes that enable evolution of novel organismal features are likely to be different from those that enable quantitative alterations of existing traits.

Therefore, to measure evolvability, we need to be more specific about the theoretical context in which the measurements will function. First, we must specify evolvability *Of* what class of organismal or population feature. Examples might be a trait, a DNA sequence, or a property of the GP map. Second, we must define the stimulus *Under* which we want to measure evolvability. Stimuli that create the opportunity for evolution include particular kinds of natural selection or genetic drift. Finally, we need to know *Over* what timescale we want to predict evolution. Attributes that predict short-term evolution might be irrelevant for long-term evolution. Specification of the *Of-Under-Over* conditions supplies the essential theoretical context for measurement.

Table 6.1 lists some of the many possible choices for *Of-Under-Over.* Note that properties in the *Of* column require further specification (i.e., which trait or GP map property, or fitness of what entity). Not all combinations of *Of-Under-Over* are sensible; for example, the evolution of novelty under strict stabilizing selection in all phenotypic dimensions is not likely.

Table 6.1
Examples of choices for *Of*, *Under*, and *Over*

Of	Under	Over
Allele	Directional selection	Cell cycle
Genome	Neutral process	One generation
Trait mean	Approach to an optimum	Hundreds of generations
Trait variance	Corridor selection	Macroevolutionary time†
Discrete state	Fluctuating selection	
Fitness	White noise motion of optimum	
GP map	Brownian motion of optimum	
Mutation rate	Mutation pressure	
Recombination		
Novel traits		
Speciation rate		
Species range		

† Time is frequently assessed in years when generation time is unknown, evolves, or when multiple organisms with different generation times are involved.

6.5 From Intrinsic Property to Attribute to Measurement: Two Examples

To illustrate the link between intrinsic properties, attributes, and measurements that are suggested by particular theoretical contexts, let's consider two cases where the measurement of evolvability is based on well-established theoretical models: short-term evolution of quantitative traits in response to directional selection, and the neutral evolution of discrete properties of genotypes.

6.5.1 Response of Quantitative Traits to Directional Selection

Additive genetic variation quantifies the evolvability *Of* the mean of a quantitative trait, z, in an outbred population *Under* directional selection *Over* one generation. The relevant theory is based on the Lande equation, $\Delta\bar{z} = V_A\beta$, where $\Delta\bar{z}$ is the change in the mean of trait z due to selection, V_A is the additive genetic variance in a trait z, and β is the selection gradient (Lande 1979). In this model, the manifestation strength is $\Delta\bar{z}$. The quantity V_A measures the evolvability of the population. Although it is often narrowly defined for random-mating diploid populations, we use V_A more broadly to encompass the inherited variation that causes offspring to resemble their parents, regardless of ploidy and mating system (Sztepanacz et al., chapter 12). It is estimated from the phenotypic resemblance of related and unrelated individuals and thus is an attribute of a population. The selection gradient β quantifies the stimulus strength, and it is the change in relative fitness w for a unit change in z. Mathematically, $\beta = COV(z, w) / V_z$, where $COV(z, w)$ is the covariance between z and relative fitness (w), and V_z is the phenotypic variance in z, the sum of V_A and the environmental and other causes of variance, which we refer to as V_R. Thus, β is not just a property of the population that evolves but the result of the interaction of the population with its environment. The derivation and significance of the Lande equation is developed in Hansen's chapter (chapter 5).

We often want to compare V_A or β across traits, populations, or species. When traits are measured in the same units, the values can be compared directly, bearing in mind that V_A

has units of the trait squared, while β has units of relative fitness/trait unit. When the traits to be compared are measured in different units, however, standardization is essential.

Since we often care about the proportional change in trait values, a natural way to standardize the Lande equation is by the trait mean, yielding response as a proportion of the trait's starting value

$$\frac{\Delta\bar{z}}{\bar{z}} = \frac{V_A}{\bar{z}^2}\bar{z}\beta = e_\mu\beta_\mu, \tag{6.1}$$

where e_μ is the mean-scaled evolvability, and β_μ is the mean-scaled selection gradient (Hansen et al. 2003b, 2011; Hereford et al. 2004. A key advantage of this standardization is that $\beta_\mu = 1$ when the trait is fitness, providing a natural marker for strong selection (Hereford et al. 2004). Mean-scaled evolvability, e_μ, offers a useful metric to compare changes in V_A when other components of the phenotypic variance are changing simultaneously, for example, when drift or environment affects both V_A and V_R (Hoffmann and Merilä 1999; Whitlock and Fowler 1999), or when the trait mean differs due to selection, environment, or inbreeding. Still, comparing evolvability using e_μ assumes that the variance can be meaningfully expressed as a proportion of the trait mean (Houle et al. 2011; Pélabon et al. 2020), which is not the case for variables on an absolute scale, such as probabilities, or for those on interval or ordinal scales. For many traits, particularly morphological traits (Gingerich 2000), variances are positively correlated with means, such that we expect that mean standardization will tend to homogenize e_μ of traits with different means. For other traits, such as clutch size in birds (Pélabon et al. 2020) or the traits described by G. Wagner (chapter 10), changing the trait mean will change e_μ and the proportional response to selection. For these traits, evolution of the trait mean causes evolution of evolvability (Hansen and Wagner, chapter 7).

Prior to Lande (1979), quantitative geneticists wrote the response equation as the "breeder's equation" $\Delta\bar{z} = V_A\beta = h^2S$, where $h^2 = V_A/V_p$ is the heritability, and $S = COV(z, w)$ is the selection differential (Falconer 1981). Because heritability is a dimensionless quantity, it may seem natural to standardize both sides of the equation by the phenotypic standard deviation, $\sqrt{V_z}$, yielding response in units of standard deviation,

$$\frac{\Delta\bar{z}}{\sqrt{V_z}} = h^2\frac{COV(z, w)}{\sqrt{V_z}} = h^2\sqrt{V_z}\,\beta = h^2\beta_\sigma, \tag{6.2}$$

where β_σ is the intensity of selection, symbolized i in the animal breeding literature. This standardization invites one to see h^2 as a measure of evolvability and β_σ as a measure of selection. However, because V_A is part of V_z, the factor that controls evolvability is part of the standardization (Houle 1992), making this variance standardization a "rubber ruler" (Houle et al. 2011). It also confounds evolvability and selection by multiplying β by a function of the evolvability. The pernicious impact of this standardization is perhaps most striking when there is a linear relationship between and z and w. In this case, by definition, an increase in V_A increases evolvability but leaves β unchanged. The Lande equation thus shows that $\Delta\bar{z}$ increases and that the reason for this is the increase in evolvability; the mean standardized equation (6.1) reflects this fact. However, in equation (6.2), when V_A increases, both h^2 and β_σ increase, obscuring the fact that selection has not changed. Furthermore, the rubber ruler effect increases the standard deviation used as a measuring stick. Comparison of response using equation (6.2) in populations with different levels of V_A will therefore understate the effect of increased evolvability on response to selection and spuriously suggest that the strength of selection has increased.

6.5.2 Neutral Evolution

The neutral model predicts the evolution of organismal features when mutation and drift are the only evolutionary forces acting (Kimura 1983). Drift is the random sampling of alleles from one generation to the next due to stochasticity in reproduction or survival. Let's further restrict our attention to the rate at which discrete genotypic differences evolve between two populations. *Of* is thus any discrete neutral feature, such as a DNA or an amino acid sequence, and *Over* is whatever timescale we care to specify. The *Under* conditions are explicitly mutation, drift, and effective absence of selection.

Modeling this neutral evolutionary process requires two key parameters, the population size N, and the rate of mutation to new neutral variants, μ. When there are alternative genotypes at a given generation, the necessary outcome of the drift process is that eventually all the individual alleles in the population will descend from a copy of just one allele, which is called a fixation event. In a diploid population of N individuals and $2N$ alleles, there will be $2N\mu$ new neutral mutations in each generation and a probability of $1/(2N)$ that each mutation eventually becomes fixed. Thus, the rate of accumulation of differences between ancestor and descendant per generation, k, is

$$k = 2N\mu \times \frac{1}{2N} = \mu. \tag{6.3}$$

The simplicity of this outcome is a key to the usefulness of the neutral theory. This equation also shows that the neutral mutation rate, μ, is the attribute that generates evolvability under the neutral theory.

This seemingly straightforward conclusion hides two complications. First, how do we know which variants are neutral? The absence of selection that makes μ a dispositional parameter is difficult to measure directly, and variants with small fitness effects can be effectively neutral when the influence of drift at rate $1/(2N)$ is much greater than the effect of natural selection (Ohta 1992). This means that μ is entangled with N; as N decreases, the effective μ increases. Thus, even if we call μ a mutation rate, it is a function of a molecular mutation rate, the GP map that determines the phenotypic effect of the mutation, the shape of the fitness landscape, and N. The second complication is that the population size parameter, N, is not the census count of individuals in the population, but the effective population size, N_e, a complex function of the structure and history of the population that usually makes N_e much less than the census size (see e.g., Charlesworth and Charlesworth 2010, chapter 5.2).

We do have strong models for some aspects of sequence data, and in these cases, separating the innate mutation rate from assumptions about selection is relatively straightforward. For example, mutations to synonymous codons are likely to be effectively neutral in many species with population sizes that are not too large. Mutation rates at the molecular level are readily measured, and, at least to a first approximation (Hodgkinson and Eyre-Walker 2011), the mutation rate does not differ depending on whether the sequence is neutral. This makes it possible to parameterize models of robustness and evolvability for protein sequences (Ancel Meyers et al. 2005) and RNA secondary structure (A. Wagner 2008). However, attempts to measure μ for classes of traits that plausibly have neutral networks (Schuster et al. 1994), such as gene regulatory networks, or morphological traits, are speculative, as we do not know when changes in those organismal features generate equivalent fitness.

The consistency of genomic evolutionary rates (Kumar 2005) suggests that μ usually evolves quite slowly, arguing that the model applies *Over* fairly long timescales, mostly because the number of generations from the lucky mutation event to when it takes over the population is of order $4N_e$ generations.

The stimulus for evolution by drift is population size, and its strength is $1/(2N_e)$. Taken literally, equation (6.3) suggests that the existence of neutral mutations ($\mu > 0$), is a sufficient cause for evolution to occur under the neutral model, similar to radioactive decay, where the disposition of an atom to decay is in itself the cause of its manifestation. This makes the typical genotype with its GP map the bearer of the mutation rate. This literal interpretation, however, ignores the fact that effective neutrality is also influenced by N_e. Taking this into account makes the population the bearer of the attribute μ. Somewhat uncomfortably, this result places N as both a stimulus and one of the background conditions that shapes μ. To place μ at the genotype level, we could assume a class of mutations that are effectively neutral unless the population size is very large. For example, vertebrates generally have effective population sizes of order 10,000 (chapter 4 in Lynch 2007), justifying a treatment of N_e as a background condition to μ for variants that will still evolve neutrally at all smaller values of N_e.

6.6 Measurement and Screening Off

We claimed that V_A and μ are the relevant dispositional attributes that dictate evolvability *Under* short-term directional selection and neutral evolution, respectively. Using the concept of screening-off (Salmon 1971), we can make the case that V_A and μ are not just relevant to evolvability *Under* directional selection and drift, but that they are also the best measures of evolvability under those scenarios.

Imagine a causal sequence where a distal cause (D) leads to a proximal cause (P) that in turn leads to a manifest change (M). P screens-off D when all you need to know to predict M is P. This is true when a prediction based on P alone is equivalent to prediction based on both P and D, and it is different from a prediction based on D alone. Screening-off is also helpful when trying to sort out correlates from causes. If we add a causally irrelevant feature C, which is correlated with P because it is also an outcome of cause D but has no direct effect on M, C is screened-off from prediction by both D and P. Screening-off has been used in evolutionary biology as a tool to identify which factors in a causal chain are "better" explanations for an outcome (Brandon 1982; Brandon et al. 1994). In particular, Brandon (1982) claimed that screening-off justifies Mayr's (1963, 184) intuition that "natural selection favors (or discriminates against) phenotypes, not genes or genotypes."

Applying the concept of screening-off to the evolvability *Of* a trait mean, *Under* directional selection, *Over* one generation, V_A is the only evolvability attribute that is appropriate in the Lande equation, because V_A screens-off the mutation rate, GP map and population history that cause V_A to have its actual value. We can build better understanding of evolvability by focusing on the distal attributes that affect V_A (Sober 1992), but we will not achieve better predictive power in the single generation considered in the Lande model. After the first generation, V_A itself can change due those very attributes that are screened off.

6.7 Evolvability Attributes

Several attributes have been hypothesized to affect population-level evolvability. We list the most important of these in table 6.2 and discuss them further in the following sections. These attributes are specific to the *Of, Under,* and *Over* conditions considered. For example, if *Of* is the transition from one base pair to another, we will measure evolvability as the probability that a mutation occurs per unit time. Alternatively, our *Of* might be the number of mutational changes in the entire genome *U*, which takes on any positive integer and is referable not just to the rate of mutation but also to the average number of base pairs involved in each mutational event. The *Under* column lists some simple evolutionary stimulus scenarios, including directional selection favoring only a change in the trait mean (D), evolution on a curved fitness landscape (C) that may favor a change in the mean, while simultaneously selecting on other aspects of variation, neutral evolution (N), or mutation pressure alone (M). The *Over* column denotes the relevant timescales from one generation to macroevolutionary trends. We do not include attributes that may apply above the population level.

6.7.1 Genetic

The first class of attributes concerns the processes of *mutation* and *recombination* on haplotypes inherited from parents to offspring. Sequencing allows the rates of these processes to be measured at the genotypic level, without any reference to phenotype, placing these processes at the far left of figure 6.1. Mutation and recombination are usually measured as the rate of discrete mutations or recombination events per generation per base

Table 6.2
Attributes hypothesized to affect population-level evolvability

Intrinsic property	Intrinsic level†	Bearer‡	*Of* Discrete	*Of* Continuous	*Under*§	*Over*‖
Genomic mutation	G	I	Rate	Bp affected	Any	I, L
Recombination	G	I	Rate	Gene conversion tract size	Any	I, L
Mutational effect	GP	I	Robustness, probability	Canalization, effect size	N, M, D	I
Conditional effect	GP	I	Rate	Conditional effect size	C	I
Pleiotropy	GP	I	Conditional probability	Angle to trait	C	I
Plasticity	GP	I	Switch points	Reaction norm	D, C	S, I
Integration	GP	I	Covariance	Covariance	C	I
Modularity/autonomy	GP	I	Modularity	Modularity	C	I
Versatility	GP	I	Dimension	Dimension	Any	I, L
Mutational impact	Vy	I	Mutation number, bias	Mutational bias, variance	N, M, D	I
Conditional impact	Vy	I	Mutation number	Mutational variance	C	I
Genetic variation	Vn	P	Diversity H, π	Additive genetic variance	N, D	S
Conditional variation	Vn	P	Conditional rate	Conditional variance	C	S

† Level: G, genetic; GP, property of GP map; Vy, variability; Vn, variation.
‡ Bearer of the property: I, Individual organisms or their components, such as genotype, gene, or trait; P, population.
§ Under: D, directional selection; C, evolution on a curved fitness landscape; N, neutral; M, mutation.
‖ Over timescale: S, short (1 to 10s of generations); I, intermediate (10^2-10^6 generations); L, long ($>10^6$ generations).

pair. They also affect a continuously distributed number of base pairs. A single mutational event, for example, can alter a variable number of base pairs (Schrider et al. 2013), as does the recombinational process of gene conversion. On a genome-wide level, the total number of mutations or recombination events may be large integer values. These rates generally evolve slowly, suggesting that measurements are most relevant over intermediate to long timescales.

6.7.2 GP Map

We identify several evolvability attributes at the level of the GP map, because they determine phenotypic effects conditional on the existence of genetic (or environmental) variation. These are: mutational effects, conditional effects, pleiotropy, integration, modularity, versatility, and plasticity. One can measure them without regard to the rate of genetic changes. For example, one could engineer novel genetic variants to quantify these properties. This makes the typical genotype the bearer of the GP map properties.

Most fundamentally, a *mutational effect* is quantified by the probability that a mutation has a discrete effect, or as the average effect of a mutation on a continuous trait. Interactions between genotypes may cause mutational effects to differ either due to dominance of allelic variation at the site of the mutation or epistatic interactions of alleles at different sites in the genome. Epistatic interactions of mutational effects are what generates evolvability of the GP map (Hansen 2006, chapter 5).

Fixation of a mutation may alter genetic robustness or canalization, affecting variability in subsequent generations. The genetic robustness of a particular trait, such as an amino acid sequence or the 3-dimensional structure of a polymer can be measured as the probability of mutational effect on the phenotype. This makes robustness difficult to measure, as the absence of an effect can never be established experimentally. Measures of discrete effects are useful in models of mutation pressure, mutation load, neutral evolution, and developmental systems drift.

The term canalization was coined to refer to the process of evolving reduced effect size (Waddington 1942). It is now often used to refer to the relative effect sizes of a mutation in two different genotypes; the genotype in which a variant has a larger effect is less canalized than the one with a smaller effect (Flatt 2005). Continuous effect sizes can be measured directly for mutations with large effects, such as gene knockouts, or indirectly from the total mutational variance, coupled with an estimate of the number of mutations that potentially cause that variation.

Pleiotropy and *conditional effects* are complementary ways to characterize mutational effects on multiple traits. Conditional effect quantifies the probability or size of phenotypic effects, conditioned on the absence of some other effect(s), whereas pleiotropy is the tendency of variants to affect more than one trait. These two ways of looking at effects differ when our a priori definition of a trait is not the same as what selection sees. For example, if we measure long bone lengths, such as the femur and tibia, it is natural to characterize pleiotropy as the proportion of mutations that affect both bones, or as the angle between the multivariate direction of an effect and the trait axes. However, if selection favors an increase in both bones, it might be more natural to see the relevant trait as size, and a mutation that affects both traits as nonpleiotropic. Conditional effects are useful in the context of models of evolution on curved fitness landscapes, where, for example, fitness is maximized by changing the relative lengths of the leg bones while holding leg length constant. Conditional effect sizes can sometimes be estimated for a given set of phenotypes, but identifying the full set of phenotypes pleiotropically affected by a given genetic variant is not currently feasible (Paaby and Rockman 2013).

Integration and *modularity* concern the collective pleiotropic and conditional effects of the genome-wide distribution of mutational effects. Integration is minimally defined as the degree of covariation among traits (e.g., Olson and Miller 1958; Cheverud 1996; Armbruster et al. 2014), and *modularity* is defined as the degree to which sets of integrated traits covary less with other sets of traits (Hendrikse et al. 2007; Klingenberg 2008). These definitions make no reference to adaptation and thus concern what Armbruster et al. (2014) term phenomenological integration and modularity, in contrast to the more restrictive sense of adaptive integration and modularity (see Pavličev et al., chapter 8). Many indices of integration have been proposed (Armbruster et al. 2014), and measures of modularity build on these measures of integration to identify clusters of integrated traits that are relatively independent of other such clusters (Zelditch and Goswami 2021). Much of the literature on integration and modularity focuses on morphology, but the concepts apply more generally to other suites of potentially correlated traits, such as gene expression or behavior.

Plasticity is the relationship between the phenotype and the environment in which an organism exists. It can be summarized as a function that relates the distribution of phenotypes to the environment, called a reaction norm. Although we presented mutational effects without explicit references to environment, these are more realistically approached as the study of the effects of genetic variation on reaction norms. In this context, genetic or environmental canalization is the evolution of those reaction norms, considering either the genetic or environmental background in which mutations occur. The importance of plasticity to the GP map and then for evolvability is evident (Sultan 2017) but extremely difficult to measure.

Vermeij (1973a,b) termed the number of dimensions in which the phenotype can vary *versatility* and further proposed that versatility enhances evolvability by providing more alternative phenotypes for selection to assay and increase the prospects for evolution of novel traits. For example, Vermeij observed that primitive mollusk shells were simple linear forms, while derived taxa evolved shells that coil in 2 and later 3 dimensions and speculated that a gain in dimensionality enabled the increased variety of forms.

6.7.3 Variability and Variation

Variability is the propensity of an individual to produce phenotypic variation due to genetic events filtered through the GP map, and is measured as the rate of increase in such variation. Variation measures the degree to which genomic or phenotypic properties differ among members of a population. Although closely related conceptually, a key distinction between variability and variation is that many evolutionary processes can alter variation on a short timescale (Pélabon et al., chapter 13), whereas properties at the mutational and GP map levels are likely to evolve more slowly, rendering estimates of variability relevant over longer timescales than variation. Attributes that we treated as part of the GP map, such as mutational and conditional effects, are quantified through their effect on variability and variation.

Mutational impact encompasses both the number and the effects of mutations. Discrete mutational impacts measure the number of mutations that have a specified effect. Measures of continuous mutational impact characterize how mutation changes the distribution of phenotypes in the population. The effect on the trait mean results from mutational bias, while the effect on the variance is a function of the mutational variance, V_M, the increase in genetic variation from a single generation of mutation. Different definitions of V_M focus either on mutational effects that increase V_A (as described in section 6.5.1) or the effects of mutations once they are fixed (Lynch and Hill 1986).

Genetic variation can be measured for either discrete properties, as the probability of genetic differences between randomly chosen individuals (H or π), or for quantitative properties measured as variance. These variability and variation attributes are frequently used in models of mutation acting alone, of response to selection, and of change in a neutral model.

Measures of *conditional impact* quantify the effects of mutations on a focal trait or set of traits when effects on other traits are held constant (Hansen 2003; Hansen et al. 2003a; Hansen and Houle 2008). Conditional effects depend on the integration and modularity of the GP map but have a more direct relationship to evolvability than their GP map counterparts. If we are interested in the evolution of trait X, holding trait Y constant, we can directly ask: What is the evolvability of trait X if selection favors a change in X, while holding Y at its current value? Despite the conceptual clarity of this connection between conditional properties and an evolutionary model combining directional and stabilizing selection, it is often doubtful that all appropriate traits to condition on have been identified, particularly when pleiotropic interactions affect traits expressed at different ages or life stages.

The measurement of mutational variability is still challenging in most species, as mutations are individually rare, and their phenotypic effects are often small. Sequencing of parents and their descendants readily generates estimates of genomic mutational variability. At the phenotypic level, a typical design is a mutation-accumulation experiment, where mutations are allowed to build up in an initially homogeneous set of lines in the absence of natural selection (Houle and Kondrashov 2006). After a substantial number of generations, the cumulative change in mean and variance are measured. This can only be accomplished in model organisms that allow such designs. In contrast, genetic variation can readily be studied in any organism. Sequencing directly measures discrete genotypic variation. Quantitative genetic variation can be estimated by quantifying the relative similarity of related and unrelated individuals.

6.8 Measurement of Evolvability Attributes versus Measurements of Evolvability

In representational measurement, we assign numbers to the attributes of entities so that the mathematical relationships among the numbers capture empirical relationships in the real world (Krantz et al. 1971; Hand 2004). In the context of evolvability, representational measurements can allow quantitative predictions of the disposition to evolve when associated *Of* and *Under* conditions are met, as in our example of changes in trait mean under directional selection (see section 6.1). In the Lande model, when the additive genetic variance in a selected trait doubles, the population will evolve twice as fast.

We could hope that the attributes listed in table 6.2 have at least a monotonic relationship with evolvability. Close consideration, however, shows that this is often not the case. The assumptions under which many of these attributes measure evolvability are not general.

For some attributes, a change of assumptions can reverse the sign of their relationship to evolvability.

A familiar example is the relationship between recombination rate and evolution that figures prominently in the literature on the evolution of sex (Otto 2009). For each set of conditions under which recombination enhances the rate of adaptation, another set of conditions exists under which the converse is true. Under the assumptions of the Fisher-Muller model for the evolution of sex, mutations that enhance fitness act additively, such that the fitness of a genotype having two such mutations is always better than those with only one. In this case, sex and recombination enhance the rate of evolution by rapidly bringing together favored mutations in the same genotype. However, if sign epistasis is common, alleles favored in isolation will only spread in the presence of a subset of other alleles. In these cases, selection will create positive gametic disequilibrium between jointly favored alleles, but recombination will tend to separate them, reducing the rate of adaptation.

Modularity and integration provide another example (Houle and Rossoni 2022). For many authors, the phenomenological definition of integration and modularity is inadequate, and they instead define integration and modularity as the tendency of functionally related parts of an organism to covary, and for functional suites to be independent of each other (e.g., Olson and Miller 1958; Cheverud 1996; G. Wagner 1996; G. Wagner and Altenberg 1996; Pavličev et al., chapter 8). Under such a definition, modularity "is expected to improve evolvability by limiting the interference between the adaptation of different functions" (G. Wagner and Altenberg 1996, 967). This positive effect of modularity on the evolvability *Of* complex characters assumes that collections of integrated traits are *Under* selective regimes that differ from those affecting traits in different functional modules. Similarly, integration is expected to increase evolvability when the pattern of covariation among traits aligns with the orientation of the fitness landscape. However, if modularity and the directions of selection are not aligned, evolvability is actually reduced (Hansen 2003; Welch and Waxman 2003). The result of these conflicting considerations is that while everyone can agree that integration and modularity have an important role in determining evolvability, they have no monotonic relationship to evolvability (Armbruster et al. 2014; Houle and Rossoni 2022).

In most systems, we have relatively few data on the types of changes favored by natural selection in suites of traits for which integration and modularity are relevant. The major source of evidence for the adaptiveness of phenomenological integration and modularity is that evolution often proceeds by alteration of integrated modules, but this may be explained either by the assumption that modularity is adaptive or that phenomenological modularity constrains evolution in other possible directions (Houle and Rossoni 2022). The morphologically distinct and developmentally integrated liver of vertebrates provides a potential example of a mismatch between integration and selection. The hepatocytes of the liver are the site of a wide variety of biochemical and physiological functions, including synthesis, storage, detoxification, and digestion. It is, however, a challenge to unite the morphological integration of the liver with a hypothesis that relies on the coordinated evolution of these diverse functions. The complexity of hepatocyte function instead suggests that the liver is a locus for trade-offs among competing functions, rather than an organ individuated to optimize compatible functions.

The relationship of robustness to evolvability is similarly contingent on assumptions. On its face, robustness would seem to reduce evolvability due to the reduced probability that any mutation will change the phenotype. Andreas Wagner (2008) pointed out that this is not necessarily so, because robust genotypes can reside in neutral networks of variants with no phenotypic effect that give access to high-fitness alternative phenotypes, enhancing evolvability. Conversely, Mayer and Hansen (2017) pointed to the possibility that robust genotypes are embedded in neutral networks that are not well connected, reducing evolvability, as suggested by the naïve expectation. These alternative assumptions about the nature of neutral networks can be tested only in exceptional circumstances (e.g., Zheng et al. 2020), leaving generalizations about the sign of the relationship between robustness and evolvability unverifiable in most cases.

6.9 Toward Better Measures of Evolvability

We have emphasized the importance of a specific theoretical context to the usefulness of measures of evolvability. Strong theory supports the relevance of measures of genomic mutation rates and of variability and genetic variance as evolvability attributes in many contexts. Other attributes, such as recombination and GP map properties, require more qualifying assumptions before their relationship to evolvability can be specified. The development of novel theoretical contexts may suggest more useful ways to connect these attributes to evolvability.

Predictive validity, the usefulness of the measure to predict some future outcome, represents the best method to understand the value of a measurement (Hand 2004), but is particularly difficult to assess for dispositional properties, for which we also need to know the stimulus strength to make such predictions. Even if we know the stimulus strength, the predictive ability can be compromised by inaccuracy of the measurements or shortcomings in the model used to relate measurements to predicted outcome. For example, the Lande equation assumes that the directional selection measured on the focal trait is the sole source of selection, and, in particular, that indirect responses of the focal trait to selection on other correlated traits are absent.

Predictive experiments have repeatedly been performed for short-term evolution by comparing realized evolutionary changes *Of* a character *Under* artificial selection *Over* a given number of generations with predictions obtained from the Lande equation. These experiments have provided somewhat inconsistent results, particularly so for indirect response to selection of genetically correlated traits, leading some authors to question the ability of additive variance and covariance to predict short-term evolution. These are readily, if unsatisfyingly, explained in principle by a wide variety of possible violations of the assumptions of the Lande model (Walsh and Lynch 2018, 504–506). In practice, limited predictive power mostly results from inaccuracies of measures of evolvability or from the effect of background conditions (e.g., small population size; see Pélabon et al. 2021). Accounting for these effects greatly improves the predictive validity of additive variance on short timescales, suggesting that the other possible violations of the Lande equation have little practical significance.

Over intermediate timescales, initial estimates of evolvability generally have diminished predictive power, particularly in small populations (Weber and Diggins 1990). This is expected, as we know that variation-based measures of evolvability may evolve on this timescale (Pélabon et al., chapter 13), for example, by loss of genetic variation.

When we are concerned with longer timescales, predictions are rarely possible (see Barrick et al. 2009 for a representative exception), and instead validation is only possible by *retrodiction*, the relationship between the properties of extant populations and the prior rate of evolutionary change among related ancestral taxa. The prime example is the retrodiction of rates of molecular evolution by genomic mutation rate (Lynch et al. 2016). This relationship is so strong that rates of evolution have frequently been used to estimate mutation rates, following the neutral theory (Nachman and Crowell 2000). Somewhat surprisingly, retrodictions of evolutionary rates of quantitative traits among populations and species using variance and variability-based measures of evolvability suggest that these measures of evolvability have greater predictive validity than their performance at intermediate times scales would suggest (Voje et al., chapter 14). Although uncertainties characterizing measurements of the *Under* and *Over* conditions call for cautious interpretation, these results suggest that additive and mutational variance may represent fairly accurate measures of evolvability far beyond a handful of generations.

We believe that the main reason for the predictive power of short-term measures of evolvability lies in the well-defined model of evolution provided by the Lande equation, where attributes to measure evolvability and those to measure evolution are simultaneously defined (a similar argument could be made for mutation and molecular evolution). This approach contrasts with measurements of evolvability using GP map attributes that are not defined in the context of an evolutionary model. The validity of those evolvability measurements has not yet been demonstrated via predictions or retrodictions.

6.10 Evolvability Is a Unifying Concept

The ability to conceive and execute useful and predictive measurements is the hallmark of a maturing scientific field. The concept of evolvability is clearly measurable in some contexts, including the neutral theory and the Lande model. In other contexts, we may know what to measure but not how to measure it. For example, if we are interested in the ability to optimize one suite of traits while holding others constant, conditional evolvability is logically what we need to measure (Hansen et al. 2003a), but we rarely know what to condition on, as both pleiotropy and the actual shape of the fitness landscape are unknown. In still other contexts, like the ability to generate novel traits, we have only speculative notions of attributes that generate such a disposition.

For some observers, the diverse status of measurement for each of these variant concepts of evolvability may serve to enhance the claim that each conception of evolvability refers to a fundamentally different phenomenon (Pigliucci 2008; Brookfield 2009; Minelli 2017). We instead find unity in the growing recognition that all definitions of evolvability conceptualize the disposition of a population (or higher-level entity) to evolve (G. Wagner and Altenberg 1996; Love 2003; Hansen 2006; Brigandt et al., chapter 4). To apply that definition to the universe of entities with the potential to evolve, we need to specify the conditions *Of*, *Under*, and *Over*. The actual properties that constitute evolvability are different, depending on whether one is interested in the evolution of DNA sequences, proteins, or organism-level phenotypes. To some extent, this viewpoint echoes the positions taken in Villegas et al. (chapter 3) and Brigandt et al. (chapter 4), although those contributions note that, in addition to being applied to a wide variety of phenomena, the concept of evolvability plays a wide diversity of roles in scientific discourse.

This combination of unity of concept and diversity of application is already inherent in the typical definitions of evolution as, for example, "change over time in the characteristics of a population of living organisms" (Charlesworth and Charlesworth 2010, xxv). Few observers would maintain that we need to discard the word "evolution," depending on which characteristics, which timescale, or which forces shape those changes. The multifarious instantiations of evolvability are reminiscent of another disposition—energy, which quantifies the potential for matter to perform work. Energy is stored in many different forms, including elastic energy, chemical energy, potential energy, and the thermal energy that the humble thermometer quantifies. Each of these other forms of energy is quantified using a different measure appropriate to its nature. So it is with evolvability: one dispositional concept with many measures.

Acknowledgments

We thank Thomas Hansen, Günter Wagner, and Mihaela Pavličev for detailed and helpful comments on the manuscript. We thank the Center for Advanced Study (CAS) at the Norwegian Academy of Science and Letters for its generous funding of this workshop. The critical enthusiasm of the many participants was key to molding this contribution into its final form. DH thanks CP and Thomas Hansen for organizing and shepherding the evolvability project from its beginnings, and Alan Love for organizing the 2017 workshop titled "Generic and Genetic Explanations of Evolvability" at the Minnesota Center for Philosophy of Science, at which many of his ideas on this topic began their incubation. Preparation of this manuscript was supported by US National Science Foundation (www.nsf.gov) Division of Environmental Biology grant 1556774 to DH and by Norwegian Research Council grant no. 287214 to CP.

References

Ancel Meyers, L., F. D. Ancel, and M. Lachmann. 2005. Evolution of genetic potential. *PLoS Computational Biology* 1: e32.

Armbruster, W. S., C. Pélabon, G. H. Bolstad, and T. F. Hansen. 2014. Integrated phenotypes: Understanding trait covariation in plants and animals. *Philosophical Transactions of the Royal Society B* 369: 20130245.

Barrick, J. E., D. S. Yu, S. H. Yoon, H. Jeong, T. K. Oh, D. Schneider, R. E. Lenski, et al. 2009. Genome evolution and adaptation in a long-term experiment with *Escherichia coli*. *Nature* 461: 1243–1274.

Brandon, R. 1982. The levels of selection. *PSA: Proceedings of the Biennial Meeting of the Philosophy of Science Association* 1982: 315–323.

Brandon, R. N., J. Antonovics, R. Burian, S. Carson, G. Cooper, P. S. Davies, C. Horvath, et al. 1994. Sober on Brandon on screening-off and the levels of selection. *Philosophy of Science* 61: 475–486.

Brookfield, J. F. 2009. Evolution and evolvability: Celebrating Darwin 200. *Biology Letters* 5: 44–46.

Chang, H. 2004. *Inventing Temperature: Measurement and Scientific Progress.* Oxford: Oxford University Press.

Charlesworth, B., and D. Charlesworth. 2010. *Elements of Evolutionary Genetics.* Greenwood Village, CO: Roberts and Company.

Cheverud, J. M. 1996. Developmental integration and the evolution of pleiotropy. *American Zoologist* 36: 44–50.

Falconer, D. S. 1981. *Introduction to Quantitative Genetics.* London: Longman.

Flatt, T. 2005. The evolutionary genetics of canalization. *Quarterly Review of Biology* 80: 287–316.

Gingerich, P. D. 2000. Arithmetic or geometric normality of biological variation: An empirical test of theory. *Journal of Theoretical Biology* 204: 201–221.

Grantham, T. A. 2007. Is macroevolution more than successive rounds of microevolution? *Palaeontology* 50: 75–85.

Haag-Liautard, C., M. Dorris, X. Maside, S. Macaskill, D. L. Halligan, D. Houle, B. Charlesworth, et al. 2007. Direct estimation of per nucleotide and genomic deleterious mutation rates in *Drosophila*. *Nature* 445: 82–85.

Hand, D. J. 2004. *Measurement Theory and Practice: The World through Quantification.* London: Arnold.

Hansen, T. F. 2003. Is modularity necessary for evolvability? Remarks on the relationship between pleiotropy and evolvability. *Biosystems* 69: 83–94.

Hansen, T. F. 2006. The evolution of genetic architecture. *AREES* 37: 123–157.

Hansen, T. F., and D. Houle. 2008. Measuring and comparing evolvability and constraint in multivariate characters. *Journal of Evolutionary Biology* 21: 1201–1219.

Hansen, T. F., W. S. Armbruster, M. L. Carlson, and C. Pélabon. 2003a. Evolvability and genetic constraint in *Dalechampia* blossoms: Genetic correlations and conditional evolvability. *Journal of Experimental Zoology B* 296: 23–39.

Hansen, T. F., C. Pélabon, W. S. Armbruster, and M. L. Carlson. 2003b. Evolvability and genetic constraint in *Dalechampia* blossoms: Components of variance and measures of evolvability. *Journal of Evolutionary Biology* 16: 754–766.

Hansen, T. F., C. Pélabon, and D. Houle. 2011. Heritability is not evolvability. *Evolutionary Biology* 38: 258–277.

Hendrikse, J. L., T. E. Parsons, and B. Hallgrímsson. 2007. Evolvability as the proper focus of evolutionary developmental biology. *Evolution & Development* 9: 393–401.

Hereford, J., T. F. Hansen, and D. Houle. 2004. Comparing strengths of directional selection: How strong is strong? *Evolution* 58: 2133–2143.

Hodgkinson, A., and A. Eyre-Walker. 2011. Variation in the mutation rate across mammalian genomes. *Nature Reviews Genetics* 12: 756–766.

Hoffmann, A. A., and J. Merilä. 1999. Heritable variation and evolution under favourable and unfavourable conditions. *TREE* 14: 96–101.

Houle, D. 1992. Comparing evolvability and variability of quantitative traits. *Genetics* 130: 195–204.

Houle, D., and A. Kondrashov. 2006. Mutation. In *Evolutionary Genetics: Concepts and Case Studies,* edited by C. W. Fox, and J. B. Wolf, 32–48. Oxford: Oxford University Press.

Houle, D., C. Pélabon, G. P. Wagner, and T. F. Hansen. 2011. Measurement and meaning in biology. *Quarterly Review of Biology* 86: 3–34.

Houle D., and Rossoni D. M. 2022. Complexity, evolvability, and the process of adaptation. *AREES.* 53: 137–159.

Jablonski, D. 2008. Species selection: Theory and data. *AREES* 39: 501–524.

Kimura, M. 1983. *The Neutral Theory of Molecular Evolution.* Cambridge: Cambridge University Press.

Klingenberg, C. P. 2008. Morphological integration and developmental modularity. *AREES* 39: 115–132.

Krantz, D. H., R. D. Luce, P. Suppes, and A. Tversky. 1971. *Foundations of Measurement.* Vol. I: *Additive and Polynomial Representations.* New York: Academic Press.

Kumar, S. 2005. Molecular clocks: Four decades of evolution. *Nature Reviews Genetics* 6: 654–662.

Lande, R. 1979. Quantitative genetic analysis of multivariate evolution applied to brain:body size allometry. *Evolution* 33: 402–416.

Love, A. C. 2003. Evolvability, dispositions, and intrinsicality. *Philosophy of Science* 70: 1015–1027.

Lynch, M. 2007. *The Origins of Genome Architecture.* Sunderland, MA: Sinauer.

Lynch, M., and W. G. Hill. 1986. Phenotypic evolution by neutral mutation. *Evolution* 40: 915–935.

Lynch, M., M. S. Ackerman, J.-F. Gout, H. Long, W. Sung, W. K. Thomas, and P. L. Foster. 2016. Genetic drift, selection and the evolution of the mutation rate. *Nature Reviews Genetics* 17: 704.

Mayer, C., and T. F. Hansen. 2017. Evolvability and robustness: A paradox restored. *Journal of Theoretical Biology* 430: 78–85.

Mayr, E. 1963. *Animal Species and Evolution.* Cambridge, MA: Belknap Press.

McGee, T. D. 1988. *Principles and Methods of Temperature Measurement.* New York: John Wiley & Sons.

Minelli, A. 2017. Evolvability and its evolvability. In *Challenges to Evolutionary Theory: Development, Inheritance and Adaptation*, edited by P. Huneman, and D. M. Walsh, 211–238. New York: Oxford University Press.

Nachman, M. W., and S. L. Crowell. 2000. Estimate of the mutation rate per nucleotide in humans. *Genetics* 156: 297–304.

Nuño de la Rosa, L. 2017. Computing the extended synthesis: Mapping the dynamics and conceptual structure of the evolvability research front. *Journal of Experimental Zoology B* 328: 395–411.

Ohta, T. 1992. The nearly neutral theory of molecular evolution. *AREES* 23: 263–286.

Olson, E. D., and R. L. Miller. 1958. *Morphological Integration.* Chicago: University of Chicago Press.

Otto, S. P. 2009. The evolutionary enigma of sex. *American Naturalist* 174: S1–S14.

Paaby, A. B., and M. V. Rockman. 2013. The many faces of pleiotropy. *Trends in Genetics* 29: 66–73.

Pélabon, C., E. Albertsen, A. Le Rouzic, et al. 2021. Quantitative assessment of observed vs. predicted responses to selection. *Evolution* 75: 2217–2236.

Pélabon, C., C. H. Hilde, S. Einum, and M. Gamelon. 2020. On the use of the coefficient of variation to quantify and compare trait variation. *Evolution Letters* 4: 180–188.

Pigliucci, M. 2008. Opinion—Is evolvability evolvable? *Nature Review Genetics* 9: 75–82.

Salmon, W. C. 1971. *Statistical Explanation and Statistical Relevance.* Pittsburgh: University of Pittsburgh Press.

Schrider, Daniel R., D. Houle, M. Lynch, and M. W. Hahn. 2013. Rates and genomic consequences of spontaneous mutational events in *Drosophila melanogaster.* *Genetics* 194: 937–954.

Schuster, P., W. Fontana, P. F. Stadler, and I. L. Hofacker. 1994. From sequences to shapes and back: A case study in RNA secondary structures. *Proceedings of the Royal Society B* 255: 279–284.

Sober, E. 1992. Screening-off and the units of selection. *Philosophy of Science* 59: 142–152.

Sultan, S. E. 2017. Developmental plasticity: Re-conceiving the genotype. *Interface Focus* 7: 20170009.

Vermeij, G. J. 1973a. Adaptation, versatility, and evolution. *Systematic Zoology* 22: 466–477.

Vermeij, G. J. 1973b. Biological versatility and earth history. *PNAS* 70: 1936–1938.

Waddington, C. H. 1942. Growth and determination in the development of *Drosophila.* *Nature* 149: 264–265.

Wagner, A. 2008. Robustness and evolvability: A paradox resolved. *Proceedings of the Royal Society B* 275: 91–100.

Wagner, G. P. 1996. Homologues, natural kinds and the evolution of modularity. *American Zoologist* 36: 36–43.

Wagner, G. P., and L. Altenberg. 1996. Perspective: Complex adaptations and the evolution of evolvability. *Evolution* 50: 967–976.

Walsh, B., and M. Lynch. 2018. *Evolution and Selection of Quantitative Traits.* Oxford: Oxford University Press.

Wang, J., H. C. Fan, B. Behr, and Stephen R. Quake. 2012. Genome-wide single-cell analysis of recombination activity and *de novo* mutation rates in human sperm. *Cell* 150: 402–412.

Weber, K. E., and L. T. Diggins. 1990. Increased selection response in larger populations. II. Selection for ethanol vapor resistance in *Drosophila melanogaster* at two population sizes. *Genetics* 125: 585–597.

Welch, J. J., and D. Waxman. 2003. Modularity and the cost of complexity. *Evolution* 57: 1723–1734.

Whitlock, M. C., and K. Fowler. 1999. The changes in genetic and environmental variance with inbreeding in *Drosophila melanogaster.* *Genetics* 152: 345–353.

Zelditch, M. L., and A. Goswami. 2021. What does modularity mean? *Evolution & Development* 23: 377–403.

Zheng, J., N. Guo, and A. Wagner. 2020. Selection enhances protein evolvability by increasing mutational robustness and foldability. *Science* 370: eabb5962.

7 The Evolution of Evolvability

Thomas F. Hansen and Günter P. Wagner

The evolution of evolvability became a topic in the 1990s, and since then, it has progressed from controversies about its radical or conventional nature to a mature research program with hypotheses motivated in evolutionary theory and theoretical population genetics. Evolvability is an outcome of a variety of organismal traits, and it evolves along with these traits. In this chapter, we first review the theoretical basis for the main modes of evolution of evolvability, including adaptation at various levels; contingent evolution based on indirect, canalizing, and congruent selection; and neutral evolution, including systems drift. We then present an overview of organismal properties that may influence evolvability and provide some selected reviews of their possible modes of evolution.

7.1 Introduction

The diversity of life testifies to the capacity of organisms to evolve into a variety of complex forms and modes of existence. On one hand, this reflects the power of natural selection to build complex adaptations. One of the main achievements of twentieth-century evolutionary biology was the theoretical and empirical demonstration of the efficacy of natural selection. Given heritable variation, selection can produce stunning changes in little time. On the other hand, the diversity of life also reflects the ability of at least some organisms to produce new variations that can fuel selection. The ability to produce and maintain potentially adaptive heritable variation is what we call *evolvability.* In their drive to demonstrate the power of natural selection, the architects of the modern synthesis took the existence of heritable variation more or less for granted. With some exceptions, variation was not seen as a property in need of explanation or studied as an evolving variable. This view started to change toward the end of the last century, when an increasing number of researchers began to study variation and the ability of organisms to produce variation. In particular, the architectural features of organisms that structure their variational properties, the genotype-phenotype map, became a subject of study (e.g., Wagner and Altenberg 1996).

What happened in evolutionary theory toward the end of the last century can be described as a case of theory expansion, in which previously external conditions and assumptions about variability became "endogenized" in the sense of being treated as something to be explained within the theory (Okasha 2021). Thus emerged the study of "the evolution of evolvability." This phrase, popularized by Dawkins (1988), refers to the evolution of the organismal or population properties that influence the ability to evolve.

As evolvability is an outcome of organismal properties, there is no mystery or paradox to its evolution. Evolvability evolves along with the traits that influence it. Such questions as "is evolvability evolvable?" and attempts to demonstrate that evolvability can evolve through simulations or experiment are therefore of limited interest. The question is not whether evolvability can evolve, but how it evolves.

In this chapter, we review principles for the evolution of evolvability and conclude that there is no simple or agreed-on answer to how this actually happens. A variety of opinions and hypotheses can be found, and the answer may depend on phylogenetic scale, what traits are considered, whether origin or maintenance is to be explained, and on whether ultimate or proximate explanation is intended. A key question is whether organisms are somehow adapted to be evolvable. Gould (2002) stated a "paradox of evolvability" as "how can something evolve that is not of immediate use?" As explained above, this is not a literal paradox, but Gould intended to challenge a perceived notion of evolvability as an individual-level adaptation, and, as selection is a population phenomenon, evolvability must materialize through population variation, making it difficult to conceive how it can be directly selected on the individual level. This problem was recognized as far back as by Dobzhansky (1937), who referred to it as the "paradox of viability."

Nevertheless, the notion that organisms are designed for evolvability is widespread and has strong antecedents. Riedl (1977, 1978) pointed to the improbability of producing functional variations through mutation in complex organisms with many interlocking parts. He inferred that organisms must be structured to vary along functional lines. Similar ideas have been expressed by Berg (1957, 1958), Waddington (1957), Olson and Miller (1958), Cheverud (1982, 1984), Conrad (1983, 1990), G. Wagner (1986, 1996, 2014), Raff (1996), Gerhart and Kirschner (1997), and A. Wagner (2005), among others. The concept of modularity has been particularly important in this line of thinking. Wagner and Altenberg (1996), for example, identified unbounded pleiotropy as a major impediment to complex adaptation and suggested that the partitioning of the organism into variationally distinct parts allowed each part to evolve in a quasi-independent manner such that adaptation in one part would not fatally interfere with adaptation in other parts. The emphasis on cis-regulatory modularity in evolutionary developmental biology (e.g., Stern 2000; Carroll 2008) reflects this idea. Because regulatory proteins function in a variety of contexts, mutations in their coding sequence have been assumed to be highly pleiotropic and therefore unlikely to improve the organism. In contrast, mutations in cis-regulatory modules may have more narrow effects due to the specificity of the regulatory modules themselves. While it is becoming clear that cis-regulatory mutations are not the only way of making modular changes in gene regulation (e.g., Lynch and Wagner 2008), modular change remains a guiding principle in evolutionary developmental biology.

Gould (2002) proposed two solutions to the paradox of evolvability. One was to give up on the idea of evolvability as a (narrow-sense) adaptation and instead view it as an exaptation, a trait that has evolved for a different purpose than its current function. In this view, evolvability may or may not be maintained as a (broad-sense) adaptation, but the focus is shifted from selection on evolvability itself to the evolution of the various traits and properties that relate to evolvability. Evolvability is likely subject to a variety of indirect selection pressures caused by adaptation in correlated traits, and these need not be related to whether evolvability is of benefit to the organism or population (Sniegowski and Murphy 2006; Hansen 2011). In particular, evolvability may be a side effect of the ways organisms structure their development and physiology to be coordinated, robust, and/or flexible in relation to environmental conditions (e.g., Gerhart and Kirschner 1997). For example, constraints on body symmetry may evolve as adaptations to ensure a functional symmetric body in the face of developmental perturbance, but as a side effect may also structure genetic variation to be more symmetric, which will facilitate the evolvability of symmetric changes and reduce the evolvability of asymmetric changes. Selection is also not all powerful, and systems drift and other forms of neutral evolution of genetic architecture are central to the evolution of evolvability (A. Wagner 2005; Lynch 2007).

Gould's (2002) other proposed solution was that evolvability evolves as a group- or lineage-level adaptation. Among those who favor an adaptive view of evolvability and explicitly consider the level of selection, some form of higher-level selection seems to be the favored solution (e.g., Gerhart and Kirschner 1997). This even includes Dawkins (1988), who went as far as describing the evolution of evolvability as "a kind of higher-level selection." Yet it is not obvious that one needs to move beyond conventional within-population selection to solve the paradox of evolvability (e. g., Wagner 1981). In this chapter, we construct some ways in which evolvability can be said to evolve by direct individual- or gene-level selection.

In any case, our position is that evolvability is an outcome of a variety of organismal characteristics, and that its evolution must ultimately be understood in terms of the evolution of these characteristics and the various traits that influence them. As there is a varied set of relevant characteristics, most known modes of evolution are relevant, and this includes adaptation through direct selection for evolvability on various levels, contingent changes due to indirect selection on traits or properties correlated with evolvability, and neutral modes of evolution, such as systems drift and accumulation of nearly-neutral changes that affect evolvability. Many aspects of evolvability are also deeply integrated with and therefore constrained by organismal structure and life cycle, which sets up historical contingencies and major transitions in body plan or inheritance systems as relevant modes of evolution.

7.2 Modes of Evolution of Evolvability

In this section, we review the theoretical basis for the main modes of evolution of evolvability. We aim to clarify motivation, conditions, problems, and arguments more than to judge general plausibility. Plausibility of different hypotheses is better considered in relation to the evolution of particular characteristics, as illustrated in section 7.3.

7.2.1 Adaptation

The motivation for viewing evolvability as an adaptation is that organisms seem designed to be evolvable, and adaptation by natural selection is the main source of design in nature. If we accept that organismal architecture generates accurate inheritance, coordinated variability, continuity, recombination, and robustness to an unexpected degree, then, to the extent that these properties facilitate evolvability, we must infer that organismal structure facilitates evolvability to an unexpected degree. An unexpected degree of functional organization is design, and if evolvability is the function, evolvability must be an adaptation.

Leaving aside the debates as to whether organisms really are unexpectedly evolvable and whether this is due to frozen constraints with ancient origin or continuously maintained in individual lineages, the problem with the above argument is that evolvability may evolve by selection without being an adaptation. Even if organisms display complex functional order that reflects a multitude of adaptations, it does not follow that every property of the organism is an adaptation for any specific function. Adaptation for property X requires direct selection being caused by property X (Sober 1984), and in the case of evolvability, this means that evolvability must cause the selection for the properties that influence evolvability. The obvious alternative is that these properties are selected for reasons unrelated to evolvability, and that the evolvability evolves as a correlated response. In this scenario, organisms may appear designed to facilitate evolvability without evolvability being the adaptation.

Adaptations can exist on different levels in the biological hierarchy. The paradox of evolvability: "How can something evolve that is not of immediate use?" is only paradoxical on the implicit assumption that the functional benefit must pertain to an individual organism. For the adaptationist, the easy way out is to assign the level of selection for evolvability to other units, such as genes, groups, or lineages. We will discuss each of these below, but first we sketch a way one may also meaningfully talk about individual-level selection for evolvability.

An episode of selection can be formally described as a mapping from a set of entities (e.g., population of individuals) to another set (Kerr and Godfrey-Smith 2009). Fitness is assigned to different types of entities to describe their expected ratio of representation after selection to representation before selection. To select *for* evolvability on the level of individuals, we need to set up the mapping in such a way that a high-evolvability type has high fitness because it has high evolvability. This is not possible if the episode of selection does not include reproduction, because there is then no possibility for the evolvability to manifest itself. It is of course possible that high-evolvability types have high fitness in this scenario, and we can have individual selection *of* evolvability, but not individual selection *for* evolvability (sensu Sober 1984). We can, however, consider a mapping from a set of adult individuals in one generation to their adult offspring in the next (or even a later) generation. In this case, a high-evolvability individual may produce a set of candidate offspring that is more adaptable (e.g., more variable in some ecologically relevant trait), and if the selection happens in an uncertain, changing, or unfavorable environment, the high-evolvability types may end up with a better representation among adults in the next generation (i.e., higher fitness), because they were more likely to produce some offspring that were well adapted to the environment they encountered.

This scenario, familiar from the literature on bet-hedging strategies and from the tangled-bank hypothesis for the evolutionary maintenance of sexual recombination, is a candidate for direct individual-level selection for evolvability. On a similar basis, candidates for individual-level adaptation for evolvability may also be found in mechanisms for stress-induced release of genetic variation in situations where the individual is likely to be maladapted. This can take the form of phenotypic expression of segregating "hidden" variation (e.g., Rutherford and Lindquist 1998) or the form of increased rates of mutation under stress (e.g., Galhardo et al. 2007).

7.2.2 Gene-Level Adaptation

Some may object that the above mechanism is selection of family groups more than of individuals, and it can also be considered in terms of selection on the level of genes. The relative stability of alleles makes it more natural to consider episodes of selection that last over more generations. Hence, if we consider a mapping from a population of alleles to a later point in time, we can see that an allele that generates more variable descendants may be better represented, because it is more likely that some of its descendants were successful in an unfavorable environment. If we consider the mapping to be composed of a series of selective episodes, the fitness over the total mapping is the product of the Wrightian fitness values of the individual episodes of selection. Formally, if W_i is the fitness of the allele in the ith episode of selection, the fitness over a sequence of episodes is $W = \Pi_i W_i$. Now consider an allele that starts out with fitness W_0 but produces offspring with variable heritable fitness. Then, by the fundamental theorem of natural selection, the mean fitness of the subpopulation carrying the allele will increase in each episode of selection with a term v that is equal to the variance of fitness divided by the mean fitness. Hence, assuming for simplicity that v stays constant over the sequence of selection episodes, the fitness of the allele over the whole sequence will be $W = \Pi_i (W_0 + iv)$. Assuming the same process operating in the population as a whole, the relative fitness of the allele after t episodes of selection will be

$$w = \frac{\prod_{i=0}^{t-1} (W_0 + iv)}{\prod_{i=0}^{t-1} (\bar{W}_0 + iV)},$$

where $\bar{W}_0$ is the intial mean fitness of the population, and V is the total variance in fitness of the population divided by the mean fitness. From this equation, and assuming also that V stays constant, we see that the allele will increase in frequency due to selection if

$$\prod_{i=0}^{t-1} (W_0 + iv) > \prod_{i=0}^{t-1} (\bar{W}_0 + iV),$$

which can be approximated as

$$\frac{v}{W_0} - \frac{V}{\bar{W}_0} > \frac{-2ln(W_0 / \bar{W}_0)}{t}.$$

The left-hand side of this equation can be interpreted as the difference in evolvability between carriers of the allele and the population at large. The term v/W_0 is the initial average opportunity for selection (i.e., the variance in relative fitness) of the subpopulation carrying the allele, and the term $V/\bar{W}_0$ is the initial opportunity for selection in the population at large. The numerator on the right-hand side is the initial fitness cost of carrying the allele. A cost is to be expected due to the likely immediate deleterious effects of variation. Hence, an allele that increases the opportunity for selection on its carriers will spread if it can overcome direct fitness costs. As the number of episodes of selection (e.g., generations) increases, the more likely it is that the high-evolvability allele will increase in frequency. Wagner (1981) described a similar mechanism in terms of selection on modifiers of Malthusian fitness.

Well-known examples that may fit this description include "mutator" alleles that increase the mutation rate of their carriers, alleles that increase the recombination or outcrossing rates of their carriers, and "modifiers" that epistatically increase the effects of other allele substitutions in the genomes in which they are situated. More generally, we can consider any allele that influences the variational properties of the organism as a putative evolvability allele, and if the variational changes caused by such an allele increase the opportunity for selection on its descendants, it may spread in the population according to the above criterion.

In this model, alleles that elevate the opportunity for selection are favored because their descendants are able to adapt in the sense of increasing their mean fitness faster than the population at large. This is direct selection for evolvability, because it is the effect of the allele on evolvability that makes the difference in the selective outcome, and the described process thus has the potential to create and maintain adaptations for evolvability. We can imagine such alleles spreading in maladapted species with more scope for changing fitness variation, and that species living in changing environments may maintain higher evolvability because they tend to be maladapted more of the time.

7.2.3 Group- or Lineage-Level Adaptation

Even if they are divided on whether adaptive evolution of evolvability occurs at all, commentators as different as Dawkins (1996), Gould (2002), and Lynch (2007) all suggest that this would require some form of higher-level selection. A shift to higher-level units with internal evolutionary processes that may differ in evolvability is indeed an obvious solution to the problem of constructing direct selection for evolvability. Populations with high evolvability will on average be better adapted to their environment, and they may tend to survive better, bud off more offspring populations, or produce more propagules of individuals that transfer the evolvability-enhancing traits to other populations.

Group selection has had a bad press, particularly when associated with naive best-for-the-species styles of argument. Evolvability may be good for the species, but this is no explanation for why species are evolvable. Nevertheless, carefully formulated models have demonstrated that group selection can be efficient in maintaining group adaptations (Okasha 2006). Although the focus of such models has been on social traits like altruism, there is also work supporting group or lineage selection as a viable hypothesis for the maintenance of sex and recombination (Maynard Smith 1978; Nunney 1989), and for emergent species-level traits and trends more generally (e.g., Lloyd and Gould 1993; Jablonski 2008).

The modern treatment of group selection is based on the Price theorem and works by splitting the evolutionary change over an episode of selection into components attributable to selection among groups and selection within groups, or from the group-selection perspective by treating lower-level selection as transmission effects (Price 1972; Okasha 2006; Frank 2012). This approach requires a recognizable group structure in which groups can be assigned a fitness based on differential contribution to the metapopulation at the end of the episode of selection. Any trait correlated with the group fitness will experience group selection, and if the correlation is causal, the trait can evolve as a group-level adaptation. There are two limitations, however, in applying this to evolvability.

The first is that group-level selection may be overcome by within-group selection. As we have seen above, the mean fitness of an evolvable population will increase with a factor equal to the genetic opportunity for selection (i.e., the evolvability of fitness) per generation. Hence, the group evolvability of fitness will be the among-group variance in the evolvability of (within-group) fitness. We do not have any estimates of group evolvability of fitness, but we can illustrate its potential impact with a thought experiment. Let us assume a metapopulation in which half the populations have an evolvability for fitness of 1% and the other half an evolvability of 3%, meaning that selection would increase the genetic value of fitness with 1% per generation in the former and by 3% in the latter. These values are consistent with current estimates of evolvability of life-time fitness (Hansen and Pélabon 2021). If we assume that the subpopulations' contributions to the metapopulation are proportional to their final mean fitness, then we can compute that group selection will increase mean evolvability of fitness by 0.5% per generation (i.e., from 2% to 2.01%). This matches the evolvability of many quantitative traits and may thus balance many processes acting to diminish within-population evolvability. Group adaptation for evolvability is thus plausible, provided a group structure with enough variation in evolvability can be maintained.

The long-term maintenance of group differences in evolvability would require a degree of remixing of groups before within-group selection removes individually deleterious evolvability-enhancing traits. This is a serious limitation for species- or clade-level selection with little opportunity for remixing. To maintain evolvability as an adaptation on these levels, it must be constrained on lower levels. This would not work if the evolvability in question is a function of quantitative polygenic traits, which inevitably are themselves evolvable on the organismal level. But it is more plausible when evolvability is the outcome of discontinuous changes in the body plan or inheritance system that may become burdened and difficult to reverse. Indeed, many discussions of evolvability from a macroevolutionary perspective concern innovations such as the evolution of new character identities, transitions to new levels of organization, or the construction of qualitatively new niches. More generally, any historical contingency may irreversibly set new evolutionary possibilities that can be conceptualized as a change in evolvability. Species selection then emerges as a potent mechanism for preserving and proliferating clades with constrained traits that provide for richer evolutionary possibilities.

The second limitation to evolvability as a group adaptation is that group selection, just like individual selection, may be indirect. Even when there is population structure facilitating group selection, traits influencing evolvability may be (group) selected for reasons unrelated to their effects on evolvability. For example, Lloyd and Gould (1993) argued that species selection may favor genetic variability, because species with a subdivided population structure will both tend to maintain more genetic variation and be more prone to speciate. Traits that generate the subdivided population structure, such as having non-planktonic larvae, may then be considered species-level adaptations, but not for (within-species) evolvability, as the increase in genetic variation is a side effect and not a cause of increased speciation rates. In contrast, Dobzhansky (1937) argued that evolvability ("evolutionary plasticity") was selected on the species level because it reduces extinction risk by increasing the ability to adapt to environmental change. This would be direct species selection for evolvability and would act to maintain evolvability as a species-level adaptation to changing environments.

7.2.4 Contingent Evolvability: Epistasis and Trait Evolution

The most obvious situation in which evolvability is favorable is when the population is under directional selection. Under directional selection on a trait, both the rate of change in the trait and the increase in mean fitness caused by the selection are proportional to the additive genetic variance in the trait. If we take the additive variance as a measure of trait evolvability, we can ask how the evolvability itself is likely to change in this situation. Carter et al. (2005) showed that the per generation change in the additive variance in a polygenic trait under linear selection with a selection gradient β is

$$\Delta V_A = 2\beta(C_3 + \varepsilon V_A^2) + o(\beta),$$

where C_3 is the additive-genetic third cumulant of the trait, ε is a measure of directional epistasis, and $o(\beta)$ designate terms that vanish under weak selection. The third cumulant is positive when there is positive skew in the distribution of genetic effects, which will happen if alleles that increase the trait tend to be rare. Hence, this term describes the leading effects of allele-frequency changes on the variance, which will increase if rare alleles tend to increase in frequency, but decrease if common alleles tend to increase in frequency. The second term describes the leading effects of epistasis. A positive ε means that allele substitutions with positive effects on the trait tend to elevate the effects of other genetic changes, and a negative ε means that allele substitutions with positive effects on the trait tend to depress the effects of other genetic changes. Hence, positive epistasis in the direction of selection will increase evolvability, while negative epistasis in the direction of selection will decrease evolvability. Note that this has nothing to do with build-up of linkage disequilibrium or hitchhiking of alleles.

An important insight from this model is that the evolution of evolvability under directional selection does not depend on whether it is favorable to the population. Whether the evolvability is increasing or decreasing under linear directional selection depends, at least to a first approximation, on the details of the variational architecture. The main factor determining whether evolvability will increase or decrease is the directionality of epistasis. This stems from the fact that directional epistasis determines the correlation between the trait and its variability. Positive directional epistasis implies a positive correlation in the sense that increasing the trait will also make it more variable. Changing a trait in a direction of positive epistasis will tend to elevate the effects of new mutations, and thus the input of new mutational variance (Hansen et al. 2006).

We can now recognize the mechanism for evolution of evolvability in this model as indirect selection. Direct selection on the trait generates indirect selection on trait variability, and this indirect selection can be positive or negative, depending on whether the correlation between the trait and its variability is positive or negative. In this light, Hansen (2011) proposed that trait evolvability mainly evolves as a correlated response to trait evolution. The argument is that such indirect selection is likely to be ubiquitous, strong, and variable, and it will tend to swamp weaker effects of alternatives, such as genetic drift and canalizing selection.

7.2.5 Adaptive, Canalizing, Conservative, and Hitchhiking Selection on Alleles

Trait selection has complex and sometimes counterintuitive effects on the underlying genes. In this section, we decompose the gene-level effects of directional and stabilizing selection on a trait into four distinct forces and discuss their interplay in the evolution of evolvability. This section is technical, and a theory-averse reader may skip to the next section. In a nutshell, we show that the effects of selection on a gene can be decomposed into (1) an *adaptive force* favoring allele substitutions that improve trait adaptation, (2) a *canalizing force* favoring alleles that epistatically reduce the effects of other alleles when the fitness function is concave (and disfavors them when it is convex), (3) a *conservative force* that acts against rare alleles when the fitness function is concave, and (4) a *hitchhiking force* favoring alleles in positive linkage disequilibrium with other favorable alleles.

Let $z = z(a_1, \ldots, a_n, y)$ be a trait that is a function of the state of a number of loci, a_1 to a_n, as well as a focal locus, y, the effects of which we will examine. We will follow Hansen and Wagner (2001) in measuring the genotypic state of all the loci on a scale set by the phenotypic effect they will have if substituted into a given reference genotype. That is, $y = z(a_1, \ldots, a_n, y) - z(a_1, \ldots, a_n, 0)$, where all the other loci are at their reference values, which by definition are $a_i = 0$ for all i. We will take the reference values of the *as* to be their population means. Let the relative fitness of the phenotype, z, be $w(z)$, and consider this as a function of the state of our focal locus. In the appendix, we show that the change in relative fitness due to substituting $y = y$ for $y = 0$ in an epistatic trait architecture is

$$s \approx \beta y - \gamma^2\left(\varepsilon_1 + \frac{1}{2}\delta_1\right)V_A y - \frac{1}{2}\gamma^2((1+\delta)^2 + \varepsilon_2 V_A)y^2 + \beta\left(\delta + \left(\varepsilon_3 + \frac{\delta_2}{2}\right)y\right)y,$$

where β and $-\gamma^2$ are the first- and second-order selection gradients on the trait, and V_A is its additive genetic variance. The epsilons and the deltas, defined in the appendix, are measures of patterns of epistasis and linkage disequilibrium. The ε_1 is a measure of the directional epistatic contribution of the change. This is the measure of y's effect on evolvability; it will be positive if the change y tends to elevate the effects of other loci, and negative if y tends to reduce the effects of other loci. The directional epistatic ε-parameter discussed in section 7.2.4 is a weighted average of the ε_1 across all loci affecting the trait. The ε_2 is a measure of the magnitude of the epistatic modifications due to y; it will usually be positive. The ε_3 measures whether the directionality of the epistatic modifications matches the linkage disequilibrium between y and the loci it modifies. The δ is the summed linkage disequilibrium between y and the *as*, so that the product δy is a measure of how much carriers of the y-genotype differs from other individuals due to disequilibrium with other loci. The δ_1 and δ_2 measure how epistasis among the a-loci matches patterns of linkage disequilibrium.

The four terms in the equation illustrate four distinct forces of selection on our focal locus. The first term, βy, describes an *adaptive* force due to trait adaptation. An allelic substitution is favored if it changes the trait in the direction of selection, and as discussed above, this may generate indirect selection on evolvability. This force is likely to dominate the dynamics if the trait is not at a fitness equilibrium, and its effect on evolvability will depend on the directionality of epistasis (ε_1). The second term describes a *canalizing* force. An allelic substitution with a net canalizing effect ($\varepsilon_1 < 0$) will be favored under stabilizing selection ($\gamma^2 > 0$), while a substitution with a net decanalizing effect (i.e., increasing evolvability ($\varepsilon_1 > 0$)) will be favored if there is positive curvature ($\gamma^2 < 0$) in the fitness landscape. There is also a canalizing effect due to directional third-order fitness epistasis between y and pairs of a-loci (the δ_1 term). Under stabilizing selection, this will favor changes in a direction opposite to the direction of epistasis between loci in positive linkage disequilibrium. The third term describes a *conservative* force acting against any change under stabilizing selection ($\gamma^2 > 0$). Because it increases with the square of the effect of the change (i.e., with y^2), the conservative force will overpower the other forces when the effect size of the mutation increases, and thus block mutations above a certain size from participating in both adaptation and systems drift under stabilizing selection. For example, ignoring linkage disequilibrium and directional selection, a canalizing mutation cannot be favorable if its phenotypic effect, y, exceeds $2\varepsilon_1 V_A$, which is quite strict. If we assume that the mean-scaled additive variance (i.e., the evolvability) is 0.1%, then a mutation with an average 10% modification of other loci would have to have a phenotypic effect less than 1% to have any possibility of being favored under stabilizing selection.

These three forces were named in Le Rouzic et al. (2013), who derived them for a general multilinear epistatic architecture but without linkage disequilibrium (see also Hermisson et al. 2003). The fourth term describes a *hitchhiking* force due to linkage disequilibrium between the focal change and favorable alleles (δ and ε_3), or favorable combinations of alleles (δ_2), at other loci. Any new mutation will necessarily appear in a particular genetic background, and its initial dynamics, and hence invasion probability, will be influenced by this association. Under free recombination, the initial association is rapidly broken down, but random linkage disequilibrium may still affect its dynamics, as in the Hill-Robertson effect (Felsenstein 1974). While these mechanisms will have indirect and haphazard effects on evolvability, we can also recognize the gene-level adaptation discussed in section 7.2.2 as deriving from this type of hitchhiking. If the focal evolvability-enhancing allele is causally involved in making its associated alleles more favorable, then the hitchhiking can be considered as direct selection for evolvability. This may happen directly through epistatic modification of the associated alleles (the ε_3 term) or through modification of mutation or recombination rates that increases the chance of the focal allele becoming associated with something adaptive. To be effective, such adaptive hitchhiking requires some mechanism for maintaining the specific association. This can come about through tight linkage, population structure, or selection. Pavličev et al. (2011) presented a model of how epistatic modifiers of multivariate genetic variation can be maintained in linkage disequilibrium with their target genotype by selection in the face of recombination, thereby generating changes in the G-matrix that match patterns of directional selection (see also Wagner and Bürger 1985).

To fully understand the evolution of genetic changes that modify evolvability, we must consider all these forces, as well as genotype-by-environment interaction and genetic drift (see sections 7.2.6 and 7.2.7 below). For example, while stabilizing selection on the trait induces canalizing selection on the underlying loci, this force is unlikely to drive evolvability to zero, because it will conflict with other forces (Hermisson et al. 2003; Le Rouzic et al. 2013). In particular, as the canalizing force weakens with reduced additive variance, while the conservative force is less affected, there will be a lower limit to the canalization that can be achieved. This lower limit will be larger if stabilizing selection is stronger, and we get the counterintuitive result that stronger stabilizing selection may lead to a less-canalized genetic architecture with larger mutational effects (Wagner et al. 1997; Le Rouzic et al. 2013).

Hence, two findings from this analysis are that evolvability in the sense of allelic (and mutational) effect sizes (1) is likely to evolve in idiosyncratic manners that depend on details of genetic architecture and patterns of selection, and (2) will be robustly maintained at a nonzero level due to haphazard indirect selection, the conservative force, and inevitable mutation bias against perfect canalization. The analysis further identifies two possible mechanisms for adaptive increase of evolvability. One is through hitchhiking with favorable alleles generated by the evolvability-enhancing mechanism (the ε_3 term), and the other is through decanalizing selection in a convex fitness landscape (the ε_1 term). As fitness landscapes are more likely to be concave when populations are well adapted (close to fitness peaks), the latter force may normally act to reduce evolvability through adaptive canalization, but there may be situations in changing or fluctuating environments in which the population is temporarily in convex (or less concave) areas of the fitness function, which may act to elevate evolvability relative to more stable environments (Le Rouzic et al. 2013; see also Layzer 1980).

7.2.6 Congruence

Genetic effects can also evolve due to associations with environmental variation. A paradigmatic example is Haldane's theory for dominance evolving as a side effect of selection for the wild-type allele to be robust against unusual environmental conditions (Bagheri 2006). This example has been generalized to the proposal that genetic canalization evolves as a side effect of environmental canalization (Wagner et al. 1997; Ancel and Fontana 2000; Meiklejohn and Hartl 2002; de Visser et al. 2003). This *congruence hypothesis* is based on the idea that genetic and environmental robustness may result from similar physiological mechanisms, and if selection favors robustness against environmental perturbations, then it will indirectly favor robustness against genetic perturbations (i.e., allele substitutions). In general, robustness will be favored in concave fitness landscapes, and environmental variation will then add a force of indirect canalizing selection on genetic effects similar to the (genetic) canalizing force discussed above.

7.2.7 Neutral Evolution of Evolvability

Michael Lynch (e.g., 2005, 2007) has argued that many aspects of genome architecture are determined by genetic drift and mutation pressure. The key to his argument is that weak selection is inefficient in small populations. A common rule of thumb is that fitness differences need to be larger than $1/4N_e$ to dominate genetic drift and mutation pressures. To see what this means, consider that the ratio between the fixation probabilities of an advantageous and a disadvantageous allele with a (heterozygous) fitness difference of s is approximately e^{4sNe} (Bürger and Ewens 1995). Hence, if $s = 1/4N_e$, then the ratio of the fixation probabilities is merely $e^1 \approx 2.71$, which would allow frequent invasions of the deleterious allele and not be sufficient to overcome even mild differences in mutation rate. Increasing either the fitness difference or the effective population size by an order of magnitude, however, would increase the ratio of the fixation probabilities by three orders of magnitude and make selection very powerful relative to drift. Lynch has argued that the relevant measures of N_e for many multicellular organisms, like plants and animals, are often quite small, usually less than 10,000. This means that genotypes with fitness differences below a few hundredths of a percent or so will be practically indistinguishable by selection. Even fitness differences of a percent or more may be dominated by drift, mutation, or rare migration in local populations of large-bodied organisms.

These facts limit the potential for fine-tuning genetic architecture in multicellular organisms. If we ignore linkage disequilibrium, we can rewrite the canalizing force discussed above as $2\varepsilon_1 yL$, where L $(=\gamma^2 V_A/2)$ is the load generated by the curvature in the fitness function, and $\varepsilon_1 y$ is the average modification of the effect of substitutions at other loci. If the trait generates a strong fitness load of 10%, and our substitution generates a large 10% average modification of other loci, we see that the canalizing force generates a fitness advantage of

$s = 2\%$ for the canalizing allele, which would only need $N_e = 12.5$ to generate a ratio of fixation probabilities equal to e^1 and be effective in generating canalizing adaptations with effective population sizes above 100 or so. These numbers are only realistic, however, if the focal locus can modify many loci in a consistent manner. An epistatic modification of 10% of one other locus out of say 100 affecting the trait, would make $\varepsilon_1 y = 0.001$, and $s = 0.02\%$, which would require $N_e = 1250$ just to generate a ratio of fixation probabilities equal to e^1. In practice, then, there is little room for adaptive canalization (or decanalization) on a locus-by-locus basis in multicellular organisms. If adaptive canalization of polygenic traits happens at all, it must be through systemwide modifications that allow the simultaneous change of many loci at once. This point has been argued by Proulx and Phillips (2005) based on related considerations, which they extend to other aspects of the evolution of genetic architecture, such as the evolution of dominance and the invasion of gene duplications.

According to Lynch (2007), the large, redundant, and complex genomes of multicellular plants and animals can be explained by the accumulation of mildly deleterious changes that slip under the resolution of selection. Such changes include the invasion and subfunctionalization of genes after duplication, the expansion and modularization of regulatory sites, the expansion of introns, and the proliferation of transposable elements. All these processes may facilitate evolvability. For example, the invasion and subfunctionalization of duplicated genes is likely to be slightly deleterious but sets up a potential for subdivision and specialization of function that can be used to provide more refined adaptation. The expansion of gene families and regulatory elements provides for a richer toolbox to be used for future evolution. In these cases, evolvability evolves as a contingent side effect of the changes in genome architecture, but unlike the mechanisms discussed above, this is not due to indirect selection deriving from other functions but is due to the near-neutral accumulation of slightly deleterious changes.

Some limitations of the efficacy of selection are also likely to hold for unicellular and small-bodied organisms, as increasing population sizes may increase the frequency of selective sweeps that cause stochastic changes on linked loci with effects similar to genetic drift. Gillespie (2000) has referred to this as genetic draft, and it puts upper limits on the (long-term) effective population size, even when the census size is practically infinite (Lynch 2007).

Systems drift is another neutral mechanism relevant to the evolution of evolvability (e.g., True and Haag 2001; Hahn et al. 2004; McCandish 2018). If a character under stabilizing selection has a polygenic architecture, then many different genotypes may generate the same optimal character state. If the subspace formed by these genotypes is connected through genetic steps that slip under the resolution of selection, then neutral evolution can proceed in this subspace. As the different genotypes in the subspace may have different variational properties (i.e., different mutational spectra), we have the potential for evolution of evolvability without changes in the trait itself.

Many have argued that the neutral exploration of such subspaces, or neutral networks, allows evolution to find or poise itself for new innovations (Kauffman 1993; Schuster et al. 1994; A. Wagner 2005, 2008; Payne and Wagner 2014). This mechanism sets up the possibility of a positive relationship between robustness and evolvability, in that increasing robustness will decrease the phenotypic differences between genotypes, which will increase the size of near-neutral subspaces. More robust characters or genotype-phenotype maps may therefore be more evolvable in the sense of being able to better explore their genotypic neighborhood. Systems drift is also the main mechanism for the evolution of postzygotic reproductive isolation through the Bateson-Dobzhansky-Muller process, and may thus facilitate local adaptation, speciation, and species selection by reducing gene flow between incipient species.

7.2.8 System-Level Contingency

Many aspects of evolvability are not character specific but outcomes of general organismal architecture or population properties. Species- or population-level properties, such as population density, distribution range and structure, mating system, reproductive rates, and modes of dispersal, will affect the evolvability of the whole organism, as will rates and accuracy of development, modes of reproduction and inheritance, and mechanisms of homeostasis and plasticity. The evolution of any such property will generate contingent changes in the evolvability of specific characters. We will discuss some of these properties in section 7.3, and many others are discussed in other chapters of this book. Here we just emphasize the distinction between character-specific and general evolvability. One fundamental aspect of the evolution of evolvability is the evolution of distinct characters in the first place. The emergence of a variationally quasi-independent character also implies the emergence of a quasi-independent character-specific evolvability, which is likely to be contingent on the developmental origin of the character (Wagner 2014).

Changes in the developmental or genetic system may also be instrumental in the maintenance of evolvability when they become integrated into the body plan or life cycle of the organism in ways that cannot be easily undone. Major transitions such as multicellularity; sexual reproduction; and the evolution of organizers, axes, or symmetries in the body plan are likely to get "burdened" when other traits are organized around them, and their effects on evolvability are then frozen and maintained, even if these effects by themselves become unfavorable to the individual organisms or populations carrying them. A genetic instantiation of this process is the maintenance of duplicated genes after subfunctionalization has rendered the duplicates complementary and both essential.

7.3 Organismal Properties Related to Evolvability and Their Evolution

Table 7.1 lists some properties that may influence evolvability with suggested modes of evolution. Many of these properties are discussed elsewhere in this book, and here we only provide a few selected reviews to illustrate the application of our theoretical concepts and perspectives.

7.3.1 Sex and Recombination

The realization that sexual recombination can produce new variation transcending current phenotypic ranges was perhaps the first major insight in evolvability in the modern synthesis, and it was instrumental in the acceptance of the efficacy of natural selection (e.g., Beatty 2016). Consequently, and in fact going back as far as Weismann (1889), the idea arose that the function of sexual reproduction was to generate variation for natural selection to act on (e.g., Dobzhansky 1937). Although sexual recombination is crucial for the maintenance of evolvability in multicellular organisms, this is not a sufficient explanation for its evolution. There are individual-level costs to sexual reproduction, and its maintenance

Table 7.1
Properties that influence evolvability with suggested modes of evolution of evolvability

Property	Effect on evolvability	Mode of evolution
Variation[5,6,12,13,14]	Allows selection	*G, MS, C, G*
Mutability[5,7]	Source of variation	
Mutation rate[7]		A_i, *S*, A_g
Mutation effect[7,11]		*C, K*, A_i
Mutational target[7]		*N, C*, A_i
Recombination[7]	Generates new variation. Allows complex adaptation	A_g, A_i, *S*
Mating system[12]	Affects maintenance of variation	*S*, A_g
Symmetry[7,10,15,17]	Constraint and facilitator	*S, K*
Modularity[7,8,15,17]	Facilitates quasi-independence	
Cis regulation[9]		*C, N*
Pleiotropy[8,9,10,16]		*C, K*, A_i
Char. identity[15]		*S*
Continuity/fidelity[7,15]	Allows complex adaptation	*C, S, K*
Robustness[8,11,16]	Allows systems drift and hidden variation	*G, K*
Plasticity[5,13]	Capacitance, Baldwin effect	*C*
Epistasis[7,8,9,10]	Allows evolution of gene effects	*C, K*
Individuality[16]	Transition of selection level, facilitates specialization	*S*, A_g

Notes: The letter *A* signifies evolving as an adaptation for evolvability, with A_i specifying individual or gene-level adaptation and A_g specifying group or species adaptation. Contingent evolution is indicated by *C* when the indirect selection stems from the trait in question and by *S* when the indirect selection stems from systems-level properties (including selective constraints and burden). Neutral evolution is indicated by *N*, canalizing selection by *K*, congruent selection by *G*, and mutation-selection balance by *MS*. Relevant chapters are indicated by superscripted numbers: 5 Hansen, 6 Houle and Pélabon, 7 this chapter, 8 Pavličev et al., 9 Hallgrímsson et al., 10 G. Wagner, 11 A. Wagner, 12 Sztepanacz et al., 13 Pélabon et al., 14 Voje et al., 15 Armbruster, 16 Galis, 17 Jablonski.

requires powerful selection pressures or constraints (Williams 1975). Given that obligate asexuals are rare and phylogenetically short-lived, the maintenance of sex has been called the queen of problems in evolutionary biology (Bell 1982). Despite much research and a multitude of hypotheses, no complete consensus has appeared (Hartfield and Keightley 2012).

The many hypotheses for the evolution of sex and recombination span most modes of evolution that we have discussed above. The most direct link to evolvability is found in the Red Queen hypothesis, which in its original formulation explains the maintenance of sex as a group- or species-level adaptation for evolvability in a changing biotic environment (Maynard Smith 1978). The hypothesis subsequently has become more focused on sex as an adaptation to deal with arms races with evolving parasites that tend to adapt to common genotypes, thus giving an advantage to rare and novel genotypes. In some of these formulations, the advantage is more on the individual than on the population level, and they would have been better labeled as cases of the tangled-bank hypothesis than as cases of the classic Red Queen hypothesis. According to the tangled-bank hypothesis (Ghiselin 1974; Bell 1982), sex is maintained as an adaptation for producing variable offspring to increase the chances that some are successful in an uncertain environment. We have outlined how this could be constructed as an individual- or gene-level adaptation for evolvability, but note that some related individual-advantage hypotheses for sex, such as reducing local or sibling competition, do not locate the advantage in evolvability, which would then evolve as a side effect. Nevertheless, some form of higher-level selection remains a plausible mechanism for the maintenance of sex (e.g., Nunney 1989). Furthermore, sex and recombination as adaptations for evolvability on some level is supported by their tendency to be more frequent in, and sometimes being induced by, stress and environmental degradation.

Some hypotheses for the maintenance of sexual recombination are based on advantages to breaking up linkage disequilibrium (Felsenstein 1974; Otto 2009; Hartfield and Keightley 2012). These advantages may include reduction of the mutation load in the presence of synergistic epistasis among deleterious alleles (the deterministic-mutation hypothesis), a reduction of the fixation load (Muller's ratchet), or a reduction of selective interference between advantageous alleles (the Fisher-Muller hypothesis). In each case, some form of group-level advantage seems plausible. For the Fisher-Muller hypothesis, the population advantage is in terms of elevated evolvability, which is then maintained as a group adaptation. For the other hypotheses, the link to evolvability is less obvious, but one can view the proposed advantages as facilitating the maintenance of complex adaptations, and to the extent that allowing the maintenance of complex adaptations is seen as an aspect of evolvability, the deterministic-mutation and Muller's-ratchet hypotheses also explain the maintenance of sex as an adaptation for evolvability. The tendency for sex to be less common in marginal environments can also be seen in this light as an adaptation to protect local adaptations from dilution due to external gene flow.

Sexual recombination is fundamentally integrated with organismal architecture. There are associations with meiosis, replication and DNA repair on the cellular level, and with dispersal and life-cycle stages on the organismal level. The evolution of evolvability may be constrained by all these factors. In some cases, this may result in absolute constraints or a "burden" that maintains sex and evolvability in the face of short-term costs on the individual level. The fact that there are no known asexual mammals may reflect a particularly severe constraint against parthenogenetic development in this clade (perhaps due to gamete-specific imprinting). Increased recombination may also evolve through near-neutral expansions of genomes rendering more complex, larger-bodied organisms more evolvable than microorganisms with leaner genomes (Lynch 2007).

7.3.2 Coordinated Variation (Continuity, Modularity, Symmetry)

Lewontin (1978) suggested that adaptive evolution requires two preconditions that he called continuity and quasi-independence. Continuity "means that small changes in a characteristic must result in only small changes in ecological relations" (i.e., fitness). Quasi-independence "means that there is a great variety of alternative paths by which a given characteristic may change, so that some of them will allow selection to act on the characteristic without altering other characteristics of the organism" (i.e., modularity). These two properties can be seen as manifestations of a more general property, which we call coordinated variation. The key point is that biological organisms are so complex and multidimensional that their variations must be organized, partitioned, and channeled to be usefully selected. Selection cannot optimize thousands of parts, genes, and traits simultaneously, and the potential for selective interference grows multiplicatively with increasing complexity. The only way to optimize many parts without eliminating variation altogether is to organize the variation along a limited number of functional lines (Wagner and Altenberg 1996). How this comes about is perhaps the most difficult question in evolvability research. The many contingent modes of evolution render it almost paradoxical, and we regard this problem as substantially unsolved.

Riedl (1978) suggested that complex organisms are evolvable because their variational and functional interdependencies are congruent rather than in conflict. He called this principle the "immitatory epigenotype," because the developmentally integrated parts of the organism (the "epigenotype") "imitates" functional interdependencies among the parts. This leads to the question of whether direct selection for evolvability is responsible for the correspondence between functional and variational constraints. This question has been explored in the context of the evolution of modularity. Variational modularity is a statement about the distribution of environmental and genetic perturbations on a set of traits. A variational pattern is said to be modular if there are sets of traits that more likely vary together and quasi-independently from other traits. From the genetic point of view, modularity can be seen as a pattern of pleiotropy, in which certain genes primarily affect a subset of phenotypic traits and others less. The nature and distribution of pleiotropic effects are themselves genetically influenced and therefore evolvable, which raises the question of what evolutionary forces shape the pattern of pleiotropy (Pavličev and Cheverud 2015). Wagner (1996) proposed that functional modularity could emerge from a combination of stabilizing and fluctuating directional selection. Later, many studies have found that pleiotropy or mutational variation can potentially evolve to align with patterns of selection/function (e.g., Hansen 2003; Kashtan and Alon 2005; Jones et al. 2007, 2014; Draghi and Wagner 2008; Pavličev et al. 2011; Melo and Marroig 2015). The generality of such results is unclear, however. The evolution of pleiotropy is a special case of the evolution of gene effects and thus subject to all the forces we have discussed above. Due to the conservative force discussed in section 7.2.5, strong stabilizing selection may block canalization, and the relationship of mutational canalization to strength of stabilizing selection is nonlinear with a minimum at intermediate strengths of selection (Wagner et al. 1997; Hermisson et al. 2003; Le Rouzic et al. 2013). As a result, we expect a nonlinear relation of evolvability to strength of stabilizing selection across directions in morphospace. In the presence of directional selection, the evolution of multivariate evolvability will be determined by multivariate patterns of directional epistasis, which makes a simple alignment with patterns of selection unlikely. In many models of the evolution of modularity, the outcome is as much influenced by internal constraints or biased sets of alternatives as by the pattern of selection acting on the phenotype (Gardner and Zuidema 2003; Hansen 2011; Guillaume and Otto 2012).

Body symmetries and metamerisms are examples of structural features coordinating variation and thus facilitating evolvability in some dimensions while constraining it in others (e.g., Jablonski 2020). Such symmetries may evolve for functional reasons, which may cause the evolution of functionally organized evolvability by indirect selection. For example, variation along a left-right axis could become canalized due to selection for increased developmental robustness if left-right differences are largely deleterious in an elongated moving organism. A likely side effect is an "immitatory" canalization of genetic and mutational effects, which will symmetrize and facilitate evolvability. A related example emphasized by Gerhart and Kirschner (1997) involves tissue organization, in which the ability of a growing organ to recruit vascularization to ensure an adequate supply of oxygen and nutrients according to need is functional in terms of allowing growth and accommodating size differences in the organ. Selection for such recruiting mechanisms will then indirectly select for evolvability, as they allow evolutionary changes in the size of the organ without the need for genetic changes in vascularization. We can recognize these examples as instances of congruent evolution, and note how congruence may allow organization of genetic variability along functional lines by means of indirect selection. Conversely, congruence may also reduce evolvability along functionally important axes due to canalizing selection in these directions.

As for the continuity or smoothness of the genotype-phenotype map, it is clear that the evolution of complex adaptations requires a supply of small-effect modifications that combine at least partly additively. Although it is an empirical fact that many genotype-phenotype maps are continuous and order-preserving in this sense (e.g., Gjuvsland et al. 2011), it is unclear why they have these properties. Kauffman (1993) argued that these properties are not expected in complex random interaction networks and inferred that some level of developmental and gene-regulatory order was necessary for evolvability. Based on the above theory, it seems likely that this would require systematic canalizing selection. Some studies have also found that epistatic interactions, and thus the ruggedness of the genotype-phenotype map, tend to be canalized under selection, thus favoring the evolution of a degree of additivity (Hermisson et al. 2003; Hansen et al. 2006; Le Rouzic et al. 2013). Continuity is also a function of the fidelity of inheritance, and it is thus preconditioned on the establishment of an accurate replication mechanism.

7.3.3 Mutability

While segregating variation in sexual populations normally holds potential for change far outside the original range, there comes a point at which new mutations are needed for further change. The amount of mutational input is variable across traits and taxa (Houle 1998) and may change either through changes in the phenotypic effects of new mutations or through their rate of appearance.

The evolution of mutational effects, as opposed to rates, is a special case of the evolution of allelic effects and is largely determined by patterns of epistasis. When the genetic background is changing, effects of new mutations will change in accordance with their epistatic relation to the genetic background (e.g., Hansen et al. 2006). A more specific mechanism for the evolution of mutational effects (including pleiotropy) is in terms of "inherited allelic effects" (Hansen 2003, 2006, 2011). This is a form of intralocus epistasis in which subsequent mutations on the same allele inherit the variational properties established by previous mutations. For example, a mutation establishing a new cis-regulatory module may not only cause expression of the gene in a novel context but may also change the mutational spectrum to make subsequent mutations in the same allele more likely to have effects in this novel context. In this case, there is an automatic association between trait effects and mutability, which facilitates the evolution of evolvability and may even lend itself to gene-level adaptation for evolvability, as outlined in section 7.2.2. The adaptation is not automatic, however, and favorable trait changes may also be achieved through elimination of regulatory elements, which may reduce evolvability.

The evolution of mutation rates is perhaps the best-studied case for potentially adaptive evolution of evolvability (Good and Desai 2016). From microbial systems, there is evidence for elevated mutation rates in stressful or novel environments (Cox and Gibson 1974; Radman et al. 1999; Metzgard and Wills 2000; Caporale 2003; Galhardo et al. 2007;

Diaz-Arenas and Cooper 2013). The extent to which this is an evolved adaptation for evolvability or just a side effect of stress on replication fidelity is debatable (e.g., Sniegowski and Murphy 2006), but our criterion for adaptive evolution can apply to modifiers of mutation rate, because a hypermutator allele will directly cause and (in bacteria) stay linked to elevated mutation rates and may thus be favored in situations in which evolvability is advantageous.

Still, there are more deleterious than advantageous mutations, and therefore selection to improve replication fidelity and reduce rates of mutation could be effective. Such canalizing selection will not drive mutation rates to zero, however, as the supply and effect of antimutator alleles must decrease with increasing canalization and hit a balance with increasing mutation bias toward increased rates (Lynch 2008). Mutation rates are also influenced by physiology and genome architecture, which expose them to indirect selection that may overwhelm direct selection for evolvability. The large patterns in the evolution of mutation rates are therefore likely to be contingent side effects of other changes. For example, transposable elements may increase mutation rates, and although it has been argued that transposable elements are maintained to favor evolvability (McClintock 1984), it is difficult to rule out alternatives, such as the elements acting as genomic parasites replicating for their own benefit and causing mutations as a side effect (Sniegowsky and Murphy 2006; Lynch 2007). This possibility is also supported by the existence of evolved systems (e.g., pi-RNAs and possibly methylation) to suppress such elements (Zemach et al. 2010; Kofler 2019). Although such systems could possibly be exapted into capacitors that could release dormant elements to boost evolvability in times of stress, it seems likely that their primary adaptive function is to suppress the elements.

The major determinant of trait-specific mutation rates may be the number of genes or genomic positions that potentially affect the trait. Houle (1998) put forward the hypothesis that trait differences in standing additive variation are largely determined by the mutational target size of the trait. This hypothesis goes a long way toward explaining why complex traits, such as fitness, life-history, and behavior, are more evolvable than simple morphological traits, and it suggests that the recruitment and elimination of genes affecting a trait are crucial to the long-term evolution of its evolvability. Genes may be recruited or eliminated by regulatory evolution, gene duplication, subfunctionalization, and pseudogenization, which are all subject to contingent changes unrelated to evolvability. For example, the maintenance of duplicated genes may depend less on future evolvability than on accidental subfunctionalization that renders both copies necessary (Force et al. 1995).

The impact of mutations on evolvability depends on how well they are maintained in the population. Selection acts on standing variation, and the relationship of standing variation to mutation is complicated and the subject of a huge literature, which we will not review here. Standing variation in mutation-selection balance not only depends on the mutation rate but also on genetic architecture in the form of mutational effect sizes, number of loci, epistasis, dominance, and pleiotropy. It is further influenced by external factors, such as the mode and strength of selection, population size and structure, recombination rates, mating system, and environmental variation. All these are prone to change for reasons unrelated to evolvability, which sets up contingent change as a major factor in the evolution of population evolvability as captured by the G-matrix.

7.4 Conclusion

After research on the evolution of evolvability started in earnest some 30 years ago (see Nuño de la Rosa 2017, and chapters 2 and 3[1]), there have been many advances in terms of hypotheses, concepts, and mathematical formalism. The field has progressed from loose verbal models and computer simulations with tenuous connections to biological facts to theory well grounded in mathematical population genetics, evolutionary theory, and molecular biology. This research has shown that there is nothing paradoxical about the evolution of evolvability and that it does not require any special higher-level selection to work. It is possible for the evolution of evolvability to proceed by conventional within-population selection at the gene or individual level. This does not exclude group- or lineage-level selection, however, which remains relevant in many specific cases (see table 7.1). It has further become clear that evolvability is susceptible to various forms of indirect selection and contingencies, and the degree to which organisms are adapted for evolvability is unsettled. Systems drift and other forms of (nearly) neutral evolution may also be important. Many aspects of evolvability are fundamentally integrated with organismal body plans and life cycles. They are thus subject to deep constraints and major transitions that must be considered for a full understanding of the subject. Finally, the genotype-phenotype map has emerged as the focal determinant of evolvability, and it has been made clear that epistasis controls the evolution of evolvability on the proximate level.

Empirical research is lagging, however. There has been progress in the empirical understanding of patterns, determinants, and consequences of evolvability through research on the developmental, physiological, and molecular basis of the genotype-phenotype map, on the quantification of genetic and mutational variation, and on molecular changes in experimental evolution. But there has been less research that directly addresses the evolution of evolvability. We only have anecdotal information about actual evolution of evolvability in nature, and selection experiments have rarely been used—and almost never set up—to test hypotheses about the evolution of evolvability (but see A. Wagner, chapter 11). There is a shortage of relevant information on crucial parameters, such as patterns of epistasis and pleiotropy. We also lack quantitative comparative studies of evolvability, which would be essential to test the theory.

7.5 Appendix

Let the relative fitness of a genotype $\{a_1, \ldots, a_m, y\}$ with phenotype z be $w(z(a_1, \ldots, a_m, y))$. The marginal fitness of the genotype y at the focal locus is then $\mathrm{E}[w(z(a_1, \ldots, a_m, y)) \mid y]$, where the expectation is taken over the values of the vector $\mathbf{a} = \{a_1, \ldots, a_n\}$. Let the variance matrix of this vector be $\mathbf{A}$, and take the mean of the a_i as reference genotype. For simplicity, assume that $\mathbf{A}$ does not depend on y, but we allow the possibility of the change y being in linkage disequilibrium with the a-vector by assuming $\mathrm{E}[\mathbf{a} \mid y] = \mathbf{b}_{\mathbf{ay}} y$, where $\mathbf{b}_{\mathbf{ay}}$ is the vector linear gradient ("regression") of $\mathbf{a}$ on y. By a second-order Taylor approximation around the reference value of $\mathbf{a} = 0$, we get

1. References to chapter numbers in the text are to chapters in this volume.

$$\begin{aligned}
\mathrm{E}[w(z(\mathbf{a},y)\,|\,y] &\approx w(z(\mathbf{0},y)) + \left(\frac{\partial w(z(0,y))}{\partial \mathbf{a}}\right)^{T} \mathrm{E}[\mathbf{a}\,|\,y] + \frac{1}{2}\mathrm{E}\left[\mathbf{a}^{T}\frac{\partial^{2} w(z(0,y))}{\partial \mathbf{a}\,\partial \mathbf{a}^{T}}\mathbf{a}\,|\,y\right] \\
&= w(z(\mathbf{0},y)) + \left(\frac{\partial w(z(0,y))}{\partial \mathbf{a}}\right)^{T} \mathbf{b}_{\mathbf{a}y}y + \frac{1}{2}\mathrm{Tr}\left(\frac{\partial^{2} w(z(0,y))}{\partial \mathbf{a}\,\partial \mathbf{a}^{T}}\mathbf{A}\right) \\
&\quad + \frac{1}{2}\mathbf{b}_{ay}^{T}\frac{\partial^{2} w(z(0,y))}{\partial \mathbf{a}\,\partial \mathbf{a}^{T}}\mathbf{b}_{ay}y^{2},
\end{aligned}$$

where Tr is the trace function, and $\frac{\partial^{2} w(z(0,y))}{\partial \mathbf{a}\,\partial \mathbf{a}^{T}}$ is the Hessian matrix of $w(z(\mathbf{a}, y))$ with respect to a evaluated at $\mathbf{a}=0$. A second-order Taylor approximation of fitness with respect to y around $y=0$ now gives

$$\begin{aligned}
\mathrm{E}[w(z(\mathbf{a}, y)\,|\,y] &\approx \mathrm{w}_{0} + \frac{\partial w}{\partial y}y + \frac{1}{2}\frac{\partial^{2} w}{\partial y^{2}}y^{2} + \left(\frac{\partial w}{\partial \mathbf{a}}\right)^{T}\mathbf{b}_{\mathbf{a}y}y + \left(\frac{\partial^{2} w}{\partial y\,\partial \mathbf{a}}\right)^{T}\mathbf{b}_{\mathbf{a}y}y^{2} \\
&\quad + \frac{1}{2}\mathrm{Tr}\left(\left(\frac{\partial^{2} w}{\partial \mathbf{a}\,\partial \mathbf{a}^{T}} + \frac{\partial^{3} w}{\partial y\,\partial \mathbf{a}\,\partial \mathbf{a}^{T}}y + \frac{1}{2}\frac{\partial^{4} w}{\partial y^{2}\partial \mathbf{a}\,\partial \mathbf{a}^{T}}y^{2}\right)\mathbf{A}\right) \\
&\quad + \frac{1}{2}\mathbf{b}_{ay}^{T}\frac{\partial^{2} w}{\partial \mathbf{a}\,\partial \mathbf{a}^{T}}\mathbf{b}_{ay}y^{2} + o(y^{2}),
\end{aligned}$$

where we have simplified the notation and all differentials are evaluated at the reference genotype $\{\mathbf{a}, y\} = \{\mathbf{0}, 0\}$. The change in marginal fitness due to substituting a $y=y$ for a $y=0$ genotype is then

$$\begin{aligned}
\Delta w &= \mathrm{E}[w(z(\mathbf{a},y)|y] - \mathrm{E}[w(z(\mathbf{a},0)|y=0] \\
&\approx \frac{\partial w}{\partial y}y + \frac{1}{2}\frac{\partial^{2} w}{\partial y^{2}}y^{2} + \left(\frac{\partial w}{\partial \mathbf{a}}\right)^{T}\mathbf{b}_{\mathbf{a}y}y + \left(\frac{\partial^{2} w}{\partial y\,\partial \mathbf{a}}\right)^{T}\mathbf{b}_{\mathbf{a}y}y^{2} \\
&\quad + \frac{1}{2}\mathrm{Tr}\left(\left(\frac{\partial^{3} w}{\partial y\,\partial \mathbf{a}\,\partial \mathbf{a}^{T}}y + \frac{1}{2}\frac{\partial^{4} w}{\partial y^{2}\partial \mathbf{a}\,\partial \mathbf{a}^{T}}y^{2}\right)\mathbf{A}\right) + \frac{1}{2}\mathbf{b}_{ay}^{T}\frac{\partial^{2} w}{\partial \mathbf{a}\,\partial \mathbf{a}^{T}}\mathbf{b}_{ay}y^{2}.
\end{aligned}$$

The relevant partial derivatives of fitness evaluated in the reference genotype are

$$\frac{\partial w}{\partial y} = \frac{\partial w}{\partial z}\frac{\partial z}{\partial y} = \frac{\partial w}{\partial z} \equiv \beta,$$

$$\frac{\partial^{2} w}{\partial y^{2}} = \frac{\partial^{2} w}{\partial z^{2}}\left(\frac{\partial z}{\partial y}\right)^{2} + \frac{\partial w}{\partial z}\frac{\partial^{2} z}{\partial y^{2}} = \frac{\partial^{2} w}{\partial z^{2}} \equiv -\gamma^{2},$$

$$\frac{\partial w}{\partial \mathbf{a}} = \beta\mathbf{1}, \qquad \frac{\partial^{2} w}{\partial y\,\partial \mathbf{a}} = -\gamma^{2}\mathbf{1} + \beta\frac{\partial^{2} z}{\partial y\,\partial \mathbf{a}}, \qquad \frac{\partial^{2} w}{\partial \mathbf{a}\,\partial \mathbf{a}^{T}} = -\gamma^{2}\mathbf{J} + \beta\frac{\partial^{2} z}{\partial \mathbf{a}\,\partial \mathbf{a}^{T}},$$

$$\frac{\partial^{3} w}{\partial y\,\partial \mathbf{a}\,\partial \mathbf{a}^{T}} = \frac{\partial^{3} w}{\partial z^{3}}\mathbf{J} - \gamma^{2}\left(\frac{\partial^{2} z}{\partial \mathbf{a}\,\partial \mathbf{a}^{T}} + \mathbf{1}^{T}\frac{\partial^{2} z}{\partial y\,\partial \mathbf{a}} + \left(\frac{\partial^{2} z}{\partial y\,\partial \mathbf{a}}\right)^{T}\mathbf{1}\right) + \beta\frac{\partial^{3} z}{\partial y\,\partial \mathbf{a}\,\partial \mathbf{a}^{T}},$$

$$\frac{\partial^4 w}{\partial y^2 \partial \mathbf{a} \partial \mathbf{a}^T} = \frac{\partial^4 w}{\partial z^4}\mathbf{J} + \frac{\partial^3 w}{\partial z^3}\left(\frac{\partial^2 z}{\partial \mathbf{a} \partial \mathbf{a}^T} + 2\mathbf{1}^T \frac{\partial^2 z}{\partial y \partial \mathbf{a}} + 2\left(\frac{\partial^2 z}{\partial y \partial \mathbf{a}}\right)^T \mathbf{1}\right)$$

$$-\gamma^2\left(2\frac{\partial^3 z}{\partial y \partial \mathbf{a} \partial \mathbf{a}^T} + 2\left(\frac{\partial^2 z}{\partial y \partial \mathbf{a}}\right)^T \frac{\partial^2 z}{\partial y \partial \mathbf{a}} + \mathbf{1}^T \frac{\partial^3 z}{\partial y^2 \partial \mathbf{a}} + \left(\frac{\partial^3 z}{\partial y^2 \partial \mathbf{a}}\right)^T \mathbf{1}\right)$$

$$+\beta \frac{\partial^4 z}{\partial y^2 \partial \mathbf{a} \partial \mathbf{a}^T},$$

where $\mathbf{J}$ is an $n \times n$ matrix of ones, and $\mathbf{1}$ is a $1 \times n$ vector of ones. As illustrated in the first two equations, we have used the fact that all first derivatives of z with respect to $\mathbf{a}$ and y are unity when evaluated in the reference genotype. For further simplification, let us ignore the third- and fourth-order selection gradients and assume bilinear epistasis, which implies that the only nonzero derivatives of the trait are with respect to linear and bilinear combinations of the y and the a_i. Using this and feeding back into the above equation yields

$$\Delta w \approx \beta y + \beta \mathbf{1}^T \mathbf{b}_{\mathbf{a}y} y - \frac{1}{2}\gamma^2 y^2 - \frac{1}{2}\gamma^2 \mathrm{Tr}\left(\left(\frac{\partial^2 z}{\partial \mathbf{a} \partial \mathbf{a}^T} + \mathbf{1}^T \frac{\partial^2 z}{\partial y \partial \mathbf{a}} + \left(\frac{\partial^2 z}{\partial y \partial \mathbf{a}}\right)^T \mathbf{1}\right)\mathbf{A}\right) y$$

$$-\frac{1}{2}\gamma^2 \mathrm{Tr}\left(\left(\left(\frac{\partial^2 z}{\partial y \partial \mathbf{a}}\right)^T \frac{\partial^2 z}{\partial y \partial \mathbf{a}}\right)\mathbf{A}\right) y^2 + \frac{1}{2}\beta\, \mathbf{b}_{ay}^T \frac{\partial^2 z}{\partial \mathbf{a} \partial \mathbf{a}^T} \mathbf{b}_{ay} y^2 - \frac{1}{2}\gamma^2 \mathbf{b}_{ay}^T \mathbf{J} \mathbf{b}_{ay} y^2$$

$$-\gamma^2 \mathbf{1}^T \mathbf{b}_{\mathbf{a}y} y^2 + \beta\left(\frac{\partial^2 z}{\partial y \partial \mathbf{a}}\right)^T \mathbf{b}_{\mathbf{a}y} y^2.$$

Define the following composite parameters:

$$\varepsilon_1 = \sum_i \frac{\partial^2 z}{\partial y \partial a_i} \frac{\sum_j A_{ij}}{V_A}, \qquad \varepsilon_2 = \sum_i \sum_j \frac{\partial^2 z}{\partial y \partial a_i} \frac{\partial^2 z}{\partial y \partial a_j} \frac{A_{ij}}{V_A}, \qquad \varepsilon_3 = \sum_i \frac{\partial^2 z}{\partial y \partial a_i} b_{a_i y},$$

$$\delta = \sum_i b_{a_i y}, \qquad \delta_1 = \sum_i \sum_{j \neq i} \frac{\partial^2 z}{\partial a_i \partial a_j} \frac{A_{ij}}{V_A}, \quad \delta_2 = \sum_i \sum_{j \neq i} \frac{\partial^2 z}{\partial a_i \partial a_j} b_{a_i y} b_{a_j y},$$

where, as shown in Hansen and Wagner (2001), $V_A = \sum_i \sum_j A_{ij}$ is the additive genetic variance (including "hidden" variation due to linkage disequilibrium). Fitting these parameters into the above equation yields

$$\Delta w \approx \beta y + \beta\left(\delta + \left(\varepsilon_3 + \frac{\delta_2}{2}\right) y\right) y - \gamma^2\left(\varepsilon_1 + \frac{1}{2}\delta_1\right) V_A y - \frac{1}{2}\gamma^2 (1 + \varepsilon_2 V_A + 2\delta + \delta^2) y^2,$$

which is rearranged to obtain the equation in the main text with the notation $s = \Delta w$.

Acknowledgments

Thanks to the members of the evolvability group at the Centre for Advanced Study in Oslo and the evolvability journal club for discussions and comments. Arnaud Le Rouzic, Scott Armbruster, and Christophe Pélabon made helpful comments on the manuscript.

References

Alberch, P. 1991. From genes to phenotype—dynamic-systems and evolvability. *Genetica* 84: 5–11.

Ancel, L. W., and W. Fontana. 2000. Plasticity, evolvability, and modularity in RNA. *Journal of Experimental Zoology B* 288: 242–283.

Bagheri, H. C. 2006. Unresolved boundaries of evolutionary theory and the question of how inheritance systems evolve: 75 years of debate on the evolution of dominance. *Journal of Experimental Zoology B* 306: 329–359.

Beatty, J. 2016. The creativity of natural selection? Part I: Darwin, Darwinism, and the mutationists. *Journal of the History of Biology* 49: 659–684.

Bell, G. 1982. *The Masterpiece of Nature.* Berkeley, CA: University of California Press.

Berg, R. L. 1959. A general evolutionary principle underlying the origin of developmental homeostasis. *American Naturalist* 93: 103–105.

Berg, R. L. 1960. The ecological significance of correlation pleiades. *Evolution* 17: 171–180.

Bürger, R., and W. Ewens. 1995. Fixation probabilities of additive alleles in diploid populations. *Journal of Mathematical Biology* 33: 557–575.

Caporale, L. H. 2003. Natural selection and the emergence of a mutation phenotype: An update of the evolutionary synthesis considering mechanisms that affect genome variation. *Annual Review of Microbiology* 57: 467–485.

Carroll, S. B. 2008. Evo-devo and an expanding evolutionary synthesis: A genetic theory of morphological evolution. *Cell* 134: 25–36.

Carter, A. J. R., J. Hermisson, and T. F. Hansen. 2005. The role of epistatic gene interactions in the response to selection and the evolution of evolvability. *Theoretical Population Biology* 68: 179–196.

Cheverud, J. M. 1982. Phenotypic, genetic, and environmental morphological integration in the cranium. *Evolution* 36: 499–516.

Cheverud, J. M. 1984. Quantitative genetics and developmental constraints on evolution by selection. *Journal of Theoretical Biology* 110: 155–171.

Conrad, M. 1983. *Adaptability: The Significance of Variability from Molecule to Ecosystem.* New York: Plenum Press.

Conrad, M. 1990. The geometry of evolution. *BioSystems* 24: 6–81.

Cox, E. C., and T. C. Gibson. 1974. Selection for high mutation rates in chemostates. *Genetics* 77: 169–184.

Dawkins, R. 1988. The evolution of evolvability. In *Artificial Life: The Proceedings of an Interdiciplinary Workshop on the Synthesis and Simulation of Living Systems,* edited by C. Langton, 202–220. Santa Fe, NM: Addison Wesley.

Dawkins, R. 1996. *Climbing Mount Improbable.* New York: Viking.

de Visser, J. A. G. M., J. Hermisson, G. P. Wagner et al . 2003. Evolution and detection of genetic robustness. *Evolution* 57: 1959–1972.

Diaz Arenas, C., and T. F. Cooper. 2013. Mechanisms and selection of evolvability: Experimental evidence. *FEMS Microbiology Reviews* 37: 572–582.

Dobzhansky, T. 1937. *Genetics and the Origin of Species.* New York: Columbia University Press.

Draghi, J., and G. P. Wagner. 2008. Evolution of evolvability in a developmental model. *Evolution* 62: 301–315.

Felsenstein, J. 1974. The evolutionary advantage of recombination. *Genetics* 78: 737–756.

Force, A., W. A. Cresko, F. B. Pickett, S. R. Proulx, C. Amemiya, and M. Lynch. 2005. The origin of subfunctions and modular gene regulation. *Genetics* 170: 433–446.

Frank, S. A. 2012. Natural selection. IV. The Price equation. *Journal of Evolutionary Biology* 25: 1002–1019.

Galhardo, R. S., P. J. Hastings, and S. M. Rosenberg. 2007. Mutation as a stress response and the regulation of evolvability. *Critical Reviews in Biochemistry and Molecular Biology* 42: 399–435.

Galis, F., and J. A. J. Metz. 2007. Evolutionary novelties: The making and breaking of pleiotropic constraints. *Integrative and Comparative Biology* 47: 409–419.

Gardner, A., and W. Zuidema. 2003. Is evolvability involved in the origin of modular variation? *Evolution* 57: 1448–1450.

Gerhart, J., and M. Kirschner. 1997. *Cells, Embryos and Evolution: Towards a Cellular and Developmental Understanding of Phenotypic Variation and Evolutionary Adaptability.* New York: Wiley.

Ghiselin, M. T. 1974. *The Economy of Nature and the Evolution of Sex.* Berkeley, CA: University of California Press.

Gillespie, J. H. 2000. Genetic drift in an infinite population: The pseudohitchiking model. *Genetics* 155: 909–919.

Gjuvsland, A. B., J. O. Vik, J. A. Wooliams, and S. W. Omholt. 2011. Order-preserving principles underlying genotype-phenotype maps ensure high additive proportions of genetic variance. *Journal of Evolutionary Biology* 24: 2269–2279.

Good, B. H., and M. M. Desai. 2016. Evolution of mutation rates in rapidly adapting asexual populations. *Genetics* 204: 1249–1266.

Gould, S. J. 2002. *The Structure of Evolutionary Theory.* Cambridge, MA: Belknap.

Guillaume, F., and S. P. Otto. 2012. Gene functional trade-offs and the evolution of pleiotropy. *Genetics* 192: 1389–1409.

Hahn, M. W., G. C. Conant, and A. Wagner. 2004. Molecular evolution in large genetic networks: Does connectivity equal constraint? *Journal of Molecular Evolution* 58: 203–211.

Hansen, T. F. 2003. Is modularity necessary for evolvability? Remarks on the relationship between pleiotropy and evolvability. *BioSystems* 69: 83–94.

Hansen, T. F. 2006. The evolution of genetic architecture. *AREES* 37: 123–157.

Hansen, T. F. 2011. Epigenetics: Adaptation or contingency? In *Epigenetics: Linking Genotype and Phenotype in Development and Evolution,* edited by B. Hallgrímsson and H. B. K. Hall, 357–376. Berkeley, CA: University of California Press.

Hansen, T. F., and C. Pélabon. 2021. Evolvability: A quantitative-genetics perspective. *AREES* 52: 153–175.

Hansen, T. F., and G. P. Wagner. 2001. Modeling genetic architecture: A multilinear model of gene interaction. *Theoretical Population Biology* 59: 61–86.

Hansen, T. F., J. M. Alvarez-Castro, A. J. R. Carter, J. Hermisson, and G. P. Wagner. 2006. Evolution of genetic architecture under directional selection. *Evolution* 60: 1523–1536.

Hartfield, M., and P. D. Keightley. 2012. Current hypotheses for the evolution of sex and recombination. *Integrative Zoology* 7: 192–209.

Hermisson, J., and G. P. Wagner. 2004. The population genetic theory of hidden variation and genetic robustness. *Genetics* 168: 2271–2284.

Hermisson, J., T. F. Hansen, and G. P. Wagner. 2003. Epistasis in polygenic traits and the evolution of genetic architecture under stabilizing selection. *American Naturalist* 161: 708–734.

Houle, D. 1998. How should we explain variation in the genetic variance of traits? *Genetica* 102/103: 241–253.

Jablonski, D. 2008. Species selection: Theory and data. *AREES* 39: 501–524.

Jablonski, D. 2020. Developmental bias, macroevolution, and the fossil record. *Evolution & Development* 22: 103–125.

Jones, A. G., S. J. Arnold, and R. Bürger. 2007. The mutation matrix and the evolution of evolvability. *Evolution* 61: 727–745.

Jones, A. G., R. Bürger, and S. J. Arnold. 2014. Epistasis and natural selection shape the mutational architecture of complex traits. *Nature Communications* 5: 3709.

Kashtan, N., and U. Alon. 2005. Spontaneous evolution of modularity and network motifs. *PNAS* 102: 13773–13778.

Kauffman, S. A. 1993. *The Origins of Order. Self-Organization and Selection in Evolution.* Oxford: Oxford University Press.

Kerr, B., and P. Godfrey-Smith. 2009. Generalization of the Price equation for evolutionary change. *Evolution* 63: 531–536.

Kofler, R. 2019. Dynamics of transposable element invasions with piRNA clusters. *Molecular Biology and Evolution* 36: 1457–1472.

Layzer, D. 1980. Genetic variation and progressive evolution. *American Naturalist* 115: 809–826.

Le Rouzic, A., J. M. Álvarez-Castro, and T. F. Hansen. 2013. The evolution of canalization and evolvability in stable and fluctuating environments. *Evolutionary Biology* 40: 317–340.

Lewontin, R. C. 1978. Adaptation. *Scientific American* 239: 212–231.

Lloyd, E. A., and S. J. Gould. 1993. Species selection on variability. *PNAS* 90: 595–599.

Lynch, M. 2005. The origins of eukaryotic gene structure. *Molecular Biology and Evolution* 23: 450–468.

Lynch, M. 2007. *The Origins of Genome Architecture.* Sunderland, MA: Sinauer Associates.

Lynch, M. 2008. The cellular, developmental, and population-genetic determinants of mutation-rate evolution. *Genetics* 180: 933–943.

Lynch, V. J., and G. P. Wagner. 2008. Resurrecting the role of transcription factor change in developmental evolution. *Evolution* 62: 2131–2154.

Maynard Smith, J. 1978. *The Evolution of Sex.* Cambridge: Cambridge University Press.

McCandish, D. M. 2018. Long-term evolution on complex fitness landscapes when mutation is weak. *Heredity* 121: 449–465.

McClintock, B. 1984. The significance of responses of the genome to challenge. *Science* 226: 792–802.

Meiklejohn, C. D., and D. L. Hartl. 2002. A single mode of canalization. *TREE* 17: 468–473.

Melo, D., and G. Marroig. 2015. Directional selection can drive the evolution of modularity in complex parts. *PNAS* 112: 470–475.

Melo, D., A. Porto, J. M. Cheverud, and G. Marroig. 2016. Modularity: Genes, development, and evolution. *AREES* 47: 463–486.

Metzgard, D., and C. Wills. 2000. Evidence for the adaptive evolution of mutation rates. *Cell* 101: 581–584.

Nunney, L. 1989. The maintenance of sex by group selection. *Evolution* 43: 245–257.

Nunney, L. 1999. Lineage selection: Natural selection for long-term benefit. In *Levels of Selection in Evolution*, edited by L. Keller, 238–252. Princeton, NJ: Princeton University Press.

Nuño de la Rosa, L. 2017. Computing the extended synthesis: Mapping the dynamics and conceptual structure of the evolvability research front. *Journal of Experimental Zoology B* 328: 395–411.

Okasha, S. 2006. *Evolution and the Levels of Selection.* Oxford: Oxford University Press.

Okasha, S. 2021. The strategy of endogenization in evolutionary biology. *Synthese* 198: 3413–3435.

Olson, E. C., and R. L. Miller. 1958. *Morphological Integration.* Chicago: University of Press.

Otto, S. P. 2009. The evolutionary enigma of sex. *American Naturalist* 174: S1–S14.

Pavličev, M., and J. M. Cheverud. 2015. Constraints evolve: Context dependency of gene effects allows evolution of pleiotropy. *AREES* 46: 413–434.

Pavličev, M., J. M. Cheverud, and G. P. Wagner. 2011. Evolution of adaptive phenotypic variation patterns by direct selection for evolvability. *Proceedings of the Royal Society B* 278: 1903–1912.

Payne, J. L., and A. Wagner. 2014. The robustness and evolvability of transcription factor binding sites. *Science* 343: 875–877.

Price, G. R. 1972. Extension of covariance selection mathematics. *Annals of Human Genetics* 35: 485–490.

Proulx, S. R., and P. C. Phillips. 2005. The opportunity for canalization and the evolution of genetic networks. *American Naturalist* 165: 147–162.

Radman, M., I. Matic, and F. Taddei. 1999. Evolution of evolvability. *Annals of the New York Academy of Science* 870: 146–155.

Raff, R. A. 1996. *The Shape of Life: Genes, Development, and the Evolution of Animal Form.* Chicago: University of Chicago Press.

Riedl, R. 1977. A systems-analytical approach to macro-evolutionary phenomena. *Quarterly Review of Biology* 52: 351–370.

Riedl, R. 1978. *Order in Living Organisms: A Systems Analysis of Evolution.* New York: Wiley.

Rutherford, S., and S. Lindquist. 1998. Hsp90 as a capacitor for morphological evolution. *Nature* 396: 336–342.

Schuster, P., W. Fontana, P. F. Stadler, and I. L. Hofacker. 1994. From sequences to shapes and back: A case-study in RNA secondary structures. *Proceedings of the Royal Society B* 255: 279–284.

Sniegowski, P. D., and H. A. Murphy. 2006. Evolvability. *Current Biology* 16: 831–834.

Sober, E. 1984. *The Nature of Selection: Evolutionary Theory in Philosophical Focus.* Cambridge, MA: MIT Press.

Stern, D. L. 2000. Evolutionary developmental biology and the problem of variation. *Evolution* 54: 1079–1091.

True, J. R., and E. S. Haag. 2001. Developmental systems drift and flexibility in evolutionary trajectories. *Evolution & Development* 3: 109–119.

Waddington, C. H. 1957. *The Strategy of the Genes: A Discussion of Some Aspects of Theoretical Biology.* London: Allen & Unwin.

Wagner A. 2005. *Robustness and Evolvability in Living Systems.* Princeton, NJ: Princeton University Press.

Wagner A. 2008. Robustness and evolvability: A paradox resolved. *Proceedings of the Royal Society B* 275: 91–100.

Wagner, A., and P. F. Stadler. 1999. Viral RNA and evolved mutational robustness. *Journal of Experimental Zoology B* 285: 11–127.

Wagner, G. P. 1981. Feedback selection and the evolution of modifiers. *Acta Biotheoretica* 30: 79–102.

Wagner, G. P. 1984. Coevolution of functionally constrained characters: Prerequisites of adaptive versatility. *BioSystems* 17: 51–55.

Wagner, G. P. 1986. The systems approach: An interface between development and population genetic aspects of evolution. In *Patterns and Processes in the History of Life,* edited by D. M. Raup and D. Jablonski, 149–165. Berlin: Springer.

Wagner, G. P. 1996. Homologues, natural kinds and the evolution of modularity. *American Zoologist* 36: 36–43.

Wagner, G. P. 2014. *Homology, Genes, and Evolutionary Innovation.* Princeton, NJ: Princeton University Press.

Wagner, G. P., and L. Altenberg. 1996. Complex adaptations and the evolution of evolvability. *Evolution* 50: 967–976.

Wagner, G. P., and R. Bürger. 1985. On the evolution of dominance modifiers II. A non-equilibrium approach to the evolution of genetic systems. *Journal of Theoretical Biology* 113: 475–500.

Wagner, G. P., G. Booth, and H. Bagheri-Chaichian. 1997. A population genetic theory of canalization. *Evolution* 51: 329–347.

Weismann, A. 1889. *Essays upon Heredity and Kindred Biological Problems.* Oxford: Oxford University Press.

Williams, G. C. 1975. *Sex and Evolution.* Princeton, NJ: Princeton University Press.

Zemach, A., I. E. McDaniel, P. Silva, and D. Zilberman. 2010. Genome-wide evolutionary analysis of eukaryotic DNA methylation. *Science* 328: 916–919.

8 The Genotype-Phenotype Map Structure and Its Role in Evolvability

Mihaela Pavličev, Salomé Bourg, and Arnaud Le Rouzic

The potential for evolutionary change is deeply anchored in the kind and amount of heritable phenotypic variation that organisms can produce, and thus in the way that genetic predispositions translate into the phenotype. That translation is the genotype-phenotype (GP) map. We first explain the two common conceptualizations of GP map: the global correspondence map and the local mechanistic map, and how they relate to each other. We focus on the structural aspects of the GP mapping, as summarized in the notions of pleiotropy and epistasis, and argue that their effect on evolvability is not captured sufficiently in the current theory. One way to approach this problem is to address mechanistic, causal mapping explicitly and explore systematically the variational properties of various well-known biochemical or regulatory processes. This may allow us not only to better account for the effects of pleiotropy and epistasis, but also to potentially complement these summarizing concepts themselves with notions that better capture the underlying mechanisms—thus adding the mechanistic aspect to the global GP map.

8.1 Introduction

The genotype-phenotype (GP) map captures the translation of genetic predispositions into phenotypic traits and is thus intimately associated with the potential to evolve (the evolvability). Despite the shared general notion, the GP map concept is used with various distinct meanings in biological literature; it is therefore paramount to explain how these meanings relate to one another and be specific about our use of the concept in the present chapter. Most generally, a GP map establishes a mere correspondence between a set of possible genotypes and a set of phenotypes (Lewontin 1974). We refer here to the ensuing theoretical space as a *global* GP map (figure 8.1). A global map is not thought to itself evolve; instead, evolution is conceptualized as movement of a population on this map. Populations and species inhabit (more or less) distinct portions of the map. Referring to the global map does not imply any intention to address the explicit mechanistic nature of mapping or its change as the population moves through the map. When studying the genotype-phenotype relations in a population in the context of the global map, the nonlinearities, or the lack of one-to-one mapping, which naturally arise in the mechanisms of development and physiology, are subsumed into variational concepts, such as pleiotropy or epistasis.

In contrast, when the concept of genotype-phenotype map is applied to individual genotypes, it refers to mapping of a single genotype to a single phenotype through all the organismal processes, including morphogenesis, growth, and physiology (figure 8.1). To

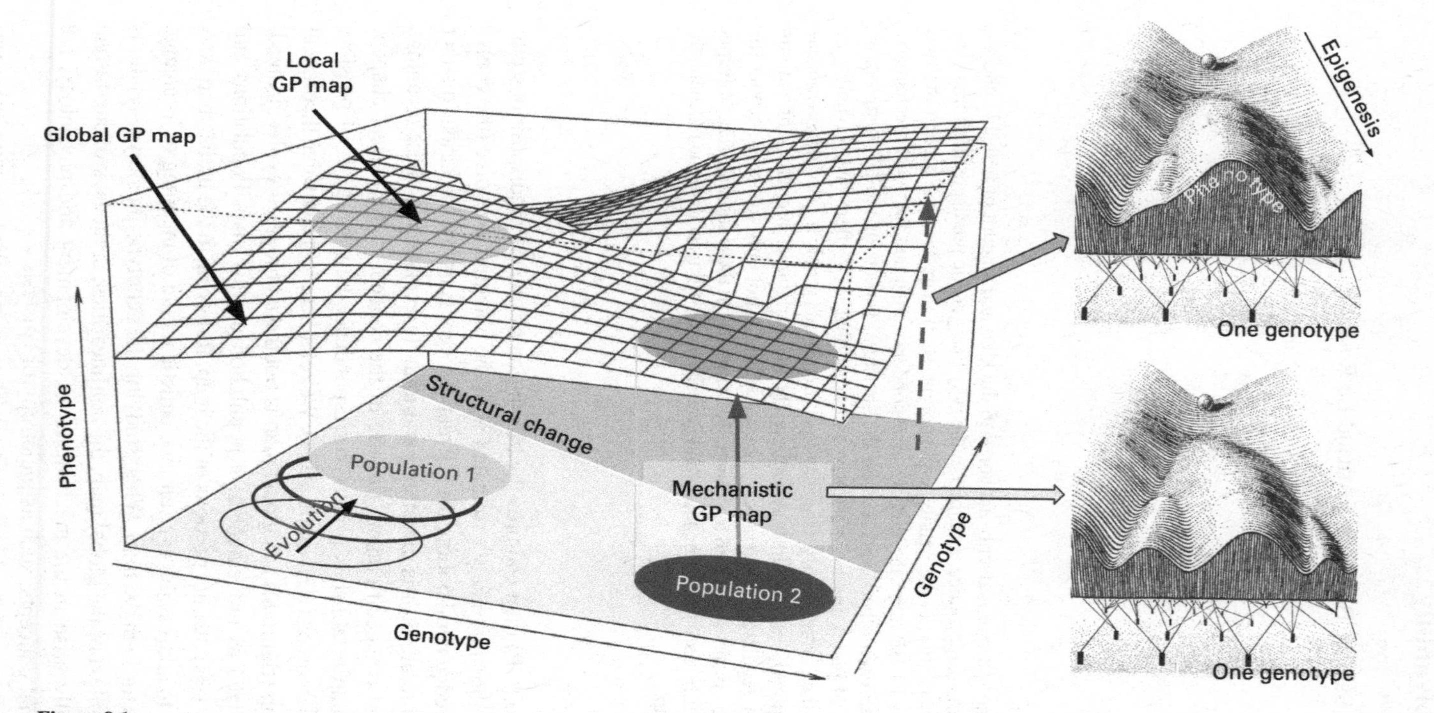

Figure 8.1
Three types of GP maps. GP maps describe the relationship between the genotype space (represented by 2 dimensions) and the phenotype space (here 1 dimensional). The entire surface represents a hypothetical global GP map, relating all possible genotypes to the phenotype. Ellipses illustrate local GP maps, which can be deduced based on variation in populations. The GP map concept can also be extended to the description of physiological, developmental, and functional mechanisms by which genotypes direct the epigenesis of phenotypes (mechanistic GP map). Populations evolve by moving in the genotype space, which may change the phenotype and the structure of the local GP map. In this figure, the global map shows a discontinuity in phenotype space, such that genotypes on either side of the boundary labeled "structural change" can show a discrete difference in phenotype. These differences are represented by the two panels on the right. In these panels, the process of epigenesis proceeds like a ball rolling downhill on each landscape (Waddington 1957). Maps in front of the 'structural change' boundary will reliably produce one of the four phenotypes at the bottom of each channel, shown in the lower right panel, while those on the other side of the boundary produce just two possible phenotypes. According to this definition, the global GP map is constant and cannot evolve, the local GP map evolves along with genetic changes in the population, and the mechanistic GP map evolves when populations cross the structural change limit.

each genotype, *a local map* is thereby attributed, represented by Waddington's (1957) epigenetic landscapes reproduced on the right side of figure 8.1. This *mechanistic* GP map explicitly aims to capture some aspect of the nature of mapping and may refer to any intermediate or final level of the phenotype (e.g., gene expression, RNA secondary structure, enzyme activity, or the adult femur length). Such mechanistic maps vary greatly in how detailed they are. This local GP map can be thought to evolve as the population moves through the global GP map. We aim to show in this chapter that the summary concepts used to capture the mapping complexity in the global map, such as pleiotropy or epistasis, often insufficiently capture the potential to evolve. We argue therefore that we would profit from rethinking the mapping in the global GP map in terms of mechanistic mapping (as proposed by Alberch 1991).

When we wish to study how any system works, we observe the consequences of perturbing it. Similarly, we learn about how nature works by observing its variation. In evolving populations, important variation arises by genetic mutations in single individuals. When sequence changes have effects that percolate through the developmental and physiological mechanisms of the individual GP map to change phenotypes of individuals, they cause heritable phenotypic differences in the population, thus constituting the raw material for selection. How the mutations affect the phenotype in individuals (i.e., which characters vary or covary across individuals) is a consequence of the mechanisms constituting GP mapping. This mechanistic GP map is thus not a description of how heritable phenotypic variation manifests, it is instead a description of the underlying processes that shape its manifestation. The observed pattern of variation is not itself a GP map, but a consequence of the variation interacting with a structure of the GP map—in the same way that a shadow of an object is not an object itself, but a consequence of light interacting with an object.

It is important to emphasize that the GP map is not dependent on the presence of variation, it applies regardless of whether all the genotypes actually exist. We thus distinguish between a GP map of traits and what we call the *genetic architecture of traits.* The latter term refers to a statistical population variation summary focusing on observed genetic variation and thus is limited to the traits and mechanisms that vary in the population. The genetic architecture is influenced by the GP map, but also by effects of allele frequency and linkage. Thus, changes to the latter factors can change the genetic architecture of traits, even when the mechanistic GP map remains constant.

We will frequently refer to the *structure* of the mechanistic GP map. Two aspects of GP map structure will be addressed: the involvement of genes in multiple traits, and the dependency of the genetic effect on the genetic background (i.e., corresponding to pleiotropy and epistasis). In both aspects, we do not refer to the statistical concepts (except when explicit; see Hansen, chapter 5)[1] but to the structures necessary for them to arise. For example, roles of two genetic sequences must depend on each other for the statistical interaction to arise when the underlying sequences vary (neglecting linkage disequilibrium, which is transient). Similarly, a gene must be involved in generating different body parts for these parts to covary when mutation arises (variational pleiotropy). We consider such mechanistic pleiotropy and epistasis to describe the general GP map structure. Within this structure, other evolutionary

1. References to chapter numbers in the text are to chapters in this volume.

changes, such as changes of mutational rate or effect size, are thought to occur. Singling out the GP structure as we do here is based on the assumption that it evolves more slowly than the changes within a given structure (note that this assumption underlies the existence of pleiotropic constraints). In other words, we consider that the global GP map consists of *structurally neutral* regions. Moving in these structurally neutral regions entails phenotypic modifications, but not changes of the structural aspects of the mechanistic GP maps. Addressing this topic thoroughly would merit a separate chapter, so to support the notion of separating conceptually structural changes from changes within a given structure, we merely point to the pattern of trait evolution being hierarchical, with structural aspects (e.g., body plan traits, homologies) evolving more slowly than modifications within given structures (e.g., relative sizes of given parts).

These clarifications of assumptions and the use and interpretation of GP maps will help us explain the full role of the (mechanistic) GP map in evolution. Because we consider GP structure to be invariant across subregions of the global GP map (i.e., the *structurally neutral* regions), it can be treated within this region as an *a priori*. In this space, the extant variation as well as the variation to be encountered by the same system due to future mutations (variability sensu Wagner and Altenberg 1996) percolate through the same structure, restricting in the same way the range of patterns of variation and covariation that can be generated by mutation, recombination, and segregation. Genetic variation thus makes the underlying system's structure visible and can be used to explore the GP map structure as well as its effects on the intermediate- and short-term responses to selection. The evolvability in this context is not based solely on the variation currently segregating, but refers to the propensity of the system itself to generate variation.

In this chapter, we focus on the mechanistic GP map structure and its consequences for the general propensity to vary, and thus for the ability or inability of a population to respond to selection on trait means. Addressing the evolution in the space of the *structurally* neutral variation of the GP map corresponds to what is often referred to as short- and intermediate-term evolution. Whether this level of change fully corresponds to within-species evolution is an empirical question, as the GP map structure may differ between species for some processes or body parts but not for others.

The chapter has the following organization. We first explain why the evolvability of complex organisms is an intriguing phenomenon, how this involves the GP map, and based on that, what GP map one might expect in evolvable organisms in section 8.2. Next, we elaborate in more detail the consequences of the GP map structure for evolvability in section 8.3. We point to the difficulties that the statistical approach to the GP map encounters when addressing evolutionary prediction. This leads us in section 8.4 to recognize the need to integrate the mechanistic detail of GP maps, rather than only distribution of extant variation, to predict evolutionary response. We explain how this detail can be integrated with the population genetic approach to model the evolution and evolvability of the phenotype, starting from the mechanistic structure of the map rather than starting from the phenotype. Finally, in section 8.5, we explain the principles of assessing GP map structure and the resulting quantitative genetic measurements useful to both quantitative and population genetic approaches.

8.2 Evolving Complex Phenotypes

8.2.1 Phenotypes Too Complex to Adapt by Chance: Fisher's Geometric Model

We can start to understand the role of the GP map structure by observing how evolvability changes as the complexity of the system increases. Fisher (1930) illustrated this in his "geometric model." Consider the two-trait situation, shown in figure 8.2. The circle of radius d centered at the optimum O, encloses all the points that would be closer to the optimum than genotype P. Given this situation, Fisher showed that the probability that an arising mutation will be advantageous approaches 1/2 when the mutational step (e) is very small relative to d and decreases with increasing mutational effect size.

Extrapolating this logic to increasing number of traits, the more traits that are affected by a mutation, the smaller the probability that a mutation will be advantageous. The expectation based on this simple intuitive model is therefore that with the increase of complexity (assuming increase in pleiotropy), the probability that each mutation has of being advantageous decreases (Orr 2000). This is the problem of complex adaptation. It should be mentioned that in population genetic dynamics, the situation is somewhat more complex: Probabilities of fixation must also be considered (Kimura 1983), and pleiotropy also introduces advantages into the system by increasing the trait's mutational target size (Hansen 2003; Pavličev and Hansen 2011; see section 8.3.3). Orr (2000) thus used Fisher's geometric model to support an important intuition about the problem of evolvability set by complex adaptation.

By exploring the implicit assumptions of this restricting model, one can understand the conditions that enabled complexity to nevertheless evolve. For example, note that Fisher uses a very particular GP map structure; namely, that of an extreme form of pleiotropy, in which each mutation affects all traits with equal probability and effect size, as represented here by a continuous circle, is replaced by various fragments of the circle (representing the mutationally accessible directions), or by an ellipse (representing asymmetrical mutational sizes), this becomes a different GP map, with differing predictions. We will address such GP maps next.

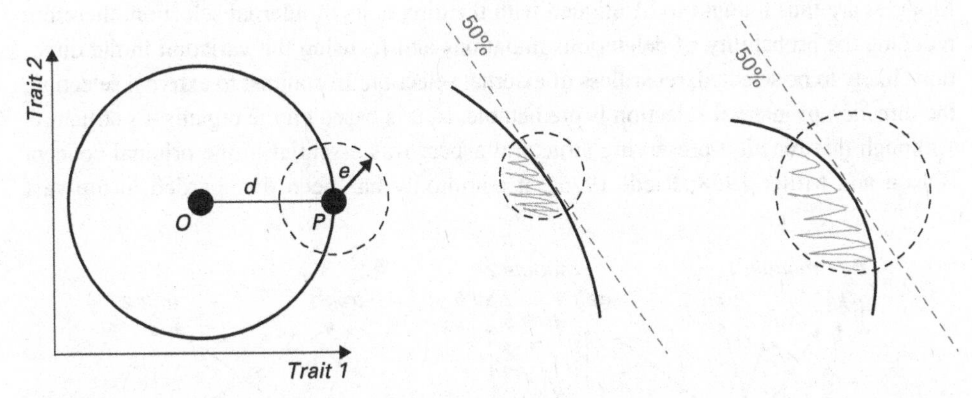

Figure 8.2
Fisher's geometric model. The two axes represent two phenotypic traits. O, optimal phenotype; P, mean phenotypic value of a genotype or population; d, distance of P from the optimum; e, mutational effect size. Note that as the mutational step size increases, the proportion of possible mutational outcomes that are closer to the optimum than the present phenotype (shaded part of the small circles in the middle and right panels) decreases. Adapted from Pavličev and Wagner (2012).

8.2.2 GP Structures Increasing Evolvability: Modularity and Robustness

Since complex organisms do evolve, what then are the GP structures that make organisms evolvable? Here we briefly explain two major structural principles: modularity and robustness (see A. Wagner, chapter 11).

In Fisher's model, universal pleiotropy leads to decreased frequency and size of beneficial mutations. Pleiotropy, the single mutation causing change in multiple phenotypic traits, causes covariation between these traits at the population level. To avoid this effect of pleiotropy, *modularity* of the GP map restricts mutational effects of single loci to sets of traits with common function or development (figure 8.3A, a module sensu Wagner and Altenberg 1996). At the population level, such a GP map will generate covarying clusters of traits, which covary less with other trait clusters.

Why is this structure considered to promote evolvability? In the short term, evolvability is proportional to the availability of heritable genetic variation in the direction of selection, which is not entangled with variation in other phenotypic traits, as these traits may encounter different selection. Note that evolvability in Fisher's model was diminished, because, with increasing complexity, the dimensionality of mutational effect also increased; it is unlikely that changing all traits simultaneously will be advantageous for all traits. Modularity allows independent selection responses of modules without interference due to correlation between modules—an aspect that increases the evolvability by reducing pleiotropic constraint.

Note, however, that the idea of modularity, as expressed most influentially by Wagner and Altenberg (1996), is not only about decreasing the dimensionality of mutational change. Instead, it also involves a functional aspect: The proposed modules are focused on structures with common function or common development. Thus the traits integrated in a module will most likely be selected together, due to *internal selection* (Schwenk and Wagner 2000). Internal selection means that only some directions of variation maintain the initial functionality of the organism, while others don't, regardless of the environment. For example, the mutations that perturb heart function or disable reproduction will be unconditionally deleterious to fitness in any environment. Such selection occurs within the organism. Modules are thus thought to be aligned with the directions of internal selection, therefore reducing the probability of deleterious mutations and focusing the variation in the directions likely to be selected, regardless of external selection. In contrast to external selection, the direction of internal selection is predictable, as it is based on the organism's structure. Although this function-preserving structural aspect was essential in the original concept (Olson and Miller 1958; Riedl 1975), it commonly has been disregarded in the vast

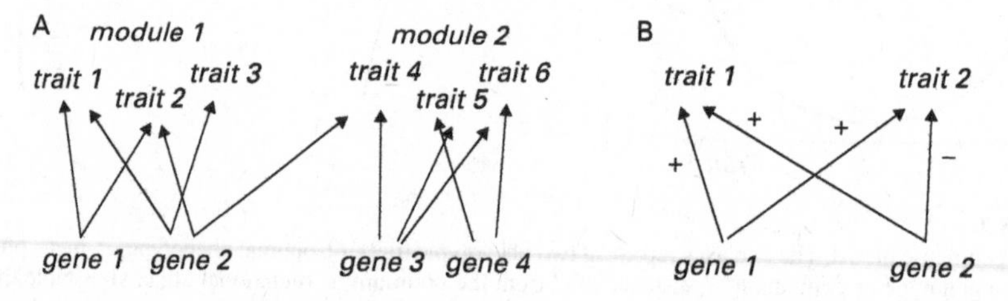

Figure 8.3
Two structures of the GP map generate reduced population covariance. (A) Modular pleiotropy. (B) GP map structure that may result in hidden pleiotropy, depending on effect sizes and allele frequencies.

subsequent literature on modularity. This led to a variety of misleading conclusions concerning the effects of modularity on evolvability (Jablonski, chapter 17).

The assessment of pleiotropy will be discussed in section 8.5. Mechanistically, we know from developmental and variational studies that while most genes are reused in development of multiple traits, each mutation does not affect all traits of an organism (e.g., G. Wagner et al. 2008; Wang et al. 2010). Note that recent work in disease genetics has drawn a different conclusion, all genes affecting all traits (*omnigenic model;* Boyle et al. 2017). Yet this discrepancy is not a contradiction. Disease is not a complex trait in the sense that the vertebrate forelimb is a complex trait; it is not an individualized, evolved, and evolvable biological unit but a consequence of a pathological variant thereof (Pavličev and Wagner 2022). It is therefore problematic to use the insights from disease genetics to draw conclusions in evolutionary genetics directly. As biological traits are embedded in an organism, there are many more possibilities to cause a defect than to generate variation relevant for evolution. With respect to biological traits, then, pleiotropy is ubiquitous but not universal, so each mutation does not contribute to variation in all biological traits. Because mutational effects percolate through the developmentally and physiologically organized GP map, the distribution of pleiotropic effects of genes on traits does not follow a random play of chance but instead affects the sets of genes from functionally interacting pathways that are active in specific cells, tissues, and organs, and are not invoked in others. Therefore, we can expect that variation will, however partially, reflect these patterns.

Note that the principle of modularity does not imply that population variation consists of entirely independently varying blocks of traits. There will be covariance even when the traits belong to different modules. This is not surprising, as many fundamental pathways are shared among traits, tissues, and cell types. To the extent that the genes in these pathways vary, they will produce covariation between modules.

Robustness is another structural property of GP maps that can increase evolvability. We will address only genetic (not environmental) robustness here. We have seen that the concept of modularity suggests a specific arrangement of mutational effects with respect to the direction in the phenotypic space. In contrast, genetic robustness is about the distribution of mutational *effect sizes*. A GP map is robust if many mutations do not change the phenotype. Therefore, those cryptic mutations may accumulate in the population. If one imagines genotypic space as a network of all possible genetic sequences that are a single mutation away from each other, then in a robust population, a subnetwork of such connected genotypes map to the same phenotype (the so-called *neutral network* of genotypes; Schuster et al. 1994; Fontana 2002). For example, an RNA molecule may be able to fold into its secondary structure and remain functional despite several mutational changes in its sequence.

The idea that genetic robustness confers evolvability is based on the observation that if the individuals are distributed across a large neutral network, many would reside at its outer edges, close to the border and just a single mutation away from the neighboring neutral network, conferring a different phenotype (A. Wagner 2008). Put differently, a particular incoming mutation in such cases occurs on a large range of different genetic backgrounds, potentially with different outcomes. The phenotypic effects of the mutation differ in the network due to epistasis, leading to heritable phenotypic variation where there was none initially (Hermisson and Wagner 2004; Richardson et al. 2013; Geiler-Samerotte et al. 2019). New phenotypes are not necessarily advantageous, but a large network increases the probability that some will be.

Note that robustness does not monotonically increase evolvability for several reasons. First, strong robustness disables the evolutionary process altogether by suppressing variation. Second, the general association between large neutral network and high evolvability depends on the population size; small populations maintain only a few genotypes and thus cannot realize the advantages of large neutral networks. More interestingly, robustness also depends on the details of the individual GP map structure and the accessibility of the alternative phenotypes (Draghi et al. 2010; Mayer and Hansen 2017). Even large neutral networks may only have access to a small number of alternative phenotypes, and vice versa. Human genetic disease may be considered as an example. Genetic variants that cause human disease by perturbing specific buffered pathways do not release random phenotypic variation but result in rather specific, recurring, disease phenotypes. Thus, the phenotypes arising due to perturbation are constrained to a specific phenotypic space by the rest of the organismal regulation. To what extent the range of the accessible phenotypes is correlated with the degree of robustness depends on the systemic structure into which the trait in question is embedded.

In summary, the intuition behind the role of specific GP map structures in increasing evolvability is that they generate the kind of variational distribution that, more likely than random distributions, allows for a viable and even advantageous response to selection. Let us next take a closer look at the general relation between the GP map and the measure of evolvability—the amount of heritable genetic variation.

8.3 Variational Consequences of GP Maps in the Short and Long Term

In this section, we explain how GP map structure affects variation in general and what consequences these effects have for evolvability in the short and long terms. For interested readers, effects and their estimations are described in detail in the section 8.5 (see also Hansen, chapter 5). The general approach used by evolutionary biologists to understand the underlying mechanisms governing evolution is that of forward modeling. We use models predicting how the patterns of variation influence future response to selection. We thus derive expectations about future phenotypes (e.g., in artificial breeding) or about processes that must have acted in the past, to explain what we see in extant species or in the paleontological record. Importantly, the discrepancies between the predictions and reality can reveal that our models do not sufficiently capture the underlying causal processes we aim to understand.

8.3.1 Response to Selection

The immediate response to directional selection is driven by the standing additive genetic variation in the population, summarized in a genetic variance matrix **G**. When several phenotypic traits are considered, the phenotypic response can be predicted by the multivariate breeder's equation $\Delta\mathbf{z}=\mathbf{G}\beta$ (Lande and Arnold 1983), where $\Delta\mathbf{z}$ is a vector of changes in mean phenotypic traits in one generation, and β is the selection gradient (strength and direction of selection). **G** measures the short-term evolvability of the population, describing the amount and the structure of the genetic variation available to respond to selection.

The standing genetic variation is fueled by mutations, whose contribution can be measured by another variance matrix, **M** (how both matrices are measured is described in section 8.5). This mutational variance matrix depends heavily on the GP map, as it quantifies the statistical distribution of phenotypes resulting from genetic mutations. The size of

M depends on the sensitivity of the phenotype to changes in the underlying genotypes, while the shape of **M** depends on the correlation of different traits to the same genetic change. The standing genetic variation **G** is conditioned both by the influx of mutations (**M**) and by the recent history of the population and is thus linked to the GP map. But genetic variation rarely accumulates generation after generation without being affected by environmental or demographic events. Stabilizing or directional selection can indeed erode the genetic variance in some specific directions of the phenotypic space; genetic drift in small populations may also affect the geometry of the genetic covariance (Chantepie and Chevin 2020).

Standing genetic variation fuels short-term response to selection. Once initial existing variation is exhausted, genetic evolution relies on new mutations and is thus more directly affected by the **M** matrix and the underlying GP map (Lande 1980; Turelli 1985; G. Wagner 1989; Slatkin and Frank 1990).

8.3.2 Evolution of Variance Matrices

The short-term evolvability of a population depends on **G**, which is largely influenced by recent patterns of selection, gene flow, and drift in the population. Nevertheless, genetic variation is ultimately produced by mutation, and the evolution of **G** is influenced, to some extent, by **M**. The structure of **M**, however, may not be constant. When GP maps are complex and nonlinear, epistatic patterns can drive the evolution of **M** along with the evolution of the genotype. For instance, some genetic backgrounds can be robust (when the local GP map is *flat*), while other backgrounds can be more sensitive to genetic change (when the map is *steep*). Consequently, depending on the structure of the GP map, the mutational pattern **M** may change when the genotypes change in the population. Ultimately, various aspects of the evolvability of a population rely directly and indirectly on the GP map: This map conditions the evolution of **M**, which contributes to the evolution of **G**, which determines the evolution of the population.

8.3.3 Correspondence between GP Map and Genetic (G) or Mutational (M) Variance Matrices

We have seen how the distribution and dynamics of variance and covariance in **G** affects evolvability in the short-term, while **M** affects evolvability in the long term. The genetic variance and covariance between phenotypic traits in the **G** and **M** matrices are sums of contributions at many polymorphic genetic locations (loci). Unlike variance, which is always positive, covariance between traits due to a single polymorphic locus can also be negative. The covariance contributions of loci thus can also cancel out, depending on the exact structure of the GP map and the allele frequencies. An example in Figure 8.3B shows a minimal map with full pleiotropy. Note that the effects of genes 1 and 2 can potentially cancel each other, resulting in no covariance between traits 1 and 2, given a specific combination of effect sizes and allele frequencies. Such pleiotropic effects that are not reflected in the variance matrix are called hidden pleiotropy (Gromko 1995; Baatz and Wagner 1997). Hidden pleiotropy may have advantages compared to modular pleiotropy, as such GP structure provides greater mutational domains per trait on average (Hansen 2003). However, Baatz and Wagner (1997) have shown that hidden pleiotropy can cause a constraint, because despite the lack of covariance, selection on single traits does affect the variance of other traits, which may be under stabilizing selection. It can be easily shown that this effect strengthens as the

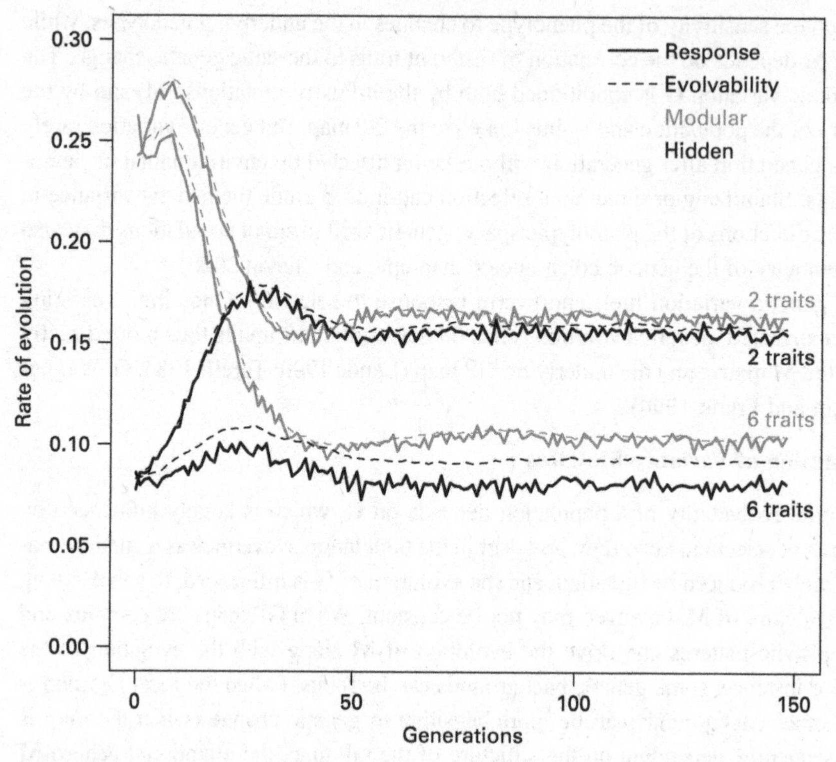

Figure 8.4
Simulated focal trait evolution given a hidden pleiotropic (black) and modular (gray) GP map. Corridor evolution (Baatz and Wagner 1997) is applied with directional selection ($\beta = 1$) on focal trait and stabilizing selection (quadratic selection gradient $\gamma = 2$) on all other traits. Both GP maps consist of 6 traits and 20 loci affecting the focal trait, with the same initial genetic variance and no covariance between the focal trait and all other traits. The solid lines show the rate of evolution of the focal trait (the mean across 10 repeats). The dashed lines track the 1-generation prediction based on Lande and Arnold (1983), updated for current **G**. For each GP map, a situation with 1 constraining trait (2 traits total) and 5 constraining traits (6 traits total) is shown. Evolvability with hidden pleiotropy is lower than for the modular pleiotropy case. The figure shows that the hidden pleiotropy imposes cost on evolvability, as the response deviates even in the short term from the predicted response (dashed), and that this deviation increases with the number of traits. Trait number also affects modular GP (due to linkage, not shown). Thus, despite the initially identical **G** values, the response, even in 1 generation, differs from the prediction, and it differs between GP maps. Starting values in this plot differ due to 1000 generations of stabilizing selection on all traits to achieve mutation-selection equilibrium prior to 150 generations of corridor selection (for simulation details, see Hansen et al. 2019).

stabilizing selection on other traits strengthens, and it also strengthens with the number of traits that share loci (Hansen et al. 2019; figure 8.4).

A related aspect to consider is that even though equivalent **G** matrices can be generated by different GP maps, they may not be equally accessible in populations with different GP maps, affecting the selection response. For example, without pleiotropy, the lack of covariance between divergently selected traits is a default (assuming no linkage disequilibrium). In contrast, lack of covariance requires coordinated modifications at many loci when it evolves by matching the single-locus covariance contributions to cancel out. Although this effect appears subtle when considering two traits, it is substantial when matching includes multiple traits and when the mutations are rare but large, leading to erratic covariance change and departures from **G**-based predictions (figure 8.4; based on Pavličev, unpublished data). Linkage, in addition, will generate covariance even in modular maps (figure 8.4).

In summary, the example of hidden and modular pleiotropy shows that no unique correspondence exists between GP structure and **G/M**. Nevertheless, the differences among GP maps generating the same **G** affect the response to selection (figure 8.4), which shows that GP map structure cannot be predicted based on the statistical summary matrices.

In addressing how GP map affects variation, we have so far only focused on one aspect of GP structure: pleiotropy. We assumed independence between effects at single loci. However, variation also arises as a consequence of yet another GP map aspect: the interdependencies between effects at different loci. This phenomenon, termed epistasis, defines the shape of the GP map, the presence of hills and valleys in the genotypic landscape, and the ruggedness of the genotype-to-phenotype relationship. Developmentally and physiologically, interdependency of mutational effects is expected (e.g., proteins do not function in isolation; they depend on conformation and abundance of many other proteins to interact with, either directly or indirectly). Effects of mutation in one protein will therefore depend on the genetic sequence of other proteins. Interdependency can affect not only the size of a mutation but also its direction (i.e., which traits are affected), so the interdependency can affect the mutation's contribution to trait variance and the covariance between traits. Analogous to contributions of pleiotropic loci to covariance across loci, the contributions of epistatic interactions to variance and covariance can add up across pairs of loci (directional epistasis; see section 8.5.2) or cancel out. In the later case, the overal contributions of interactions to variance (and covariance) are reduced and effects of single interactions invisible in the final statistical matrices. This aspect of GP structure and its effects on long-term evolvability, especially in multivariate settings, is difficult to assess systematically and has therefore received limited attention. From the above, it can be assumed that the existence of pleiotropy and epistasis is not captured fully in the statistical matrices. When it comes to long-term predictions, or to explaining the past, organismal details not captured in the statistical concepts (**G** and **M**) start to matter. How can we properly integrate them into evolutionary theory?

8.4 Turning the Question Bottom-Up: What Variation *Can* Real GP Maps Generate?

The prevailing approach of evolutionary theory is to study the effects of allele changes on statistical population-level assessment of phenotypes (including fitness). As explained above, the differences in structural details of the GP map can affect evolution even when they are not reflected in summary measurements of variation. Therefore, we should turn our attention to these differences to understand how and what kind of mutational variation can arise in the first place, even when the developmental and physiological detail and their evolution may not be our main interest. Placing genotypes at the bottom, and phenotype at the top, as in figure 8.1, we thus ask the bottom-up question: What kind of variation can the encountered physiological and developmental processes generate?

8.4.1 Integrating Mechanistic Detail: What Type, How Much, and How?

Mechanistic detail conferring the structure of the genotype-phenotype map should thus be integrated into the existing theory. But what kind of detail matters, how fine-grained does it need to be, and how do we integrate it? In section 8.3, we showed that pleiotropy matters, yet the GP map affects other aspects of variation besides pleiotropy that may be necessary to predict long-term evolutionary trajectories. The kind and amount of organismal detail needed to better understand and predict evolution are empirical questions. Instead of trying to interpret the variation pattern observed (which can have many causes); we ask: *What kind of variation can familiar organismal processes generate?* In other words, what variational properties do those organismal processes engender?

There are clear precedents for investigating variational properties in earlier attempts to integrate organismal processes (metabolism, enzymatic reactions, and gene regulation) with evolutionary theory. Most prominently, Kacser and Burns (1981) have used this general approach to show why mutations introduced into long enzymatic pathways are mostly recessive, explaining the dominance of the wild type as a systemic property. This work prompted the development of Metabolic Control Theory, which constitutes a major source of explicit GP map models. Metabolic Control Theory paved a way for attempts to bridge the organismal theory of metabolism to population genetic theory (e.g., Keightley and Kacser 1987; Keightley 1989, 1996; Clark 1991; Szathmáry 1993; Frank 1999, 2019a; 2019b; Bagheri et al. 2003; Bagheri and Wagner 2004). Recent use of metabolic GP models enables a plethora of mechanistic insights on sensitivity of metabolic circuits to mutations, on the network structure underlying systemic properties such as homeostasis, as well as prediction of the frequent forms of pathology (e.g., Nijhout et al. 2015, 2019; Reed et al. 2017). The connection of this work to quantitative genetics theory suggests that this work is a highly promising path forward.

Other approaches to explicit GP map models are the RNA secondary structure models (Schuster et al. 1994; Ancel and Fontana 2000) and transcriptional regulatory models. Among the latter, variational properties have been studied for some of the recurring transcriptional regulatory motifs (Gjuvsland et al. 2007, 2013; Widder et al. 2012). Yet another model of a GP map is the gene network model introduced by A. Wagner (1996; reviewed in Fierst and Philipps 2015). Furthermore, G. Wagner (chapter 10) proposes a similar kind of bottom-up approach for exploring the evolvability properties of various growth scenarios.

In short, the field of explicit GP mapping may not have been at the center stage of evolution and evolvability research, but it is well populated and growing, and we suggest that it carries a high potential for system-level understanding of evolvability. In section 8.4.2, we discuss one set of recent bottom-up approaches to asking the question of the role of the GP map in predicting long-term evolution.

8.4.2 System-Level Approaches in Physiology and Development

Population and quantitative genetic models focus either on genotype or phenotype space, respectively. However, the substantial effort to model one of these spaces generally leads to very naive consideration of the other (Lewontin 1974). Omholt (2013) argued that the mainstream framework of evolutionary theory so far lacks causal relations between genotype and phenotype. However, the realistic connections between genotype and phenotype can nowadays be integrated, thanks to intense work on physiological and developmental mechanisms, into the classical modeling framework, thus explicitly integrating causal links. From the 2000s, such models started to emerge under the name of *causally cohesive genotype-phenotype models* (hereafter cGP models; Rajasingh et al. 2008). Most cGP models are individual-based and follow a structure described in figure 8.5, where individuals are

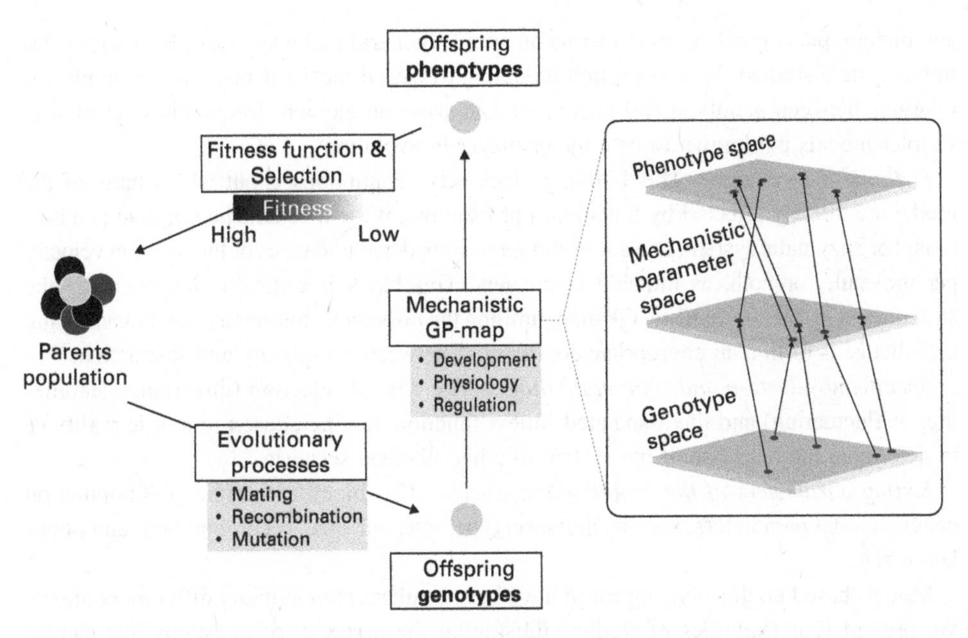

Figure 8.5
Building blocks of a cGP model. See text for explanation. This figure is inspired by figures in Omholt (2013) and Milocco and Salazar-Ciudad (2020).

characterized by a genotype determining a physiological mechanism (i.e., a mechanistic GP map) generating the expression of a phenotype. If the model involves evolution, then a fitness function is added, which maps phenotype to fitness. One of the key features of cGP models is the use of explicit physiological parameters, in which genetic variation can be assumed to exist. The variation thus arising in this model is anchored in specific causal biological hypotheses. In the next paragraphs, we first outline steps for developing a cGP model. We then illustrate the application with four studies.

To build a cGP model, five blocks of information must be provided, as listed below. These blocks are reflected on the left side of the flow diagram in figure 8.5 and are constitutive of the subsequent selection step. For this chapter, we consider these blocks as independent, whereas in practice their generation is a strongly interdependent process, with development of one block having consequences for all others, with no unique correct order of addressing them.

Defining the phenotype: The goal here is to specify traits that are sufficiently simple that their causal basis is understood and still address the question of interest. Defining a phenotype remains a challenging exercise. A global phenotype can always be segmented into more traits, but beyond a certain point, this will not improve our understanding of evolution (Houle 2001). Moreover, it is highly recommended to study traits produced by well-described underlying mechanisms.

Determining the mechanistic GP map: The joint consequence of pleiotropy and polygeny is that many interconnected pathways participate in the expression of a single trait, what Houle (1991) called functional architecture. Because the full complexity of mechanisms involved in the production of the phenotype usually cannot be integrated, the traits to be implemented in the model must efficiently capture the functional link between genotype and phenotype. A good start is to focus on simple, general pathways that play major roles in the system studied. Due to explicit modeling of causal mechanisms, realistic nonlinear relations between genotype and phenotype can arise; an element frequently neglected in simpler models but known to be a major player in evolution

Defining the genotype: This building block sets assumptions about which parts of the model are directly affected by mutation. For example, if the mechanistic GP map is a hormonal or enzymatic system, one could choose to consider abundances or the reaction velocity per molecule, or both, as mutable parameters. This block is critically dependent on the structure set by the mechanistic GP map, limiting the possible evolutionary outcomes. Again, a challenge is to find an appropriate compromise between complexity and abstraction.

Specifying selection and computing fitness: The type of selection (directional, stabilizing, or fluctuating) and the associated fitness function may be chosen to mimic reality or to determine the response of the system to a hypothetical scenario.

Setting parameters of the evolutionary process: This block determines the population genetic model parameters, such as mutation type, rate, and size, recombination, and population size.

Models based on this cGP approach have been implemented in many different contexts. We present four examples of studies, illustrating the diversity of questions that can be considered using this framework.

To our knowledge, Rajasingh et al. (2008) presented the first cGP model. They studied a species of chinook salmon (*Oncorhynchus tshawytscha*) whose subpopulations exhibit two alternative phenotypes: white-fleshed and red-fleshed. To reproduce empirical data obtained in the crosses, they implemented a model GP map with a system of ordinary differential equations (ODEs) describing the uptake and deposition of carotenoids, a metabolic pathway responsible for the flesh color in salmonids. They compared two genetic architectures (two-locus two-alleles versus single-locus three-alleles) and found that a standard single locus model was best able to explain their observations.

The mechanistic GP map incorporated into a cGP model can be a single physiological pathway or an extremely complex system, such as that used by Vik et al. (2011) to simulate in detail the functioning of a whole cell. They modeled a mouse heart cell, drawing on a large body of empirical literature and previously developed mathematical models. Existing partial models were combined to constitute a complex GP map consisting of 35 ODE with 175 parameters. Resulting action potentials and ion concentrations represent phenotypes. The model was able to reproduce ion circulation into and out of the cells. Deviations in genetic bases for ion currents (as genotypic values) were examined for their ability to reproduce disease phenotypes. Such heavily parameterized models that describe cellular processes at a fine-grained level represent promising tools to predict the proximate determinants responsible for disease symptoms.

The aim of Milocco and Salazar-Ciudad (2020) was to compare predictions based on the multivariate linear breeder's equation to those based a nonlinear mechanistic GP map modeled as a cGP model. To build the cGP, they associated an existing developmental model of mammalian teeth with 21 genetically variable developmental parameters with a population genetics model. The resulting individual phenotypes were characterized by a complex 3-dimensional structure (the tooth morphology). Phenotypes evolved for 400 generations under stabilizing selection that selected for an optimal tooth shape different from the initial tooth shape. They demonstrated that the bias between predictions from the breeder's equation and their results arose when populations were located in a nonlinear region of the global GP map. Hence, the study of models considering a realistically complex model GP map is justified by the limits of classic quantitative genetics tools, which cannot fully account for variation in development, as argued by Polly (2008).

Bourg et al. (2019) built an evolutionary model to study the evolution of the shape of a trade-off between life-history traits. Their mechanistic GP map incorporated a dynamic hormonal system that determines allocation of resources. They coupled this system with a classic population genetic model. Individuals were described by genotypes composed of genes coding for the expression of hormones and their receptors. Depending on the hormone-receptor affinities and their respective concentrations, a limited energetic resource was distributed differentially between two abstract vital functions (or traits). As a result, they obtained individual phenotypes corresponding to two abstract trait values that represented an internal energetic compromise. They let populations evolve under directional selection over 100,000 generations. Trade-offs expressed at the level of the population were not necessarily linear and could evolve due to a change in the local GP map.

These models illustrate how modeling explicit GP maps in mechanistic detail can be a useful avenue to better understand the potential for evolutionary change. We next turn to a brief overview of approaches to detect and assess GP map structure.

8.5 Detecting and Measuring Structure

In the previous sections, we described the structural aspects of the GP map and the statistical parameters that are estimated at the population level. Both these aspects are still relevant when the question is asked bottom up, as we have seen. Here, we want to briefly describe how these aspects of structure are detected and measured, and the limitations that measurement may face.

8.5.1 Direct Measurement of GP Maps

The most straightforward way to access the GP map structure is to measure an exhaustive set of genotypes directly (i.e., to generate many genotypes differing by one or a very few mutations and then measure the corresponding phenotypes). In most species, such an approach would be highly impractical, very costly, or simply not feasible. Yet combining recent technology (e.g., nucleic acid synthesis, new generation sequencing) with high-throughput phenotyping methods in micro-organisms makes it possible to explore the complexity of the GP map by generating thousands of mutants in the vicinity of wild-type sequences for single proteins (Jacquier et al. 2013; Bank et al. 2015) or noncoding RNAs (Li et al. 2016). At the molecular level, the GP map exploration generally reveals complex epistatic interaction patterns. Most combinations of single mutations do not interact, but some have strong interactions. These strong interactions can have more than two loci (Domingo et al. 2018; Poelwijk et al. 2019). In these cases, interactions are often strong enough to create sign epistasis, where the identity of the variant favored by selection depends on the genetic background it is in. This generates a fitness landscape with multiple fitness peaks. It is usually unclear whether a succession of nondeleterious mutants exists that would allow natural selection to push a population from one peak to another.

Quantitative trait locus (QTL) detection has been exploited for decades to identify molecular markers associated with quantitative trait differences among individuals. QTL detection can be attempted on specific populations generated by a controlled experimental cross design or on a sample from natural populations, including humans. With affordable and efficient new sequencing technologies, the latter have become increasingly popular under the term GWAS (for genome-wide association studies). In principle, estimating the statistical effect of marker genotypes on quantitative traits could lead to a satisfactory approximation of the underlying GP map. Whether this is achievable in practice remains unclear (Manolio et al. 2009; Young 2019; Uricchio 2020).

Detecting QTLs that display epistatic interactions is, in theory, a straightforward extension of the classical QTL detection methods: Instead of looking for markers with a significant association to the phenotype, the focus needs to be on pairs of markers with a significant interaction effect on the phenotype (Carlborg and Haley 2004; Carlborg et al. 2006). The statistical power of such analyses remains limited, even in large samples, as the number of marker pairs to test is huge. A particularly interesting form of epistasis creates variation in pleiotropy. When two phenotypic traits are affected by the same genetic variants, they share a (partially) common genetic basis, and a QTL affecting one trait is expected to affect the other trait. Genetic variants that modify pleiotropy will change the relative magnitude of those effects, which makes it possible to detect pleiotropy modifier QTLs (referred to as relationship QTLs, or rQTLs) by the presence of an interaction term between their effects on both traits (Cheverud et al. 2004). Here again, the statistical power to detect such interactions is lower than for traditional QTL mapping, and detection issues will rapidly increase with the number of phenotypic traits.

8.5.2 Measuring Statistical Properties of Genetic Architectures

Quantitative genetics aims to describe genetic architectures of traits from a statistical point of view rather than via the specific influences of identified biochemical or regulatory pathways. Genetic and phenotypic diversity on which selection acts, fueling evolution, is then measured as the additive genetic variance of quantitative traits. Here we briefly review how the structure of genetic architectures is described statistically through two distinct (although subtly related) kinds of interactions: the fact that genes may influence several characters (pleiotropy), which translates into statistical covariances among traits (in **G** and **M** matrices), and the fact that the effect of single substitutions does not add up to produce total variation (epistasis), which arises because the genetic effects depend on the genetic background in which the mutation takes place.

Mutational variances and covariances, reflecting the rate, size, and pleiotropic effects of mutations before selection, are key features of any long-term quantitative genetics theoretical prediction (Jones et al. 2007). Yet they are notoriously challenging to estimate from empirical data: Mutations are individually rare, and their effects are usually small enough that they cannot be recognized against the background of existing phenotypic variation. Typical mutational heritabilities (mutational variance relative to the phenotypic variance) are in the range of 10^{-4} to 10^{-3}, usually less than 1% of the heritable genetic variance. The most common experimental design to measure **M** matrices consists of deriving a highly inbred genotype and then maintaining replicate copies as mutation accumulation lines. The increase in phenotypic (co)variance among lines is the estimator of the **M** matrix.

M matrices have been estimated in a small number of species, most of which are short-lived organisms. Mutational covariance studies consistently report large positive correlations among fitness components and life-history traits (in *C. elegans:* Estes et al. 2005; Keightley et al. 2000, in *Daphnia:* Lynch 1985, or in *Drosophila:* Houle et al. 1994), and moderate correlations among morphological traits (Houle and Fierst 2013). There is also consistent evidence that **M** matrices may differ among close genotypes or populations (Fernández and López-Fanjul 1996), suggesting that the structure of **M** matrix is evolvable.

Mutational correlations due to pleiotropy may induce genetic correlations, which condition the evolvability of populations. As genetic drift and selection can also cause genetic correlations through linkage disequilibrium, the **G** variance matrix cannot be deduced directly from **M** and must be measured independently. The concept of additive genetic variance (and covariance) is rooted in the decomposition of genetic variances (Fisher 1918), in which heritable and environmental components can be distinguished based on the phenotypic covariance among relatives; genetically related individuals will share part of their genotype, while the residual effects will remain independent (Hansen, chapter 5). Progress in statistical methods has made it possible to use information from all related individuals in a multigeneration pedigree simultaneously and to estimate genetic variance and covariance components underlying the structure of the genetic correlations for many traits in the population.

G matrices for different traits are known from a wide variety of organisms (Roff 1996). Positive correlations among life-history traits and morphological traits are regularly reported, but they are not strong or systematic enough to define clear general rules (Pélabon et al., chapter 13). Knowledge of **M** in addition to **G** enables us to understand which portion of segregating variational pattern is due to inherent structure of the GP map, and which portion may be a consequence of the current selection. When both **G** and **M** matrices are measured on the same population, they sometimes match convincingly (Houle et al. 2017), which has two possible explanations: (1) natural selection is weak, and **G** is mostly shaped by **M**; and (2) **M** has evolved to match the pattern favored by selection. This latter hypothesis remains controversial, due to the lack of clear theoretical and empirical support (Jones et al. 2014).

In contrast to **G**, the phenotypic variance matrix **P** is considerably easier to estimate, as it can be estimated from phenotypic measurements in the population. As the environmental (nongenetic) sources of variation are often expected to dominate the phenotypic structure, evolutionary predictions from phenotypic covariances are theoretically dubious. Yet in practice, environmental covariances are often remarkably similar to the **G** matrix. This observation, sometimes referred to as the Cheverud's conjecture (after Cheverud 1988), has received a substantial amount of empirical confirmation (Roff 1996; Kruuk et al. 2008; Dochtermann 2011), although the reasons that phenotypic covariances match genetic covariances need to be clarified (Noble et al. 2019; Chevin et al 2021).

When measured in a reference genotype, epistatic effects are usually referred to as *functional* (or *physiological*) effects (Cheverud and Routman 1995). They correspond to a local description of the curvature of the GP map. When averaged over all genotypes in a population, weighted by genotype frequencies, epistatic effects are called *statistical.* Statistical and functional estimates of epistasis are complementary descriptions of the structure of the GP map, and each can be calculated from the other if we have a detailed understanding of which loci are involved (Álvarez-Castro and Carlborg 2007). The question of whether an evolutionarily relevant, global measurement of epistasis exists is not straightforward to answer. A traditional way to quantify epistasis in populations is based on an extension of the decomposition of phenotypic variance (Cockerham 1954; Lynch and Walsh 1998). Yet the epistatic contribution to the phenotypic variance carries little information about the underlying genetic architecture (Álvarez-Castro and Le Rouzic 2015). When it comes to predicting the response to directional selection, directional epistasis, a measurement of the average curvature of the GP map in the population, is a better alternative (Hansen and Wagner 2001). Measurement of directionality of epistasis attempts to capture the evolutionarily relevant part of epistasis—that is, which has the potential to speed up (positive epistasis) or slow down (negative epistasis) the evolution in the context of directional selection (Carter et al. 2005). Positive directional (or synergistic) epistasis indicates that allelic effects tend to reinforce one another, while negative (or antagonistic) epistasis describes a situation where allelic effects cancel one another. Directional epistasis can be estimated from various datasets, including line crosses, targeted mutations, or selection responses (Pavličev et al. 2010; Le Rouzic et al. 2011; Le Rouzic 2014). Epistasis is thus another structural aspect of the GP map which shows that different structures can produce the same variational distributions.

Directional epistasis has seldom been measured directly on quantitative traits, in spite of its theoretical relevance. Systematic directional epistasis is expected for traits that do not scale linearly with an underlying physiological function. For instance, growth traits may scale exponentially and thus display positive epistasis for trait increases. However, the few existing empirical measurements are not consistent. For example, chicken body size shows positive epistasis (Le Rouzic 2014), while mouse size-related traits show negative epistasis (Pavličev et al. 2010). Directional epistasis on multiplicative fitness has important theoretical consequences for the evolution of sex, recombination, and mutation rate (Phillips et al. 2000), and has thus been under intense scrutiny. Overall, there is solid empirical evidence for negative epistasis on fitness (the effects of beneficial mutations tend to cancel out), at least in microorganisms (Martin et al. 2007).

8.6 Conclusion

The way that epistemic entities are conceptualized in any theory constrains the range of problems that can be addressed using the theory—and this is no different for evolutionary theory. The specific statistical concepts of evolutionary quantitative genetics conveniently lower the dimensionality of data, yet thereby obscure important aspects of organismal complexity, which matter in particular for long-term evolvability. But to address evolvability, which aspects of real organisms and which types of detail do we need to include? Can those be integrated into existing approaches? These are, foremost, empirical questions. The nascent field of GP map studies cannot yet answer them.

Here we suggest (not for the first time) that the answers may require us to complement the existing theory by turning to the relevant epigenetic (sensu Waddington 1957) processes and address the question from the bottom-up: What mutational variation *can* these processes generate? What is the space of **M** matrices that can arise? How does this **M** space change during the evolution of the processes? What is the structure of the GP map at its various intermediate levels? Are there important principles on a less coarse-grained scale than variational modularity and robustness?

The field of modeling explicit physiological/ developmental processes and their influence on evolution poses challenges. Access to the full GP map is likely infeasible for most complex non-model organisms—but an exhaustive map is hopefully not necessary. Defining informed numerical models of aspects of GP maps (e.g., metabolic and regulatory networks or developmental cascades), around which theory can be built, may be the most feasible path to practical use of the GP map. Deciphering the consequences of the GP map structure on evolvability thus relies heavily on experimental, conceptual, and theoretical progress. This long path will require input from a wide variety of disciplines, including systems biologists; physiologists; developmental biologists; and molecular, quantitative, and population geneticists.

In practical terms, some of the empirical questions to ask are:

- What are the variational properties of specific organismal structures and processes?
- What are the consequences of those properties for evolution?
- Are there dynamical principles with common consequences for evolutionary change, and could these replace the current statistical concepts?
- How do variational properties differ between species?

The expectation is that the study of the structure of GP maps will allow us to see the systemic and molecular changes that contribute to the change in the phenotype and its variation in a context richer than linearly associating specific single nucleotide polymorphisms (SNPs) with specific phenotypic change. Long-term predictions may appear hard to validate, but using comparative studies of physiological and developmental parameters, we can start to understand how our predictions of evolvability based on GP maps correspond to long-term evolutionary change. The comparative approach could thus bridge the existing populational questions to inform the long-term, macroevolutionary change (Jablonski, chapter 17).

Acknowledgments

We acknowledge The Center for Advanced Studies at the Norwegian Academy of Science and Letters for enabling this creative workshop type, and the organizers, T. F. Hansen and C. Pélabon, who—despite obstacles—created a dynamic environment fostering deep intellectual engagement. Finally, thanks to all participants that have improved this chapter by their feedback. SB is supported by Norwegian Research Council grant 287214.

References

Alberch, P. 1991. From genes to phenotype—dynamic-systems and evolvability. *Genetica* 84: 5–11.

Álvarez-Castro, J. M., and Ö. Carlborg. 2007. A unified model for functional and statistical epistasis and its application in quantitative trait loci analysis. *Genetics* 176: 1151–1167.

Álvarez-Castro, J. M. and A. Le Rouzic. 2015. On the partitioning of genetic variance with epistasis. In *Epistasis. Methods in Molecular Biology (Methods and Protocols)*, vol 1253, edited by Moore J., S. Williams. New York: Humana Press.

Ancel, L. W., and W. Fontana. 2000. Plasticity, evolvability, and modularity in RNA. *Journal of Experimental Zoology B* 288: 242–283.

Baatz, M., and G. P. Wagner. 1997. Adaptive inertia caused by hidden pleiotropic effects. *Theoretical Population Biology* 51: 4966.

Bagheri, H. C., and G. P. Wagner. 2004. The evolution of dominance in metabolic pathways. *Genetics* 168: 1713–1735.

Bagheri, H. C., J. Hermisson, J. R. Vaisnys, and G. P. Wagner. 2003. Effects of epistasis on phenotypic robustness in metabolic pathways. *Mathematical Biosciences* 184: 27–51.

Bank, C., R. T. Hietpas, J. D. Jensen, and D. N. Bolon. 2015. A systematic survey of an intragenic epistatic landscape. *Molecular Biology and Evolution* 32: 229–238.

Bourg, S., L. Jacob, F. Menu, and E. Rajon. 2019. Hormonal pleiotropy and the evolution of allocation trade-offs. *Evolution* 73: 661–674.

Boyle, E. A., I. L. Yang, and J. K. Pritchard. 2017. An expanded view of complex traits: From polygenic to omnigenic. *Cell* 169: 1177–1186.

Carlborg, Ö., and C. S. Haley. 2004. Epistasis: Too often neglected in complex trait studies? *Nature Reviews Genetics* 5: 618–625.

Carlborg, Ö., L. Jacobsson, P. Åhgren., P. Siegel, and L. Andersson. 2006. Epistasis and the release of genetic variation during long-term selection. *Nature Genetics* 38: 418–420.

Carter, A. J., J. Hermisson, and T. F. Hansen. 2005. The role of epistatic gene interactions in the response to selection and the evolution of evolvability. *Theoretical Population Biology* 68: 179–196.

Chantepie, S., and L. M. Chevin. 2020. How does the strength of selection influence genetic correlations? *Evolution Letters* 4: 468–478.

Cheverud, J. M. 1988. A comparison of genetic and phenotypic correlations. *Evolution* 42: 958–968.

Cheverud, J. M., and E. J. Routman 1995. Epistasis and its contribution to genetic variance components. *Genetics* 139: 1455–1461.

Cheverud, J. M., T. H. Ehrich, T. T. Vaughn, S. F. Koreishi, R. B. Linsey, and L. S. Pletscher 2004. Pleiotropic effects on mandibular morphology II: Differential epistasis and genetic variation in morphological integration. *Journal of Experimental Zoology B* 302: 424–435.

Chevin, L. M., C. Leung, A. Le Rouzic, and T. Uller. 2021. Using phenotypic plasticity to understand the structure and evolution of the genotype–phenotype map. *Genetica*, in press.

Clark, A. G. 1991. Mutation-selection balance and metabolic control theory. *Genetics* 129: 909–923.

Cockerham, C. C. 1954. An extension of the concept of partitioning hereditary variance for analysis of covariances among relatives when epistasis is present. *Genetics* 39: 859–882.

Dochtermann, N. A. 2011. Testing Cheverud's conjecture for behavioral correlations and behavioral syndromes. *Evolution* 65: 1814–1820.

Domingo, J., G. Diss, and B. Lehner. 2018. Pairwise and higher-order genetic interactions during the evolution of a tRNA. *Nature* 558: 117–121.

Draghi, J. A., T. L. Parsons, G. P. Wagner, and J. B. Plotkin. 2010. Mutational robustness can facilitate adaptation. *Nature* 463: 353–355.

Estes, S., C. C. Ajie, M. Lynch, and P. C. Phillips. 2005. Spontaneous mutational correlations for life-history, morphological and behavioral characters in *Caenorhabditis elegans. Genetics* 170: 645–653.

Fernández, J., and C. López-Fanjul.1996. Spontaneous mutational variances and covariances for fitness-related traits in *Drosophila melanogaster. Genetics* 143: 829–837.

Fierst, J. L., and P. C. Phillips. 2015. Modeling the evolution of complex genetic systems: The gene network family tree. *Journal of Experimental Zoology B* 324: 1–12.

Fisher, R. A. 1918. The correlation between relatives on the supposition of Mendelian inheritance. *Transactions of the Royal Society of Edinburgh* 53: 399–433.

Fisher, R. A. 1930 *The Genetical Theory of Natural Selection.* Oxford: Oxford University Press.

Fontana, W. 2002. Modeling 'evo-devo' with RNA. *Bioessays* 24: 1164–1177.

Frank, S. A. 1999. Population and quantitative genetics for regulatory networks. *Journal of Theoretical Biology* 197: 281–294.

Frank, S. A. 2019a. Evolutionary design of regulatory control. I. A robust control theory analysis of tradeoffs. *Journal of Theoretical Biology* 463: 121–137.

Frank, S, A. 2019b. Evolutionary design of regulatory control. II. Robust error-correcting feedback increases genetic and phenotypic variability. *Journal of Theoretical Biology* 468: 72–81.

Geiler-Samerotte, K., F. M. O. Sartori, and M. L. Siegal. 2019. Decanalizing thinking on genetic canalization. *Seminars in Cell and Developmental Biology* 88: 5–66.

Gjuvsland, A. B., B. J. Hayes, S. W. Omholt, and Ö Carlborg. 2007. Statistical epistasis is a generic feature of gene regulatory networks. *Genetics* 175: 411–420.

Gjuvsland, A. B., Y. Wang, E. Plahte, and S. W. Omholt. 2013. Monotonicity is a key feature of genotype-phenotype maps. *Frontiers Genetics* 4: 216.

Gromko, M. H. 1995. Unpredictability of correlated response to selection: Pleiotropy and sampling interact. *Evolution* 49: 685–693.

Hansen, T. F. 2003. Is modularity necessary for evolvability? Remarks on the relationship between pleiotropy and evolvability. *Biosystems* 69: 83–94.

Hansen, T. F., and G. P. Wagner. 2001. Modeling genetic architecture: A multilinear theory of gene interaction. *Theoretical Population Biology* 59: 61–86.

Hansen, T. F., T. M. Solvin, and M. Pavličev. 2019. Predicting evolutionary potential: a numerical test of evolvability measures. *Evolution* 73: 689–703.

Hermisson, J., and G. P. Wagner. 2004. The population genetic theory of hidden variation and genetic robustness. *Genetics* 168: 2271–2284.

Houle, D. 1991. Genetic covariance of fitness correlates: What genetic correlations are made of and why it matters. *Evolution* 45: 630–648.

Houle, D. 2001. Characters as the units of evolutionary change. In *The Character Concept in Evolutionary Biology*, edited by Günter P. Wagner, 109–140. Cambridge, MA: Academic Press.

Houle, D., and J. Fierst. 2013. Properties of spontaneous mutational variance and covariance for wing size and shape in *Drosophila melanogaster. Evolution* 67: 1116–1130.

Houle, D., K. A. Hughes, D. K. Hoffmaster, J. Ihara, S. Assimacopoulos, and B. Charlesworth. 1994. The effects of spontaneous mutation on quantitative traits. I. Variances and covariances of life history traits. *Genetics* 138: 773–785.

Houle, D., G. H. Bolstad, K. van der Linde, and T. F. Hansen. 2017. Mutation predicts 40 million years of fly wing evolution. *Nature* 548: 447–450.

Jacquier, H., A. Birgy, H. Le Nagard, et al. 2013. Capturing the mutational landscape of the beta-lactamase TEM-1. *PNAS* 110: 13067–13072.

Jones, A. G., S. J. Arnold, and R. Bürger. 2007. The mutation matrix and the evolution of evolvability. *Evolution* 61: 727–745.

Jones, A. G., R. Bürger, and S. J. Arnold. 2014. Epistasis and natural selection shape the mutational architecture of complex traits. *Nature Communications* 5: 1–10.

Kacser, H., and J. A. Burns. 1981. The molecular basis of dominance. *Genetics* 97: 639–666.

Keightley, P. D. 1989. Models of quantitative variation of flux in metabolic pathways. *Genetics* 121: 869–876.

Keightley, P. D. 1996. Metabolic models in selection response. *Journal of Theoretical Biology* 182: 311–316.

Keightley, P. D., and H. Kacser. 1987. Dominance, pleiotropy and metabolic structure. *Genetics* 117: 319–329.

Keightley, P. D., E. K. Davies, A. D. Peters, and R. G. Shaw 2000. Properties of ethylmethane sulfonate-induced mutations affecting life-history traits in *Caenorhabditis elegans* and inferences about bivariate distributions of mutation effects. *Genetics* 156: 143–154.

Kimura, M. 1983. *The Neutral Theory of Molecular Evolution*. Cambridge: Cambridge University Press.

Kruuk, L. E. B., J. Slate, and A. J. Wilson. 2008. New answers for old questions: The evolutionary quantitative genetics of wild animal populations. *AREES* 39: 525–548.

Lande, R. 1980. The genetic covariance between characters maintained by pleiotropic mutations. *Genetics* 94: 203–215.

Lande, R., and S. J. Arnold. 1983. The measurement of selection on correlated characters. *Evolution* 37: 1210–1226.

Le Rouzic, A. 2014. Estimating directional epistasis. *Frontiers in Genetics* 5: 198.

Le Rouzic, A., D. Houle, and T. F. Hansen. 2011. A modeling framework for the analysis of artificial-selection time series. *Genetics Research* 93: 155–173.

Lewontin, R. C. 1974. *The Genetic Basis of Evolutionary Change*. New York: Columbia University Press.

Li, C., W. Qian, C. J. Maclean, and J. Zhang. 2016. The fitness landscape of a tRNA gene. *Science 352*: 837–840.

Lynch, M. 1985. Spontaneous mutations for life-history characters in an obligate parthenogen. *Evolution* 39: 804–818.

Lynch, M., and B. Walsh. 1998. *Genetics and Analysis of Quantitative Traits*. Sunderland, MA: Sinauer.

Manolio, T. A., F. S. Collins, N. J. Cox, et al. 2009. Finding the missing heritability of complex diseases. *Nature* 461: 747–753.

Martin, G., S. F. Elena, and T. Lenormand. 2007. Distributions of epistasis in microbes fit predictions from a fitness landscape model. *Nature Genetics* 39: 555–560.

Mayer, C., and T. F. Hansen. 2017. Evolvability and robustness: A paradox restored. *Journal of Theoretical Biology* 430: 78–85.

Milocco, L., and I. Salazar-Ciudad. 2020. Is evolution predictable? Quantitative genetics under complex genotype-phenotype maps. *Evolution* 74: 230–244.

Nijhout, H. F., J. A. Best, and M. C. Reed. 2015. Using mathematical models to understand metabolism, genes and disease. *BMC Biology* 13: 79.

Nijhout, H. F., J. A. Best, and M. C. Reed. 2019. Systems biology of robustness and homeostatic mechanisms. *Wiley Interdisciplinary Reviews. Systems Biology and Medicine* 11: e1440.

Noble, D. W., R. Radersma, and T. Uller. 2019. Plastic responses to novel environments are biased towards phenotype dimensions with high additive genetic variation. *PNAS* 116: 13452–13461.

Olson, E. C., and R. L. Miller. 1958. *Morphological Integration.* Chicago: University of Chicago Press.

Omholt, S. W. 2013. From sequence to consequence and back. *Progress in Biophysics and Molecular Biology* 111: 75–82.

Orr, H. A. 2000. Adaptation and the cost of complexity. *Evolution* 54: 13–20.

Pavličev, M., and T. F. Hansen. 2011. Genotype-phenotype maps maximizing evolvability: Modularity revisited. *Evolutionary Biology* 38: 371–389.

Pavličev, M., and G. P. Wagner. 2012. Coming to grips with evolvability. *Evolution: Education and Outreach* 5: 231–244.

Pavličev, M. and G. P. Wagner. 2022. The value of broad taxonomic comparisons in evolutionary medicine: Disease is not a trait but a state of a trait! *MedComm* 3: e174.

Pavličev, M., A. Le Rouzic, J. M. Cheverud, G. P. Wagner, and T. F. Hansen. 2010. Directionality of epistasis in a murine intercross population. *Genetics* 185: 1489–1505.

Phillips, P. C., S. P. Otto, and M. C. Whitlock. 2000. Beyond the average: The evolutionary importance of gene interactions and variability of epistatic effects. In *Epistasis and the Evolutionary Process,* edited by J. B. Wolf, E. D. Brodie, and M. J. Wade, 20–38. New York: Oxford University Press.

Poelwijk, F. J., M. Socolich, and R. Ranganathan. 2019. Learning the pattern of epistasis linking genotype and phenotype in a protein. *Nature Communications* 10: 1–11.

Polly, P. D. 2008. Developmental dynamics and G-matrices: Can morphometric spaces be used to model phenotypic evolution? *Evolutionary Biology* 35: 83–96.

Rajasingh, H., A. B. Gjuvsland, D. I. Våge, and S. W. Omholt. 2008. When parameters in dynamic models become phenotypes: A case study on flesh pigmentation in the Chinook salmon (*Oncorhynchus tshawytscha*). *Genetics* 179: 1113–1118.

Reed, M. C., J. A. Best, M. Golubitsky, I. Stewart, and H. F. Nijhout. 2017. Analysis of homeostatic mechanisms in biochemical networks. *Bulletin of Mathematical Biology* 79: 2534–2557.

Richardson, J. B., L. D. Uppendahl, M. K. Traficante, S. F. Levy and M. L. Siegal. 2013. Histone variant HTZ1 shows extensive epistasis with, but does not increase robustness to, new mutations. *PLOS Genetics* 9: e1003733.

Riedl, R. 1975. *Die Ordnung des Lebendigen. Systembedingungen der Evolution.* Berlin: Paul Parey Verlag.

Roff, D. A. 1996. The evolution of genetic correlations: An analysis of patterns. *Evolution* 50: 1392–1403.

Schuster, P., W. Fontana., P. F. Stadler, and I. L. Hofacker. 1994. From sequences to shapes and back: A case study in RNA secondary structures. *Proceedings of the Royal Society B* 255: 279–284.

Schwenk, K., and G. P. Wagner. 2000. Function and the evolution of phenotypic stability. *American Zoologist* 41: 552–563.

Slatkin, M., and S. A. Frank. 1990. The quantitative genetic consequences of pleiotropy under stabilizing and directional selection. *Genetics* 125: 207–213.

Szathmáry, E. 1993. Do deleterious mutations act synergistically? Metabolic control theory provides a partial answer. *Genetics* 133: 127–132.

Turelli, M. 1985. Effects of pleiotropy on predictions concerning mutation-selection balance for polygenic traits. *Genetics* 111: 165–195.

Uricchio, L. H. 2020. Evolutionary perspectives on polygenic selection, missing heritability, and GWAS. *Human Genetics* 139: 5–21.

Vik, J. O., A. B. Gjuvsland, L. Li, K. Tøndel, et al. 2011. Genotype-phenotype map characteristics of an *in silico* heart cell. *Frontiers in Physiology* 2: 106.

Waddington, C. H. 1957. *The Strategy of the Genes; A Discussion of Some Aspects of Theoretical Biology.* London: Allen & Unwin.

Wagner A. 1996. Does evolutionary plasticity evolve? *Evolution* 50: 1008–1023.

Wagner, A. 2008. Robustness and evolvability: A paradox resolved. *Proceedings of the Royal Society B* 275: 91–100.

Wagner, G. P. 1989. Multivariate mutation-selection balance with constrained pleiotropic effects. *Genetics* 122: 223–234.

Wagner, G. P., and L. Altenberg. 1996. Perspective: Complex adaptations and the evolution of evolvability. *Evolution* 50: 967–976.

Wagner, G. P., J. P. Kenney-Hunt, M. Pavličev, J. R. Peck, D. Waxman, and J. M. Cheverud. 2008. Pleiotropic scaling of gene effects and the "cost of complexity." *Nature* 452: 470–472.

Wang, Z., B.-Y. Liao, and J. Zhang. 2010 Genomic patterns of pleiotropy and the evolution of complexity. *PNAS* 107: 18034–18039.

Widder, S., R. Sole, and J. Macia. 2012. Evolvability of feed-forward loop architecture biases its abundance in transcription networks. *BMC Systems Biology* 6: 7.

Young, A. I. 2019. Solving the missing heritability problem. *PLOS Genetics* 15: e1008222.

9 The Developmental Basis for Evolvability

Benedikt Hallgrímsson, J. David Aponte, Marta Vidal-García, Heather Richbourg, Rebecca Green, Nathan M. Young, James M. Cheverud, Anne L. Calof, Arthur D. Lander, and Ralph S. Marcucio

The developmental basis for evolvability is the central concern of evolutionary developmental biology. This is because the ways in which development structures the generation of phenotypic variation can influence evolution at both micro- and macro-evolutionary scales. Despite long-standing interest in the question of how development might structure phenotypic variation, general insights have been slow to emerge, largely because the mechanistic basis for quantitative variation has not been a strong focus in the field of developmental biology. Building on a body of work concerning the developmental genetics of quantitative variation in the morphology of the vertebrate face, we argue for three central mechanistic insights that are useful in relating development to evolutionary change: (1) Genetic influences on phenotypic variation are often highly nonlinear. (2) Despite the high dimensionality of both phenotypic and genomic variation, the two are often linked by latent factors, which—by capturing the influence of variation-determining developmental processes—act to drive integrated phenotypic variation. (3) Developmental systems tend to exhibit stochastic metastability, which ensures robustness but at the same time brings about the potential for discontinuous change. Taken together, these insights have implications for understanding long-standing issues in the developmental basis for evolvability, in particular, questions concerning the dynamics of continuous versus discontinuous evolutionary change.

Difficulty has hitherto arisen because variation is not studied for its own sake.
—Bateson, 1894. vii

By examining a great number of individuals of the same form of life, we find the types, variabilities and correlations of as many of these organs or characters as we choose.
—Pearson, 1900, 402

9.1 Introduction

The origin of evolutionary developmental biology is often cast as a project of synthesis between evolutionary biology on the one hand and developmental biology on the other. In this view, this project arose as a conscious effort in the mid-1970s, driven by a realization that development had been left out of evolutionary explanations following the "resolution" of the debate between the Mendelians and the biometricians via the modern synthesis (Gould 1977). This is, of course, an oversimplification, and it overlooks many important contributions to the integration of development and evolutionary explanations that occurred between the 1920s and the mid-1970s: these include the seminal works of Waddington and Schmalhausen on canalization (Waddington 1942, 1953, 1957; Schmalhausen 1949), as well as DeBeer's masterful integration of experimental embryology and comparative anatomy (DeBeer 1937). This narrative is also unfortunate because it implies a resolution of the divide between the Mendelian and biometrical schools of thought that did not actually occur. Fisher (1918) and others developed a coherent theory of quantitative genetics that was consistent with Mendelian inheritance. However, the modern synthesis did not actually engage with the work of such Mendelians as Bateson (1894) on the complexities of phenotypic variation and, in particular, on the prevalence of continuous versus discontinuous patterns of phenotypic variation.

The Mendelian-biometric debate was not resolved at the inception of the modern synthesis, because not enough was known about the developmental-mechanistic basis for phenotypic variation. Bateson's ideas were caricatured, and Goldschmidt's ridiculed, not because their conjectures on the "material basis for variation" could be empirically dismissed, but rather because they did not fit the assumptions of the infinitesimal model and so were orthogonal to the conceptual framework of the modern synthesis. It is ironic that while the architects of the modern synthesis harnessed the analysis of phenotypic variation (principally via analysis of variance) as a means to understand the mechanics of evolution at the population level, the field that they created tended to be dismissive of the study of variation for its own sake.

With the benefit of hindsight, it is easy to see continuity between Bateson's (1894) focus on discontinuity in patterns of variation, Goldschmidt's (1940) idea of "systemic mutations" creating discrete alternative states ("hopeful monsters"), Waddington's (1957) chreods and their relation to Rene Thom's (1983) mathematical catastrophe theory, and "tipping points" and "attractor states" in modern systems biology (Mojtahedi et al. 2016; Richard et al. 2016; Brackston et al. 2018; Yang et al. 2021). Despite this apparent continuity, however, a conceptual gulf persists between evolutionary biology and population genetics on the one hand and experimental developmental biology on the other, and this has hampered progress toward mechanistic understanding of the developmental basis for phenotypic variation. Rather than studying variation for its own sake, experimental developmental biology's focus tends to be on mechanisms or pathways and related concepts that are abstracted and disembodied from the individual organisms that provide the subjects of study. For most experimental studies in developmental biology, differences among genotypes or treatments are the units of analysis. Variation among individuals in such groups is rarely dwelt on and is usually regarded as a nuisance that reduces statistical power or complicates the ability to derive insight into the mechanisms or pathways that are the focus of study. By contrast, in genetics as well as in evolutionary biology, variation is the primary focus of study: Differences among individuals in populations form the fundamental units for most questions of interest. *For experimental developmental biology to provide insight into the developmental basis for evolvability, the critical question is: What is the mechanistic basis for variation among individuals within populations?* This is the same question that Bateson asked but lacked the technology to answer. To advance our understanding of how developmental mechanisms interact with population dynamics to produce evolutionary change, it is critical to harness the power of contemporary experimental developmental biology to address the "study of variation for its own sake."

9.2 Development and the Structure of Genotype-Phenotype Maps

The genotype-phenotype (G-P) map is a heuristic device, first explicitly used by Lewontin (1974), that describes the relationship between genetic and phenotypic variation. G-P "maps" have taken many forms (figure 9.1), but all share the basic notion that there are potential spaces, elements, or parameters of genetic and phenotypic variation in which one can be mapped on the other. Importantly, this "mapping" of genetic to phenotypic variation occurs through organismal development and physiology. The simplest kind of G-P map relates the 3 genotypes at a particular locus to a single continuous phenotypic outcome. Here, the assumption is that there is some kind of connection to degree of gene activity or function that maps to a phenotypic trait (Wright 1934; see figure 9.1C). (A specific variation of this general idea is the gene expression to phenotype (GxP) map, in which phenotypic variation is plotted against either the quantified or predicted level of gene expression (Green et al.

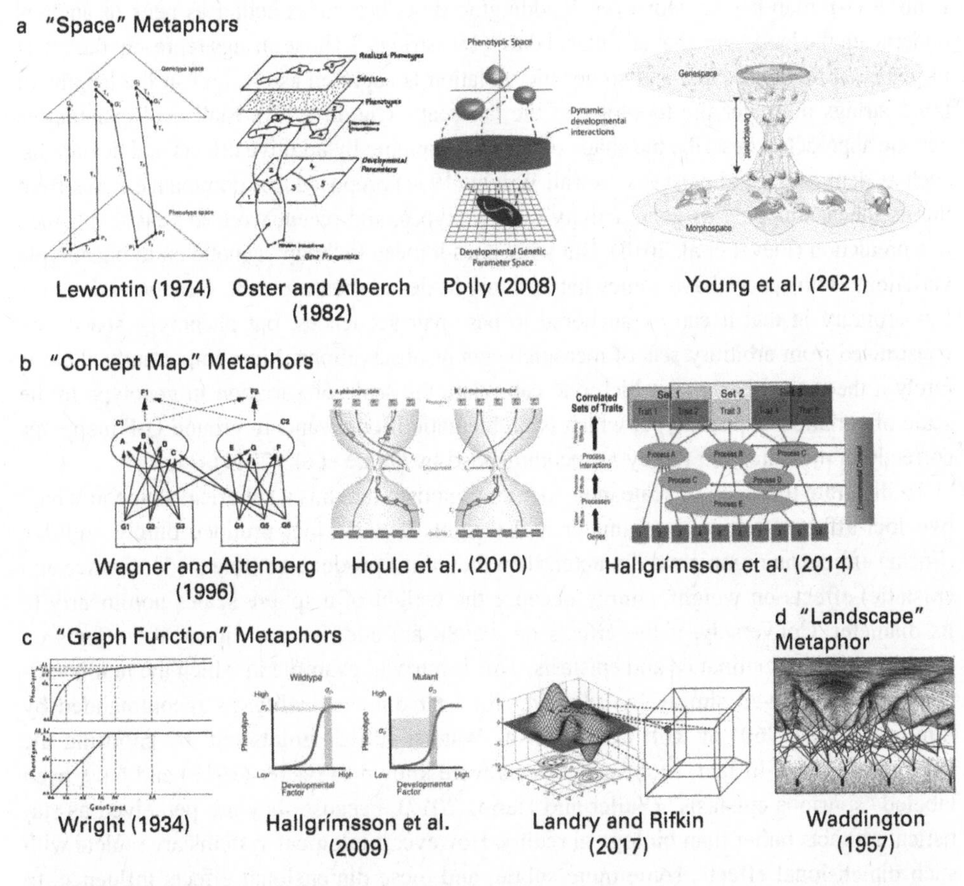

Figure 9.1
Various forms of "genotype-phenotype maps." (A) Depictions of relations of genetic and phenotypic variation on 2- or 3-dimensional planes generally follow the depiction of Lewontin (1974). (B) Various forms of concept mappings take inspiration from depictions of modularity in development by Wagner and Altenberg (1996) and Wagner (1996). There are many examples of graph function analogies that relate genotypes, gene expression, or developmental parameters to phenotypes; to our knowledge, the first example of this is in Wright's (1934) influential discussion of the developmental basis for dominance. (D) Waddington's "guidewire" metaphor for the influence of genes on the epigenetic landscape.

2017). Lewontin's (1974) G-P map (figure 9.1A) depicts genetic and phenotypic variation on separate genotype and phenotype spaces. Concept mapping has also been used to convey the G-P map idea. The most prominent example of this is the modular G-P map of Wagner and Altenberg (1996) (figure 9.1B), which conveys the modular organization gene effects on traits. The genotype-phenotype map concept has also been extended to apply to gene-regulatory network topologies and protein interactomes (Ahnert 2017).

Development is added to G-P maps as either intermediate spaces (Oster and Alberch 1982; Polly 2008; Courtier-Orgogozo et al. 2015) or as intermediate sets of connections among conceptual elements (Hallgrímsson et al. 2014) (figure 9.1B). Alternatively, development is treated as a "map" of parameter states that has some relationship to a corresponding map of phenotypes, as in Alberch (1991) or Kacser and Burns (1981). These intermediate levels are variously conceived such that they modulate (increase or decrease) and structure (create covariation or modular organization of) phenotypic variation.

Waddington's epigenetic landscape (1957), another well-known metaphor for development, is not a G-P map per se. However, Waddington describes genes acting as pegs or anchors underneath the landscape that are attached to it by "strings." Those strings represent the effect of genes on the landscape, and so genetic variation is captured as changes in the lengths of those strings that alter the topology of the landscape (figure 9.1D). Statistical quantitative genetic approaches describe the shape of the G-P mapping by additive effects and deviations, such as dominance and epistasis. Sewall Wright (1934) proposed that dominance arises from the nonlinear mapping of gene activity to phenotypes, and recent work in yeast confirmed this prediction (Fiévet et al. 2018). But what does it mean to "map" genetic on to phenotypic variation? Both spaces have somewhat arbitrary scales of measurement. Genetic variation is less arbitrary in that it can be anchored to base-pair sequences, but phenotype spaces are constructed from arbitrary sets of measurements or observations. More importantly, there is rarely a theoretical basis on which one can relate the scale of variation in genotype to the scale of variation in phenotype, which is problematic if one wants to ground G-P mappings correctly in measurement theory as recommended by Houle et al. (2011).

To illustrate the issue of scale, consider an organism that has a spherical shape in which two loci affect variation in diameter and weight. If these loci produce purely additive (linear) effects on organismal diameter, then they must produce nonlinear (dominance and epistatic) effects on weight, simply because the weight of a sphere scales nonlinearly to its diameter. Conversely, if the effects on weight are additive and linear, the effects on diameter exhibit dominance and epistasis. This is a trivial example in which the nonlinearities can be removed simply by adjusting for trait dimensionality, as recommended by Simpson et al (1960) or Roff (2012), or as Wagner (2015) points out, by choosing the appropriate scale. In fact, such scale effects were known to Fisher (1918) and have been labeled "spurious epistasis" (Sailer and Harms 2017), because they are perceived as statistical artifacts rather than biological reality. However, biological systems are replete with such dimensional effects, some quite subtle; and these dimensional effects influence, to varying degrees, many anatomical and physiological traits. Except in the most obvious situations, they are rarely considered and are often unknown. For example, Genome-Wide Association studies for body mass in production animals such as sheep are run on weight rather than on its cube root, along with linear measures of body and limb lengths (Cao et al. 2020; Tao et al. 2020). These geometric effects dictate that loci with additive effects in one measure must have nonadditive effects in the other unless the dimensionality effects are considered. For shape variation in geometric morphometrics, or for traits such as fractal dimensionality, functional outputs of organs with complex surface area/volume relationships, or pairwise voxel-based shape distances in registered brain images, scale and dimensionality effects are often difficult to disentangle in an informed manner based on principles and measurement theory.

From a statistical-genetic perspective, epistasis is a statistical deviation from additivity that is explained by an interaction term (Cheverud and Routman 1995b; Hansen and Pélabon 2021). From a developmental biology perspective, epistasis refers to mechanistic interactions, such as when the function or effect of one gene depends on the genotype at another or on genetic background. The complicating effects of dimensionality and scale aside, it is clear that epistasis is common, although how epistasis contributes to phenotypic variation is debated (Carlborg and Haley 2004; Mackay 2014). One reason for this debate may be the conflation of the statistical and biological concepts. As Hansen points out, biological epistasis does not only result in epistastic variance but can also modify additive variance (Hansen and Pélabon 2021). These differing views of epistasis have a long history in quantitative genetics, extending back to the opposing arguments of Ronald Fisher and Sewall Wright on the subject (Hansen 2015).

9.2.1 The Developmental Origins of Nonlinearities

But how can investigation of developmental mechanisms advance our understanding of the shape of genotype-phenotype relationships? Here we run straight into the conceptual divide that separates quantitative genetics on one hand from developmental biology on the other, a gulf that is remarkably similar today to the one that prevented Pearson and Bateson from understanding each other while agreeing, in principle, on the central importance of variation. For quantitative genetics, and much of evolutionary biology, epistasis and dominance are statistical concepts. They are defined in terms of deviations from linearity, as initially outlined by Fisher (1918). In developmental biology, however, gene interactions or dominance effects are mechanistic concepts that are premised on some understanding of developmental systems. In fact, this distinction is parallel to the one that motivated the contrasting pleas of Bateson and other Mendelians to study the "material basis" for variation, as opposed to the biometric argument that insight would emerge from ever more measurements and larger samples, and which famously led to the initial rejection of Fisher's foundational paper in 1918 (Norton and Pearson 1976).

This same conceptual divide impedes applying our growing understanding of developmental mechanisms to the question of how development influences genotype-phenotype relationships to influence evolutionary change (Hansen 2021). This is because to answer the question of how development is relevant to understanding epistasis and dominance, you essentially have to turn the problem around. Nonlinear genotype-phenotype maps, from a developmental perspective, are not caused by dominance or epistasis. Instead, nonlinearities are thought to emerge from developmental mechanisms, and these, in turn, produce dominance and epistasis. Thus, understanding these phenomena emerges from understanding the "material basis" of variation, rather than from solely studying the statistical patterns they create. Wright's suggestion that dominance arises from nonlinear mappings of gene activity onto phenotypic effects (Wright 1934) is an early example of this line of thought.

The concept of global or nonspecific epistasis is one approach to operationalizing this idea in the framework of statistical genetics (Cheverud and Routman 1995). In this view, epistasis and dominance arise indirectly, due to nonlinearities in development, even when genetic variation may produce additive effects on underlying developmental variables (Otwinowski et al. 2018). In other words, genetic effects map onto latent variables that capture the effects of developmental processes. It is the nonlinearities inherent in these developmental processes that result in genetic effects having nonlinear relationships to phenotypic variation.

But where do nonlinearities in development arise in the first place? The activity or expression level of a gene rarely translates directly to variation in phenotype. Instead, genetic effects act through multiple intermediate levels that are usually defined relative to an ontology or mechanistic model of a developmental system. Relations among these various levels can take a variety of forms, so a G-P function that describes the relationship between gene activity and phenotype is effectively an aggregate of the effects at these various levels of organization (Klingenberg 2004; Hallgrímsson et al. 2014; Green et al. 2017; see figure 9.2A). It turns out that many developmental processes are inherently nonlinear. Enzyme kinetics exhibit complex, nonlinear dynamics (Kacser and Burns 1981; Reuveni et al. 2014). The regulation of gene expression is generally driven by feedback control mechanisms that are highly nonlinear, sometimes producing bistable responses to continuous variation in the underlying parameters (Becskei et al. 2001; Jiménez et al. 2015). Cellular behavior can also relate nonlinearly to phenotypic outcomes. Relationships between the spatiotemporal dynamics of cell proliferation to morphogenesis are poorly understood. However, complex, nonlinear interactions can occur in growing tissues due to such factors as mechanical interactions with extracellular matrix or basement membranes as well as the interactions between cell polarity and proliferation (Wyczalkowski et al. 2012; Glen et al. 2019). Given this context, when genotype-phenotype maps are linear, this may occur when combined effects of multiple levels of process that are each nonlinear combine to approximate linearity (figure 9.2A). Perhaps more commonly, G-P maps may appear linear because the realized variation in a natural population falls within a portion of the potential G-P map, and that portion may well be linear.

Gene expression to phenotype maps represent compounding effects not just across levels of development but also across cells within organisms. The recent emergence of single-cell RNAseq is revolutionizing our understanding of how individual cell gene expression patterns and resulting cell behavior relate to morphogenetic processes in developing tissues. This has revealed, for example, that random monoallelic expression is common in mammalian cells (Deng et al. 2014). Thus in organisms heterozygous for a loss-of-function allele, some cells will express the nonfunctional allele, while others express the wildtype allele at the same level as wildtype cells. At the level of individual cells, therefore, the gene expression-phenotype map is not the hyperbolic curve envisioned by Wright (1934) but instead is a discrete function. When dominance occurs in such settings, the phenotypic outcome is either robust or disproportionately sensitive to altered behavior in 50% of cells rather than to altered average behavior or gene expression of all cells. Even in homozygotes, single-cell RNAseq analyses are also revealing the surprising extent of among-cell variation in gene expression, even for clonal populations of the same cell type, the significance of which is a subject of current interest (Pelkmans 2012; Shi et al. 2019).

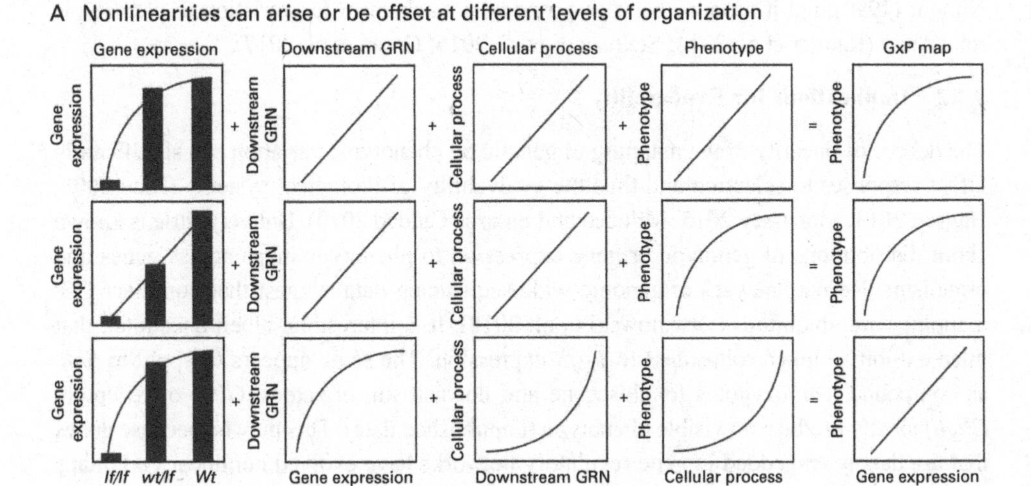

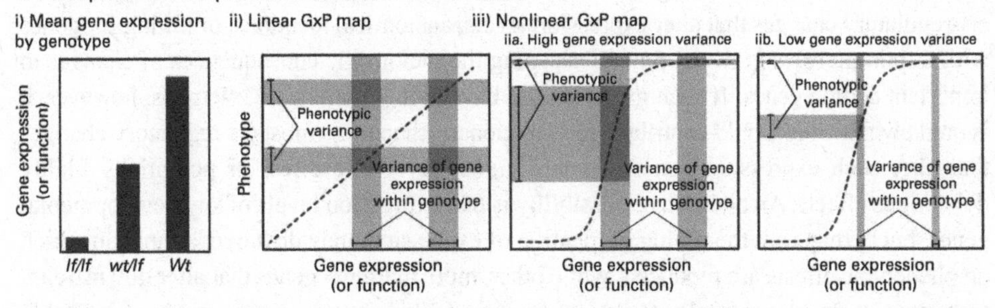

Figure 9.2
(A) The shape of a GxP map is a composite of multiple levels of organization, each of which will have a different form. At each level, nonlinearities may be counteracted or augmented from lower levels. (B) The shape of a G-P map interacts with variation within a genotype in order to modulate phenotypic variance. High levels of gene expression variance in a genotype will accentuate the tendency for a nonlinear G-P map to modulate phenotypic variance.

In prior work, our group has investigated the developmental underpinnings of G-P maps for mouse craniofacial morphology. In a chick model, we quantified a protein expression to morphology map for Sonic Hedgehog and craniofacial shape (Young et al. 2010). In subsequent work, we quantified bulk GxP maps for an *Fgf8* allelic series as well as for *Wnt9b* in a mouse model with variably penetrant cleft-lip (Green et al. 2017, 2019). In all three cases, the map is highly nonlinear. For the *Fgf8* allelic series, we found that variation in *Fgf8* expression did not really produce an effect on morphology unless it dropped below 50% of the wildtype level. Beyond that point, however, small changes in *Fgf8* expression are associated with large changes in morphology. We also quantified the genome-wide gene expression patterns in the developing face using bulk-RNAseq and found that genes downstream of *Fgf8* showed a similar nonlinear relationship to morphology (Green et al. 2017). One consequence of this finding is that the amount of phenotypic variation that corresponds to variation in gene or protein expression varies along the curve (figure 9.2B). This phenomenon was proposed as an explanation for variation in developmental stability by Klingenberg and

Nijhout (1999), but it has also been proposed as a mechanism for variation in phenotypic robustness (Ramler et al. 2014; Steinacher et al. 2016; Green et al. 2017).

9.2.2 Implications for Evolvability

The degree of linearity of the mapping of genetic on phenotypic variation can significantly affect responses to selection and thus the evolvability of biological systems (Rice 2008; Hansen 2013; Morrissey 2015; Milocco and Salazar-Ciudad 2020). But very little is known about distributions of genotype or gene expression to phenotype maps across genes and organisms. Recent analyses of genome-wide sequencing data suggest that nonlinear G-P mappings are ubiquitous (Otwinowski et al. 2018). It is interesting, albeit anecdotal, that mice exhibit so much robustness to *Fgf8* expression. The same appears to apply to *Shh*, as compound heterozygotes for this gene and downstream effectors (*Gli1*) or receptors (*Ptch*) mostly produce no visible phenotype (unpublished data). This may be because genes that are deeply embedded in gene regulatory networks have evolved nonlinear GxP maps that create robustness to variation in their expression. This is a reasonable but untested hypothesis. Stern (2000) suggested that evolutionary change most commonly involves cis-regulatory changes that alter expression levels, anatomical locations, or timing for genes with important roles in development, escaping the pleiotropic consequences of changes to important coding genes. If such genes tend to have highly nonlinear GxP maps, however, it is unlikely that they would contribute to evolutionary change, as most cis-regulatory changes that alter such expression patterns would either have little effect or potentially highly deleterious effects. An alternative possibility is that expression levels of key developmental genes, but perhaps not the timing or location of expression, may drift over a range in which no phenotypic effects are produced, while other, more marginal genes that alter downstream responses or developmental interactions are more likely to serve as sources of heritable variation in phenotypes. This would explain the apparent conservation of so many central pathways in development in the face of remarkable phenotypic diversification (Parikh et al. 2010; Galis chapter 16).[1]

The idea that evolutionary change occurs primarily at the margins of gene-regulatory networks is a hypothesis that needs to be tested systematically across developmental systems and phylogeny. This hypothesis is difficult to test, however. There are methods for quantifying network centrality that have been applied to gene regulatory networks (Jalili et al. 2016). The problem is that few gene-regulatory networks are really known with sufficient resolution and certainty to support this kind of analysis on a global scale. Clues to this question, though, can be gleaned from studies of modifier genes for structural birth defects. In the case of the Shh pathway and holoprosencephaly, for example, genes such as *Cdo* and *Boc* appear to modify the penetrance spectrum for *Shh* pathway mutations (Zhang et al. 2011). Evolutionary change for morphology that depends on the *Shh* pathway, such as early brain morphogenesis of face formation, may well be more likely to involve such peripheral "modifiers" or even small, cumulative effects of genes with even more distant connection to the pathway itself. It is suggestive that in a multivariate genotype-phenotype mapping (Aponte et al. 2021) of the effects of the *Shh* pathway on craniofacial shape in mice, the central genes such as *Shh* and *Gli1* contribute little, while more periph-

1. References to chapter numbers in the text are to chapters in this volume.

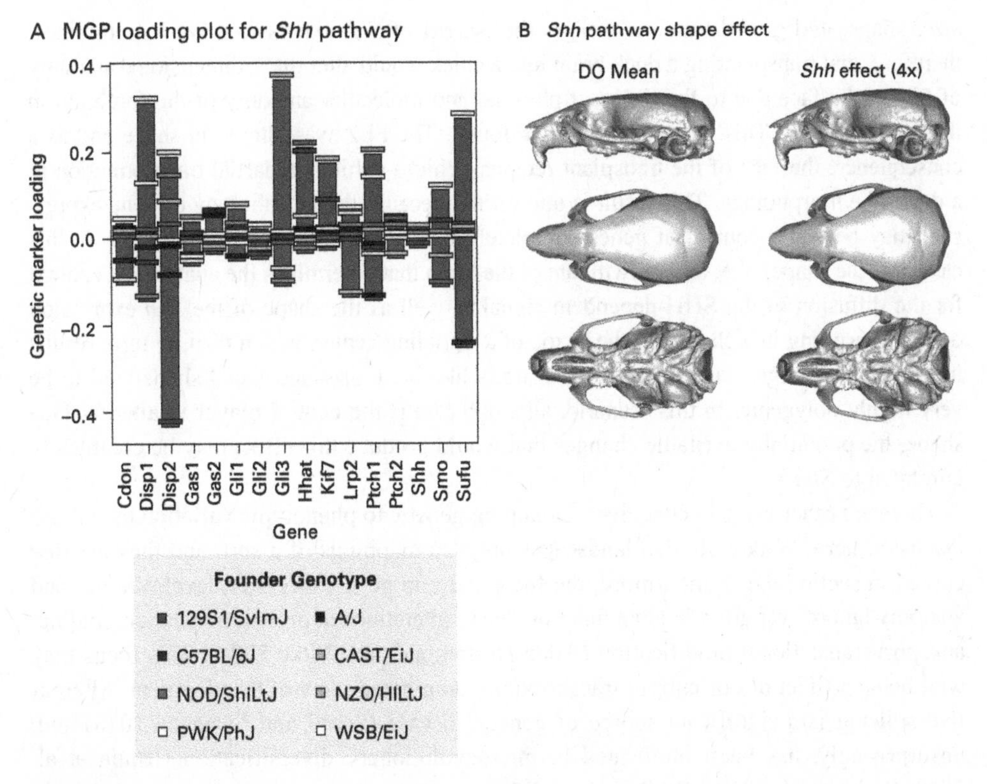

Figure 9.3
Multivariate genotype-phenotype mapping for the *Shh* pathway and adult craniofacial shape in mice. (A) The loadings by gene for members of this pathway are shown on the principal shape axis associated with the pathway as a whole. (B) The craniofacial shape effect of the pathway, as defined by the the following gene list: *Shh, Hhat, Displ, Disp2, Ptchl, Ptch2, Gasl, Gas2, Gas3, Glil, Gli2, EHZF, Gli2A, Gli3A, Lrp2, Hhrp, Gli2, Gpc3, Gli3, Smo, Cdon, Kif7, Sufu*. Genes that do not appear on the graph do not have annotated markers in the diversity outbred mouse megaMUGA panel.

eral members of the pathway such as *Ptch2, Disp2,* and *Sufo* account for the majority of the phenotypic variation that corresponds to this pathway (figure 9.3). However, this prediction needs to be tested systematically with available data for phenotypic traits with well-characterized developmental underpinnings, particularly as gene interactions can vary across developmental contexts (Greenspan 2009).

Genes with central roles in development can influence morphology in ways other than their level of expression. A neglected dimension of this problem is the morphology of gene expression (Xu et al. 2015; Martínez-Abadías et al. 2016). It turns out that the anatomical context in which a gene is expressed may significantly influence how its expression influences subsequent morphogenesis. We have demonstrated this by experimentally inducing an alteration in the shape of *Shh* expression in a duck-chick transplant model (Hu et al. 2015b). This model leverages the work of Ralph Marcucio and colleagues that established the role of a Sonic Hedgehog signaling center (the FEZ) in the ectoderm of the midface in determining the growth and morphogenesis of the facial prominences during face formation. *Shh* expression in the FEZ is induced by an SHH-dependent signal that diffuses from the developing forebrain. Duck and chick development is characterized by forebrains that differ in size, shape, and growth rate, as well as the pattern of *Shh* expression. We hypothesized, therefore, that transplanting a duck brain into a chick would alter the 3-dimensional anatomy of *Shh* in the face due to the different physical and molecular anatomy of the forebrain in these two species. This is exactly what we found. The FEZ was altered in shape and as a consequence, the face of the transplant recipient chicks exhibits a partial transformation to a duck-like morphology. This result is interesting, because the morphology of gene expression may be under control of genes completely unrelated to the gene in question. In this case, it is the shape, size, and growth rate of the brain that determines the anatomical context for the diffusion of the SHH-dependent signal as well as the shape of the *Shh* expression domain, resulting in a change in the shape of a signaling center, which then, in turn, results in altered morphogenesis. Morphological traits like brain size and facial shape tend to be very highly polygenic. In this scenario, although *Shh* is the central player in altering face shape, the potentially heritable changes that would produce this effect may be completely unrelated to *Shh*.

There are other ways to conceive of mapping genetic to phenotypic variation than those discussed here. Waddingtonian landscapes are also mappings of a sort, and they are discussed in section 9.4. Furthermore, the focus here on gene expression level, timing, and anatomy ignores variation in gene function due to alterations in protein translation, folding, and post-translational modification (Allan Drummond and Wilke 2009). This focus may well be an artifact of our current transcriptome-dominated view of development. Alternative splicing is a significant source of genetic disease (Scotti and Swanson 2016) and, unsurprisingly, has been implicated in microevolutionary diversification (Irimia et al. 2009; Smith et al. 2018). While it is well known that tertiary protein structure is highly conserved compared to both amino acid and DNA sequence variation (Konaté et al. 2019), such changes must contribute to evolutionary change at some level and frequency (Bajaj and Blundell 1984). Mapping genetic variation that alters protein synthesis to phenotypic variation is rare, so it is difficult to generalize about the shape of such maps.

Accordingly, although the exercise of mapping genetic to phenotypic variation makes assumptions that can be uncomfortable, this heuristic device is still a very useful way to frame our understanding of how development influences phenotypic variation and, thus, evolutionary change. Such maps, however constructed, are likely to be nonlinear, especially for genes with important roles in development. This fundamental aspect of biological systems needs to be better incorporated into quantitative-genetic theory, as Hansen (2013), Morrisey (2015), Rice (2008), and others have also argued.

9.3 Integration and the Dimensionality of G-P Maps

From one perspective, phenotypic variation has a higher dimensionality than genetic variation, because even though genomes consist of finite numbers of base-pair sequences, phenotypes can be quantified in infinitely many ways. However, closer examination of variation at levels from protein structure to morphology reveals a different pattern—one that shows that the variation structure at these higher levels is simpler than that of genomic sequences. We argue that this reduction in effective dimensionality is caused by the effects of integration on phenotypic variation (Hallgrímsson and Lieberman 2008; Hallgrímsson et al. 2009). If this is generally true, the implication is that developmental systems tend to be integrated, such that large potential spaces of genomic variation funnel down to a smaller, more circumscribed set of phenotypic outcomes.

The funneling of variation to a smaller set of possibilities is trivially true for synonymous mutations and holds somewhat less trivially for amino-acid sequence variation in proteins that does not alter their functional properties. At higher levels of organization, however, this pattern is much more complex, as it derives from the channeling of variation through developmental processes that produce structured influences on phenotypic variation. Study of this phenomenon has a long and complex history in evolutionary and developmental biology and is central to most constructions of how development influences evolutionary change. Darwin's discussion of correlation of growth assumes the existence of developmental factors that influence more than one trait (Darwin 1859). This is the central idea behind D'Arcy Thompson's work on the relationship between growth and shape (Thompson 1942), Huxley's pioneering work on allometric variation (Huxley 1932), and the factor-based conceptualization of correlated sets of traits by Sewall Wright (1932) and Olson and Miller (1958). This is also the idea behind the quantitative dissection of covariance structure in complex morphological traits (Wagner 1984, 1989).

Complementary but also orthogonal to this line of thinking is the concept of modularity. Modularity is orthogonal to integration because, at least as articulated by Wagner and colleagues (Wagner 1995, 2001a, 2001b; Wagner and Altenberg 1996; Wagner et al. 2007), it is a much more fundamental ontological statement about how life is structured. The module concept is much more than the idea that some traits covary more with each other than with others. Instead, it conveys the idea that there is a modular structure to developmental systems, from molecular pathways to gene and protein interactomes to tissue level interactions and anatomical features. In other words, there are units that are in part dissociated from others and often reused in different contexts (Schlosser and Wagner 2004). In its strongest form, this concept underpins homology (the correspondence of modules) and evolutionary novelty (the origin of new modules), and it aligns closely with the character concept (Wagner 2014).

So where do the concepts of integration and modularity meet within the conceptual landscape of evolutionary and developmental biology? The central tension that makes the answer to this question less obvious than one might think is one of ontology. A module is a thing—an element in an ontology of developmental systems. Integration, by contrast, is a property of a system. As Wagner has argued for canalization (Wagner et al. 1997), integration is a dispositional concept that refers to the *tendency* of phenotypic traits to covary (Hallgrímsson et al. 2009). In the palimpsest model for integration, covariation patterns are driven by variation in developmental processes. Since such processes occur at different times, scales, and locations in development, they have overlapping effects on the covariance structure of a set of phenotypic traits (Hallgrímsson et al. 2009). The palimpsest model allows for mechanistic dissection of covariation structure using the methods of developmental biology, because one can perturb specific processes and make predictions about changes to phenotypic covariance structure (Hallgrímsson et al. 2006, 2007, 2009; Jamniczky and Hallgrímsson 2009). Beyond agreeing with the obvious statement that fully integrated systems are not evolvable (Schlosser and Wagner 2004), the palimpsest model does not actually require the existence of modules in the natural-kind sense. Instead, it flows from an ontological model that is continuous with the older conceptions of trait correlation, from Wright's (1932) factors or latent variables to

Olson and Miller's (1958) correlation sets. This is not to say that the palimpsest model of integration contradicts the modularity concept, but rather that it is based on a different ontology of evolution and development.

This does not mean that the palimpsest model of integration is necessarily built on a firmer conceptual foundation than the modularity concept is. Instead of modules as natural kinds, the palimpsest reifies a different concept—that of the "developmental process." What is a developmental process? Although there are many things that we recognize clearly as "processes," such as cell proliferation, epithelium to mesenchyme transformation, or cell differentiation, there is actually no agreed-on definition of this commonly used phrase. Interestingly, in his book *Evolution of Developmental Pathways,* Wilkins (2002, 9) defined developmental pathways as "the underlying causal chain of gene activities that propels a particular developmental process." As for what is meant by "developmental process," he left that bit undefined. The Atchley and Hall (1991) model posits developmental units that correspond roughly to hierarchically arranged processes and cell populations underlying the generation of morphological variation in the mandible. But they also provide no general definition of a developmental process.

In its weakest sense, a developmental process is simply a series of events that together produce some outcome within a developmental system. What counts as a process in development, therefore, is in the eye of the beholder and depends on judgments about the relative importance of some events relative to others. When constructing an explanation of a developmental phenomenon, such as neural-crest migration, "important" events may be all those that are necessary for neural-crest migration to occur and particularly those that are specific to neural-crest migration as opposed to some other phenomenon of interest. In the context of the developmental basis for evolutionary change, importance may have more to do with the potential to contribute to the generation of heritable variation.

In his pitch for "variational structuralism," Wagner (2014) argues that there is a tension in biology between structuralism and functionalism in which the former focuses on organismal form and developmental architecture, while the latter focuses on adaptation and the functions of traits. The functionalism-structuralism distinction is well explored in the social sciences, but much less so in evolutionary biology, with notable exceptions such as Amundson (2005). By Wagner's (2014) definition, the entire project of integrating evolution and development is structuralist, but it may be more accurate to characterize concepts in evolutionary and evolutionary developmental biology as falling along a continuum between these two extremes. Concepts like modularity and novelties along with *bauplans* and characters are clearly on the structuralist end. However, concepts that deal with how development influences the generation of variation, such as integration or even developmental constraints, are closer to the functionalism of the modern synthesis. This is because they make minimal reference to abstract organization of developmental systems but instead deal with the flow of variation from genomes to phenotypes to populations. In this view, development is not unstructured, but exactly in which way it is structured is defined by the observer in a manner that is partly dependent on the context. We make this abstract point because this is where integration and the modularity concept meet. Both make strong statements about how development affects evolution via some form of dimensionality reduction. In the case of modularity, complex systems can evolve because everything is not equally connected to everything else. In the case of integration and the palimpsest model, phenotypic variation is structured by variation-generating developmental processes. In both cases, there is a general sense that we observe that development reduces the effective dimensionality of variation, whether it is by packaging it via units of some kind (modularity) or by creating axes of covariation that correspond to the influences of developmental processes on multiple traits (integration and the palimpsest).

9.3.1 Evidence for Dimensionality Reduction via Integration

Is it necessarily the case that development reduces the effective dimensionality of variation? Patterns of covariation for phenotypic traits obviously provide one kind of evidence for this. The degree to which variation in a set of covarying traits is explained by a few underlying factors is one measure of the extent to which development reduces the dimensionality of variation, as formalized in Wagner's variance of eigenvalues metric for integration (Wagner 1990; Pavličev et al. 2009). For morphological traits, such as limb element lengths, 3-dimensional shapes of skeletal elements in vertebrates, human faces, or outlines and vein configurations of fly wings, it is common to observe that the vast majority of phenotypic variation is captured by a few underlying factors (Young and Hallgrímsson 2005; Hallgrímsson et al. 2009; Haber and Dworkin 2017; Larson et al. 2018; Pitchers et al. 2019). This may not always be the case. Human brain morphology appears to have an unusually complex variation structure, for example (Naqvi et al. 2021). It is also not known to what extent reticular anatomical structures, such as vascular networks, nephrons, or lymphatic vessels exhibit this form of dimensionality reduction, as such morphologies are rarely quantified in a multivariate statistical framework.

Another line of evidence for dimensionality reduction via integration comes from the patterns of pleiotropic effects from mutations. Of course, if mutations have randomly varying patterns of pleiotropic effects, pleiotropy alone does not result in dimensionality reduction. But if mutations in different genes tend to converge on some limited set of patterns of pleiotropy, then this reflects the channeling of variation via developmental processes (or modular structure to development). There are many examples where this channeling clearly occurs. From our work, for example, we observe that mutations in mice that affect growth at the cranial synchondroses produce very similar patterns of shape change across the mouse skull, even if they involve very different mechanisms or even different kinds of perturbations to the synchondroses at the cellular level (Parsons et al. 2015). Similarly, mutations that affect overall growth are often associated with similar patterns of allometric change that would count as shared patterns of pleiotropy (Hallgrímsson et al. 2019).

A related source of evidence comes from observations of mutations with major effect. Bateson (1894) argued, as did both Goldschmidt (1940) and Alberch (1989), that much can be learned about the relationship between development and evolution from the study of mutations with large, often deleterious effect. Interestingly, Alberch (1989) referred to the "logic of monsters," by which he meant that such mutations are not random but rather follow rules that reflect constraints or regularities in developmental architecture. This pattern can also be seen in the patterns of morphological variation associated with genetic syndromes in humans (Hallgrímsson et al. 2020). Here, such mutations tend to produce effects that follow the same axes of covariation that characterize background variation in unrelated, unaffected individuals but to degrees that are more phenotypically extreme (unpublished data).

A more direct, statistical approach to the question of how development structures genetic influences on phenotypic variation is to search for latent variables that link multivariate genomic and phenotypic data. Mitteroecker et al. (2016) proposed a partial least-squares method, called multivariate genotype-phenotype mapping (MGP), which implements this approach. Applying it to mouse sample genotyped for 353 SNPs and phenotyped for 11 traits, they determined that 3 latent variables accounted 90% of genetic variance, which suggests a fairly low dimensionality for the genotype-phenotype map for these traits. Building on this work, we have implemented a process-centered MGP approach in which the joint effects of biologically coherent sets of genes can be related to multivariate phenotypic variation (Aponte et al. 2021). We have adapted this approach to the 8-way cross Diversity Outbred mouse population (Churchill et al. 2012) and implemented it on a large (N=1,145) dataset of 3-dimensional craniofacial landmarks (Percival et al. 2017; Katz et al. 2020). An important advantage of geometric morphometric datasets is that directions of variation in shape space are meaningful (Pitchers et al. 2019). This crucial property allows us to compare the directions of shape change associated with different gene-ontology sets both to each other and to the shape changes associated with mutations of known effect.

Our MGP analyses of craniofacial shape show that gene lists associated with processes such as chondrocyte differentiation or Fgf signaling correlate with large amounts of variation in craniofacial shape (Aponte et al. 2021). Interestingly, the largest three axes (PCs) of background shape variation tend to covary strongly with process effects, while lower order PCs tend to exhibit lower correlations with MGP process effects. Thus the major axes (PCs) of shape variation capture the conjoint effects of many underlying processes, while lower order axes exhibit weak correlations with process effects. This is not necessarily expected. One could imagine, for example, that each major axes of shape variation is driven by a very small number of specific processes of large effect, while processes of smaller effect associate strongly with lower order axes. This is not the case, however. The major axes of shape variation capture the conjoint effects of many processes that essentially move the phenotype in the same direction in morphospace. Furthermore, mutants with known mutations tend to have directions of effect that covary with those of gene lists for associated pathways or processes (Aponte et al. 2021). These results suggest that development structures the flow of variation such that its effective dimensionality is reduced and that this occurs via the channeling of variation through processes that drive axes of covariation among phenotypic traits. This would also imply that genotype-phenotype maps tend to be characterized by patterns of pleiotropy determined by covariance-generating developmental processes, which is a prediction of the palimpsest model for the developmental basis for integration (Hallgrímsson et al. 2009).

To what extent are genotype-phenotype maps generally structured in this way? This is very difficult to test systematically for many reasons, not the least of which is the inherent arbitrariness in the definitions of phenotypic traits (Wagner and Zhang 2011). Wagner and Zhang (2011), however, observed from analyses of QTL data that SNPs with larger effects on phenotypic variation also had higher tendency for pleiotropic effects. This observation is borne out in a systematic analysis of pleiotropic effects in GWAS studies of various human diseases (Chesmore et al. 2018). This is exactly what one would predict from the palimpsest model. It is also what one observes from analyses of mutations of major effect in model organisms (Hallgrímsson et al. 2009). This tendency for large mutational effects to associate positively with pleiotropy may occur because as development is perturbed to a greater degree, there will be a greater tendency for knock-on effects on other aspects of development. Such effects can be nonlinear, which would further exacerbate the tendency for mutations of small effect to be more local and those of large effect to be more pleiotropic. A further prediction of the palimpsest model is that the degree of penetrance for disease-causing mutations would correlate with the degree of multi-system involvement, with more severe manifestations exhibiting a greater tendency for systemic disruption.

To illustrate this, imagine the distribution of genetic effects on cell proliferation in the maxillary prominences of developing vertebrate face. Mutations with small effects on this process might result in a locally confined alteration in facial shape. Mutations with larger effects might secondarily produce displacements of surrounding tissues, producing more widespread changes in facial shape. Finally, mutations with large negative effects might result in prominences too small to allow normal formation of the primary palate, resulting in a global transformation of facial shape, including altered nasal morphology.

Many mutations affect ubiquitous processes rather than something highly specific in terms of location and developmental stage, as in the previous example. Genetic diseases such as cohesinopathies (Santos et al. 2016; Newkirk et al. 2017), ribosomopathies (Dauwerse et al. 2011) and spliceosomopathies (Bernier et al. 2012) affect ubiquitous, global processes. Yet they often combine highly specific manifestations, such as the upturned nose in Cornelia de Lange Syndrome, with varied phenotypic features. These varied but oddly specific phenotypic manifestations of mutations to ubiquitous or global processes suggest that aspects of developmental systems often vary in sensitivity to perturbations to such processes. These specific manifestations may often represent "tip of the iceberg" phenomena, in that they are visible manifestations of widespread underlying dysregulation. Fusion of the facial prominences to form a primary palate may be particularly sensitive to mutations that perturb cell adhesion or epithelium to mesenchymal transformation, for example, while bone growth at growth plates and synchondroses may be particularly sensitive to mutations that affect extracellular matrix organization (e.g., collagen). In this case, mild perturbations to such processes may result in a highly localized, specific phenotype like cleft lip or altered stature while more severe (larger effect) mutations that affect those same processes will produce other phenotypes as other aspects of development reach a threshold at which these perturbations come to matter. Such scenarios are also consistent with the palimpsest model and with the observation that mutations of larger effects on a particular trait also tend to be more pleiotropic.

Multiple lines of evidence converge on the view that the observed dimensionality of phenotypic variation is reduced as variation flows from genetic influences through developmental processes. This has important implications for complex trait genetics, in the sense that gene- or SNP-level explanations of phenotypic variation may be missing the forest for the trees. Meaningful explanations of phenotypic variation must address this structuring role of developmental processes. In fact, it is difficult to imagine how else to make sense of the increasingly long lists of genes that are associated with many complex traits. What this means for evolvability of developmental systems, however, is a more complex question, as modeling the impact of pleiotropy patterns on the evolvability of specific traits has shown (Pavličev and Hansen 2011; Pavličev and Wagner 2012).

9.4 Canalization, Developmental Noise, and Continuous versus Discontinuous Variation

Waddington (1957) described phenotypes as resulting from canalized trajectories on a probability landscape formed by the interacting elements of developmental systems. In this metaphor, when perturbations exceed the tendency of the system to proceed along such trajectories, discontinuous change in the phenotype may result. For many years, work on canalization has focused more on the genetics and mechanisms underlying continuous variation—or the modulation of the magnitude of phenotypic variance (Scharloo 1991; Wagner et al. 1997; Hallgrímsson et al. 2006, 2009). Much less attention in evolutionary developmental biology has been paid to what is arguably a more interesting aspect of Waddington's metaphor—the existence of cusps or "tipping points" between relatively discrete alternative phenotypic outcomes. However, this aspect of Waddington's epigenetic conceptual framework inspired catastrophe theory in mathematics (Thom 1975) and is foundational to modern systems biology (Fagan 2012).

A central idea behind Waddington's epigenetic view of development, or the "cybernetics of development" in his words (Waddington 1957), is that developmental systems exhibit *metastability.* This means that biological systems have a tendency to resist the effects of perturbations within some range, but when that range is exceeded, a transition to an alternative metastable state occurs. Another key element in Waddington's epigenetic view of development is the presence of developmental noise or, in his words, "looseness of play in the epigenetic machine" (Waddington 1957, 39). The result is a view of development as a stochastic system, resistant to perturbation, but with the potential to change dramatically given an insult of sufficient magnitude.

The extent to which developmental systems are stochastic and exhibit metastability as envisioned by Waddington and at what levels of organization is an important question with direct implications for the developmental basis for evolvability. It is well known that variation can be either continuous or discontinuous and when continuous, various shapes of distributions are possible. Polygenic traits that are discontinuous are normally modeled as thresholds on underlying liability distributions that are continuous. As Davidson (1982, 2002, 2010) argued, metastability may be adaptive. Life-cycle transitions, for example, should involve rapid and irreversible transitions from one metastable state to another, simply because piecemeal transitions are likely to be maladaptive. The same could be claimed for cell differentiation. Cells with a phenotype intermediate between epithelium and mesenchyme, for example, would be nonfunctional in most developmental contexts. Davidson described development as a "progression of states of spatially defined regulatory gene expression" (Davidson et al. 2002, 1670), implying a series of discontinuous transitions between metastable states. All this is to say that the existence of discontinuities in variation in both evolution and development is not particularly novel or controversial. In fact, there are strong theoretical reasons to regard metastability as a fundamental feature of developmental systems.

That being said, actual evidence for metastability has tended to be very indirect until recently. In humans, it is well known that disease-causing mutations are often associated with a range of phenotypic outcomes. People with the same holoprosencephaly associated mutation, for example, can exhibit a phenotypic range from normal facial development to various forms of holoprosencephaly, including cyclopia (Roessler and Muenke 2010). Remarkably, analyses of large reference genome datasets also reveal multiple genetic variants associated with rare Mendelian disease in healthy individuals (Chen et al. 2016; Tarailo-Graovac et al. 2017). While such phenotypic heterogeneity effects are often attributed to gene or gene-environment interaction effects (Girirajan et al. 2012), variable phenotypic outcomes also occur in monogenic mice where such interaction effects are likely to be minimal. In fact, this tendency for mutations of major effect to be associated with increased phenotypic variance was a major focus of early work on the mechanistic basis for canalization (Mather 1953; Thoday 1958; Rendel 1959, 1967). We have shown that in both mice and humans, the magnitude of the effect of mutations on craniofacial shape correlates positively with their effect on phenotypic variance within genotypes (Hallgrímsson et al. 2009, 2020). We argued in section 9.1 that nonlinear mapping of developmental processes to phenotypic outcomes is one explanation for such effects. Even this explanation, however, assumes that there is some "looseness of play" in the system. Such nonlinear mappings only modulate robustness if there is some magnitude of variance in a developmental process that can produce differing magnitudes of phenotypic change depending on the location along the curve (figure 9.2B).

So what is the source of this looseness of play in development? Recent advances in the ability to quantify gene expression at the level of individual cells in developing embryos combined with theoretical advances in systems biology are breathing new life into Waddington's epigenetic landscape metaphor for development. Individual cells, it turns out, have surprisingly variable patterns of gene expression (Chen et al. 2012; Osorio et al. 2019). More importantly, this variance is itself tied to the progression of cells along cell-fate trajectories. Gene expression patterns can be viewed as a noisy, dynamical system, and the "landscape" of gene expression is given explicit meaning as a quasipotential surface (Ferrell 2012). On this landscape, cell fates are "attractors," or wells into which cells progress, and from which they may exit through transitions driven by stochastic fluctuations in gene expression patterns. As cells move from one attractor to another, they pass through a transition state, or "tipping point," (Brackston et al. 2018; Coomer et al. 2020; Guillemin and Stumpf 2021). A key characteristic of tipping points between cell fates is an increase in among-cell variance and a decrease in among-cell covariance of gene expression patterns (Chen et al. 2012). When cells are at such a transitional state, individual cell behaviors are less stable and more easily perturbed to transition along alternative paths. Studies have confirmed the existence of such patterns during cell differentiation in various developmental systems (Mojtahedi et al. 2016; Richard et al. 2016; Brackston et al. 2018; Yang et al. 2021), suggesting that cell differentiation processes exhibit metastability. At the cellular level, therefore, development does appear to exhibit the "looseness of play" envisioned by Waddington.

But is stochastic behavior of individual cells during development relevant to understanding the developmental basis for organismal-level variation or in evolutionary contexts, or even phenotypic heterogeneity in human disease? The answer to that question likely depends on context. There is likely genetic variation for the "shape" of the quasipotential landscape of cellular differentiation, and selection presumably acts to reduce the deleterious consequences of this variation for normal development. There may also be heritable variation in the magnitudes of gene expression variance or the covariance of gene expression among cells, as is suggested by the recent finding of heritable variation in single-cell gene expression variance in cultured human lymphoblastoid cell lines (Osorio et al. 2019). The presence of heritable variation in the variability of cell-specific gene expression is also suggested by the fact that this occurs in some human diseases that involve global dysregulation of gene expression, such as Cornelia de Lange Syndrome (Kawauchi et al. 2009; Santos et al. 2016).

The phenotypic consequences of global dysregulation of gene expression provide an important clue to the question of whether the stochastic metastability of cellular behavior matters for understanding the developmental basis for evolutionary change. In Cornelia de Lange syndrome, global dysregulation of gene expression does not result in random patterns of dysmorphology but rather quite specific but anatomically distributed features, such as altered midfacial and nasal morphology and congenital heart defects (Santos et al. 2016). To return to the "tip of the iceberg" metaphor, this pattern suggests that some developmental processes are more sensitive to global dysregulation of cell-specific gene expression patterns than others. Calof et al. (2020) argue that a general feature of such disorders, or "transcriptomopathies" is a constellation of neurodevelopmental, craniofacial, and somatic growth features. This may be because for some reason, these processes are less canalized (shallower wells in a quasipotential landscape). Such patterns can also occur, however, when a developmental process depends critically on a pattern of differentiation in a very small number of cells. In such cases, increased variance of gene expression or a change in the shape of the landscape could mean that an insufficient number of such critical cells do whatever it is that is required for development to proceed normally.

Dependence on the activity of a small number of cells occurs fairly frequently in development, but particularly during early stages of pattern formation and morphogenesis. At the cardiac crescent stage in mice, for example, a small group of ectodermal cells located in the ventral midline of the forebrain are critical for specification of the ventral midline and septation of the forebrain into right and left lobes (Hallonet et al. 2002; Corbin et al. 2003). The primitive heart (endocardial tubes) must function at a very early stage in vertebrates with a very small number of cells. Similarly, facial prominence outgrowth and the fusion of the facial prominences during face formation depend on specific gene expression patterns in relatively small, single cell layer patches of ectoderm (Chai and Maxson 2006; Hu and Marcucio 2009; Hu et al. 2015a). Growth and proper timing of fusion at cranial sutures depends on surprisingly complex interactions of specific and small cell populations in developing sutures (Farmer et al. 2021). Similarly, pattern formation in vertebrate limbs depends on gene expression patterns in small, single cell layer patches of ectoderm (Farmer et al. 2021). There are many other such examples. However, these examples mentioned are interesting, as they also represent some of the most common locations for structural birth defects in humans, many of which have diverse and sometimes polygenic causes.

Developmental systems may also exhibit sensitivity for reasons other than dependence on some small, critical cell populations. The preponderance of neurodevelopmental symptoms in human genetic disease, for instance, suggests that higher level cognitive functions in humans can be perturbed in many different ways. Human growth as measured by stature is similarly sensitive to a wide range of perturbation. Using data from Hallgrímsson et al. (2020) and Hammond (2007) on stature for 6,580 human subjects with 529 different genetic syndromes and 15,000 unaffected unrelated subjects (Cole et al. 2016; Shaffer et al.

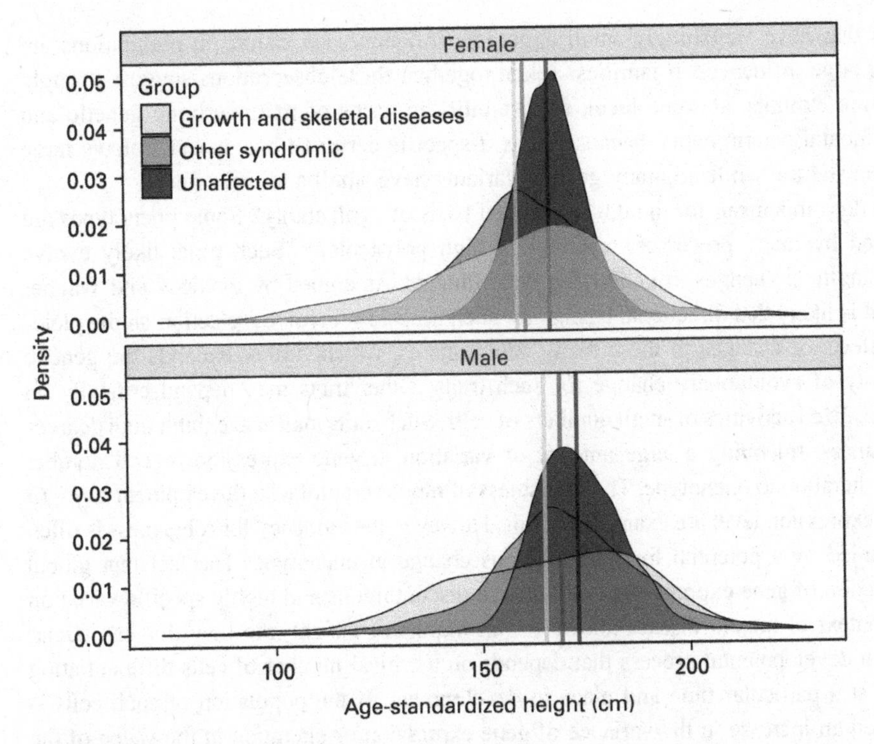

Figure 9.4
Distributions of stature for humans with genetic disease compared to unaffected, unrelated subjects. These distributions show that although the variance of stature is increased in syndromic subjects generally, this is particularly true for genetic diseases with known effects on somatic growth or musculoskeletal development.

2016; Hallgrímsson et al. 2020; Liu et al. 2021), we compared age-standardized distributions of stature for syndromic subjects to unaffected unrelated controls (figure 9.4). Syndromic subjects overall have reduced mean stature but also an increased variance of stature. Most mutations appear to reduce stature while a few have the opposite effect. This trend is accentuated when you look only at those syndromes with known effects on growth, cartilage or bone development (figure 9.4). Human stature is extraordinarily polygenic, influenced by thousands of genomic variants (Wood et al. 2014). It is also sensitive to environmental effects, such as nutritional status (Prentice et al. 2013). The finding that so many genetic diseases affect stature is consistent with this picture. Clearly, variation in many metabolic and developmental processes funnel their effects toward changes in stature. Interestingly, this diversity of determinants does not result in phenotypic robustness, which contradicts the prediction of Soulé (1982) and Lande (1977), who argued that traits that represent the composite of many underlying factors should have low variance. Stature is clearly a trait with many and diverse influences, and yet it is highly variable and also so sensitive to perturbation that it is a commonly used sentinel indicator of environmental stress in children. How this occurs is not known, but it must involve heterogeneity in the potential effects of different developmental influences on stature as well as complex and potentially nonlinear interactions among these potential influences. A related conundrum is the fact that stature is both extraordinarily polygenic and highly heritable. Yet it frequently varies dramatically among family members. This must mean that genetic variants that have vanishingly small aggregate influences on stature in populations are exerting large influences in families. Taken together, these observations suggest strongly that stature exhibits, at some level, metastability in terms of its underlying genetic and developmental determinants, because context-specific perturbations appear to have large effects on stature, while so many genetic variants have small additive effects.

What does this mean for the developmental basis of evolvability? Some phenotypes are influenced by many processes, resulting in high polygenicity. Such traits likely evolve through multiple changes in underlying determinants. As argued by Pavličev and Wagner (2012), it is likely that directional change for such traits also involves selection on the pleiotropic effects of changes to these many determinants, which further expands the genetic complexity of evolutionary change for such traits. Other traits may depend critically on highly localized activities of small numbers of cells. Such traits may also exhibit high degrees of robustness, tolerating a large amount of variation in gene expression or cell number without alteration to phenotype. The robustness of mouse craniofacial development to *Fgf8* and *Shh* expression level are examples of this. However, the tendency for robustness is often accompanied by a potential for discontinuous change in phenotype. The fact that global dysregulation of gene expression contributes to discontinuous and highly specific variation in the context of human disease allows for an important insight into how this can occur. Imagine a developmental process that depends on a critical number of cells differentiating properly at a particular time and place in development. If the population of such cells is small, then an increase in the variance of gene expression or alteration in the shape of the developmental quasipotential landscape might push development over a critical threshold in some individuals and not in others, purely as a consequence of sampling variation among individual embryos (figure 9.5). Such situations reveal the existence of thresholds or "tipping points" in development. Where such tipping points exist, a change in the mean gene expression pattern may also result in a discontinuous change in phenotype. As so elegantly argued by Alberch (1989), structural birth defects reveal underlying regularities in development that are relevant to evolution. The fact that specific malformations occur as a consequence of global dysregulation, for example, shows that there are processes in development that are more sensitive to perturbation than others are. Such processes are likely to be nexuses of discontinuous change, both in the generation of variation in populations and in the developmental basis for evolutionary change.

9.5 Conclusion: Implications for the Developmental Basis of Evolvability

The central question of evolutionary developmental biology is the developmental basis for evolvability (Hendrikse et al. 2007). Since natural selection acts on phenotypic variation, addressing this question revolves around the ways in which development biases the structure of phenotypic variation. Some of us (Hallgrímsson et al. 2002, 2009; Hendrikse et al. 2007) have argued previously that this structuring of variation takes two forms. These are the packaging of traits into patterns of covariation via the influence of shared developmental effects on one hand and the modulation of the magnitude of variance expressed for a given genetic or environmental effect on the other. Here, we build on this argument to develop three somewhat interwoven threads that further flesh out how development structures phenotypic variation. In the first, we argue that nonlinearities are common in

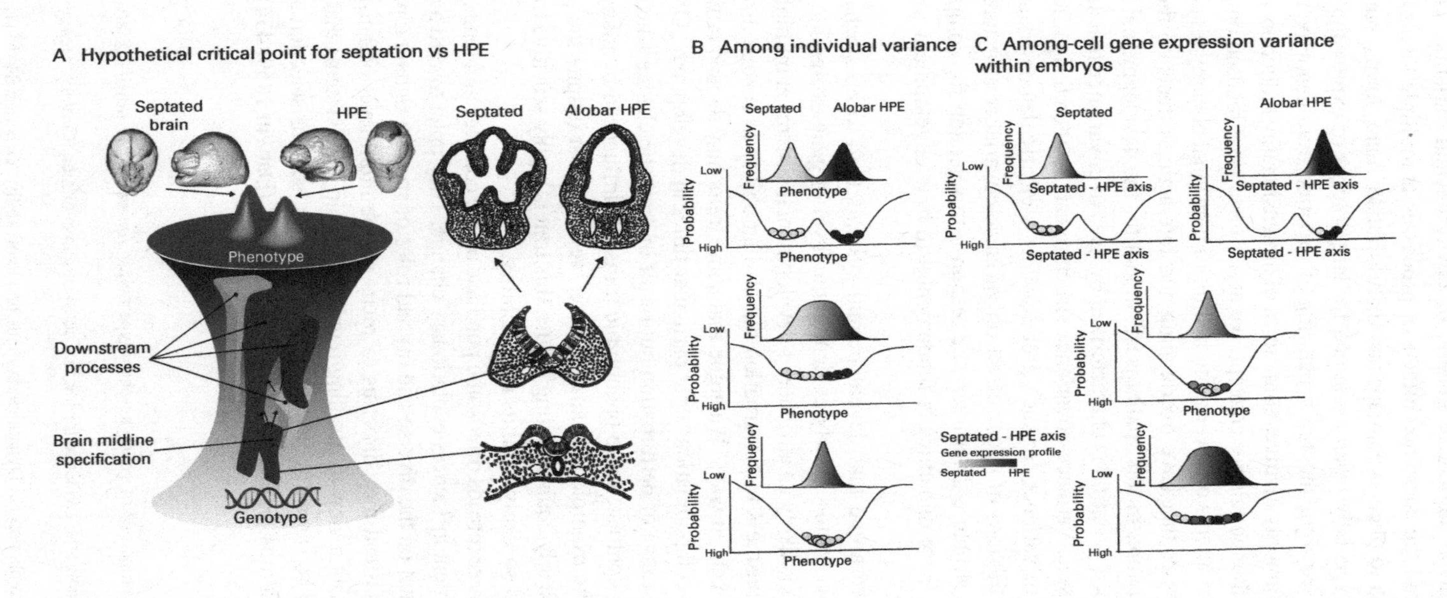

Figure 9.5
Metastability in development and discontinuous phenotypic variation. (A) A metaphorical genotype-phenotype map showing an early developmental process with knock-on effects on later development. Here the critical process is specification of the midline of the brain, which depends on a small number of ventral midline neuroepithelial cells. (B) Here it is shown among individual-phenotypic variation across developmental stages. (C) Variation in gene expression among cells in this critical small population of cells and their descendants shows how sampling variation at early stages can lead to individual embryos pursuing one of two distinct developmental trajectories.

development, and that this can result in modulation of the amount of phenotypic variance that corresponds to a given amount of genetic or environmental variation. Genotype-phenotype maps represent the combined effects of processes at multiple levels of organization. Gene expression to phenotype maps can therefore take many forms, but they are likely to be nonlinear, especially when examined for large ranges of expression.

In the second thread, we argue that the structuring of variation from genotype to phenotype by development commonly involves some form of dimensionality reduction. Although this can be cast around the modularity concept, we argue for a view focused on developmental process, in which patterns of covariation in multidimensional phenotypes are driven by developmental processes that have overlapping and interacting effects on the overall variance-covariance structure. In this "palimpsest" model for the developmental basis for variation in complex traits, variation is characterized by axes of covariance that are driven by developmental processes. Variance-covariance structure can be altered significantly by modulating the amount of variance for such processes, as is clearly demonstrated by our work on the effect of mutations of major effect on variance-covariance structures. Since the variance-covariance structure determines the response to selection for complex phenotypes, understanding these dynamics is fundamental to how development influences evolvability.

In the third thread, we argue that developmental systems tend to be stochastic and exhibit metastability. This is most clearly evident at the level of gene expression patterns for individual cells, but it is likely an attribute of developmental systems at multiple levels. Such systems exhibit robustness to perturbation but also an ability to change discontinuously when perturbed in certain ways. We argue that some traits may depend critically on what happens in specific cell populations or at particular times in development. Often such processes are robust to a range of perturbation but result in dramatic change when pushed past some threshold. By contrast, some phenotypic traits are influenced by many underlying processes. Using the example of human stature, we show that despite the presence of many, diverse and probably countervailing inputs, this trait is highly sensitive to diverse genetic perturbations and so likely exhibits metastability.

Taken together, these three threads offer some purchase on the complex and long-standing question of how development influences evolutionary change. They point toward a research direction that is focused not on the diverse and myriad roles of specific signaling pathways or transcription factors. Instead they direct us to focus on the multi-level organization of developmental systems with a view to gleaning regularities from such systems that help us understand how they can be shaped by evolutionary processes, such as selection and drift to produce "endless forms most beautiful and most wonderful" (Darwin 1859, 434).

References

Ahnert, S. E. 2017. Structural properties of genotype-phenotype maps. *Journal of the Royal Society Interface* 14: 20170275.

Alberch, P. 1989. The logic of monsters: Evidence for internal constraint in evolution. *Geobios, Memoire Special* 12: 21–57.

Alberch, P. 1991. From genes to phenotype: Dynamical systems and evolvability. *Genetica* 84: 5–11.

Allan Drummond, D., and C. O. Wilke. 2009. The evolutionary consequences of erroneous protein synthesis. *Nature Reviews Genetics* 10: 715724.

Amundson, R. 2005. *The Changing Role of the Embryo in Evolutionary Thought: Roots of Evo-Devo*. Cambridge: Cambridge University Press.

Aponte, J. D., D. C. Katz, D. M. Roth, et al. 2021. Shapes and genescapes: Mapping multivariate phenotype-biological process associations for craniofacial shape. *bioRxiv*: 2020.2011.2012.378513.

Atchley, W. R., and B. K. Hall. 1991. A model for development and evolution of complex morphological structures. *Biological Review* 66: 101–157.

Bajaj, M., and T. Blundell. 1984. Evolution and the tertiary structure of proteins. *Annual Review of Biophysics and Bioengineering* 13: 453–492.

Bateson, W. 1894. *Materials for the Study of Variation Treated with Especial Regard to Discontinuity in the Origin of Species (reprinted in 1992)*. London: Johns Hopkins University Press.

Becskei, A., B. Séraphin, and L. Serrano. 2001. Positive feedback in eukaryotic gene networks: Cell differentiation by graded to binary response conversion. *EMBO Journal* 20: 2528–2535.

Bernier, F. P., O. Caluseriu, S. Ng, et al. 2012. Haploinsufficiency of SF3B4, a component of the pre-mRNA spliceosomal complex, causes Nager syndrome. *American Journal of Human Genetics* 90: 925–933.

Brackston, R. D., E. Lakatos, and M. P. H. Stumpf. 2018. Transition state characteristics during cell differentiation. *PLOS Computational Biology* 14: e1006405.

Calof, A. L., R. Santos, L. Groves, C. Oliver, and A. D. Lander. 2020. Cornelia de Lange Syndrome: Insights into neural development from clinical studies and animal models. In *Neurodevelopmental Disorders,* edited by J. Rubenstein, B. Chen, P. Rakic, K. Y. Kwan, 129–157, London: Elsevier.

Cao, Y., X. Song, H. Shan, et al. 2020. Genome-wide association study of body weights in Hu sheep and population verification of related single-nucleotide polymorphisms. *Frontiers in Genetics* 11: 588.

Carlborg, Ö., and C. S. Haley. 2004. Epistasis: Too often neglected in complex trait studies? *Nature Reviews Genetics* 5: 618–625.

Chai, Y., and R. E. Maxson, Jr. 2006. Recent advances in craniofacial morphogenesis. *Developmental Dynamics* 235: 2353–2375.

Chen, L., R. Liu, Z.-P. Liu, M. Li, and K. Aihara. 2012. Detecting early-warning signals for sudden deterioration of complex diseases by dynamical network biomarkers. *Scientific Reports* 2: 342.

Chen, R., L. Shi, J. Hakenberg, et al. 2016. Analysis of 589,306 genomes identifies individuals resilient to severe Mendelian childhood diseases. *Nature Biotechnology* 34: 531–538.

Chesmore, K., J. Bartlett, and S. M. Williams. 2018. The ubiquity of pleiotropy in human disease. *Human Genetics* 137: 39–44.

Cheverud, J. M., and E. J. Routman. 1995. Epistasis and its contribution to genetic variance components. *Genetics* 139: 1455–1461.

Churchill, G. A., D. M. Gatti, S. C. Munger, and K. L. Svenson. 2012. The diversity outbred mouse population. *Mammalian Genome* 23: 713–718.

Cole, J. B., M. Manyama, E. Kimwaga, et al. 2016. Genomewide Association Study of African children identifies association of SCHIP1 and PDE8A with facial size and shape. *PLOS Genetics* 12: e1006174.

Coomer, M. A., L. Ham, and M. P. H. Stumpf. 2020. Shaping the epigenetic landscape: Complexities and consequences. *bioRxiv*: 2020.2012.2021.423724.

Corbin, J. G., M. Rutlin, N. Gaiano, and G. Fishell. 2003. Combinatorial function of the homeodomain proteins Nkx2.1 and Gsh2 in ventral telencephalic patterning. *Development* 130: 4895–4906.

Courtier-Orgogozo, V., B. Morizot, and A. Martin. 2015. The differential view of genotype–phenotype relationships. *Frontiers in Genetics* 6: 179.

Darwin, C. 1859. *The Origin of Species* (reprinted 1975). New York: Avenel.

Dauwerse, J. G., J. Dixon, S. Seland, et al. 2011. Mutations in genes encoding subunits of RNA polymerases I and III cause Treacher Collins syndrome. *Nature Genetics* 43: 20.

Davidson, E. H. 1982. Evolutionary change in genomic regulatority organization: Speculations on the origins of novel biological structure. In *Evolution and Development,* edited by J. T. Bonner, 65–84. New York: Springer-Verlag.

Davidson, E. H. 2010. *The Regulatory Genome: Gene Regulatory Networks in Development and Evolution*. Amsterdam: Elsevier.

Davidson, E. H., J. P. Rast, P. Oliveri, et al. 2002. A genomic regulatory network for development. *Science* 295:1669–1678.

DeBeer, G. 1937. *The Development of the Vertebrate Skull.* London: Oxford University Press.

Deng, Q., D. Ramsköld, B. Reinius, and R. Sandberg. 2014. Single-Cell RNA-Seq reveals dynamic, random monoallelic gene expression in mammalian cells. *Science* 343: 193–196.

Fagan, M. B. 2012. Waddington redux: Models and explanation in stem cell and systems biology. *Biology & Philosophy* 27: 179–213.

Farmer, D. J. T., H. Mlkochova, Y. Zhou, et al. 2021. The developing mouse coronal suture at single-cell resolution. *Nature Communications* 12: 4797.

Ferrell, J. E., Jr. 2012. Bistability, bifurcations, and Waddington's epigenetic landscape. *Current Biology* 22: R458–466.

Fiévet, J. B., T. Nidelet, C. Dillmann, and D. de Vienne. 2018. Heterosis is a systemic property emerging from nonlinear genotype-phenotype relationships: Evidence from in vitro genetics and computer simulations. *Frontiers in Genetics* 9: 159.

Fisher, R. A. 1918. The correlation between relatives on the supposition of Mendelian inheritance. *Transactions of the Royal Society of Edinburgh* 52: 399–433.

Girirajan, S., J. A. Rosenfeld, B. P. Coe, et al. 2012. Phenotypic heterogeneity of genomic disorders and rare copy-number variants. *New England Journal of Medicine* 367: 1321–1331.

Glen, C. M., M. L. Kemp, and E. O. Voit. 2019. Agent-based modeling of morphogenetic systems: Advantages and challenges. *PLOS Computational Biology* 15: e1006577.

Goldschmidt, R. 1940. *The Material Basis of Evolution.* New Haven, CT: Yale University Press.

Gould, S. J. 1977. *Ontogeny and Phylogeny.* Cambridge, MA: Belknap Press of Harvard University Press.

Green, R. M., J. L. Fish, N. M. Young, et al. 2017. Developmental nonlinearity drives phenotypic robustness. *Nature Communication* 8:1970.

Green, R. M., C. L. Leach, V. M. Diewert, et al. 2019. Nonlinear gene expression-phenotype relationships contribute to variation and clefting in the A/WySn mouse. *Developmental Dynamics* 248: 1232–1242.

Greenspan, R. J. 2009. Selection, gene interaction, and flexible gene networks. *Cold Spring Harbor Symposia on Quantitative Biology* 74: 131–138.

Guillemin, A., and M. P. H. Stumpf. 2021. Noise and the molecular processes underlying cell fate decision-making. *Physical Biology* 18: 011002.

Haber, A., and I. Dworkin. 2017. Disintegrating the fly: A mutational perspective on phenotypic integration and covariation. *Evolution* 71: 66–80.

Hall, B. K., and T. Miyake. 1995. Divide, accumulate, differentiate: Cell condensation in skeletal development revisited. *International Journal of Developmental Biology* 39: 881–893.

Hallgrímsson, B., and D. E. Lieberman. 2008. Mouse models and the evolutionary developmental biology of the skull. *Integrative and Comparative Biology* 48: 373–384.

Hallgrímsson, B., K. Willmore, and B. Hall. 2002. Canalization, developmental stability, and morphological integration in primate limbs. *American Journal of Physical Anthropology* S35: 131–158.

Hallgrímsson, B., J. J. Y. Brown, A. F. Ford-Hutchinson, D. Sheets, M. L. Zelditch, and F. R. Jirik. 2006. The brachymorph mouse and the developmental-genetic basis for canalization and morphological integration. *Evolution and Development* 8: 61–73.

Hallgrímsson, B., D. E. Lieberman, N. M. Young, T. Parsons, and S. Wat. 2007. Evolution of covariance in the mammalian skull. *Novartis Foundation Symposia* 284: 164–185.

Hallgrímsson, B., H. Jamniczky, N. M. Young, C. Rolian, T. E. Parsons, J. C. Boughner, and R. S. Marcucio. 2009. Deciphering the palimpsest: Studying the relationship between morphological integration and phenotypic covariation. *Evolutionary Biology* 36: 355–376.

Hallgrímsson, B., W. Mio, R. S. Marcucio, and R. Spritz. 2014. Let's face it—complex traits are just not that simple. *PLOS Genetics* 10: e1004724.

Hallgrímsson, B., D. C. Katz, J. D. Aponte, et al. 2019. Integration and the developmental genetics of allometry. *Integrative and Comparative Biology* 59: 1369–1381.

Hallgrímsson, B., J. D. Aponte, D. C. Katz, et al. 2020. Automated syndrome diagnosis by three-dimensional facial imaging. *Genetics in Medicine* 22: 1682–1693.

Hallonet, M., K. H. Kaestner, L. Martin-Parras, H. Sasaki, U. A. Betz, and S. L. Ang. 2002. Maintenance of the specification of the anterior definitive endoderm and forebrain depends on the axial mesendoderm: A study using HNF3beta/Foxa2 conditional mutants. *Developmental Biology* 243: 20–33.

Hammond, P. 2007. The use of 3D face shape modelling in dysmorphology. *Archives of Disease in Childhood* 92: 1120.

Hansen, T. F. 2013. Why epistasis is important for selection and adaptation. *Evolution* 67: 3501–3511.

Hansen, T. F. 2015. Measuring gene interactions. In *Epistasis: Methods and Protocols,* edited by J. H. Moore and S. M. Williams, 115–143. New York: Springer.

Hansen, T. F. 2021. Epistasis. In *Evolutionary Developmental Biology: A Reference Guide,* edited by L. Nuño de la Rosa and G. B. Müller, 1097–1110. Cham, Switzerland: Springer.

Hansen, T. F., and C. Pélabon. 2021. Evolvability: A quantitative-genetics perspective. *AREES* 52: 153–175.

Hendrikse, J. L., T. E. Parsons, and B. Hallgrímsson. 2007. Evolvability as the proper focus of evolutionary developmental biology. *Evolution & Development* 9: 393–401.

Houle, D., C. Pélabon, G. P. Wagner, and T. F. Hansen. 2011. Measurement and meaning in biology. *Quarterly Review of Biology* 86: 3–34.

Hu, D., and R. S. Marcucio. 2009. A SHH-responsive signaling center in the forebrain regulates craniofacial morphogenesis via the facial ectoderm. *Development* 136: 107–116.

Hu, D., N. M. Young, X. Li, Y. Xu, B. Hallgrímsson, and R. S. Marcucio. 2015a. A dynamic Shh expression pattern, regulated by SHH and BMP signaling, coordinates fusion of primordia in the amniote face. *Development* 142: 567–574.

Hu, D., N. M. Young, Q. Xu, et al. 2015b. Signals from the brain induce variation in avian facial shape. *Developmental Dynamics* 244: 1133–1143.

Huxley, J. S. 1932. *Problems of Relative Growth.* London: Methuen.

Irimia, M., J. L. Rukov, S. W. Roy, J. Vinther, and J. Garcia-Fernandez. 2009. Quantitative regulation of alternative splicing in evolution and development. *BioEssays* 31: 40–50.

Jalili, M., A. Salehzadeh-Yazdi, S. Gupta, O. Wolkenhauer, M. Yaghmaie, O. Resendis-Antonio, and K. Alimoghaddam. 2016. Evolution of centrality measurements for the detection of essential proteins in biological networks. *Frontiers in Physiology* 7: 375.

Jamniczky, H. A., and B. Hallgrímsson. 2009. A comparison of covariance structure in wild and laboratory muroid crania. *Evolution* 63: 1540–1556.

Jiménez, A., J. Cotterell, A. Munteanu, and J. Sharpe. 2015. Dynamics of gene circuits shapes evolvability. *PNAS* 112: 2103–2108.

Kacser, H., and J. A. Burns. 1981. The molecular basis of dominance. *Genetics* 97: 639–666.

Katz, D. C., J. D. Aponte, W. Liu, et al. 2020. Facial shape and allometry quantitative trait locus intervals in the Diversity Outbred mouse are enriched for known skeletal and facial development genes. *PLOS One* 15: e0233377.

Kawauchi, S., A. L. Calof, R. Santos, et al. 2009. Multiple organ system defects and transcriptional dysregulation in the Nipbl(+/–) mouse, a model of Cornelia de Lange Syndrome. *PLOS Genetics* 5: e1000650.

Klingenberg, C. P. 2004. Dominance, nonlinear developmental mapping and developmental stability. In *The Biology of Genetic Dominance,* edited by R. A. Veitia, 37–51. New York: CRC Press.

Klingenberg, C. P., and H. F. Nijhout. 1999. Genetics of fluctuating asymmetry: A developmental model of developmental instability. *Evolution* 53: 358–375.

Konaté, M. M., G. Plata, J. Park, D. R. Usmanova, H. Wang, and D. Vitkup. 2019. Molecular function limits divergent protein evolution on planetary timescales. *eLife* 8: e39705.

Lande, R. 1977. On comparing coefficients of variation. *Systematic Zoology* 26: 214–217.

Larson, J. R., M. F. Manyama, J. B. Cole, et al. 2018. Body size and allometric variation in facial shape in children. *American Journal of Physical Anthropology* 165: 327–342.

Lewontin, R. C. 1974. *The Genetic Basis of Evolutionary Change.* New York: Columbia University Press.

Liu, C., M. K. Lee, S. Naqvi, et al. 2021. Genome scans of facial features in East Africans and cross-population comparisons reveal novel associations. *PLOS Genetics* 17: e1009695.

Mackay, T. F. C. 2014. Epistasis and quantitative traits: Using model organisms to study gene-gene interactions. *Nature Reviews Genetics* 15: 22–33.

Mäki-Tanila, A., and W. G. Hill. 2014. Influence of gene interaction on complex trait variation with multilocus models. *Genetics* 198: 355–367.

Martínez-Abadías, N., R. Mateu, M. Niksic, L. Russo, and J. Sharpe. 2016. Geometric morphometrics on gene expression patterns within phenotypes: A case example on limb development. *Systematic Biology* 65: 194–211.

Mather, K. 1953. Genetical control of stability in development. *Heredity* 7: 297–336.

Milocco, L., and I. Salazar-Ciudad. 2020. Is evolution predictable? Quantitative genetics under complex genotype-phenotype maps. *Evolution* 74: 230–244.

Mitteroecker, P., J. M. Cheverud, and M. Pavličev. 2016. Multivariate analysis of genotype-phenotype association. *Genetics* 202: 1345–1363.

Mojtahedi, M., A. Skupin, J. Zhou, et al. 2016. Cell fate decision as high-dimensional critical state transition. *PLOS Biology* 14: e2000640.

Morrissey, M. B. 2015. Evolutionary quantitative genetics of nonlinear developmental systems. *Evolution* 69: 2050–2066.

Naqvi, S., Y. Sleyp, H. Hoskens, et al. 2021. Shared heritability of human face and brain shape. *Nature Genetics* 53: 830–839.

Newkirk, D. A., Y. Y. Chen, R. Chien, et al. 2017. The effect of Nipped-B-like (Nipbl) haploinsufficiency on genome-wide cohesin binding and target gene expression: Modeling Cornelia de Lange syndrome. *Clinical Epigenetics* 9: 89.

Norton, B., and E. S. Pearson. 1976. A note on the background to, and refereeing of, RA Fisher's 1918 paper 'On the correlation between relatives on the supposition of Mendelian inheritance.' *Notes and Records of the Royal Society of London* 31: 151–162.

Olson, E. C., and R. L. Miller. 1958. *Morphological Integration*. Chicago: University of Chicago Press.

Osorio, D., X. Yu, Y. Zhong, et al. 2019. Extent, heritability, and functional relevance of single cell expression variability in highly homogeneous populations of human cells. *bioRxiv*: 574426.

Oster, G., and P. Alberch. 1982. Evolution and bifurcation of developmental programs. *Evolution* 36: 444–459.

Otwinowski, J., D. M. McCandlish, and J. B. Plotkin. 2018. Inferring the shape of global epistasis. *PNAS* 115: E7550.

Parikh, A., E. R. Miranda, M. Katoh-Kurasawa, et al. 2010. Conserved developmental transcriptomes in evolutionarily divergent species. *Genome Biology* 11: R35.

Parsons, T. E., C. M. Downey, F. R. Jirik, B. Hallgrímsson, and H. A. Jamniczky. 2015. Mind the gap: Genetic manipulation of basicranial growth within synchondroses modulates calvarial and facial shape in mice through epigenetic interactions. *PLOS One* 10: e0118355.

Pavličev, M., and T. F. Hansen. 2011. Genotype-phenotype maps maximizing evolvability: Modularity revisited. *Evolutionary Biology* 38: 371–389.

Pavličev, M., and G. P. Wagner. 2012. A model of developmental evolution: selection, pleiotropy and compensation. *TREE* 27: 316–322.

Pavličev, M., J. M. Cheverud, and G. P. Wagner. 2009. Measuring morphological integration using eigenvalue variance. *Evolutionary Biology* 36: 157–170.

Pearson, K. 1900. *The Grammar of Science*. Second edition. London: Adam and Charles Black.

Pelkmans, L. 2012. Using cell-to-cell variability—A new era in molecular biology. *Science* 336: 425–426.

Percival, C. J., R. Green, D. M. Gatti, D. Pomp, et al. 2017. QTL analysis of a trade-off in bone length within the mouse zygomatic arch. *Faseb Journal* 31, 387.3.

Pitchers, W., J. Nye, E. J. Márquez, A. Kowalski, I. Dworkin, and D. Houle. 2019. A multivariate genome-wide association study of wing shape in *Drosophila melanogaster*. *Genetics* 211: 1429–1447.

Polly, P. D. 2008. Developmental dynamics and G-Matrices: Can morphometric spaces be Used to model phenotypic evolution? *Evolutionary Biology* 35: 83–96.

Prentice, A. M., S. E. Moore, and A. J. Fulford. 2013. Growth faltering in low-income countries. *World Review of Nutrition and Dietetics* 106: 90–99.

Ramler, D., P. Mitteroecker, L. N. Shama, K. M. Wegner, and H. Ahnelt. 2014. Nonlinear effects of temperature on body form and developmental canalization in the threespine stickleback. *Journal of Evolutionary Biology* 27: 497–507

Rendel, J. M. 1959. Canalization of the scute phenotype of *Drosophila*. *Evolution* 13: 425–439.

Rendel, J. M. 1967. *Canalization and gene control*. London: Logos Press.

Reuveni, S., M. Urbakh, and J. Klafter. 2014. Role of substrate unbinding in Michaelis–Menten enzymatic reactions. *PNAS* 111: 4391–4396.

Rice, S. H. 2008. Theoretical approaches to the evolution of development and genetic architecture. *Annals of the New York Academy of Sciences* 1133: 67–86.

Richard, A., L. Boullu, U. Herbach, et al. 2016. Single-cell-based analysis highlights a surge in cell-to-cell molecular variability preceding irreversible commitment in a differentiation process. *PLOS Biology* 14: e1002585.

Roessler, E., and M. Muenke. 2010. The molecular genetics of holoprosencephaly. *American Journal of Medical Genetics Part C: Seminars in Medical Genetics* 154C: 52–61.

Roff, D. A. 2012. *Evolutionary Quantitative Genetics*. Dordrecht: Springer Science & Business Media.

Sailer, Z. R., and M. J. Harms. 2017. Detecting high-order epistasis in nonlinear genotype-phenotype maps. *Genetics* 205: 1079–1088.

Santos, R., S. Kawauchi, R. E. Jacobs, et al. 2016. Conditional creation and rescue of nipbl-deficiency in mice reveals multiple determinants of risk for congenital heart defects. *PLOS Biology* 14: e2000197.

Scharloo, W. 1991. Canalization—Genetic and developmental aspects. *AREES* 22: 65–93.

Schlosser, G., and G. P. Wagner. 2004. Introduction: The modularity concept in developmental and evolutionary biology. In *Modularity in Development and Evolution,* edited by G. Schlosser, and G. P. Wagner, 1–11. Chicago and London: University of Chicago Press.

Schmalhausen, I. I. 1949. *Factors of Evolution.* Chicago: University of Chicago Press.

Scotti, M. M., and M. S. Swanson. 2016. RNA mis-splicing in disease. *Nature Reviews Genetics* 17:19–32.

Shaffer, J. R., E. Orlova, M. K. Lee, et al. 2016. Genome-Wide Association Study reveals multiple loci influencing normal human facial morphology. *PLOS Genetics* 12: e1006149.

Shi, J., T. Li, L. Chen, and K. Aihara. 2019. Quantifying pluripotency landscape of cell differentiation from scRNA-seq data by continuous birth-death process. *PLOS Computer Biology* 15: e1007488.

Simpson, G. G., A. Roe, and R. C. Lewontin. 1960. *Quantitative Zoology.* New York: Harcourt, Brace & World.

Smith, C. C. R., S. Tittes, J. P. Mendieta, E. Collier-Zans, H. C. Rowe, L. H. Rieseberg, and N. C. Kane. 2018. Genetics of alternative splicing evolution during sunflower domestication. *PNAS* 115: 6768–6773.

Soulé, M. E. 1982. Allomeric variation 1. The theory and some consequences. *American Naturalist* 120: 751–764.

Steinacher, A., D. G. Bates, O. E. Akman, and O. S. Soyer. 2016. Nonlinear dynamics in gene regulation promote robustness and evolvability of gene expression levels. *PLOS One* 11: e0153295.

Stern, D. L. 2000. Evolutionary developmental biology and the problem of variation. *Evolution* 54: 1079–1091.

Tao, L., X. Y. He, L. X. Pan, J. W. Wang, S. Q. Gan, and M. X. Chu. 2020. Genome-wide association study of body weight and conformation traits in neonatal sheep. *Animal Genetics* 51: 336–340.

Tarailo-Graovac, M., J. Y. A. Zhu, A. Matthews, C. D. M. van Karnebeek, and W. W. Wasserman. 2017. Assessment of the ExAC data set for the presence of individuals with pathogenic genotypes implicated in severe Mendelian pediatric disorders. *Genetics in Medicine* 19: 1300–1308.

Thoday, J. 1958. Homeostasis in a selection experiment. *Heredity* 12: 401–415.

Thom, R. 1975. *Structural Stability and Morphogenesis.* Reading, MA: Benjamin Cummings Publishing Company.

Thom, R. 1983. *Mathematical Models of Morphogenesis.* Chichester, England: Elliz Norwood.

Thompson, A. 1942. *On Growth and Form.* Cambridge: Cambridge University Press.

Waddington, C. H. 1942. Canalisation of development and the inheritance of acquired characters. *Nature* 150: 563–565.

Waddington, C. H. 1953. The genetic assimilation of an acquired character. *Evolution* 7: 118–126.

Waddington, C. H. 1957. *The Strategy of the Genes.* New York: Macmillan.

Wagner, G. P. 1984. On the eigenvalue distribution of genetic and phenotypic dispersion matrices: Evidence for a non-random origin of quantitative genetic variation. *Journal of Mathematical Biology* 21: 77–95.

Wagner, G. P. 1989. A comparative study of morphological integration in *Apis mellifera* (Insecta, Hymenoptera). *Zeitschrift für zoologische Systematik und Evolutionsforschung* 28: 48–61.

Wagner, G. P. 1995. Adaptation and the modular design of organisms. In *Advances in Artificial Life,* edited by A. M. F. Morán, J. J. Merelo, and P. Chacón. 317–328. Berlin: Springer Verlag.

Wagner, G. P. 1996. Homologues, natural kinds and the evolution of modularity. *American Zoologist* 36: 36–43.

Wagner, G. P., ed. 2001a. *The Character Concept in Evolutionary Biology.* New York: Academic Press.

Wagner, G. P. 2001b. Characters, units and natural kinds. In *The Character Concept in Evolutionary Biology,* edited by G. P. Wagner, 1–13. New York: Academic Press.

Wagner, G. P. 2014. *Homology, Genes, and Evolutionary Innovation.* Princeton, NJ: Princeton University Press.

Wagner, G. P. 2015. Two rules for the detection and quantification of epistasis and other interaction effects. In *Epistasis: Methods and Protocols,* edited by J. H. Moore and S. M. Williams, 145–157. New York: Humana Press.

Wagner, G. P., and L. Altenberg. 1996. Complex adaptations and the evolution of evolvability. *Evolution* 50: 967–976.

Wagner, G. P., and J. Zhang. 2011. The pleiotropic structure of the genotype-phenotype map: The evolvability of complex organisms. *Nature Reviews Genetics* 12: 204–213.

Wagner, G. P., G. Booth, and H. Bagheri-Chaichian. 1997. A population genetic theory of canalization. *Evolution* 51: 329–347.

Wagner, G. P., M. Pavličev, and J. Cheverud. 2007. The road to modularity. *Nature Genetics* 8: 921–931.

Wilkins, A. S. 2002. *The Evolution of Developmental Pathways.* Sunderland, MA: Sinauer Associates.

Wood, A. R., T. Esko, J. Yang, et al. 2014. Defining the role of common variation in the genomic and biological architecture of adult human height. *Nature Genetics* 46: 1173–1186.

Wright, S. 1932. General, group and special size factors. *Genetics* 17: 603–619.

Wright, S. 1934. Physiological and evolutionary theories of dominance. *American Naturalist* 68: 24–53.

Wyczalkowski, M. A., Z. Chen, B. A. Filas, V. D. Varner, and L. A. Taber. 2012. Computational models for mechanics of morphogenesis. *Birth Defects Research. Part C, Embryo Today: Reviews* 96: 132–152.

Xu, Q., H. Jamniczky, D. Hu, R. M. Green, R. S. Marcucio, B. Hallgrímsson, and W. Mio. 2015. Correlations between the morphology of sonic hedgehog expression domains and embryonic craniofacial shape. *Evolutionary Biology* 42: 379–386.

Yang, X. H., Z. Wang, A. Goldstein, et al. 2021. Tipping-point analysis uncovers critical transition signals from gene expression profiles. *bioRxiv*: 668442.

Young, N., and B. Hallgrímsson. 2005. Serial homology and the evolution of mammalian limb covariation structure. *Evolution* 59: 2691–2704.

Young, N. M., H. J. Chong, D. Hu, B. Hallgrímsson, and R. S. Marcucio. 2010. Quantitative analyses link modulation of sonic hedgehog signaling to continuous variation in facial growth and shape. *Development* 137: 3405–3409.

Zhang, W., M. Hong, G. U. Bae, J. S. Kang, and R. S. Krauss. 2011. Boc modifies the holoprosencephaly spectrum of Cdo mutant mice. *Disease Model and Mechanisms* 4: 368–380.

10 Models of Contingent Evolvability Suggest Dynamical Instabilities in Body Shape Evolution

Günter P. Wagner

In this chapter, three questions are considered: What is the relationship between simple models of trait development and Hansen and Houle (H&H) evolvability of a quantitative trait? What is the relationship between trait variation and trait mean across mammalian species? And what is the effect of the evolution of evolvability on body shape evolution? It is shown that the H&H evolvability of a trait depending on the developmental interaction of two underlying traits is the sum of the H&H evolvabilities of these traits. Empirically, it is shown that the standard deviation of body size characters increases on average across species, proportional to the mean body size, suggesting that the H&H evolvability could be constant across species. Finally, a model of constant H&H evolvability predicts that the evolution of body proportions could face runaway dynamics, leading to alternative nonadaptive outcomes. In this chapter, I outline research questions that follow from the idea of contingent evolvability, that is, the idea that evolvability evolves as coincidental side effects of the evolution of other characters, for instance, mean body size.

10.1 Introduction

In this chapter, I address three focused questions: (1) Considering the evolution of a quantitative trait, how do assumptions about the developmental underpinnings of that trait affect the evolutionary dynamics of trait evolvability? (2) What is the relationship between trait mean and evolvability? And (3) What are the consequences of systematic changes in trait evolvability for the evolution of body shape?

Although this chapter is ostensibly about the evolution of evolvability, I will only briefly mention broader issues related to evolvability in general, such as its conceptual status; what biological factors affect evolvability; and how, in general, evolvability changes during evolution. All these questions are also addressed in other chapters in this book and do not need to be covered here in any detail. Instead, this chapter is meant to point to a path forward, accepting that evolvability is an important and measurable biological property, and that evolution of evolvability can be *contingent,* that is, the result of an interaction between natural selection on traits and the underlying developmental processes. The range of possible models for the evolution of evolvability are discussed by Hansen and Wagner (chapter 7).[1] Of these, contingent evolution of evolvability is a likely mode for evolvability evolution. The foundational results for this view of the evolution of evolvability are the pioneering papers by Thomas

1. References to chapter numbers in the text are to chapters in this volume.

Hansen and colleagues (Hansen 2006; Hansen et al. 2006) modeling the evolution of variational properties under natural selection due to epistatic interactions among genes.

In brief, the work by Hansen and colleagues has shown that, as a quantitative trait changes under natural selection, epistatic interactions among the underlying genes lead to changes in the variational properties of the trait (i.e., the mutational variance). This idea in itself was not unexpected, as it has been clear for a long time that evolution of variational properties, relevant for evolvability, requires epistatic interactions, that is, context-dependent gene effects (e.g., Wagner et al. 1997). What was unexpected is the finding that the variational consequences of natural selection on a trait are entirely dependent on the statistical distribution of interaction effects of the segregating loci in that population. In other words, the form of natural selection (i.e., whether it is directional or stabilizing) does not determine the nature of changes in the variational properties that shape evolvability. For instance, there are scenarios where stabilizing selection, instead of leading to reduced mutational variability of the trait, as expected under Waddington's (1957, 1959) theory of canalization, is predicted to increase the mutational variance (Hermisson et al. 2003). Furthermore, directional selection on a quantitative trait can either increase the mutational variance or decrease it, depending on whether epistatic interactions are, on average, positive or negative (Hansen et al. 2006). *The effect of natural selection on evolvability is contingent on the pattern of epistatic interactions.* This result is the basis of the *theory of contingent evolvability.*

The conclusion outlined above raises the question of how to proceed from there. This chapter is meant as a preliminary answer to how to turn this conclusion into a positive research agenda, rather than just a rejection of an adaptive explanation for the evolution of evolvability. The answer given here is necessarily a narrow one, because any answer has to rely on and build on a body of existing mathematical theory. For the present study this theory pertains to the evolution of quantitative traits under natural selection as it has been developed in the field of evolutionary quantitative genetics by Russ Lande, Michael Lynch, Michael Turelli, Nick Barton, and Reinhard Bürger, as well as many others.

The foundational results of the theory of contingent evolvability were based on a class of models that was designed to be as general as possible while still retaining mathematical tractability, in other words, the model of multilinear epistasis (Hansen and Wagner 2001). As such this class of models is both phenomenological (i.e., captures gene interactions given some assumptions) and quite flexible (and thus complex). On the flip side, multilinear models are not mechanistic: Their connection to underlying molecular and developmental processes is not explicit. In this study, I want to start at the opposite end. I make simple, although plausible mechanistic assumptions and investigate their consequences for contingent evolvability (see also Pavličev et al., chapter 8). This approach is painfully limited. But scientific progress, if it can be achieved at all, relies on defining problems that can be solved, however limited the immediate implications are.

10.2 Evolvability of a Quantitative Trait

Measuring evolvability, or to be precise, devising a measurement scale of evolvability, requires a mathematical model for the evolution of the trait we are considering. The best understood model for the short-term evolution of a phenotypic trait is evolutionary quantitative genetics (Lande 1976). For a single quantitative trait in continuous time and overlapping generations, the response to selection is described by a modification of the breeder's equation, a.k.a. the "Lande equation":

$$\dot{\bar{z}} = \frac{\partial \bar{m}}{\partial \bar{z}} V_A,$$

where z is the phenotypic value of a quantitative trait, $\bar{z}$ is the population mean value of z, $\dot{\bar{z}}$ is the time derivative of the mean value (i.e., the instantaneous rate of evolutionary change), $\bar{m}$ is the mean Malthusian fitness, and V_A is the additive genetic variance of z. Models for molecular sequence evolution or macro-molecular shape space evolution have also been developed (Schuster et al. 1994; Fontana and Schuster 1998; Stadler et al. 2001; Fontana 2002); they belong to another class of mathematical models.

From the Lande equation, a formal measure of evolvability has been devised by Houle (1992) and developed further by Hansen and Houle (2008),

$$I_A = \frac{V_A}{\bar{z}^2},$$

which is the squared additive genetic coefficient of variation, $cv = \frac{\sqrt{V}}{\bar{z}}$. One can also consider this measure as the mean-squared normalized additive genetic variance. The justification for this measure has been extensively discussed in Hansen and Houle (2008) and elsewhere in this book (Hansen, chapter 5; Houle and Pélabon, chapter 6; and A. Wagner, chapter 11).

The property measured by the Hansen & Houle (H&H) evolvability is best described as *segregational evolvability,* as it is based on the genetic variation that is segregating in a population, V_A. In many theoretical contexts, evolvability is considered a variational property of a genotype or a class of genotypes rather than a population property. For a single quantitative character, the relevant variational property is the mutational variance, V_M, if the average mutational effect is zero. It is easy to extend the basic idea of the H&H evolvability to measure the variational property of a class of genotypes by replacing V_A with V_M,

$$I_M = \frac{V_M}{\bar{z}^2},$$

which could be called the *variational H&H evolvability.* For the purpose of general discussions of evolvability I_M might be more useful, since it is independent of a variety of population biological variables, such as effective population size, the strength of natural selection, and inbreeding. The relevance of V_M as a measure of medium term evolvability derives from the fact that V_M determines the rate of replenishment of genetic variation under longer term evolution. The asymptotic amount of additive genetic variance under sustained directional selection is proportional to V_M (Hill 1982), as is the rate of neutral evolution of a quantitative trait (Lynch 1990). Hence V_M is a predictor of the medium-term evolutionary potential.

Taking the H&H evolvability as our starting point, let us note three things: (1) by its definition, the H&H evolvability depends on the relationship between the additive or mutational variance of a trait and its mean value, (2) by this token, the evolution of evolvability is described as a change of the variance relative to the change of the mean, and finally (3), to be meaningful, evolvability requires some fixed scale type for the quantification of the trait. The first two points define the agenda for the study of the evolution of contingent evolvability of a quantitative trait as framed here: How does the mutational variance of a trait change with evolutionary changes in the mean value of the trait? And how do these changes impact patterns of evolution? The third point amounts to a prohibition of certain transformations to the scale of the trait values. What is required is that any allowable scale transformation keeps the H&H measure invariant. Otherwise, the H&H evolvability would depend on the arbitrary choice of measurement scale. In this case, it implies that H&H is only meaningful for ratio-scale variables (i.e., variables that are defined up to a multiplicative constant).

Theoretical studies of how the additive variance changes under directional selection have shown that the outcome depends on the detailed distributions of allelic effects at each locus (Turelli and Barton 1990; Bürger 1991), and it thus seems to have no general solution. In contrast, the mutational variance is a property of the genotype (in a fixed environment, to obviate genotype environment interaction) and is primarily determined by the developmental and physiological underpinnings of the trait in addition to the genomic mutation rate. As mentioned in section 10.1, the relationship between changes in the trait mean and mutational variance can be driven by epistatic interactions among genes (Hansen 2006). In this chapter, I do not analyze a general epistatic model but instead explore the implications of a few very simple models of growth and development.

Evolution of evolvability in the H&H perspective happens when there is a systematic relationship between the mean and the variance of a trait, called mean-variance coupling. In statistics, mean-variance coupling has been observed in many data types and has precipitated efforts to eliminate it through so-called variance-stabilizing transformations. There is a straightforward mathematical theory that determines which transformation is removing which form of mean-variance coupling. The most widely used variance-stabilizing transformation is the log transformation, which specifically eliminates an increase in variance proportional to the square of the mean value of the variable. Other transformations remove other kinds of mean-variance coupling (e.g., the arcsine square root transformation for ratios). In statistics, these transformations are performed to force the data to conform with the assumptions of statistical models, which is justified to answer certain statistical questions. However, there is also a tendency in theoretical modeling to use such transformations to achieve mathematical tractability. This latter practice is more problematic. For one, a nonlinear transformation also affects the shape of the fitness function, which rarely is taken into account. Nevertheless, we need to be aware that the mean variance coupling reflects the biology of the situation that is relevant for the evolution of evolvability.

In the current context, we have to assume that the quantitative trait is measured on a ratio scale (Hand 2004), for example, length or weight. The only permissible transformation on a ratio scale variable is multiplication with a constant, as is done when we change the scale from centimetrs to inches or meters. It is easily shown that under these transformations, the H&H evolvability measure remains invariant.

Below I consider two simple types of developmental models and how they affect H&H evolvability. First, I consider the effect of developmental interactions between traits and how it affects their evolvability (section 10.3). Then I consider quantitative traits that grow according to some model, like exponential or linear in ontogenetic time, and derive how their H&H evolvability changes with the mean of the character (section 10.4). And finally, I consider how contingent evolvability affects the evolution of body shape.

10.3 How Developmental Interdependencies Affect Evolvability

During development, many characters arise from physical interactions between different cell populations, in particular, in vertebrates. For instance, most distinct organs of vertebrates develop form an epithelial-mesenchymal interaction, where a population of mesenchymal cells aggregates underneath a patch of an epithelium and then induce one another's growth and development via reciprocal signaling. For instance, all skin derivatives form in this way, such as teeth, hair, feathers, and mammary glands. But internal organs also follow this form of development, for instance, the metanephros (kidney) of amniotes, which arises through the interaction between a part of the Wolffian duct and the nephrogenic mesenchyme. It is thus of interest to investigate the evolvability of a trait x_1 that depends on signaling from another character x_2. Let us consider a rather simple model of this sort:

$$x_1 = kx_2.$$

This model assumes that the size of character 1, x_1, depends on the signaling strength from character 2, which is proportional to the size of character 2, x_2. Then k is a factor that determines the rate at which character 1 responds to signals from character 2. It is therefore the slope of a "developmental reaction norm" of character 1 in response to the signals coming from character 2.

A possible example is the well-known regulation of liver size by the amount of muscle mass present in the body of a mammal (Michalopoulos 1990). Liver size has to match the metabolic demand of the rest of the body, which is largely determined by the amount of skeletal muscle. To match demand for detoxifying capacity, the liver grows in response to endocrine signals from the skeletal muscle cells.

The multiplicative relationship between k and x_2 in determining x_1 has important implications. The most important one is that character 1 can only develop if character 2 is present and active. Also, if character 1 cannot respond to or receive the signal from character 2, then character 1 also cannot develop. For instance, this would be the case if a mutation rendered the receptor for the signal from character 2 nonfunctional or if this receptor is not expressed in the cells of character 1. An example is the loss of penal spines in the human lineage, which is caused by loss of expression of androgen receptors in the skin of the phallus (Reno et al. 2013).

When deriving a model of the H&H evolvability of character 1, I assume that k and x_2 are quantitative traits with their own mutational variance. Also note that the multiplicative dependency of x_1 on k and x_2 leads to bilinear epistasis among the genes affecting k and x_2, with respect to x_1. I can thus directly use the results from Hansen and Wagner (2001) to derive the evolvability of x_1 as a function of the evolvabilities of k and x_2. The result is pleasingly simple:

$$I(x_1) = I(x_2) + I(k).$$

The evolvability of x_1 is the sum of the evolvabilities of x_2 and k. Intuitively, x_1 can change because of changes in k or x_2 or both, and thus the evolvability of x_1 is the sum of the evolvabilities of k and x_2. The somewhat lengthy derivation of this result is presented in the appendix to this chapter.

As an informal generalization, I suggest that if during evolution, new developmental interactions arise, the resulting phenotype will increase in evolvability by the sum of the evolvabilities of the participating characters.

10.4 Contingent Evolvability Due to Mean-Variance Coupling

As mentioned above, the H&H evolvability is the ratio of the mutational variance of a character and its squared mean. Hence H&H evolvability evolves when there is a systematic relationship between the mutational variance and the mean of the trait. Here I consider two simple models of mean-variance coupling. First, let us assume that the character x grows through the division of cells, where each daughter cell can itself divide. This leads to an exponential increase in the size of x during growth. The final size, in this model, is then determined by some maturation signal that stops growth, for instance, through the endocrine activity of the gonads.

10.4.1 Exponentially Growing Trait

Let T be the age at which the growth of an organ x stops, and λ be the instantaneous growth rate. Then the size of the organ at the end of the growth phase will be

$$x_T = x_0 e^{\lambda T},$$

where x_0 is the size of the rudiment before exponential growth, for instance, the size of a mesenchymal condensation. Now let us assume that T is constant, for instance, due to an environmental cue that coordinates the maturation of the individual in a local population and thus there is no within-population variation in T. Further let us assume that there is genetic variation among individuals in the growth rate λ. Then the mean and the variance of x_T is approximately:

$$\bar{x}_T \approx x_0 e^{\bar{\lambda} T},$$
$$V_m(\bar{x}_T) \approx x_0^2 T^2 (e^{\bar{\lambda} T})^2 V_m(\lambda) = T^2 V_m(\lambda) \bar{x}_T^2.$$

From these expressions, it is clear that in this model of organ growth, there is a strong mean-variance coupling, where the variance is proportional to the square of the mean value. Inserting this expression into the formula for the H&H evolvability, we obtain the following result:

$$I_m(x_T) = \frac{T^2 V_m(\lambda) \bar{x}_T^2}{\bar{x}_T^2} = T^2 V_m(\lambda) = const.$$

In this model, the H&H evolvability of the trait is constant.

10.4.2 Stem Cell Growth Model

Many cells in the organism do not originate from an exponential growth process where both daughter cells each can again divide. Instead, cells are often added to the body by a stem cell growth process. In the case of a stem cell, each division creates two unequal daughter cells. One continues dividing, and the other differentiates and does not divide again.

It is easy to see that the stem cell model leads to linear growth and constant variance:

$$x_T = x_0 + \lambda T,$$
$$\bar{x}_T = x_0 + \bar{\lambda} T,$$
$$V_m(x_T) = T^2 V_m(\lambda) = const.$$

Consequently, the H&H evolvability becomes

$$I_m(x_T) = \frac{T^2 V_m(\lambda)}{\bar{x}_T^2},$$

and thus, the H&H evolvability decreases with the inverse square of the mean.

These examples lead to the proposition for which no formal proof is possible as it stands, namely that the H&H evolvability in general is non-increasing with increasing trait mean:

$$\frac{\partial I}{\partial \bar{x}} \leq 0.$$

Of course, it is mathematically possible to write an equation for V_m so that the derivative of evolvability is positive. What I mean, however, is that I did not find a mechanistic model that would predict that the H&H evolvability I can increase with trait mean. Of course, I am open to any counterexamples that actually makes biological sense.

10.5 Empirical Patterns of Mean-Variance Coupling

Patterns of phenotypic variation have been studied over many decades, and the most notable contributions are the books by Bateson (1894), Olson and Miller (1958), and Yablokov (1974) as well as the paper by Hallgrímsson and Maiorana (2000). To my knowledge, the relationship between the mean and the variance of a trait has not been systematically investigated, in particular for data that were not subjected to log-transformation.

As outlined above (section 10.4), the questions are (1) whether variance is changing with the mean of the character, and (2) if the variance is changing, what is the functional shape of this increase? The models presented in section 10.4 predicted that variance will increase with the square of the mean under the exponential growth model, while under the stem-cell-like growth model, the variance should be independent of the mean.

For the mean-variance coupling to be evolutionarily relevant, one would need evidence that at least the additive variance is changing in a systematic way with the mean, or ideally, that the mutational variance is behaving in this way. For that to be useful, one would need accurate genetic data from a large number of closely related species. Needless to say, gathering this data would be a huge undertaking, and I am not aware of any such data. For now, we have to rely on data about phenotypic variance. These data are relevant based on the *Cheverud conjecture.* That conjecture states that the phenotypic variance is roughly proportional to the additive genetic variance and thus trends of phenotypic variance might be indicative of parallel trends in genetic variance (Cheverud 1988). Furthermore, the equilibrium additive variance under stabilizing selection is strongly influenced by the mutational variance (e.g., Turelli 1984; Bürger 1986, 1988; Bürger and Hofbauer 1994). Under sustained directional selection, the genetic variance is proportional to the mutational variance (Hill 1982).

To test for a mean-variance relationship, the relationship between the standard deviation (Std) and the mean phenotype was investigated. Both these measures are linear with respect to their original measurement scale. A linear relationship between standard variation and mean implies a quadratic relationship between variance and the mean, which is consistent with the exponential growth model.

Hallgrímsson and Maiorana (2000) have assembled phenotypic variation data for body mass, body length, and tail length of 353 mammalian (therian) species with a median sample size per species of 92 and a minimum of 8 for one species and a maximum sample size of 3,727. A regression of standard deviation on mean for body mass shows clear signs of heterogeneity, which likely reflects differences in the biology of different groups. For that reason, we focus here on some larger taxonomic groups in the hope that within clades, the mean-Std relationship is more homogenous. The largest groups in those data are the Rodentia ($N=146$), and within Rodentia, the Murinae ($N=45$), the Chiroptera ($N=143$), and the Eulipotyphla (remains of the former "Insectivora," $N=22$).

The question at hand is: What is the empirical relationship between the variance (or standard deviation, *std*) and the mean (m) of a quantitative character among species? Hence, we can ask whether a power equation of the sort

$$std = k\, m^b$$

can describe the relationship between standard deviation and mean, and what the exponent b is for these data. For statistically evaluating these questions, it is convenient to log-transform the power equation:

$$\log std = b \log m + \log k.$$

Note that this transformation is done only to statistically assess the functional relationship between the mean and the standard deviation of the quantitative characters, which themselves are not log transformed. This treatment is different from standard practice, where the observational variable is subjected to transformations.

If the relationship between standard deviation and mean follows a power law, then the relationship between the log *std* and log m should be linear, regardless of b and k. We can assess that by considering the residual plot for the linear regression of log *std* on log m and other tests for deviations from linearity. Furthermore, the slope b of the linear regression is an estimate of the exponent of the power relationship. If $b=1$, then the standard deviation is proportional to the mean, and the variance is proportional to the square of the mean. If we assume that the additive variance is proportional to the phenotypic variance, then $b=1$ would mean that the H&H evolvability is, on average, constant across species. Of course, there can be variation from species to species, as reflected in the residuals of the regression.

Applying a linear regression to the log-log data of mean and standard deviation of body weight to all 353 therian species in this data set, we see a strongly linear relationship (figure 10.1b, c) consistent with a power equation relationship between standard deviation and mean across species. A runs test for deviations from linearity yields a p-value of 0.089. The slope and therefore the exponent of the power function is estimated to be 1.087 and is significantly, but only slightly, different from 1 (table 10.1).

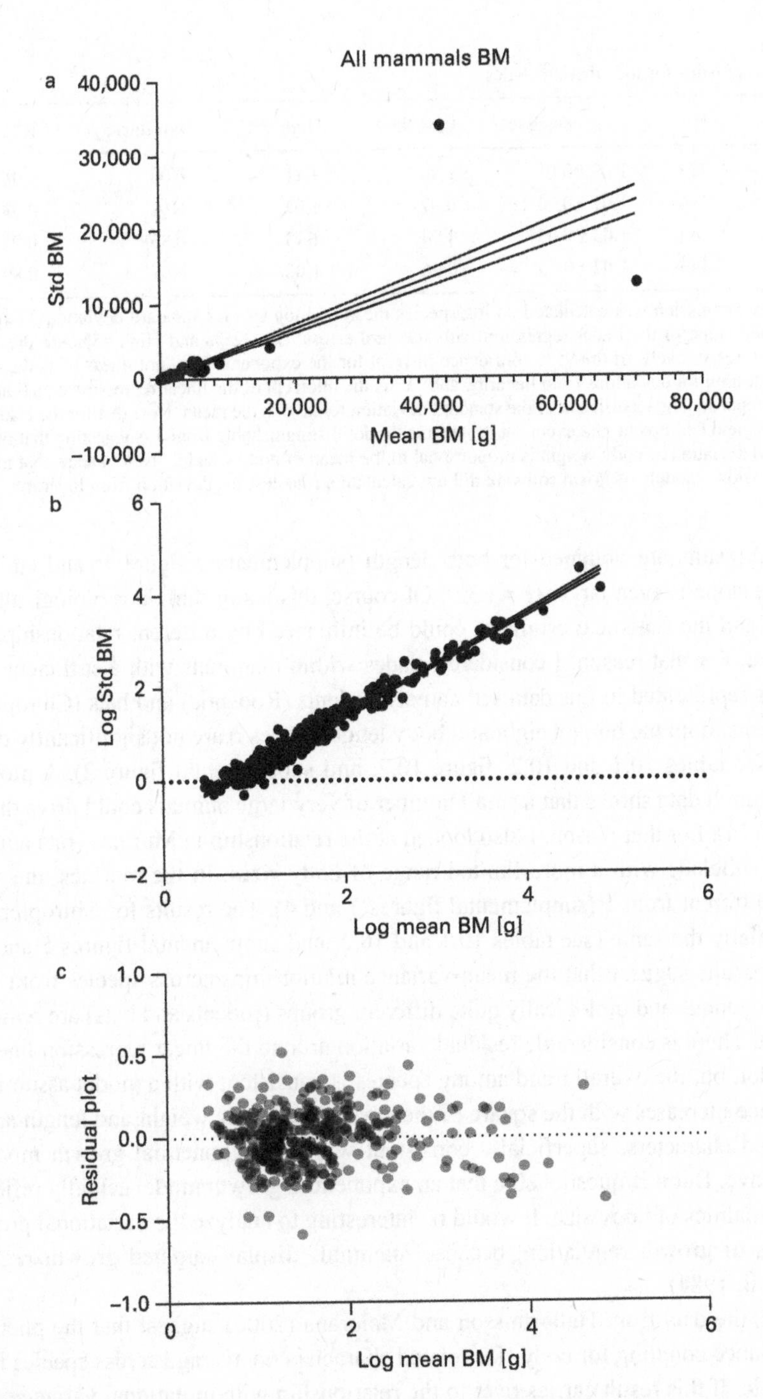

Figure 10.1
The relationship between species average body mass [g] and the standard deviation for 353 mammalian species in the Hallgrímsson and Maiorana (2000) data set.

Table 10.1
Body mass statistics for four therian clades

Taxon	*N*	Exponent ± se	Low 95%	High 95%	not linear *p*	R^2	*k*
All Theria	353	1.09±0.01	1.06	1.11	0.09	0.96	0.16
Rodentia	146	1.01±0.02	0.97	1.06	N/A	0.93	0.25
Murinae	45	1.03±0.05	0.94	1.13	0.86	0.92	0.26
Chiroptera	143	1.02±0.03	0.96	1.08	N/A	0.89	0.16

Note: Linear regression was calculated on log species mean and log species standard deviation. "Exponent" is the estimated slope of the linear regression with standard errors. Low 95% and High 95% are the lower and upper limits, respectively, of the 95% confidence interval for the exponent. The "not linear *p*" is the *p*-value of a runs test, testing for deviations from linearity, and "*k*" is the intercept of the linear regression equation, and thus 10^k is the proportionality coefficient of the standard deviation relative to the mean. Note that for the clades Rodentia, Murinae, and Chiroptera, the exponent is statistically not distinguishable from 1, suggesting that on average, the standard deviation of body weight is proportional to the mean of body weight. "N/A" means that the "simple linear regression" module of Prism software did not calculate a runs test for deviation from linearity.

Similar results are obtained for body length (supplementary figure 1[2] and table 10.2), where the slope is even larger ($b=1.32$). Of course, this taxon sample is biologically heterogenous, and the statistical estimates could be influenced by different relationships in different taxa. For that reason, I considered clades within mammals with a sufficient number of species represented in this data set, namely, rodents (Rodentia) and bats (Chiroptera).

In rodents, both the body weight and body length slopes b are not significantly different from 1 (see tables 10.1 and 10.2, figure 10.2, and supplemental figure 2). A plot of the untransformed data shows that a small number of very large animals could drive this trend (figure 10.2a). For that reason, I also looked at the relationship in Murinae (rats and mice), a sizable subclade with a more limited range of body sizes. In the murines, the slope is also not different from 1 (supplemental figures 3 and 4). The results for Chiroptera (bats) are essentially the same (see tables 10.1 and 10.2, and supplemental figures 5 and 6).

These results suggest that the mean-variance relationships across species from reasonably homogenous and biologically quite different groups (rodents and bats) are remarkably consistent. There is considerable residual variation around the linear regression line on the log-log plot, but the overall trend among species is consistent with a model assuming that the variance increases with the square of the mean. Both body weight and length are body size related characters, superficially consistent with the exponential growth model discussed above. But it is questionable that an exponential growth model actually reflects the growth dynamics of body size. It would be interesting to analyze the variational properties of models of growth regulation, because mammals display targeted growth regulation (Riska et al. 1984).

Overall, the data from Hallgrímsson and Maiorana (2000) suggest that the phenotypic mean-variance coupling for body size related characters on average across species is close to quadratic. If this result carries over to the relationship with mutational variance, it suggests that H&H evolvability might remain constant as size evolves. In section 10.6, I explore the consequences of this mean-variance relationship. Naturally it is to be expected

2. The supplementary figures, regression results and original data are available on my lab website: https://campuspress.yale.edu/wagner/.

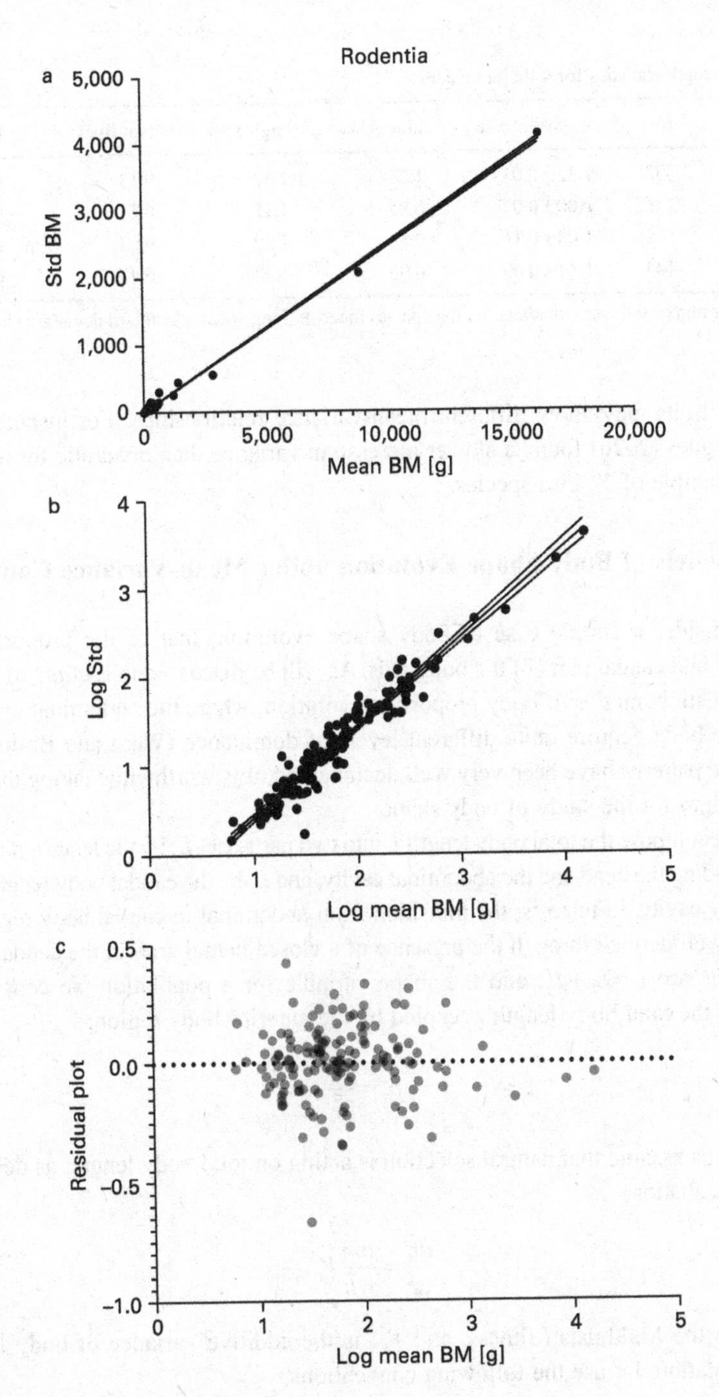

Figure 10.2
The relationship between species average body mass [g] and the standard deviation for 146 rodent species in the Hallgrímsson and Maiorana (2000) data set.

Table 10.2
Total body length statistics for 4 therian clades

Taxon	N	Exponent ± se	Low 95%	High 95%	Not linear p	R^2	k
All Theria	353	1.32 ± 0.03	1.26	1.37	0.13	0.85	0.02
Rodentia	146	1.00 ± 0.07	0.85	1.11	0.57	0.57	0.11
Murinae	45	1.08 ± 0.11	0.85	1.30	0.63	0.68	0.08
Chiroptera	143	1.14 ± 0.09	0.96	1.32	0.43	0.53	0.03

Note: Linear regression was calculated on log species mean and log species standard deviation (for explanation, see table 10.1).

that other traits may show different mean-variance relationships. For instance, Pélabon and colleagues (2020) found a slower increase in variance than quadratic for mean clutch size in a sample of 32 bird species.

10.6 Models of Body Shape Evolution under Mean-Variance Coupling

Let us consider a simple case of body shape evolution, that of the proportion of the abdominal and caudal parts of the body axis. As will be discussed in section 10.7, in fishes there are distinct modes of body proportion evolution, where the abdominal or the caudal part of the body acquire quite different levels of dominance (Ward and Brainerd 2007). Since these patterns have been very well documented, it is worthwhile taking this example as a paradigm for the study of body shape.

Let us decompose the total body length L into two parts. Let L_a be the length of the anterior body, including the head and the abdominal cavity, and L_c be the caudal body length posterior of the body cavity. In teleosts, the transition from abdominal to caudal body region is anatomically well defined through the presence of a closed hemal arch in the caudal region.

Our traits are $L = L_a + L_c$, and the shape variable for a population we consider is the fraction of the total body length occupied by the anterior body region:

$$p_a = \frac{\bar{L}_a}{\bar{L}}.$$

Now let us assume that natural selection is acting on total body length, as described by the Lande equation,

$$\frac{d\bar{L}}{dt} = \frac{\partial \bar{m}}{\partial \bar{L}} V_{A,L},$$

where m is the Malthusian fitness, and $V_{A,L}$ is the additive variance of body length. To simplify notation, let use the following conventions,

$$\frac{d\bar{L}}{dt} = \dot{\bar{L}}$$

$$\frac{\partial \bar{m}}{\partial \bar{L}} = m'.$$

which simplifies the Lande equation to

$$\dot{\bar{L}} = m' V_{A,L}.$$

If selection is in fact only on the total body length, the evolutionary change of each body region is proportional to their contribution to the additive variance of body length

$$\dot{\bar{L}}_a = m' V_{A,a},$$
$$\dot{\bar{L}}_c = m' V_{A,c},$$

assuming an absence of covariance between the anterior and the caudal body region. From this equation, we can easily derive the evolution equation for the body proportion:

$$\dot{p}_a = m' \frac{\dot{\bar{L}}_a - p_a \dot{\bar{L}}}{\bar{L}}.$$

Substituting the Lande equations for the body length and the length of the body regions, we obtain

$$\dot{p}_a = m' \frac{V_{A,a} - p_a V_{A,L}}{\bar{L}}.$$

From this expression, it is clear that whether the proportion of a body region increases or decreases due to natural selection on total body length depends on (1) the additive variance of the body region, $V_{A,a}$, (2) the current relative size of the body region, p_a, and (3) the variance of total body length, $V_{A,L}$. The anterior body region increases in relative size compared to the caudal region if

$$\dot{p}_a > 0 \leftrightarrow V_{A,a} > p_a V_{A,L}.$$

Note that the direction of the change in body shape depends on the current body shape, p_a, which suggests that history may play a role. Furthermore, since the variance of a character generally increases with its size, the outcome in terms of body shape will depend on how fast the variance is changing as a consequence of change of size, for total body length as well as for the size of each body region. Hence the prediction of how body proportions will change under selection depends on how the evolvability of the traits evolves. To further discuss this topic, we have to make assumptions about the mean-variance relationship.

10.6.1 Exponential Growth Model (Constant H&H Evolvability)

As argued in the previous section, under an exponential growth model, the H&H evolvability is predicted to be constant, because the variance remains proportional to the square of the mean. Furthermore, the empirical data on mammals discussed in section 10.5 supports a quadratic mean-variance relationship, at least for phenotypic variation.

Let us write the constant H&H evolvability as $I_A = const. = i$, and thus the mean-variance relationship as $V_A(\bar{L}) = i\bar{L}^2$ for the total length and for the body regions analogously as $V_{A,a}(\bar{L}_a) = i_a \bar{L}_a^2$ and $V_{A,c}(\bar{L}_c) = i_c \bar{L}_c^2$, respectively. Substituting these expressions into the predicted response of body shape leads to the following result:

$$\dot{p}_a = m' p_a [i_a \bar{L}_a - (p_a i_a \bar{L}_a + p_c i_c \bar{L}_c)].$$

Interpretation of this equation is as follows: m', the slope of the fitness function, determines the time scale for the change in body proportion, but it does not determine the qualitative outcome. The term $i_a\bar{L}_a$ is the product of the H&H evolvability of the anterior body region times the size of that region. The term in parentheses is the average of the evolvability of the two body regions, weighted by their relative contribution to total body length, which we can simplify as $(p_a i_a \bar{L}_a + p_c i_c \bar{L}_c) = M(a,c)$. The differential equation for p_a has three equilibrium points assuming nonzero evolvabilities of the two body regions (i.e., a point where $\dot{p}_a = 0$ while $i_a > 0$, and $i_c > 0$. These are $\hat{p}_a = 0$, $\hat{p}_a = 1$, and

$$\hat{p}_a = \frac{i_c}{i_a + i_c}.$$

Of these equilibria, the first is nonbiological, because it implies that the animal consists only of a tail, and the second assumes that there is no tail, which is possible but also unlikely. The third equilibrium is between these extremes, since the evolvabilities are all positive (figure 10.3A).

An inspection of the three equilibria reveals that the trivial fixed points, $\hat{p}_a = 0$, and $\hat{p}_a = 1$, are stable and thus the third internal equilibrium must be unstable. Thus, if p_a is larger than this fixed point, then p_a will continue to increase, and if p_a is smaller, then it will continue to decrease. The outcome of evolution in terms of body shape depends on whether, at the onset of selection for increased body length, the animal has a body proportion larger or smaller than the internal fixed point determined by the relative magnitudes of their evolvabilities (figure 10.3B). This behavior amounts to a kind of "evolvability runaway" process, where a population with an anterior body region larger than a threshold will continue to increase the relative size of its anterior body region, and a population under the same selection regime, but with an anterior body region smaller than that threshold will continue to decrease the relative size of the anterior body region. Note that these

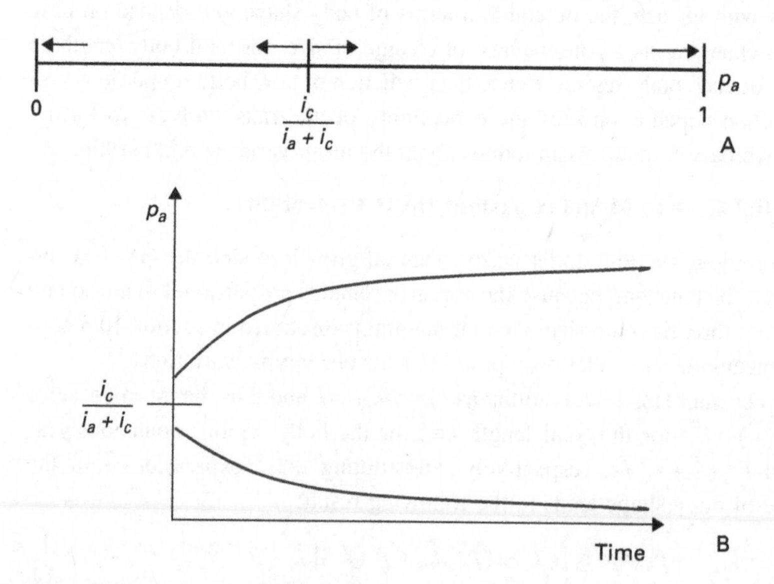

Figure 10.3
Dynamic behavior of a model of body shape evolution (see text for details): (A) fixed points of the body proportion; (B) example of bifurcation of body proportion under directional selection.

divergent outcomes will be attained under the same selection regime (i.e., selection for increased overall body size). The outcome of evolution in terms of body shape depends only on the relative evolvabilities of the anterior and posterior body regions and the initial body shape. We do not predict that if the relative size of the abdomen is decreasing, it will literally lead to animals without an abdomen, as there must be selection on the size of each body region that is not included in this model.

At this point it is worth considering an intuitive argument that may make the mathematical result accessible. The core observation is that the rate of change of a quantitative character is proportional to the additive genetic variance of that character. Furthermore, we have shown that in many cases, the variance of a quantitative trait increases with the square of the mean. Thus if selection of the same strength is acting on two characters, the one that has higher mean value to start with will also evolve faster than the smaller character. The rate of evolution of this character will further accelerate with increasing size. In that way, a small discrepancy in size between two characters under the same intensity of natural selection will grow into a major discrepancy as evolution proceeds. A small difference in the starting conditions can lead to extremely divergent outcomes. This is the logic of an "evolvability runaway process." Both outcomes, dominance of one character (e.g., abdomen) over the other (e.g., tail), are well documented in ray finned fishes. Examples are given in the next section.

10.7 Empirical Patterns of Body Shape Evolution

Body elongation has been described as the dominant mode of body shape diversification in fishes. A systematic large scale morphospace exploration based on data from 2,939 species of tropical reef fishes was published by Claverie and Wainwright (2014). In their study, the authors found that 32% of the body shape variance in their sample is accounted for by body elongation. This confirms the previous observations that body elongation is a major mode of teleost body shape evolution based on more limited samples (Friedman 2010; Sallan and Friedman 2012).

Body proportions are less well studied. The best study I am aware of is the paper by Ward and Brainard (2007). In their paper, these authors analyzed data from 867 species of ray finned fishes and found that the numbers of abdominal and caudal vertebrae are evolving independently of each other. However, the ratio of length to width of vertebrae in these body regions is highly correlated (i.e., differences in the overall body shape are due to differences in the number of abdominal and caudal vertebrae, rather than the length of abdominal and caudal vertebrae). The authors document extreme cases of body proportion (figure 10.4): The bichir, *Polypterus bichir* and *P. ornatipinnis*, as well as *Erpetoichthys calabricus* (a.k.a. reed fish or rope fish) consist of "almost only abdomen" with a very short tail region (figure 10.4 A). The other extreme, "almost only tail," with a short abdomen is exemplified by *Chitala chitala* (Bangladesh knife fish), and *Notopterus notopterus* (bronze featherback) (figure 10.4 B). Hence extreme divergence in body proportion evolved repeatedly in actinopterygian fishes.[3] These cases of extreme divergence are consistent with a model of an "evolvability runaway," as presented in section 10.6.

3. Note that figure 10.4 shows only part of the teleost phylogeny.

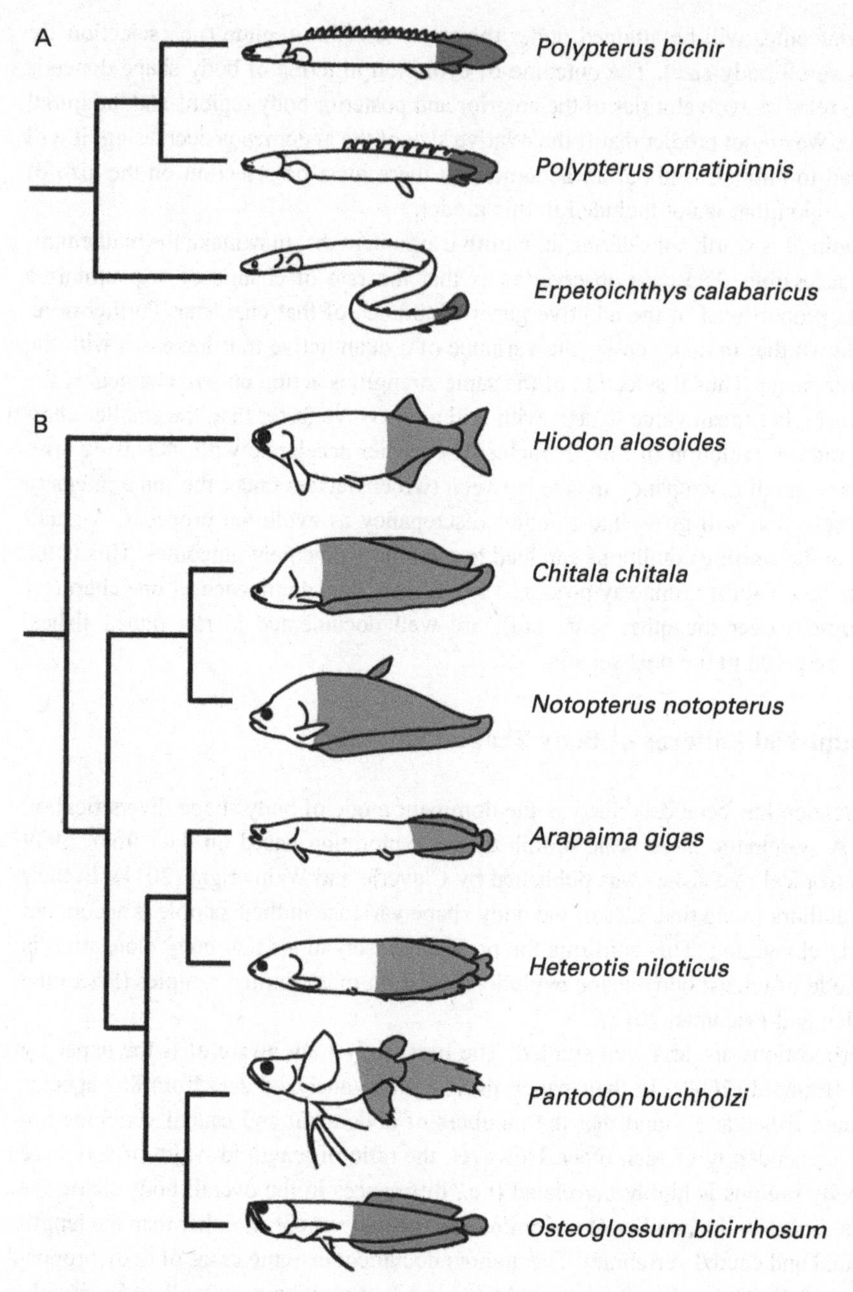

Figure 10.4
Examples of extreme body proportion among teleost fishes; gray shading indicates the tail of the body axis: (A) Reed fishes, which have a minimal tail length, and the body elongation is almost entirely due to abdominal elongation. (B) Osteoglossomorpha, which contains examples of extreme cases of tail elongation with short abdomen. (Part of Figure 3 from Ward and Brainerd (2007), reproduced with permission from the publisher).

Of course, at this point, it is not possible to decide whether the extreme body proportions evolved because of adaptive elongation of the one or the other body region, or whether they are arbitrary (i.e., the result of an "evolvability runaway process," as suggested by the mathematical model and the variation data presented above). What the theoretical work does show is that nonadaptive evolution has to be considered as a possible cause of extreme body proportions.

10.8 Conclusion and Perspective

Evolvability, or the propensity of a population to respond to directional natural selection, is itself evolving, if the mechanisms underlying the production of heritable variation are evolving (Hansen and Wagner, chapter 7). Here I have called such changes *contingent evolution of evolvability* (or "contingent evolvability" for short), that is, the evolution of evolvability is a side effect of the evolution of other traits and characters. To explore the consequences of contingent evolvability, one needs a class of cases where measures of evolvability are well defined and mathematical models of evolution by natural selection are available. This is most prominently the case for quantitative phenotypic traits, such as body size, which comes with a deep body of theoretical and empirical work. Based on this body of work, Thomas Hansen and David Houle (Hansen and Houle 2008; Hansen, chapter 5; and Houle and Pélabon, chapter 6) have established a measure of evolvability, which tells us that contingent evolution of evolvability for a quantitative trait consists of the change of the genetic variance of a trait as a consequence of changes in the mean of the character. Here I explore how evolvability is likely to change, based on mathematical models of trait development and empirical data about the variation of two quantitative traits (body weight and body length).

For this program to work, we have to observe a constraint that results from measurement theory on the mathematical models of trait evolution. In particular, we should only apply scale transformations to quantitative phenotypic variables that are consistent with their scale type (e.g., Hand 2004; Houle et al. 2011). Many quantitative traits in biology are on the ratio scale, including length and weight. Ratio scale variables can be transformed by a multiplicative constant without losing their empirical meaning, but not by a nonlinear transformation, such as the widely used log transformation. A log transformation will destroy the biological meaning of the variable and thus lead to meaningless models. The models discussed here assume that the evolution of a trait is modeled on its natural scale type.

The most surprising implication of modeling the contingent evolvability of quantitative traits is that the outcome of natural selection on body proportions can be dominated by the evolutionary dynamics of the variance. Under constant H&H evolvability, the evolutionary outcome can be arbitrary, rather than adaptive, because it is the result of a runaway process analogous to Fisher's runaway selection process for sexually selected traits (Pomiankowski and Iwasa 1998). These results are driven by natural selection, but the specific outcome depends on the contingent pattern of evolvability and the evolution of evolvability. A similar result has recently been published by Jeremy Draghi (2021) in a model of ecological specialization. It is thus conceivable that paying close attention to the contingent evolutionary dynamics of evolvability leads us to a class of evolutionary explanations that complement the well-understood adaptive optimization models. These are driven by natural selection, but natural selection neither predicts nor explains the specific outcome without reference to the contingent effects of selection on trait evolvability.

Appendix

Let us derive the additive variance of a quantitative character x_2 which depends on the interaction with another character x_1, $x_2 = kx_1$, where k is the slope of the developmental reaction norm of x_2 in response from signals from x_1. This relationship is linear in each independent variable, so it should be covered by the multilinear model, and we can use the results from Hansen and Wagner (2001) to calculate the additive variance. The genotypic value of x_2 after two substitutions at loci i and j, in general is:

$$x_{2,ij} = x_{2,0} + {}^{i}_{2}y + {}^{j}_{2}y + {}^{ij}_{222}\varepsilon\, {}^{i}_{2}y\, {}^{j}_{2}y + {}^{ij}_{221}\varepsilon\, {}^{i}_{2}y\, {}^{j}_{1}y + {}^{ij}_{212}\varepsilon\, {}^{i}_{1}y\, {}^{j}_{2}y + {}^{ij}_{211}\varepsilon\, {}^{i}_{1}y\, {}^{j}_{1}y,$$

where $x_{2,0}$ is the genotypic value of character 2 in the reference genotype. The reference effect of substitution at locus i is ${}^{i}_{2}y$, and ${}^{ij}_{2ab}\varepsilon\, {}^{i}_{a}y\, {}^{j}_{b}y$ is the interaction among the effects of locus i on character a, and of locus j on character b (see Hansen and Wagner 2001 for a detailed derivation). If the locus i is only affecting x_1 and j is only affecting k, then we have the following substitution effects.

The locus i substitution effects are:

$$x_{1,i} = x_{1,0} + {}^{i}_{1}y,$$
$$x_{2,i} = x_{2,0} + {}^{i}_{2}y,$$

where

$$x_{2,i} = k_0 x_1 = k_0(x_{1,0} + {}^{i}_{1}y) = k_0 x_{1,0} + k_0\, {}^{i}_{1}y,$$
$$x_{2,i} = x_{2,0} + k_0\, {}^{i}_{1}y,$$
$${}^{i}_{2}y = k_0\, {}^{i}_{1}y.$$

The locus j substitution effect on x_2 (x_1 is not affected per assumption) is:

$$x_{2,j} = x_{2,0} + {}^{j}_{2}y,$$
$$x_{2,j} = (k_0 + \kappa_j)x_{1,0},$$
$$x_{2,j} = x_{2,0} + \kappa_j x_{1,0},$$
$${}^{j}_{2}y = \kappa_j x_{1,0}.$$

Under the assumptions above, the equation simplifies to

$$x_{2,ij} = x_{2,0} + {}^{i}_{2}y + {}^{j}_{2}y + {}^{ij}_{222}\varepsilon\, {}^{i}_{2}y\, {}^{j}_{2}y + {}^{ij}_{212}\varepsilon\, {}^{i}_{1}y\, {}^{j}_{2}y,$$

because ${}^{j}_{1}y = 0$, ${}^{ij}_{211}\varepsilon\, {}^{i}_{1}y\, {}^{j}_{1}y = 0$, and ${}^{ij}_{221}\varepsilon\, {}^{i}_{2}y\, {}^{j}_{1}y = 0$.

Based on the developmental plasticity model (see section 10.3), we have

$$x_{2,ij} = (k_0 + \kappa_j)(x_{1,0} + {}^{i}_{1}y) = k_0 x_{1,0} + \kappa_j x_{1,0} + k_0\, {}^{i}_{1}y + \kappa_j\, {}^{i}_{1}y.$$

The first three terms on the right-hand side are easily identified as $x_{2,0}$, ${}^{j}_{2}y$, and ${}^{i}_{2}y$, respectively. The last term can either be ${}^{ij}_{222}\varepsilon\, {}^{i}_{2}y\, {}^{j}_{2}y$, (in which case ${}^{ij}_{222}\varepsilon = x_{2,0}^{-1}$) or ${}^{ij}_{212}\varepsilon\, {}^{i}_{1}y\, {}^{j}_{2}y$, (in which case ${}^{ij}_{212}\varepsilon = x_{2,0}^{-1}$). This ambiguity seems to reflect an overdetermination of the multilinear model in the case of the developmental reaction norm model. It seems that ${}^{ij}_{222}\varepsilon = x_{2,0}^{-1}$ is the more natural choice. Note that the interaction coefficient ${}^{ij}_{222}\varepsilon > 0$, and thus the model predicts overall positive epistasis. Also, the interaction coefficient decreases with $x_{2,0}$ (i.e., in large character states, the interaction becomes less important).

Now let us calculate the additive variance of x_2. Let I be the index set of loci that directly affect x_1, and J the index set of loci that affect k. Per assumption, $I \cap J = \varnothing$. Based on Hansen and Wagner (2001), the additive variance of a character with epistatic genetic architecture can be written as

$$V_{A2} = \sum_{i \in I} \langle f_{g \to i} \rangle^2 V({}^{i}_{2}y) + \sum_{j \in J} \langle f_{g \to j} \rangle^2 V({}^{j}_{2}y),$$

where $f_{g \to i}$ and $f_{g \to j}$ are the epistasis factors as defined in Hansen and Wagner (2001). These factors quantify the influence of genetic background g on the effects of substitutions at locus i and j, respectively. In our case, they reduce to

$$f_{g \to i} = 1 + \sum_{j \in J} \varepsilon_{ij} \, {}^{j}_{2}y.$$

As shown above, the epistasis coefficient is ${}^{ij}_{222}\varepsilon = x_{2,0}^{-1}$, which is in fact a constant for all i and j, We call this coefficient ε_{IJ}, so

$$f_{g \to i} = 1 + \varepsilon_{IJ} \sum_{j \in J} {}^{i}_{2}y.$$

By substituting the above identities, the average epistasis factor $\langle f_{g \to i} \rangle$ averaged over all genotypes becomes

$$\langle f_{g \to i} \rangle = \frac{\bar{k}}{k_0},$$

where $\bar{k}$ is the population average of k. Analogously, it is easy to show that

$$\langle f_{g \to j} \rangle = \frac{\bar{x}_1}{x_{1,0}}.$$

Substituting these factors into the equation for V_{A2} and observing that $V({}^{i}_{2}y) = k_0^2 V({}^{i}_{1}y)$ and $V({}^{j}_{2}y) = x_{1,0}^2 V(k)$, we have

$$V_{A2} = \bar{k}^2 V(x_1) + \bar{x}_1^2 V(k).$$

This equation is isomorphic with the approximate expression for error propagation. In the case of error propagation, this relationship is only approximately valid for small variances, but here it is exact for the additive variance within the limits of the multilinear model of gene interaction. This is not surprising, because the approximate formula of error propagation is obtained through linearization around the mean, and the additive variance is the part of variation that is explained by linear effects of gene substitutions.

The above expression for the additive variance can be rewritten in terms of H&H evolvabilities if x_1 and k are stochastically independent (i.e., there is no linkage disequilibrium between the i and j loci) by noting that then, $\bar{x}_2 = \bar{k}\bar{x}_1$, which yields

$$\frac{V_{A2}}{\bar{x}_2^2} = \frac{V(x_1)}{\bar{x}_1^2} + \frac{V(k)}{\bar{\bar{k}}},$$

which in terms of H&H evolvabilities is

$$I_2 = I_1 + I_k.$$

Acknowledgments

I thank Thomas Hansen for leading the effort to develop the evolvability concept through the project at the Centre for Advanced Studies at the Norwegian Academy of Sciences and Letters. My thanks also to the editors of this volume and to Philip Mitterröcker for reviewing my manuscript and for helpful suggestions. And I am grateful to Benedikt Hallgrímsson for giving me access to his data on body size analyzed in this chapter.

References

Bateson, W. 1894. *Materials for the Study of Variation Treated with Especial Regard to Discontinuity in the Origin of Species.* London: Macmillan.

Bürger, R. 1986. On the maintenance of genetic variation: Global analysis of Kimura's continuum-of-alleles model. *Journal of Mathematical Biology* 24: 341–351.

Bürger, R. 1988. Mutation-selection balance and continuum-of-alleles models. *Mathematical Biosciences* 91: 67–83.

Bürger, R. 1991. Moments, cumulants, and polygenic dynamics. *Journal of Mathematical Biology* 30: 199–213.

Bürger, R., and J. Hofbauer. 1994. Mutation load and mutation-selection-balance in quantitative genetic traits. *Journal of Mathematical Biology* 32: 193–218.

Cheverud, J. M. 1988. A comparison of genetic and phenotypic correlations. *Evolution* 42: 958–968.

Claverie, T., and P. C. Wainwright. 2014. A morphospace for reef fishes: Elongation is the dominant axis of body shape evolution. *PLOS One* 9: e112732.

Draghi, J. A. 2021. Asymmetric evolvability leads to specialization without trade-offs. *American Naturalist* 197: 644–657.

Fontana, W. 2002. Modelling 'evo-devo' with RNA. *Bioessays* 24: 1164–1177.

Fontana, W., and P. Schuster. 1998. Continuity in evolution: On the nature of transitions. *Science* 280: 1451–1455.

Friedman, M. 2010. Explosive morphological diversification of spiny-finned teleost fishes in the aftermath of the end-Cretaceous extinction. *Proceedings of the Royal Society B* 277: 1675–1683.

Hallgrímsson, B., and V. Maiorana. 2000. Variability and size in mammals and birds. *Biological Journal of the Linnean Society* 70: 571–595.

Hand, D. L. 2004. *Measurement Theory and Practice: The World through Quantification.* London: Arnold.

Hansen, T. F. 2006. The evolution of genetic architecture. *AREES* 37: 123–157.

Hansen, T. F., and D. Houle. 2008. Measuring and comparing evolvability and constraint in multivariate characters. *Journal of Evolutionary Biology* 21: 1201–1219.

Hansen, T. F., and G. P. Wagner. 2001. Modeling genetic architecture: A multilinear theory of gene interaction. *Theoretical Population Biology* 59: 61–86.

Hansen, T. F., J. M. Alvarez-Castro, A. J. R. Carter, J. Hermisson, and G. P. Wagner. 2006. Evolution of genetic architecture under directional selection. *Evolution* 60: 1523–1536.

Hermisson, J., T. F. Hansen, and G. P. Wagner. 2003. Epistasis in polygenic traits and the evolution of genetic architecture under stabilizing selection. *American Naturalist* 161: 708–734.

Hill, W. 1982. Rates of change in quantitative traits from fixation of new mutations. *PNAS* 79: 142–145.

Houle, D. 1992. Comparing evolvability and variability of quantitative traits. *Genetics* 130: 195–204.

Houle, D., C. Pélabon, G. P. Wagner, and T. F. Hansen. 2011. Measurement and meaning in biology. *Quarterly Review of Biology* 86: 3–34.

Lande, R. 1976. Natural selection and random drift in phenotypic evolution. *Evolution* 30: 314–334.

Lynch, M. 1990. The rate of morphological evolution in mammals from the standpoint of the neutral expectation. *American Naturalist* 136: 727–741.

Michalopoulos, G. K. 1990. Liver regeneration: Molecular mechanisms of growth control. *FASEB Journal* 4: 176–187.

Olson, E. C., and R. L. Miller. 1958. *Morphological Integration.* Chicago: University of Chicago Press.

Pélabon, C., C. H. Hilde, S. Einum, and M. Gamelon. 2020. On the use of the coefficient of variation to quantify and compare trait variation. *Evolution Letters* 4: 180–188.

Pomiankowski, A., and Y. Iwasa. 1998. Runaway ornament diversity caused by Fisherian sexual selection. *PNAS* 95: 5106–5111.

Reno, P. L., C. Y. McLean, J. E. Hines, T. D. Capellini, G. Bejerano, and D. M. Kingsley. 2013. A penile spine/vibrissa enhancer sequence is missing in modern and extinct humans but is retained in multiple primates with penile spines and sensory vibrissae. *PLOS One* 8: e84258.

Riska, B., W. R. Atchley, and J. J. Rutledge. 1984. A genetic analysis of targeted growth in mice. *Genetics* 107: 79–101.

Sallan, L. C., and M. Friedman. 2012. Heads or tails: Staged diversification in vertebrate evolutionary radiations. *Proceedings of the Royal Society B* 279: 2025–2032.

Schuster, P., W. Fontana, P. F. Stadler, and I. Hofacker. 1994. From sequences to shapes and back: A case study in RNA secondary structure. *Proceedings of the Royal Society B* 255: 279–284.

Stadler, B. M. R., P. F. Stadler, G. P. Wagner, and W. Fontana. 2001. The topology of the possible: Formal spaces underlying patterns of evolutionary change. *Journal of Theoretical Biology* 213: 241–274.

Turelli, M. 1984. Heritable genetic variation via mutation-selection balance: Lerch's Zeta meets abdominal bristle. *Theoretical Population Biology* 25: 138–193.

Turelli, M., and N. H. Barton. 1990. Dynamics of polygentic characters under selection. *Theoretical Population Biology* 38: 1–57.

Waddington, C. H. 1957. Experiments on canalizing selection. *Genetics Research* 1: 140–150.

Waddington, C. H. 1959. Canalization of development and genetic assimilation of acquired characters. *Nature* 183: 1654–1655.

Wagner, G. P., G. Booth, and H. Bagheri-Chaichian. 1997. A population genetic theory of canalization. *Evolution* 51: 329–347.

Ward, A. B., and E. L. Brainerd. 2007. Evolution of axial patterning in elongate fishes. *Biological Journal of the Linnean Society* 90: 97–116.

Yablokov, A. V. 1974. *Variability of Mammals.* New Dehli: Amerind Publishing Co.

11 Mutational Robustness and Evolvability

Andreas Wagner

Organisms are to some extent robust to DNA mutations: Their phenotypes do not change in the face of some DNA mutations that affect the gene(s) encoding these phenotypes. Robustness can facilitate evolvability—the ability of a biological system to produce phenotype variation that is both heritable and adaptive. Here I first introduce some concepts that are necessary to understand why robustness can entail evolvability. I then discuss empirical evidence that speaks to the relationship between robustness and evolvability, focusing on systems where the molecular foundations of both robustness and evolvability can be studied in detail. Finally, I discuss empirical evidence that robustness can itself evolve, and that evolvability mediated by robustness can itself be subject to adaptive evolution.

11.1 Introduction

Organisms are to some extent robust to DNA mutations. That is, their phenotypes do not change in the face of some DNA mutations that affect the gene(s) encoding these phenotypes, or the regulatory DNA driving the expression of these genes (Wagner 2005; Masel and Siegal 2009; Fares 2015). This robustness can vary among organisms, among phenotypes, and among the genotypes encoding any one phenotype (Giver et al. 1998; Lynch and Conery 2000; Salazar et al. 2003; Bloom et al. 2006; Fasan et al. 2008; Jiménez et al. 2013; Keane et al. 2014; Payne and Wagner 2014, 2019; Najafabadi et al. 2017; Starr et al. 2017; Payne et al. 2018; Vaishnav et al. 2021). Such mutational or genetic robustness is closely linked to evolvability—the ability to bring forth phenotypic variation that is both heritable and adaptive (Wagner 2005, 2008; Draghi et al. 2010; Mayer and Hansen 2017; Payne and Wagner 2019). At first sight, this relationship might seem straightforward: High robustness implies that a given number of mutations generate little phenotypic variation—adaptive or otherwise—and because natural selection requires phenotypic variation, high robustness should imply low evolvability. The argument is simple, but it is also misleading. In fact, high robustness often entails high evolvability. In this chapter, I will first explain why, and then discuss pertinent empirical evidence.

This is not an exhaustive review of the relevant literature, which could easily fill an entire book (Wagner 2005). For example, I do not discuss the role of recombination in the evolution of robustness, nor do I say much about the role of robustness to environmental change. I also omit scenarios where robustness is not adaptive, because selection favors genotypic and phenotypic diversity. Examples include the antibody diversity that helps the adaptive immune system combat pathogens. They also include the antigenic diversity that numerous pathogens create through targeted recombination or mutation processes, which help them evade host immune responses (Deitsch et al. 2009). Furthermore, I do not discuss the burgeoning theoretical literature on robustness. Instead, I provide a few key ideas that link mutational robustness and evolvability, and I discuss the empirical evidence supporting this link. More specifically, I first introduce some concepts that are necessary to understand why robustness can entail evolvability. I then discuss pertinent empirical evidence, focusing on systems where the molecular foundations of both robustness and evolvability can be studied in great detail. Finally, I discuss what we know about the evolution of robustness and evolvability.

11.2 Concepts

To understand the relationship between robustness and evolvability, it is essential to know that the same phenotype is usually encoded by many different genotypes in a *genotype space.* Such a space is typically defined as the set of all DNA or amino acid sequences of a given length. If these sequences are short, genotype space comprises a modest number of genotypes. Consider, for example, the regions of regulatory DNA known as transcription factor binding sites. Such sites are typically shorter than 16 base pairs (bps; Stewart and Plotkin 2012) and thus exist in a genotype space of fewer than $4^{16} \approx 4 \times 10^9$ molecules. When a transcriptional activator binds to such a site, its binding can help turn on a nearby gene's transcription in proportion to its binding affinity. The phenotype of such a site is its ability to bind the activator, which depends on its DNA sequence (genotype). Any one transcription factor can bind dozens to hundreds of such DNA sequences with similar affinity, and thus activate a nearby gene to a similar extent (Badis et al. 2009; Weirauch et al. 2014).

For more complex biomolecules, both genotype space and the number of genotypes encoding the same phenotype can be much larger (Reidhaar-Olson and Sauer 1990; Schuster et al. 1994; Reidys et al. 1997; Keefe and Szostak 2001). To give an example, consider RNA molecules of length $L = 30$ nucleotides, which constitute a genotype space of $4^{30} \approx 10^{18}$ RNA sequences. Their minimum free energy secondary structure phenotypes—the planar folds they can form through internal base pairing—are biologically important, because they are essential for the biological functions of many RNA molecules (Baudin et al. 1993; Powell et al. 1995). Most such RNA phenotypes are encoded by multiple RNA genotypes, and the number of genotypes encoding the same phenotype varies by several orders of magnitude among phenotypes (Wagner 2008).

Analogous observations hold for proteins. For example, it has been estimated experimentally that $\approx 10^{93}$ amino acid sequences of length 80 amino acids are able to bind ATP (Keefe and Szostak 2001). Likewise, more than 10^{56} amino acid sequences of length $L = 93$ encode the λ repressor, a transcriptional regulator of bacteriophage λ (Reidhaar-Olson and Sauer 1990). These numbers are unimaginably large, but they still constitute a vanishing fraction of genotype space. For example, in the genotype space of 20^{93} proteins with $L = 93$ amino acids, the 10^{56} λ repressors constitute a fraction $\approx 10^{-63}$ of genotype space.

Robustness does not just require that multiple genotypes encode the same phenotype. It also requires that any one genotype G has multiple *1-mutant neighbors* with this phenotype. A 1-mutant neighbor of G is a genotype that can be created from it by a single DNA mutation, such as a single nucleotide change. I will refer to the collection of all 1-mutant neighbors of a genotype G as G's (1-mutant) *neighborhood.* Such a neighborhood comprises $3L$ genotypes for DNA or RNA molecules of length L, because each of the 4 possible nucleotides can mutate into 3 other nucleotides, and $19L$ genotypes for proteins of length L amino acids, if each of the 20 proteinaceous amino acid can mutate into 19 others.

The smaller a genotype's fraction of 1-mutant neighbors with the same phenotype is, the smaller will be the robustness of this genotype. Figure 11.1a illustrates this idea in a highly simplified schematic of a hypothetical genotype G whose phenotype is indicated by the black circle in the center. This genotype has 8 1-mutant neighbors (connected to it by thick black lines), all of which are assumed to encode a different phenotype (shapes at the tip of each line) than G itself. Thus, this genotype is minimally robust to mutations. Figure 11.1b shows another hypothetical genotype G with 8 neighbors, but only 3 of these neighbors have a different phenotype. The other 5, connected to G by gray lines, encode the same phenotype (not shown) as G itself. The genotype in figure 11.1b is more robust than that in figure 11.1a. Under the assumption that among all neighboring phenotypes, some (possibly small) fraction of them is adaptive, high robustness of a genotype implies low evolvability. I will refer to this notion of robustness and evolvability as *genotypic* robustness and evolvability, because they are properties of a specific genotype encoding a phenotype (Wagner 2008). I emphasize that figure 11.1 is an abstraction to illustrate general ideas with simplifications chosen for the purpose of explanation. For example, many phenotypes are continuous rather than categorical quantities, and one genotype may encode more than one phenotype.

Although the negative association between robustness and evolvability appears inevitable from a theoretical perspective, it is reassuring that it also has empirical support. Pertinent evidence comes from an experiment that measured the ability of 20 million yeast regulatory regions (genotypes) of $L = 80$ bp to activate the expression of a yeast gene in a massively parallel assay. The phenotype of any one such sequence is the expression level of the regulated gene. The experimenters then used a deep learning neural network to predict this phenotype for those sequences whose regulatory activity they had not measured. Subsequently, they synthesized and tested thousands of further regulatory sequences and showed that the network's predictions are in excellent agreement with experimental data. With this tool in hand, the authors defined a regulatory sequence's (genotypic) evolvability as a mutation's tendency to change the expression level of the regulated gene. Not surprisingly, mutationally robust sequences were less evolvable (Vaishnav et al. 2021).

To see the limitations of these genotype-centered concepts, consider again some genotype G encoding a phenotype, such as a protein's ability to catalyze a chemical reaction. Consider a single 1-mutant neighbor of G with the same phenotype, such as genotype G_2 in figure 11.1c. This 1-mutant neighbor may itself have multiple neighbors that preserve this phenotype (one of which is shown as G_2 in figure 11.1c), which in turn may themselves have multiple neighbors with the same phenotype, and so on. In other words, the genotypes encoding the same phenotype may form a network in genotype space.

For such a network to be large and extend far through genotype space, it is sufficient that the genotypes encoding any one phenotype must have a modest nonzero fraction of neighbors that encode the same phenotype (Reidys et al. 1997; Wagner 2011). Such networks have first been described in computational models of protein and RNA folding

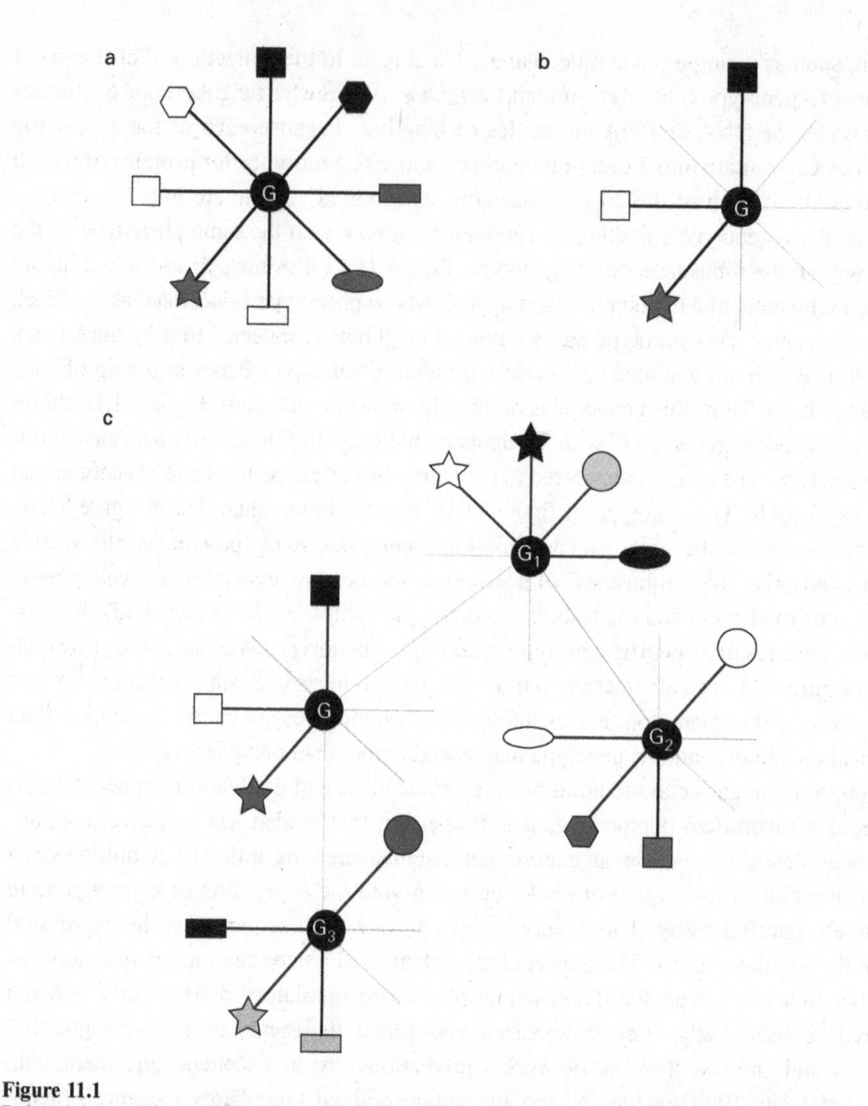

Figure 11.1
Robustness, genotype networks, and evolvability illustrated with a highly simplified hypothetical example. The figure shows genotypes as nodes in a graph, where neighboring genotypes are connected by straight lines. Different phenotypes are indicated by different shapes and their shading. a) Hypothetical minimally robust genotype *G*, whose 8 1-mutant neighbors all have a different phenotype. b) As in panel a, but only 3 neighbors have the same phenotype, whereas the remaining 5 neighbors (connected to *G* by gray lines) have the same phenotype (not shown) as *G* itself. c) As in panel b, but now the phenotypes in the neighborhood of 2 1-mutant neighbors of *G* (G_1 and G_3), as well as of 1 2-mutant neighbor (G_2) are also shown. Neighbors with the same phenotype are again connected by gray lines. Even though *G* is to some extent robust, and thus only 3 novel phenotypes are accessible in its immediate 1-neighborhood, 14 novel phenotypes are accessible from it through $G_1 - G_3$, because these networks form a phenotype-preserving genotype network. This simple schematic neglects the high-dimensional nature of genotype space, the continuous nature of many phenotypes, as well as the fact that many genotypes encode multiple phenotypes, but the key principles hold for more complex scenarios as well (Wagner 2014).

(Lipman and Wilbur 1991; Schuster et al. 1994). However, they exist on all levels of biological organization, not just for proteins and RNA molecules, but also for regulatory circuits and their gene expression phenotypes (Ciliberti et al. 2007; Schaerli et al. 2014), as well as for the chemical reaction networks encoded by metabolic genes and their ability to metabolize specific nutrients (Rodrigues and Wagner 2009).

A well-studied example among proteins is oxygen-binding globins. These ancient proteins share a common ancestor that existed many hundreds of million years ago, and they exist in both animals and plants. They have preserved their protein structure and their biochemical, oxygen-binding phenotype. At the same time, phylogenetic analysis shows that they have dramatically diverged in genotype through single amino acid changes, such that 2 globins may share less than 5% amino acid identity along their coding sequence (Goodman et al. 1988; Hardison 1996). Proteins with highly diverged genotypes and conserved phenotypes are the rule rather than the exception among biological macromolecules (Thornton et al. 1999; Rost 2002; Bastolla et al. 2003).

A population that evolves under mutation and selection acting to preserve an adaptively important phenotype explores genotype space along the kind of network illustrated in figure 11.1c. Based on computational models, such networks have first been called neutral networks (Schuster et al. 1994), suggesting that their exploration involves only neutral mutations. However, this need not be the case. For example, even though a mutation may preserve a globin's oxygen-binding ability, the mutation may increase or decrease this ability, and thus not be neutral with respect to fitness. As long as the mutation is not highly deleterious, however, it may not be eliminated from an evolving population, and it may provide a stepping-stone toward further mutation (Ohta 1992; Eyre-Walker et al. 2002; Kern and Kondrashov 2004; Kulathinal et al. 2004; Weinreich and Chao 2005; Sawyer et al. 2007). Because strict neutrality is not required for the exploration of a network of genotypes, I prefer to call such networks more generically *genotype networks*.

Genotype networks—a consequence of robustness—can facilitate evolvability. They allow an evolving population to explore a broad region of genotype space through mutations that preserve its phenotype, which may be important if the phenotype is vital for survival. During this process, the population's members also explore the mutational neighborhoods of multiple genotypes on a genotype network. In the hypothetical example shown in figure 11.1c, a total of 14 different novel phenotypes (shapes) are accessible to G via G_1, G_2, and G_3, even though genotype G is quite robust (i.e., the 1-neighborhood of G contains, just like in figure 11.1b only 3 novel phenotypes). Many more novel phenotypes may be accessible through further neighbors of these genotypes.

If different neighborhoods contain different novel phenotypes, the chances of encountering an adaptive novel phenotype can be much greater than through the exploration of just a single neighborhood. To illustrate the dramatic increase in the number of novel phenotypes that can become accessible through a genotype network, consider a guide RNA from *Trypanosoma brucei* with $L=40$ nucleotides and its minimum free energy secondary structure phenotype (accession number L25590 of the functional RNA database (https://dbarchive.biosciencedbc.jp/en/frnadb). The 1-neighborhood of this genotype G comprises $3L=120$ genotypes and could thus encode at most 120 different phenotypes. However, computational predictions of RNA secondary structures show that only 40 of these neighbors encode a novel phenotype (Wagner 2012). In other words, G is to some extent robust to mutations, and this robustness reduces the number of novel phenotypes that are accessible in its immediate (1-mutant) neighborhood. However, the 1-mutant neighborhoods of *G*'s 1-mutant neighbors encode many more novel phenotypes, 746 to be precise (Wagner 2012). In other words, just 2 mutations away from *G* 746, new phenotypes become accessible. Furthermore, the 1-mutant neighborhoods of all 2-mutant neighbors of *G* contain an even greater number of 1,174 distinct new phenotypes (Wagner 2012). Thus in just a few mutational steps away from a focal genotype, the total number of accessible novel phenotypes escalates rapidly. The total number of genotypes forming this guide RNA's secondary structure can be computed, and it is greater than 10^{17} (Jörg et al. 2008). It is not currently possible to compute the total number of novel phenotypes in the neighborhoods of all these genotypes, but this number is likely to be astronomically large as well.

These considerations show that it is shortsighted to just consider the robustness and evolvability of *genotypes*. Instead, the robustness and evolvability of *phenotypes* may be more useful. One can define the *robustness of a phenotype* as the average fraction of a genotype's neighbors with this phenotype, where the average is taken over all genotypes encoding this phenotype. Likewise, one can define the average *evolvability of a phenotype* as the total number of novel phenotypes that can be found in the neighborhoods of all genotypes encoding this phenotype. Some fraction of these novel phenotypes will be adaptive. Because more robust phenotypes have larger genotype networks, an evolving population with such a phenotype thus can access more genotype neighborhoods, which contain more novel phenotypes than the accessible neighborhoods of a less robust phenotype. In other words, phenotypic robustness can entail phenotypic evolvability.

This has first been shown computationally for RNA secondary structure phenotypes (Wagner 2008), but relevant empirical evidence exists for other systems (Ferrada and Wagner 2008; Payne and Wagner 2014). For example, consider the DNA binding sites of eukaryotic transcription factors (TF), where genotype networks have been studied for 104 mouse and 89 yeast TFs (Payne and Wagner 2014). A typical TF binds multiple DNA sequences with high affinity, and this number varies among factors, from dozens to hundreds of sites (Badis et al. 2009; Weirauch et al. 2014). For 99% of the examined factors, the majority of a factor's binding sites formed a single connected genotype network. The average robustness of these sites varied broadly among factors, ranging between 7% and 48% of a site's 1-mutant neighbors that were bound by the same factor (Payne and Wagner 2014). Larger genotype networks are formed by factors with more robust binding sites. The neighborhood of a TF's genotype network—the collection of all binding sites that are 1 nucleotide change away from at least one of the network's genotypes—harbors binding sites for multiple other TFs. If one uses the number of such novel binding sites as a measure of evolvability, high robustness entails high evolvability, that is, a larger repertoire of new binding sites that are only 1 nucleotide change away from a given genotype network (Payne and Wagner 2014).

I emphasize that the concepts introduced so far abstract from a complex reality and are subject to several caveats (De Visser et al. 2003; Meyers et al. 2005; Manrubia and Cuesta 2015; Mayer and Hansen 2017). For example, they statically enumerate genotypes with specific phenotypes and ignore the dynamics of evolving populations. Most genotype networks are much larger than any one population evolving on it, so that such a population will only be able to explore a tiny region of such a network, even on time scales of millions of years. Thus, viewing robustness and evolvability as averages over all genotypes in a network may arguably be less important than examining them in a region that an evolving population can explore. This is especially important if there is substantial variation in robustness and evolvability among different regions of a genotype network. Such variation indeed exists. For example, whereas on average, 37% of DNA sequences that bind the mouse TF Foxa2 are robust to single nucleotide changes, this percentage varies enormously among individual binding sites and ranges from 3% to 72% (Payne and Wagner 2014). Where such local variation is extreme, the relationship between robustness and evolvability may change. For example, it has been proposed that evolving populations may become entrapped in regions of genotype space where any one genotype is so highly robust that most of its neighbors have the same genotype (Manrubia and Cuesta 2015). Such entrapment can lead to low evolvability. This possibility is to date only theoretical, but it illustrates that local or regional properties of a genotype space have the potential to affect the relationship between robustness and evolvability. The starting point and duration of an evolutionary process, together with other parameters, such as the mutation rate and population size, can all potentially influence the relationship between robustness and evolvability.

For these reasons, it is important to study this relationship with empirical evidence derived from evolving populations. Two principal approaches provide such evidence. The first is experimental evolution, where whole organisms or molecules are evolved in the laboratory or in vitro. Such experiments, combined with high-throughput genotyping as well as physiological and biochemical analyses of evolved genotypes and their phenotypes, can examine the evolutionary process in real time and in exquisite molecular detail. One of their limitations is that they are restricted to short evolutionary time scales and to populations that explore only small regions of a genotype space. This time limitation imposes further constraints, such as the necessity to work at high mutation rates or at selection pressures that may be stronger than in the wild.

The second approach comprises comparative and phylogenetic studies that infer past evolutionary processes from extant organisms. It can be aided by the reconstruction of ancestral and extinct genotypes, and by biochemical analyses of these phenotypes (Bridgham et al. 2006; Dean and Thornton 2007; Ortlund et al. 2007; Eick et al. 2012; McKeown et al. 2014; Anderson et al. 2015; Nocedal et al. 2017; Starr et al. 2017, 2018). This approach can overcome the limitations of experimental evolution, but it has its own limitations, which come from its need to infer the past from the present. In the next section, I discuss data from both approaches, which show that robustness can facilitate evolvability.

11.3 Empirical Data Show that Robustness Facilitates Evolvability

One fundamental consequence of robustness is that evolving populations can accumulate *cryptic genetic variation*. This is genetic variation that does not cause phenotypic variation while it accrues but is not phenotypically neutral under all circumstances. It can give rise to novel phenotypic variation when the environment changes, or when further mutations arise (Rutherford and Lindquist 1998; True and Lindquist 2000; Masel and Bergman 2003;

True et al. 2004; Masel 2006; Jarosz and Lindquist 2010). Evolution experiments have been used to ask whether cryptic variation can facilitate adaptive evolution. They can help explain the role of robustness in adaptive evolution (Tokuriki and Tawfik 2009; Hayden et al. 2011; Rigato and Fusco 2016; Zheng et al. 2019). In one pertinent experiment, my colleagues and I used directed evolution to accumulate cryptic variation in a yellow fluorescent protein (YFP). The experiment employed repeated cycles ("generations") of mutation and selection on the yellow fluorescent light emitted by YFP. Specifically, we evolved 4 populations of YFP under strong stabilizing selection on the native yellow fluorescence phenotype, which allowed the population to accumulate cryptic genetic variation with minimal effect on the light-emitting phenotype. After 4 generations of stabilizing selection, we continued for another 4 generations but under strong directional selection for a new color phenotype, namely, green fluorescence. In parallel, we evolved 4 additional populations toward green fluorescence, but these populations had not been given the opportunity to accumulate cryptic genetic variation. We found that populations with cryptic variation evolved green fluorescence more rapidly than those without it (Zheng et al. 2019). In addition, populations with cryptic variation evolved a higher intensity of green fluorescence (Zheng et al. 2019). Moreover, populations with cryptic evolution evolved a greater diversity of green-fluorescing genotypes. In sum, this experiment not only shows that cryptic variation facilitates evolutionary adaptation. It also shows that robustness can help evolving populations find diverse (and superior) solutions to the adaptive problems they face. The reasons are easy to understand from the visual metaphor of figure 11.1c: Robustness implies that evolving populations can diversify in multiple directions from a starting genotype, and each of these directions may lead to different high-fitness genotypes.

Cryptic variation can also facilitate evolvability in other systems, and most notably in whole organisms. In one pertinent experiment, Rigato and Fusco (2016) used chemical mutagenesis to introduce a modest number (<30) mutations into the genome of *E. coli* cells. They then subjected populations of the mutagenized cells for 56 generations to strong stabilizing selection on their ability to grow on glucose. The purpose of this procedure was to eliminate phenotypic variation that may have been caused by mutagenesis and thus to preserve only cryptic variation in the populations. In addition, Rigato and Fusco also exposed populations without prior mutagenesis to strong stabilizing selection for the same number of generations. At the end of this preparatory experiment, both kinds of populations showed the same (low) amount of phenotypic variation in their growth rate on glucose. If any additional genetic variation that the mutagenized populations harbored was cryptic, then its phenotypic effects should be revealed in the right environment or genetic background. This was indeed the case, as the researchers' next experiment showed. In this experiment, the researchers subjected both the mutagenized and nonmutagenized populations to directional selection for different phenotypes, namely, high growth on lactate or glycerol (in separate experiments). The mutagenized populations adapted faster to both glycerol and lactate (Rigato and Fusco 2016). Furthermore, the researchers showed through mutagenesis that growth on glycerol is more robust to mutations than is growth on lactate, and that populations adapted more rapidly to glycerol than to lactate. In other words, the more robust phenotype was also more evolvable (Rigato and Fusco 2016). Note that the experiments discussed here involved large populations and a single change of the selective environment (Rigato and Fusco

2016; Zheng et al. 2019). Cryptic variation may affect adaptive evolution differently in smaller populations and frequently changing environments.

Other experimental studies focused on comparing the evolvability of systems with high and low robustness (Bloom et al. 2006, 2010; McBride et al. 2008; Igler et al. 2018; Zakrevsky et al. 2021). One such experiment created populations of RNA bacteriophage φ6, whose ability to survive and reproduce was either robust or sensitive to mutations (Montville et al. 2005; McBride et al. 2008). The experiment created these populations by serially passaging viral populations through the bacterium *Pseudomonas syringae* under conditions where bacteria were either infected by a single virus or simultaneously by multiple viruses. (Multiple co-infections can help a defective virus reproduce, because its defects can be complemented by other, intact co-infecting virus. During the course of multiple passages through a host, such complementation causes viral populations to become more sensitive to mutations.) The experimenters then evolved both types of populations toward increased survivorship after heat shocks of 45°C. Specifically, they passaged these viruses through bacteria for 50 viral generations and exposed them to a heat shock every 5 generations. The more robust populations adapted more rapidly to the heat shock treatment (McBride et al. 2008).

A completely different kind of experiment revolves around cytochrome P450, a class of enzymes that can catalyze reactions with multiple substrates. In this experiment, Bloom and collaborators engineered a cytochrome P450 enzyme for higher thermodynamic stability by introducing a specific stabilizing mutation into the enzyme (Bloom et al. 2006). This mutation also increases the robustness of this enzyme's activity to mutations. The researchers then asked whether the modified enzyme could more easily evolve the ability to catalyze reactions with new substrates. To find out, they introduced random mutations into the enzyme variants with high and low robustness, at an average incidence of 4.5 nucleotide changes per P450-coding gene. They then monitored the ability of both variants to catalyze reactions with multiple substrates. The more robust variants showed higher activity after mutagenesis on several substrates (Bloom et al. 2006).

All of these experiments rely on the short time scales of laboratory evolution. Phylogenetic analyses can help elucidate the relationship between robustness and evolvability on much longer time scales (Ferrada and Wagner 2008; Najafabadi et al. 2017; Nocedal et al. 2017; Starr et al. 2017). One such analysis focused on the 3-dimensional folds (tertiary structure) of ancient and well-studied enzymes (Ferrada and Wagner 2008). Because the fold of an enzyme is essential for its ability to catalyze chemical reactions, a robust fold is more likely to preserve this ability in the face of mutations. The robustness of an enzyme's fold can be estimated through at least 2 complementary approaches. The first determines the number of amino acid changes that a fold has accumulated in its evolutionary history—more robust folds tolerate more such changes, and their amino acid sequences thus change more rapidly in evolution. The second determines robustness directly from the contact density matrix of the fold, which is a descriptor of the amino acid contacts that occur in the fold (England and Shakhnovich 2003; Shakhnovich et al. 2005). As a measure of evolvability, one can estimate the number of different biochemical or biological functions that enzymes with a given fold have evolved since their evolutionary origin, for example, by examining all chemical reactions that are catalyzed by such enzymes. Such an analysis, conducted for 112 ancient enzymes, shows that enzymes with highly robust folds have evolved more diverse biochemical and biological functions (Ferrada and Wagner 2008).

Broad analyses of many proteins like this one are also supported by more focused studies of individual proteins, such as steroid hormone receptors (McKeown et al. 2014; Starr et al. 2017). These are transcription factors that bind DNA and regulate gene expression in response to steroid hormones. They are ancient proteins whose most recent common ancestor dates to more than 450 million years ago (Eick et al. 2012). This ancestor bound DNA sequences known as estrogen responsive elements (EREs), which mediate gene regulation by estrogen. The ancestor duplicated, and the duplicate evolved the ability to bind specifically to steroid-responsive elements (SREs), which differ from EREs and mediate regulation by different steroid hormones. A combination of phylogenetic analysis, mutant engineering, and biochemical experiments showed that 11 mutations in the duplicated receptor were crucial for the evolution of this new regulatory phenotype (Starr et al. 2017). These mutations occurred outside the DNA-binding domain of the protein. They left the binding specificity of the receptor unchanged, but they changed the general affinity of the receptor to DNA. In doing so, they also increased the mutational robustness of the receptor's ability to bind DNA. As a consequence, they increased by more than 20-fold the proportion of further receptor mutations that bind SREs. For example, among 160,000 variants of the ancestral receptor that lacked these 11 robustness-enhancing mutations, only 41 specifically bound SREs. In contrast, among 160,000 variants of the protein with the 11 mutations, 829 specifically bound SREs, and these variants could be reached by a smaller number of individual amino acid changes (Starr et al. 2017). In sum, mutations that increased robustness also increase the evolvability of this TF's new DNA binding and gene regulatory phenotype.

A biochemical explanation for the positive relationship between robustness and evolvability exists for an unrelated and somewhat more anecdotal example. It involves the zinc finger domain, a protein fold that is part of many TFs and helps them bind specific DNA sequences. The zinc finger domain is exceptionally robust to amino acid changes. For example, all but 7 of its 26 amino acids can be replaced by alanine without destroying its 3-dimensional structure (Michael et al. 1992). In part because of this robustness, zinc finger domains can be engineered toward a great variety of DNA binding specificities (Durai et al. 2005). They are also the most abundant protein domains in the human proteome, occurring in 500 different proteins (Venter et al. 2001). Zinc finger domains fall into different classes. One of them is the C2H2 zinc finger, so named because it contains 2 cysteines and 2 histidines. This motif has evolved much greater DNA binding diversity in metazoans than in other eukaryotes, which can be explained by the greater robustness of the metazoan C2H2 zinc finger. To see why, it is useful to know that the DNA binding affinity of a TF can be determined both by amino acids that contact specific bases and by amino acids that contact the DNA backbone. In non-metazoan C2H2 zinc fingers, DNA affinity is determined by base-contacting amino acids. However, in metazoan C2H2 zinc fingers, affinity is partly determined by backbone contacts. As a result, individual base-contacting amino acids are free to vary without complete loss of DNA binding, which facilitates variation in these amino acids and thus allows variation in DNA binding specificity. In other words, DNA binding is more robust to amino acid changes in metazoans, which also allows DNA binding specificity to vary more broadly (Najafabadi et al. 2017).

In sum, empirical evidence ranging from macromolecules to whole organisms and viruses support the notion that robustness can facilitate the adaptive evolution of new phenotypes on both short and long evolutionary time scales.

11.4 Evolution of Robustness and Evolvability

Robustness can itself evolve, and so can the associated evolvability. If this is the case, the ability of a biological system itself may be an evolving property. I will next discuss empirical evidence that speaks to this possibility.

The general question of whether evolvability itself evolves has recently been reviewed elsewhere (Payne and Wagner 2019). I will thus view this question here in the context of the evolution of robustness. The question can be subdivided into 3 parts. First, *can* robustness (and the associated evolvability) evolve? In other words, is there heritable genetic variation for these properties? Second, *do* they evolve, either in laboratory evolution experiments, or in nature? Third, do they evolve adaptively? That is, can they provide a sufficiently strong benefit that their evolution is driven by this benefit?

The first question is easy to answer. Robustness can evolve. It is subject to heritable variation on all levels of biological organization, from macromolecules and their interactions to whole organisms (Lynch and Conery 2000; Jiménez et al. 2013; Keane et al. 2014; Payne and Wagner 2014, 2019; Najafabadi et al. 2017; Starr et al. 2017; Payne et al. 2018). Several examples come from experiments to engineer specific amino acids into enzymes to increase their robustness (Giver et al. 1998; Salazar et al. 2003; Bloom et al. 2006; Fasan et al. 2008). Likewise in nature, past amino acid changes have increased the robustness of the steroid hormone receptors discussed in section 11.3 (Starr et al. 2017). I also discussed that zinc finger TFs vary between metazoan and other eukaryotes in how they contact DNA, which causes differences in the robustness of their DNA affinity to mutations (Najafabadi et al. 2017). Unrelated examples that I did not discuss include the bacterial transcription factor LexA, whose ability to regulate gene expression can be more or less robust to DNA mutations, depending on whether LexA negatively autoregulates its own expression (Marciano et al. 2014). On a higher level of biological organization, the ability of bacteriophage φ6 to survive and reproduce can vary in its robustness to DNA mutations (McBride et al. 2008). Gene duplications can increase the robustness of an organism to mutations in the duplicated genes (Lynch and Conery 2000), which is associated with increased evolvability in organisms as different as flowering plants (Theissen et al. 1996; Irish and Litt 2005; Hernandez-Hernandez et al. 2007) and vertebrates (Carroll et al. 2001; Olson 2006).

These and other examples also answer the second question: Robustness does evolve (Montville et al. 2005; Borenstein and Ruppin 2006; Bloom et al. 2007; McBride et al. 2008; Zheng et al. 2020). Experimental evolution has increased the mutational robustness of proteins, such as cytochrome P450 (Bloom et al. 2007) and yellow fluorescent protein (Zheng et al. 2020). It also succeeded in increasing the robustness of bacteriophage φ6 (Montville et al. 2005; McBride et al. 2008) and of vesicular stomatitis virus (Codoñer et al. 2006; Sanjuán et al. 2007). More importantly, evolution in the wild has also changed the robustness of various systems. For example, in their distant evolutionary history, steroid hormone receptors have accrued mutations that increase the robustness of their ability to bind DNA (Starr et al. 2017). More generally, the robustness of a protein's fold tends to increase with the evolutionary age of the protein (Toll-Riera et al. 2012). Stabilizing selection on yeast gene expression has increased the robustness of gene expression to mutations (Vaishnav et al. 2021). Many gene duplications in eukaryotic genomes have increased the robustness of a gene's function to mutations (Lynch and Conery 2000).

The third question regards the forces that drive the evolution of robustness and the associated evolvability. This question does not have a single answer. It depends on the kind of evolving system and on the conditions of its evolution, such as its population size and the mutation rate. For example, mutations are usually rare and therefore do not cause strong selection pressure for increased robustness to mutations. As a consequence, theory predicts that mutational robustness can evolve as an adaptation to mutations only when mutations are sufficiently frequent or when populations are sufficiently large (G. Wagner et al. 1997; van Nimwegen et al. 1999; De Visser et al. 2003; Wagner 2005). When these conditions are met, however, it has been shown that experimental evolution can readily increase robustness to mutations as an adaptation to mutations itself, both in proteins (Bloom et al. 2007) and in RNA viruses (Codoñer et al. 2006; Sanjuán et al. 2007).

Robustness to mutations can also evolve for at least two other reasons. First, it can emerge as a by-product of robustness to environmental change. Robustness to environmental change often entails robustness to mutations (Ancel and Fontana 2000; Meiklejohn and Hartl 2002; Bloom et al. 2006; Domingo-Calap et al. 2010; Butković et al. 2020), and environmental change is usually much more frequent than DNA mutation. As a consequence, populations experience stronger selection to become robust to environmental change (G. Wagner et al. 1997; Meiklejohn and Hartl 2002; Wagner 2005). Second, robustness may sometimes increase for no adaptive reason at all. For example, robustness often increases at least temporarily after a gene duplication, because 2 duplicate gene copies are usually redundant, such that the second copy can compensate for a deleterious mutation in the first copy (Lynch and Conery 2000). Gene duplications are frequent by-products of DNA repair and recombination processes that cells need to maintain their genomic integrity. They can be adaptive (Conant and Wolfe 2008; Nasvall et al. 2012), but they may also be maladaptive, because they carry a cost in terms of the energy needed to express them (Wagner 2007; Lynch and Marinov 2015). Thus, a gene duplication may entail high genetic robustness without necessarily being adaptive.

Such evidence shows that robustness evolves, but it does not answer the question of whether robustness can evolve because it facilitates evolvability. The problem is that evolvability, much like robustness, is a dispositional trait—it affects the potential for future evolution but need not convey immediate benefits to an organism. This indirect benefit of evolvability suggests that selection favoring evolvability is weaker than selection for other traits with direct benefits.

Such indirect selection, however, can still enhance evolvability, as a recent evolution experiment shows (Zheng et al. 2020). In this experiment, we studied the evolvability of the phenotype green fluorescence from the ancestral phenotype of yellow fluorescence in a population of evolving fluorescent proteins. More specifically, the experiment consisted of 2 separate phases. In the first phase, we evolved populations of yellow fluorescent protein toward increased *yellow* fluorescence during 4 generations of random mutation and directional selection. We subjected 4 populations to strong directional selection for this ancestral phenotype, 4 populations to weak directional selection, and 4 populations to no selection. After these 4 generations, we subjected each of the 12 populations to 4 more generations of equally strong selection for *green* fluorescence. We found that the populations that had been under strong selection for the ancestral yellow phenotype evolved green fluorescence more rapidly and to a higher level than did the other populatons. A combination of high-throughput population sequencing and mutant engineering showed that they accumulated mutations that increased mutational robustness by increasing the protein's foldability—the ability to form a correctly folded protein. These same mutations also increased protein fitness, but they did so predominantly through their effect on robustness, which hindered deleterious mutations from slowing down adaptive evolution, and which facilitated the spreading of mutations beneficial for the new phenotype (Zheng et al. 2020). This experiment shows that under the right conditions, most notably strong selection and a sufficiently high mutation rate, mutational robustness can evolve adaptively, because it enhances evolvability. An important task for future work is to find out whether such adaptive evolution of evolvability mediated by robustness also exists in the wild.

11.5 Summary and Outlook

In sum, a growing body of experimental evidence shows that robustness can facilitate evolvability. What is more, robustness can and does evolve, and it can even evolve adaptively to enhance evolvability. Genotype networks provide a unifying framework that can help explain experiments like those described in this chapter. This framework can help us understand that mutational robustness can facilitate evolvability, because it allows evolving populations to explore a wider region of a genotype space, in which the chances of finding novel and adaptive phenotypes are greater than in a smaller region.

Some of these experiments also illustrate the limitations of the genotype network framework in its simple form sketched in figure 11.1c. For example, the framework abstracts phenotypes into categories, which allows an enumeration of novel phenotypes for the purpose of mathematical and computational analyses (Schuster et al. 1994; Reidys et al. 1997; Ciliberti et al. 2007; Rodrigues and Wagner 2009). However, many phenotypes that are amenable to experimentation are continuous quantities, such as increased antibiotic resistance or enzyme activity. Thus, any one genotype network is embedded in an adaptive landscape, where genotypes with the same qualitative phenotype need not be neutral in fitness. They may differ quantitatively in the phenotype they encode, which means that this landscape's topography affects their evolutionary fate and their ability to discover novel phenotypes. To understand how robustness affects evolvability through its evolutionary dynamics on an adaptive landscape remains an exciting task for future theoretical work.

An additional complication is that many genotypes encode multiple phenotypes, even at thelowest levels of biological organization, where a promiscuous enzyme can catalyze multiple biochemical reactions (O'Brien and Herschlag 1999; Khersonsky and Tawfik 2010; Wagner 2014). An evolving genotype may thus encode an entire spectrum of phenotypes that may change with each step of evolution on and near a genotype network. Thus, although the framework of figure 11.1c serves to communicate a key principle, expanding and adapting it for different purposes will be essential to understanding why robustness often facilitates evolvability in the complex world of biological evolution.

Acknowledgments

I acknowledge funding from the European Research Council under Grant Agreement No. 739874, as well as from Swiss National Science Foundation grant 31003A_172887.

References

Ancel, L. W., and W. Fontana. 2000. Plasticity, evolvability, and modularity in RNA. *Journal of Experimental Zoology B* 288: 242–283.

Anderson, D. W., A. N. McKeown, and J. W. Thornton. 2015. Intermolecular epistasis shaped the function and evolution of an ancient transcription factor and its DNA binding sites. *eLife* 4: e07864.

Badis, G., M. F. Berger, A. A. Philippakis, et al. 2009. Diversity and complexity in DNA recognition by transcription factors. *Science* 324: 1720–1723.

Bastolla, U., M. Porto, H. E. Roman, and M. Vendruscolo. 2003. Connectivity of neutral networks, overdispersion, and structural conservation in protein evolution. *Journal of Molecular Evolution* 56: 243–254.

Baudin, F., R. Marquet, C. Isel, J. Darlix, B. Ehresmann, and C. Ehresmann. 1993. Functional sites in the 5′ region of human-immunodeficiency-virus type-1 RNA form defined structural domains. *Journal of Molecular Biology* 229: 382–397.

Bloom, J. D., S. T. Labthavikul, C. R. Otey, and F. H. Arnold. 2006. Protein stability promotes evolvability. *PNAS* 103: 5869–5874.

Bloom, J. D., Z. Lu, D. Chen, A. Raval, O. S. Venturelli, and F. H. Arnold. 2007. Evolution favors protein mutational robustness in sufficiently large populations. *BMC Biology* 5: 1–21.

Bloom, J. D., L. I. Gong, and D. Baltimore. 2010. Permissive secondary mutations enable the evolution of influenza oseltamivir resistance. *Science* 328: 1272–1275.

Borenstein, E., and E. Ruppin. 2006. Direct evolution of genetic robustness in microRNA. *PNAS* 103: 6593–6598.

Bridgham, J. T., S. M. Carroll, and J. W. Thornton. 2006. Evolution of hormone-receptor complexity by molecular exploitation. *Science* 312: 97–101.

Butković, A., R. González, I. Cobo, and S. F. Elena. 2020. Adaptation of turnip mosaic potyvirus to a specific niche reduces its genetic and environmental robustness. *Virus Evolution* 6: veaa041.

Carroll, S. B., J. K. Grenier, and S. D. Weatherbee. 2001. *From DNA to Diversity. Molecular Genetics and the Evolution of Animal Design*. Malden, MA: Blackwell.

Ciliberti, S., O. C. Martin, and A. Wagner. 2007. Innovation and robustness in complex regulatory gene networks. *PNAS* 104: 13591–13596

Codoñer, F. M., J.-A. Darós, R. V. Solé, and S. F. Elena. 2006. The fittest versus the flattest: Experimental confirmation of the quasispecies effect with subviral pathogens. *PLOS Pathogens* 2: e136.

Conant, G. C., and K. H. Wolfe. 2008. Turning a hobby into a job: How duplicated genes find new functions. *Nature Reviews Genetics* 9: 938–950.

Dean, A. M., and J. W. Thornton. 2007. Mechanistic approaches to the study of evolution: The functional synthesis. *Nature Reviews Genetics* 8: 675–688.

Deitsch, K. W., S. A. Lukehart, and J. R. Stringer. 2009. Common strategies for antigenic variation by bacterial, fungal and protozoan pathogens. *Nature Reviews Microbiology* 7: 493–503.

De Visser, J. A. G., J. Hermisson, G. P. Wagner, et al. 2003. Perspective: Evolution and detection of genetic robustness. *Evolution* 57: 1959–1972.

Domingo-Calap, P., M. Pereira-Gómez, and R. Sanjuán. 2010. Selection for thermostability can lead to the emergence of mutational robustness in an RNA virus. *Journal of Evolutionary Biology* 23: 2453–2460.

Draghi, J. A., T. L. Parsons, G. P. Wagner, and J. B. Plotkin. 2010. Mutational robustness can facilitate adaptation. *Nature* 463: 353–355.

Durai, S., M. Mani, K. Kandavelou, J. Wu, M. H. Porteus, and S. Chandrasegaran. 2005. Zinc finger nucleases: Custom-designed molecular scissors for genome engineering of plant and mammalian cells. *Nucleic Acids Research* 33: 5978–5990.

Eick, G. N., J. K. Colucci, M. J. Harms, E. A. Ortlund, and J. W. Thornton. 2012. Evolution of minimal specificity and promiscuity in steroid hormone receptors. *PLOS Genetics* 8: e1003072.

England, J. L., and E. I. Shakhnovich. 2003. Structural determinant of protein designability. *Physical Review Letters* 90: 218101.

Eyre-Walker, A., P. D. Keightley, N. G. C. Smith, and D. Gaffney. 2002. Quantifying the slightly deleterious mutation model of molecular evolution. *Molecular Biology and Evolution* 19: 2142–2149.

Fares, M. A. 2015. The origins of mutational robustness. *Trends in Genetics* 31: 373–381.

Fasan, R., Y. T. Meharenna, C. D. Snow, T. L. Poulos, and F. H. Arnold. 2008. Evolutionary history of a specialized P450 propane monooxygenase. *Journal of Molecular Biology* 383: 1069–1080.

Ferrada, E., and A. Wagner. 2008. Protein robustness promotes evolutionary innovations on large evolutionary time scales. *Proceedings of the Royal Society B* 275: 1595–1602.

Giver, L., A. Gershenson, P. O. Freskgard, and F. H. Arnold. 1998. Directed evolution of a thermostable esterase. *PNAS* 95: 12809–12813.

Goodman, M., J. Pedwaydon, J. Czelusniak, et al. 1988. An evolutionary tree for invertebrate globin sequences. *Journal of Molecular Evolution* 27: 236–249.

Hardison, R. C. 1996. A brief history of hemoglobins: Plant, animal, protist, and bacteria. *PNAS* 93: 5675–5679.

Hayden, E. J., E. Ferrada, and A. Wagner. 2011. Cryptic genetic variation promotes rapid evolutionary adaptation in an RNA enzyme. *Nature* 474: 92–95.

Hernandez-Hernandez, T., L. P. Martinez-Castilla, and E. R. Alvarez-Buylla. 2007. Functional diversification of B MADS-Box homeotic regulators of flower development: Adaptive evolution in protein-protein interaction domains after major gene duplication events. *Molecular Biology and Evolution* 24: 465–481.

Igler, C., M. Lagator, G. Tkacik, J. P. Bollback, and C. C. Guet. 2018. Evolutionary potential of transcription factors for gene regulatory rewiring. *Nature Ecology & Evolution* 2: 1633–1643.

Irish, V. F., and A. Litt. 2005. Flower development and evolution: Gene duplication, diversification and redeployment. *Current Opinion in Genetics & Development* 15: 454–460.

Jarosz, D. F., and S. Lindquist. 2010. Hsp90 and environmental stress transform the adaptive value of natural genetic variation. *Science* 330: 1820–1824.

Jiménez, J. I., R. Xulvi-Brunet, G. W. Campbell, R. Turk-MacLeod, and I. A. Chen. 2013. Comprehensive experimental fitness landscape and evolutionary network for small RNA. *PNAS* 110: 14984–14989.

Jörg, T., O. Martin, and A. Wagner. 2008. Neutral network sizes of biological RNA molecules can be computed and are atypically large. *BMC Bioinformatics* 9: 464.

Keane, O. M., C. Toft, L. Carretero-Paulet, G. W. Jones, and M. A. Fares. 2014. Preservation of genetic and regulatory robustness in ancient gene duplicates of *Saccharomyces cerevisiae*. *Genome Research* 24: 1830–1841.

Keefe, A. D., and J. W. Szostak. 2001. Functional proteins from a random-sequence library. *Nature* 410: 715–718.

Kern, A. D., and F. A. Kondrashov. 2004. Mechanisms and convergence of compensatory evolution in mammalian mitochondrial tRNAs. *Nature Genetics* 36: 1207–1212.

Khersonsky, O., and D. S. Tawfik. 2010. Enzyme promiscuity: A mechanistic and evolutionary perspective. *Annual Review of Biochemistry* 79: 471–505.

Kulathinal, R. J., B. R. Bettencourt, and D. L. Hartl. 2004. Compensated deleterious mutations in insect genomes. *Science* 306: 1553–1554.

Lipman, D., and W. Wilbur. 1991. Modeling neutral and selective evolution of protein folding. *Proceedings of the Royal Society B* 245: 7–11.

Lynch, M., and J. S. Conery. 2000. The evolutionary fate and consequences of duplicate genes. *Science* 290: 1151–1155.

Lynch, M., and G. K. Marinov. 2015. The bioenergetic costs of a gene. *PNAS* 112: 15690–15695.

Manrubia, S., and J. A. Cuesta. 2015. Evolution on neutral networks accelerates the ticking rate of the molecular clock. *Journal of the Royal Society Interface* 12: 20141010.

Marciano, D. C., R. C. Lua, P. Katsonis, S. R. Amin, C. Herman, and O. Lichtarge. 2014. Negative feedback in genetic circuits confers evolutionary resilience and capacitance. *Cell Reports* 7: 1789–1795.

Masel, J. 2006. Cryptic genetic variation is enriched for potential adaptations. *Genetics* 172: 1985–1991.

Masel, J., and A. Bergman. 2003. The evolution of the evolvability properties of the yeast prion [PSI+]. *Evolution* 57: 1498–1512.

Masel, J., and M. L. Siegal. 2009. Robustness: Mechanisms and consequences. *Trends in Genetics* 25: 395–403.

Mayer, C., and T. F. Hansen. 2017. Evolvability and robustness: A paradox restored. *Journal of Theoretical Biology* 430: 78–85.

McBride, R. C., C. B. Ogbunugafor, and P. E. Turner. 2008. Robustness promotes evolvability of thermotolerance in an RNA virus. *BMC Evolutionary Biology* 8: 231.

McKeown, A. N., J. T. Bridgham, D. W. Anderson, M. N. Murphy, E. A. Ortlund, and J. W. Thornton. 2014. Evolution of DNA specificity in a transcription factor family produced a new gene regulatory module. *Cell* 159: 58–68.

Meiklejohn, C., and D. Hartl. 2002. A single mode of canalization. *TREE* 17: 468–473.

Meyers, L. A., F. D. Ancel, and M. Lachmann. 2005. Evolution of genetic potential. *PLOS Computational Biology* 1: 236–243.

Michael, S. F., V. J. Kilfoil, M. H. Schmidt, B. T. Amann, and J. M. Berg. 1992. Metal binding and folding properties of a minimalist Cys2His2 Zinc finger peptide. *PNAS* 89: 4796–4800.

Montville, R., R. Froissart, S. K. Remold, O. Tenaillon, and P. E. Turner. 2005. Evolution of mutational robustness in an RNA virus. *PLOS Biology* 3: 1939–1945.

Najafabadi, H. S., M. Garton, M. T. Weirauch, S. Mnaimneh, A. Yang, P. M. Kim, and T. R. Hughes. 2017. Non-base-contacting residues enable kaleidoscopic evolution of metazoan C2H2 zinc finger DNA binding. *Genome Biology* 18: 1–15.

Nasvall, J., L. Sun, J. R. Roth, and D. I. Andersson. 2012. Real-time evolution of new genes by innovation, amplification, and divergence. *Science* 338: 384–387.

Nocedal, I., E. Mancera, and A. D. Johnson. 2017. Gene regulatory network plasticity predates a switch in function of a conserved transcription regulator. *eLife* 6: e23250.

O'Brien, P. J., and D. Herschlag. 1999. Catalytic promiscuity and the evolution of new enzymatic activities. *Chemistry & Biology* 6: R91–R105.

Ohta, T. 1992. The nearly neutral theory of molecular evolution. *AREES* 23: 263–286.

Olson, E. N. 2006. Gene regulatory networks in the evolution and development of the heart. *Science* 313: 1922–1927.

Ortlund, E. A., J. T. Bridgham, M. R. Redinbo, and J. W. Thornton. 2007. Crystal structure of an ancient protein: Evolution by conformational epistasis. *Science* 317: 1544–1548.

Payne, J. L., F. Khalid, and A. Wagner. 2018. RNA-mediated gene regulation is less evolvable than transcriptional regulation. *PNAS* 115: E3481–E3490.

Payne, J. L., and A. Wagner. 2014. The robustness and evolvability of transcription factor binding sites. *Science* 343: 875–877.

Payne, J. L., and A. Wagner. 2019. The causes of evolvability and their evolution. *Nature Reviews Genetics* 20: 24–38.

Powell, D., M. Zhang, D. Konings, P. Wingfield, S. Stahl, E. Dayton, and A. Dayton. 1995. Sequence specificity in the higher-order interaction of the rev protein of HIV with its target sequence, the RRE. *Journal of Acquired Immune Deficiency Syndromes and Human Retrovirology* 10: 317–323.

Reidhaar-Olson, J. F., and R. T. Sauer. 1990. Functionally acceptable substitutions in two α-helical regions of λ repressor. *Proteins: Structure, Function, and Bioinformatics* 7: 306–316.

Reidys, C., P. Stadler, and P. Schuster. 1997. Generic properties of combinatory maps: Neutral networks of RNA secondary structures. *Bulletin of Mathematical Biology* 59: 339–397.

Rigato, E., and G. Fusco. 2016. Enhancing effect of phenotype mutational robustness on adaptation in *Escherichia coli. Journal of Experimental Zoology B* 326: 31–37.

Rodrigues, J. F. M., and A. Wagner. 2009. Evolutionary plasticity and innovations in complex metabolic reaction networks. *PLOS Computational Biology* 5: e1000613.

Rost, B. 2002. Enzyme function less conserved than anticipated. *Journal of Molecular Biology* 318: 595–608.

Rutherford, S. L., and S. Lindquist. 1998. Hsp90 as a capacitor for morphological evolution. *Nature* 396: 336–342.

Salazar, O., P. C. Cirino, and F. H. Arnold. 2003. Thermostabilization of a cytochrome P450 peroxygenase. *ChemBioChem* 4: 891–893.

Sanjuán, R., J. M. Cuevas, V. Furió, E. C. Holmes, and A. Moya. 2007. Selection for robustness in mutagenized RNA viruses. *PLOS Genetics* 3: e93.

Sawyer, S. A., J. Parsch, Z. Zhang, and D. L. Hartl. 2007. Prevalence of positive selection among nearly neutral amino acid replacements in *Drosophila. PNAS* 104: 6504–6510.

Schaerli, Y., A. Munteanu, M. Gili, J. Cotterell, J. Sharpe, and M. Isalan. 2014. A unified design space of synthetic stripe-forming networks. *Nature Communications* 5: 4905.

Schuster, P., W. Fontana, P. Stadler, and I. Hofacker. 1994. From sequences to shapes and back—a case-study in RNA secondary structures. *Proceedings of the Royal Society B* 255: 279–284.

Shakhnovich, B. E., E. Deeds, C. Delisi, and E. Shakhnovich. 2005. Protein structure and evolutionary history determine sequence space topology. *Genome Research* 15: 385–392.

Starr, T. N., L. K. Picton, and J. W. Thornton. 2017. Alternative evolutionary histories in the sequence space of an ancient protein. *Nature* 549: 409–413.

Starr, T. N., J. M. Flynn, P. Mishra, D. N. A. Bolon, and J. W. Thornton. 2018. Pervasive contingency and entrenchment in a billion years of Hsp90 evolution. *PNAS* 115: 4453–4458.

Stewart, A. J., and J. B. Plotkin. 2012. Why transcription factor binding sites are ten nucleotides long. *Genetics* 192: 973–985.

Theissen, G., J. T. Kim, and H. Saedler. 1996. Classification and phylogeny of the MADS-box multigene family suggest defined roles of MADS-box gene subfamilies in the morphological evolution of eukaryotes. *Journal of Molecular Evolution* 43: 484–516.

Thornton, J., C. Orengo, A. Todd, and F. Pearl. 1999. Protein folds, functions and evolution. *Journal of Molecular Biology* 293: 333–342.

Tokuriki, N., and D. S. Tawfik. 2009. Chaperonin overexpression promotes genetic variation and enzyme evolution. *Nature* 459: 668–673.

Toll-Riera, M., D. Bostick, M. M. Albà, and J. B. Plotkin. 2012. Structure and age jointly influence rates of protein evolution. *PLOS Computational Biology* 8: e1002542.

True, H. L., and S. L. Lindquist. 2000. A yeast prion provides a mechanism for genetic variation and phenotypic diversity. *Nature* 407: 477–483.

True, H. L., I. Berlin, and S. L. Lindquist. 2004. Epigenetic regulation of translation reveals hidden genetic variation to produce complex traits. *Nature* 431: 184–187.

Vaishnav, E. D., C. G. de Boer, M. Yassour, et al. 2021. A comprehensive fitness landscape model reveals the evolutionary history and future evolvability of eukaryotic cis-regulatory DNA sequences. *bioRxiv*.

van Nimwegen, E., J. Crutchfield, and M. Huynen. 1999. Neutral evolution of mutational robustness. *PNAS* 96: 9716–9720.

Venter, J. C., M. D. Adams, E. W. Myers, et al. 2001. The sequence of the human genome. *Science* 291: 1304–1351.

Wagner, A. 2005. *Robustness and Evolvability in Living Systems.* Princeton, NJ: Princeton University Press.

Wagner, A. 2007. Energy costs constrain the evolution of gene expression. *Journal of Experimental Zoology B* 308: 322–324.

Wagner, A. 2008. Robustness and evolvability: A paradox resolved. *Proceedings of the Royal Society B* 275: 91–100.

Wagner, A. 2011. The molecular origins of evolutionary innovations. *Trends in Genetics* 27: 397–410.

Wagner, A. 2012. The role of robustness in phenotypic adaptation and innovation. *Proceedings of the Royal Society B* 279: 1249–1258.

Wagner, A. 2014. Mutational robustness accelerates the origin of novel RNA phenotypes through phenotypic plasticity. *Biophysical Journal* 106: 955–965.

Wagner, G. P., G. Booth, and H. Bagherichaichian. 1997. A population genetic theory of canalization. *Evolution* 51: 329–347.

Weinreich, D. M., and L. Chao. 2005. Rapid evolutionary escape by large populations from local fitness peaks is likely in nature. *Evolution* 59: 1175–1182.

Weirauch, M. T., A. Yang, M. Albu, et al. 2014. Determination and inference of eukaryotic transcription factor sequence specificity. *Cell* 158: 1431–1443.

Zakrevsky, P., E. Calkins, Y.-L. Kao, G. Singh, V. L. Keleshian, S. Baudrey, and L. Jaeger. 2021. In vitro selected GUAA tetraloop-binding receptors with structural plasticity and evolvability towards natural RNA structural modules. *Nucleic Acids Research* 49: 2289–2305.

Zheng, J., J. L. Payne, and A. Wagner. 2019. Cryptic genetic variation accelerates evolution by opening access to diverse adaptive peaks. *Science* 365: 347–353.

Zheng, J., N. Guo, and A. Wagner. 2020. Selection enhances protein evolvability by increasing mutational robustness and foldability. *Science* 370: eabb5962.

12 Evolvability, Sexual Selection, and Mating Strategies

Jacqueline L. Sztepanacz, Josselin Clo, and Øystein H. Opedal

This chapter considers how variation in mating systems affects evolvability in populations and how we should estimate it. Most models considered in evolutionary quantitative genetics assume random mating and identical evolvability across sexes. In this chapter, we discuss some ways in which variation in mating systems leads to a violation of these assumptions, and what this means for evolvability. We focus on two major axes of mating system variation: variation in outcrossing rate and variation in reproductive success. We present population and quantitative genetic theory specific to mating systems and review the empirical evidence to support the hypotheses put forth.

12.1 Introduction

Biologists have long been fascinated by the remarkably diverse sexual and mating systems of both plants and animals, especially since Darwin (1871, 1876). In plants, mating systems are highly variable and range from functional asexuality to obligate outcrossing enforced by genetic self-incompatibility systems (Stebbins 1974). A prominent axis of variation in plant mating systems is the outcrossing rate, or the proportion of offspring resulting from mating between genetically distinct individuals (Goodwillie et al. 2005; Moeller et al. 2017). In animals, most species have two distinct sexes and reproduce through obligate outcrossing. Mixed mating systems do occur in some hermaphroditic species, such as certain snails (Jarne and Charlesworth 1993) and nematodes (Picard et al. 2021), but are rarer than in plants. The lack of diversity in outcrossing rate for animal systems compared to plants is well compensated by the diversity of mating strategies and behaviors. The prominent axis of variation in animal mating systems is variation in the reproductive behaviors and success among males and females (Bateman 1948).

Most models considered in evolutionary quantitative genetics (see Hansen, chapter 5)[1] assume random mating and identical evolvability across sexes. In this chapter, we discuss some ways in which variation in mating systems leads to a violation of these assumptions and what this means for evolvability. Through our discussion, we aim to answer the following questions:

1. Is there a universal measuring stick for evolvability that applies meaningfully across mating systems?

1. References to chapter numbers in the text are to chapters in this volume.

2. Does selfing reduce evolvability?
3. Are males and females equally evolvable?
4. Is the heterogametic sex more evolvable than the homogametic sex?
5. How do cross-sex covariances redistribute genetic variation and evolvability?
6. Does sexual selection reduce the effective size of populations and thus impact short-term evolvability?

12.2 Evolvability Defined for Mating Systems

Any measure of evolvability derives its relevance from the theoretical context in which it will function (Houle and Pélabon, chapter 6). Our focus in this chapter is on the evolvability *of* quantitative traits in populations, *under* directional selection, *over* timescales of tens of generations. While we might expect this constant set of conditions to engender a single best measurement stick for evolvability, as we will see, the same conditions in different mating systems are likely to necessitate different measures.

Under these conditions, the heritable component of phenotypic variation in a population, the additive genetic variance, determines the response to selection, or evolvability, of that population (see Hansen, chapter 5). This is shown by the Lande equation, $\Delta\bar{z} = \sigma_A^2 \beta$, where the predicted response to selection in trait mean ($\Delta\bar{z}$) is equal to the additive genetic variance in the trait (σ_A^2) multiplied by the strength of selection on that trait (β) (Lande 1979). The *under* conditions represented by β implicitly read "under directional selection acting on a randomly mating population," which may have to be modified under different mating systems. Given the diversity of mating systems across plants and animals, we must also consider whether σ_A^2 will be representative of the true capacity for traits to evolve in a given mating system. In particular, σ_A^2 may not always reflect the genetic variance that would allow a response to selection, and sex-averaged or population-level estimates of σ_A^2 may not capture genetic variation that is entangled between the sexes as a consequence of their largely shared genome. For these reasons, the interpretation of σ_A^2 as a measure of evolvability may differ, depending on the mating system of the population under study.

For additive variance to be a useful measure of evolvability, it also must be scaled in a meaningful way that we can benchmark against a prediction and compare across traits or across populations. The most common scaling of additive variance is heritability (h^2), σ_A^2 as a proportion of total phenotypic variance (σ_P^2). As highlighted by Houle (1992), h^2 is a unitless quantity, but it cannot be meaningfully interpreted or compared without knowing the standard deviation of the trait in the focal population(s). Indeed, due to the strong positive correlations between additive-genetic, nonadditive, and environmental variance components, heritabilities are poorly correlated with the additive genetic variance (Hansen et al. 2011) and thus may be a questionable measuring stick for evolvability in many circumstances. The widespread use of h^2 as a currency of evolutionary potential may be particularly problematic when comparing inbred and outcrossing species (Charlesworth and Charlesworth 1995). Heritabilities confound differences in genetic variance with differences in environmental variance, and for a fixed environmental variance, h^2 scales nonlinearly with increasing σ_A^2. Genetic variances are expected to decrease under inbreeding, while environmental variances may increase due to reduced developmental stability of inbred individuals (Fowler and Whitlock 1999; Kelly and Arathi 2003; Noel et al. 2017). Comparing h^2 between the sexes may also be problematic, because males and females can have different environmental variances for the same trait (Wyman and Rowe 2014).

Houle (1992) proposed mean-scaled additive variance as a comparative measure of evolvability, which can be expressed as (Hansen et al. 2003):

$$e = \frac{\sigma_A^2}{\bar{z}^2}, \tag{12.1}$$

where σ_A^2 is the additive variance, and $\bar{z}$ is the trait mean. This standardization has since been adopted as a predominant metric of evolvability used in quantitative genetics when the trait mean of interest has a meaningful and nonzero value. Evolvability, when quantified in this way, can be interpreted as the proportional change in the trait mean per generation when selection on the trait is as strong as selection on fitness (Hansen et al. 2003, 2011). Comparisons of heritabilities and evolvability for over a thousand published estimates showed no correlation between the two measures (Houle 1992; Hansen et al. 2011), highlighting the importance of choosing a meaningful measure to quantify evolvability. Although heritabilities and evolvabilities may be more closely correlated within specific groups of homogeneous traits, such as livestock (Hoffmann et al. 2016), reanalyses of these data found that correlations are generally less than 0.5 and explain less than 8% of the variation in evolvability (Hansen and Pélabon 2021). Instead of a measuring stick for evolutionary potential, heritability may be better viewed as an indicator of the reliability of estimated breeding values in a population.

12.3 The Effect of Outcrossing Rate on Evolvability

In his influential series of papers on systems of mating, Sewall Wright (1921a–d) laid the foundation for much of the population-genetic theory related to mating systems (see Hill 1996 for a historical perspective). In the classic model of Wright (1921b, 1952), assuming no mutation and no selection on a focal quantitative trait, the genetic variance among groups of inbred individuals increases by a factor of $1+f$, where f is the proportional decrease in heterozygosity due to inbreeding. Hence, a population in which all individuals are completely inbred (with $f=1$) will have twice the genetic variation compared to its ancestral randomly mating population (with $f=0$), assuming that the allele frequencies of ancestral and inbred populations are the same. In natural populations subject to selection and mutation, the situation is more complex. Lande (1977) found that, when explicitly modeling the generation of variation by mutation and loss of variation through stabilizing selection for a quantitative trait, mating systems had no effect on the level of standing genetic variation. This is notably because the increase in variance due to excess homozygosity is canceled out by the more efficient purging of recessive deleterious mutations in inbred populations. In addition, he also showed that the decrease in genetic variance due to linkage disequilibrium in selfing populations is compensated for by the less efficient selection against mutations in linkage disequilibrium. In contrast, with the same assumption of a quantitative trait under stabilizing selection, Charlesworth and Charlesworth

(1995) found that genetic variation maintained by mutation-selection balance decreased under complete selfing and should be equal to one quarter of the variance maintained in an obligately outcrossing population. These contrasting results were reconciled by Lande and Porcher (2015) and Abu Awad and Roze (2018), who modeled the maintenance of genetic variation for any selfing rate and found that mating system has no or only a weak effect on genetic variance for moderate selfing rates, but that genetic variance declines abruptly for high selfing rates. The mechanism of reduced genetic variance for high selfing rates depends on the mutation rate of the trait under selection (Abu Awad and Roze 2018). If the mutation rate is low, the mutations appear so slowly that purifying selection acts efficiently to remove deleterious mutations, allowing populations to stay close to the phenotypic optimum. Because selfing populations are more efficient at purging the deleterious mutations (because they are recessive at the fitness scale; Manna et al. 2011), these populations maintain less genetic diversity than do outcrossing ones (Abu Awad and Roze 2018; Clo et al. 2020). If the mutation rate is high, however, mutations appear too quickly for purifying selection to purge them efficiently. In such cases, genetic associations emerge between deleterious mutations of opposite signs, thus reducing the deleterious effects of mutations on fitness and allowing populations to remain close to the phenotypic optimum. Due to their reduced effective recombination rates, selfing populations are better at maintaining these genetic associations, which decrease the phenotypic and genetic variance due to the negative contributions of these associations to variation (Lande and Porcher 2015; Abu Awad and Roze 2018; Clo et al. 2020).

12.3.1 Additive Variance Stored in Linkage Disequilibrium

Selfing populations are theoretically expected to be less evolvable than are outcrossing or mixed-mating populations. However, the mechanism of reduced genetic variance should play a key role in determining the short and mid-term evolvability of selfing populations. Indeed, if reduced evolvability of selfing populations is due to more efficient purging of deleterious mutations, and if adaptation depends on this previously deleterious diversity, the capacity of populations to respond to directional selection will necessarily be smaller in selfing populations (Clo et al. 2020). However, if the reduction in genetic variance is due to genetic associations and the associated linkage disequilibrium, the diversity stored within genetic associations can theoretically increase the response to selection (Lande and Porcher 2015; Abu Awad and Roze 2018). Recently, Clo et al. (2020) and Clo and Opedal (2021) showed that recombination events between phenotypically similar but genetically distinct inbred lines, within a population, led to an increase of the additive variance in the early generations of response to directional selection, because the stored diversity becomes expressed in recombining individuals. This increase in genetic diversity allows selfing populations to respond as quickly, and sometimes even quicker than their outcrossing counterparts (Clo et al. 2020; Clo and Opedal 2021). Quantifying the amount of diversity potentially stored in genetic associations in natural populations is complicated. Based on neutral diversity, it is known that selfing populations are organized into sets of repeated multilocus genotypes (Siol et al. 2008; Jullien et al. 2019), reflecting different combinations of alleles for a given quantitative trait (Clo and Opedal 2021). A straightforward way to search for cryptic diversity is to make crosses between these different inbred lines and to look for transgressive segregation in the F_2 generation (the appearance of phenotypes outside the range of phenotypes that are present in the parental generation). Transgressive segregation has been widely observed in selfing populations (Rieseberg et al. 1999, 2003), suggesting that increased evolvability of highly selfing species during directional selection could be a common phenomenon.

12.3.2 Limits of Quantitative-Genetics Models: Integrating Directional Dominance into Mutation-Selection Models

One limitation of the models reviewed above is that they assume that alleles act additively on the phenotypic scale. This assumption is potentially problematic, because it is well known that dominance contributes to the genetic architecture of quantitative traits, and hence to the genetic variance (Wolak and Keller 2014). When inbreeding and directional dominance (the fact that heterozygotes are, on average, not intermediate to homozygote values) occurs simultaneously, at least three additional terms contribute to the genetic variance (table 12.1; Cockerham and Weir 1984). The genetic variance of a population with an average inbreeding coefficient of f is then equal to

$$\sigma^2_{G,f} = (1+f)\sigma^2_A + (1-f)\sigma^2_{DR} + f\sigma^2_{DI} + 4f\sigma_{ADI} + f(1-f)H^* \quad (12.2)$$

where σ^2_A (the variance of average effects of alleles) and σ^2_{DR} (the variance of dominance deviations of a randomly mating population) are the terms typically inferred in quantitative genetics. With inbreeding, an individual may have common ancestors through both parental lines, such that nonadditive effects can contribute to the covariance between parents and offspring, which does not occur under random mating. The additional terms arising under partial inbreeding include the variance of dominance deviations of a fully inbred population σ^2_{DI}, and the covariance between the additive effect of an allele and its homozygous dominance deviation σ_{ADI} (table 12.1). This covariance arises because increased homozygosity in inbred populations increases the probability that an allele occurs in its homozygote state, which creates covariance between the contribution of an allele to the average effect and the dominance deviation of the homozygote genotype (Wolak and Keller 2014). Unlike

Table 12.1
The average effect of alleles (α) and dominance deviations (δ) associated with a genotype and the different components of genetic variance that can be inferred

Term	Symbol	Value
Average effect of allele A	α_A	$p_A G_{AA} + p_B G_{AB} - \mu = -0.42$
Average effect of allele B	α_B	$p_A G_{AB} + p_B G_{BB} - \mu = 0.18$
Dominance deviation of AA	δ_{AA}	$G_{AA} - 2\alpha_A - \mu = -0.98$
Dominance deviation of AB	δ_{AB}	$G_{AB} - \alpha_A - \alpha_B - \mu = 0.42$
Dominance deviation of BB	δ_{BB}	$\delta_{BB} = G_{BB} - 2\alpha_B - \mu = -0.18$
Additive variance	σ^2_A	$p_A 2\alpha^2_A + p_B 2\alpha^2_B = 0.15$
Dominance variance	σ^2_{DR}	$p^2_A \delta^2_{AA} + p^2_B \delta^2_{BB} + 2 p_A p_B \delta^2_{AB} = 0.18$
Inbred dominance variance	σ^2_{DI}	$p_A \delta^2_{AA} + p_B \delta^2_{BB} + (p_A \delta_{AA} + p_B \delta_{BB})^2 = 0.15$
Additive-dominance covariance	σ_{ADI}	$p_A \alpha_A \delta_{AA} + p_B \alpha_B \delta_{BB} = 0.13$
Inbreeding depression	H^*	$(p_A \delta_{AA} + p_B \delta_{BB})^2 = 0.18$

Notes: Based on Falconer and Mackay (1996). Consider a locus with two alleles A and B found at frequency $p_A = 0.3$ and $p_B = 0.7$, respectively. The genotypic values are $G_{AA} = 2$, $G_{AB} = 4$ and $G_{BB} = 4$ (B is completely dominant over A). The weighted mean genotypic value of the population is equal to $\mu = 3.82$.

the other terms in equation (12.1), this term is a covariance and can thus be negative. The third additional component, H^*, is the squared per-locus inbreeding depression, summed over all loci. While σ^2_{DR} does not affect the response to selection, the additional dominance components (notably σ^2_{DI} and σ_{ADI}) contribute to the adaptive potential, because they contribute to the covariance between parents and offspring and thus response to selection (A. Wright and Cockerham 1985; Kelly 1999a,b).

By exploring the effect of directional dominance on a fitness-related trait under stabilizing selection for populations that differ in selfing rates, Clo and Opedal (2021) showed that dominance components can explain a substantial part of the genetic variance of inbred populations. They also showed that ignoring these components leads to an upward bias in the predicted response to selection (notably because the covariance term σ_{ADI} is often substantially negative), and that when considering the effect of directional dominance, the evolutionary potential of populations remains comparable across the entire gradient in outcrossing rates.

The additional variance components discussed above complicates the measurement of evolvability for inbreeding species, and empirical estimates requires complex experimental designs (Shaw et al. 1998; Kelly and Arathi 2003). A reasonable question to ask is: Are these additional dominance terms contributing substantially to the genetic variance of quantitative traits? Given the abovementioned difficulties in estimating all the dominance (co)variance terms, it is not surprising to find very few estimates in the literature. So far, only a handful of studies have estimated all the components of variance for morphological, fitness, and agronomic traits in the plants *Zea mays* (Edwards and Lamkey 2002; Wardyn et al. 2007), *Mimulus guttatus* (Kelly 2003; Kelly and Arathi 2003; Marriage and Kelly 2009), *Nemophila menziesii* (Shaw et al. 1998), *Eucalyptus globulus* (Costa E Silva et al. 2010), and in the animals *Bos taurus* (Hoeschele and Vollema 1993), *Ovies aries* (Shaw and Woolliams 1999) and *Homo sapiens* (Abney et al. 2000). From these studies, it is possible to extract all the components of the genetic variance (ignoring epistasis) for 40 quantitative traits. For each trait, we computed the ratio of the dominance variances (σ^2_{DR}, σ^2_{DI}, σ_{AD}, and H^*) over the sum of the additive variance and the focal dominance variance component. As σ_{ADI} can be negative or positive, we used the absolute value of σ_{ADI}. All the ratios are detectably different from zero (table 12.2), suggesting that the dominance terms contribute to the genetic variance but generally to a smaller extent than the additive variance, except for σ^2_{DI} that on average tends to contribute as much to σ^2_G as the additive component. Other forms of nonadditive genetic effects, like epistasis, can also contribute to the evolutionary potential of selfing species, but in the same manner as in random mating populations (see Hansen, chapter 5).

Table 12.2
Mean, standard deviation (s.d.), and minimum and maximum values of the ratios of the different nonadditive over additive variance components

Measure	n	Mean (s.d.)	Min	Max
$\sigma^2_{DR}/(\sigma^2_A+\sigma^2_{DR})$	40	0.257 (0.226)	0	0.915
$\sigma^2_{DI}/(\sigma^2_A+\sigma^2_{DI})$	40	0.441 (0.252)	0	0.910
$\lvert\sigma_{ADI}\rvert/(\sigma_A+\lvert\sigma_{ADI}\rvert)$	40	0.236 (0.186)	0	0.737
$H^*/(\sigma^2_A+H^*)$	40	0.347 (0.359)	0	0.915

Note: Data and code for the results presented in the table are available at https://github.com/JosselinCLO/Book_Chapter_Evolvability.

12.3.3 Empirical Relationship between Outcrossing Rate, Heritability, and Evolvability

Charlesworth and Charlesworth (1995) compiled estimates of outcrossing rates, heritabilities, and coefficients of genetic variation from published studies of plants. This first quantitative survey suggested a weak overall relationship between outcrossing rate and heritability. Later, Geber and Griffen (2003) reported lower heritabilities for highly selfing species compared to outcrossing and mixed-mating ones, and Ashman and Majetic (2006) reported lower heritabilities of self-compatible compared to self-incompatible species. Recently, Clo et al. (2019) compiled the largest dataset of mating systems and heritabilities to date; they found that although overall patterns were weak, heritabilities tended to decline with increasing selfing rates. In all these studies, when an effect of the mating system was detected, it always explained limited fractions of the variation in heritabilities (<20% in Geber and Griffen 2003; Clo et al. 2019). In addition, Clo et al. (2019) found that narrow-sense heritabilities of outcrossing and mixed-mating species (selfing rate <0.8) did not differ detectably from broad-sense heritabilities of predominantly selfing species (selfing rate ≥0.8), which is expected to be a more accurate predictor of evolutionary potential for this mating category.

Charlesworth and Charlesworth (1995) also analyzed the relationship between outcrossing rates and coefficients of genetic variation. When they restricted the analysis to additive-genetic coefficients of variation, they did detect a weak positive relationship between outcrossing rate and evolvability ($R^2 \approx 9\%$). Opedal et al. (2017) reported a similar pattern for floral traits functionally related to mating systems (figure 12.1).

Although the data shown in figure 12.1 are heterogeneous, which complicates formal analysis, they suggest that the overall relationship between outcrossing rate and evolvability in plants is weak. Most comparative analyses to date have combined population-specific and species-mean outcrossing rates. Because mating systems often differ among populations of mixed-mating species (Whitehead et al. 2018), these naïve analyses could be biased due to the inclusion of species-mean outcrossing rates. Because species-mean outcrossing rates may not be representative for the evolvability measures of the populations, it will lead to a downward bias in the relationship. A more powerful approach to test associations between evolvability and mating system is to compare conspecific populations differing in outcrossing rate. Of the few studies that have attempted this exercise, Bartkowska and Johnston (2009) reported reduced nuclear genetic variance in a highly selfing population compared to mixed-mating populations of *Amsinckia spectabilis,* while the data reported by Herlihy and Eckert (2007) for *Aquilegia canadensis* and by Charlesworth and Mayer (1995) for *Collinsia heterophylla* yield weak negative correlations between population-specific outcrossing rate and evolvability. Although single estimates of outcrossing rates are not necessarily reliable estimates of the mating histories of populations (Opedal 2018), these results underline the apparently weak overall relationship between outcrossing rate and evolvability.

Associations between evolvability and mating system can also be tested by phylogenetic comparative approaches, such as comparing species pairs differing in mating systems. Across four *Oenothera* species pairs, each comprising one sexual and one functionally asexual species, Godfrey and Johnson (2014) found no consistent effect of sexuality on evolvability, although suggestive patterns were detected for some traits.

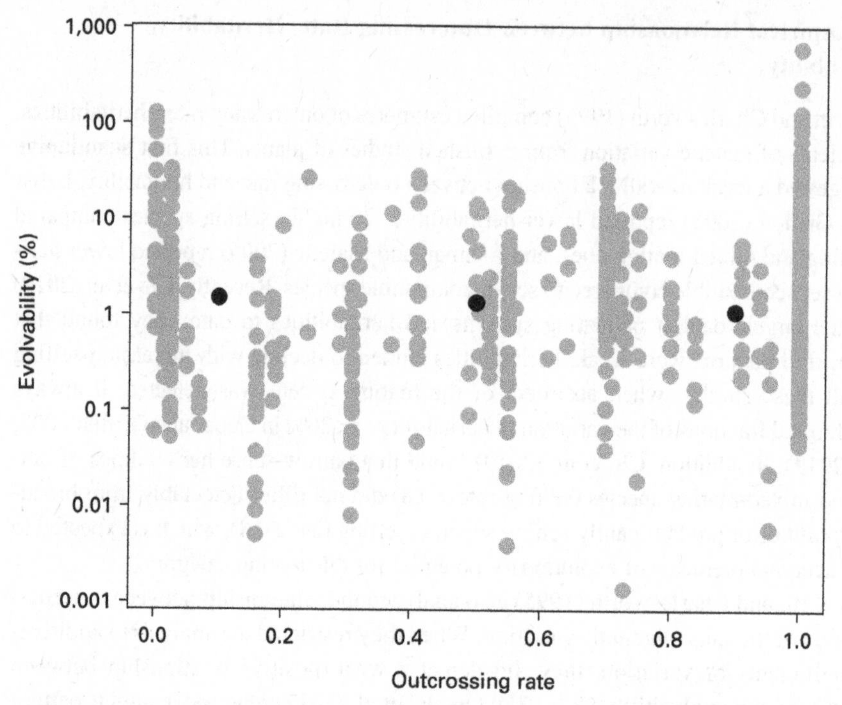

Figure 12.1
An extended version of the data compiled by Opedal et al. (2017), illustrating the apparent lack of a detectable relationship between evolvability and outcrossing rate across 763 single-trait evolvability estimates from 46 species (median $n = 12$ evolvability estimates per species, range = 1–65). Outcrossing rates plotted are population specific when available, and otherwise species means or, for self-incompatible species, assumed to be 1. The solid black symbols indicate the median for selfing (outcrossing rate < 0.2), mixed-mating (outcrossing rate 0.2–0.8), and outcrossing (outcrossing rate > 0.8 or self-incompatible) species. $R^2 < 1\%$ on either logarithmic or arithmetic scale.

12.3.4 Response to Selection as a Function of Outcrossing Rate

The empirical results reviewed above are based on the effect of the selfing rate on the additive variance, for which the effect of mating system is expected to be weak (Clo and Opedal 2021), and which is not an accurate predictor of adaptive potential of inbred populations, because the additional dominance terms contribute to the response to selection (Kelly and Williamson 2000; Clo and Opedal 2021). No trivial measures of the evolvability of populations as a function of their selfing rate are available in the literature (Kelly 1999a,b). One way to estimate the "realized" heritable variance or evolvability of a trait, as a function of the mating system, is to perform directional selection experiments with a known selection differential. Few studies have applied directional selection on quantitative traits for populations differing in their mating systems. However, in a study of the yellow monkeyflower *Mimulus guttatus,* Holeski and Kelly (2006) showed that the mean and variance in response to artificial selection differed among populations maintained under selfing, mixed mating, and outcrossing. Noel et al. (2017) used experimental evolution to generate inbred and outbred lines of the hermaphroditic snail *Physa acuta,* which were subsequently subjected to artificial selection under selfing and outcrossing. Their study provided clear evidence that inbred lines responded more slowly to selection than did outbred lines. Furthermore, when the selected lines were maintained under selfing, response to selection was initially fast but quickly decayed. Although these experiments yielded important insights into the effect of shifting from outcrossing to selfing on adaptive potential, they did not fully reconcile theory and data concerning the effect of mating system on evolvability. First, in these studies, species that are mixed mating in natural populations were subject to experimental manipulation of their mating system. Shifting from outcrossing to selfing increases the adaptive potential of populations by a factor $1+f$ (S. Wright 1921b), or higher if dominance contributes to the genetic architecture of the studied trait (Clo and Opedal 2021), explaining why selfing populations generally responded faster in these studies, at least during the first generations of selection. Second, outcrossing and mixed mating prevents the storage of diversity in genetic associations (Clo et al. 2020), and these investigators imposed complete selfing in their study populations, while residual outcrossing is always found in natural populations (Kamran-Disfani and Agrawal 2014). Maintaining a low level of outcrossing in selection experiments may allow a continued selection response by retaining cryptic diversity that is exposed under outcrossing. Directional selection experiments may remain the easiest way to estimate the "realized" evolvability of inbreeding populations, particularly for mixed-mating populations. Studies exploiting natural variation in mating systems would be particularly valuable. If predominantly selfing populations are used, due to their organization into different repeated multilocus genotypes, the immediate response to selection is likely to arise from selection among those natural purely inbred lines. In such a case, the immediate evolvability can be summarized by the among-lines genetic variance. In addition to estimating the realized evolvability for the first generations of response to selection, it seems important to study the potential patterns of transgressive segregations in recombinant lines of fully inbred genotypes. This will determine how selfing populations will evolve in the mid-term. It is nevertheless important to note that the entire variance of transgressive segregation will not predict the response to selection, as the formation of new genotypes based on the diversity stored through linkage is a stochastic process, leading to the formation of more or less well adapted new genotypes (Clo et al. 2020; Clo and Opedal 2021).

12.4 Separate Sexes and Evolvability

Some plants and the vast majority of animals have separate sexes in different individuals and reproduce through obligate outcrossing. Most estimates of additive genetic variation in these populations are obtained for a single sex or are sex-averaged to provide a population-level measure of evolvability. This is potentially problematic, because males and females can differ in the magnitude of their additive genetic variances. Sex-specific genetic variances can arise from differences in the strength and form of selection between the sexes, shaping patterns of genetic variation (Rowe and Houle 1996), sex differences in allele frequencies (Lynch and Walsh 1998; Reinhold and Engqvist 2013), sex-specific mutational effects (Mallet et al. 2012; Sharp and Agrawal 2012), sex-specific dominance (Fry 2010; Grieshop and Arnqvist 2018), sex differences in gene expression (Massouras et al. 2012), and the effect of sex chromosomes on the additive genetic variance (Husby et al. 2013). Therefore, single-sex or sex-averaged estimates of σ_A^2 may yield misleading estimates of evolvability. In the following sections, we discuss how having separate sexes can affect evolvability through the distribution of genetic variation between males and females, and through interactions between the sexes that affect how genetic variation evolves and is maintained in populations.

12.4.1 Sexual Conflict

Males and females share most of their genome and must interact in order to reproduce. The evolutionary interests of the sexes, however, are often in conflict (Trivers 1972). Sexual conflict occurs when adaptations arise that are beneficial for one sex but maladaptive for the other (Bonduriansky and Chenoweth 2009), and they may play an important role in the maintenance of genetic variation (Connallon and Clark 2012), directly influencing the evolvability of traits.

There are two mechanistically distinct forms of sexual conflict, both of which are caused by sex differences in selection and may affect evolvability. Interlocus sexual conflict refers to conflict over the outcome of interactions between males and females, such as over optimal mating rates and behaviors (Arnqvist and Rowe 2005). Sexually antagonistic coevolution caused by these interactions often results in sexual "arms races," which can lead to the rapid evolution of traits involved in the interaction (Arnqvist and Rowe 2005). One clear example is the coevolution between male grasping structures in water striders, which are used to immobilize females during mating, and female spines, which are used to hold the male away (Arnqvist and Rowe 2002a,b). In contrast, intralocus sexual conflict results from sexually antagonistic selection on the same traits expressed in both sexes, whose expression is determined by shared genetics. This form of sexual conflict arises when the direction of selection on a given allele depends on which sex that allele resides in (Arnqvist and Rowe 2013). Many traits are expected to experience intralocus sexual conflict, which may eventually be resolved by the evolution of sexual dimorphism (Bonduriansky and Chenoweth 2009). However, negative genetic covariances for fitness between males and females are often observed (e.g., Foerster et al. 2007; Wolak et al. 2018), suggesting that intralocus sexual conflict remains unresolved in many species.

12.4.2 Sex-Linked Variation

In many species, biological sex is determined by sex chromosomes. There are two primary types of sex-chromosome systems, male (XX/XY) and female (ZW/ZZ) heterogamety. Loci on these chromosomes often encode for both sex-limited and shared traits, some of which are dimorphic in trait expression (Dean and Mank 2014). However, quantitative-genetics methods for estimating σ_A^2 often ignore the potential for sex-linked loci to generate differences in genetic variance between the sexes. Estimates of genetic variation obtained from standard analyses of half-sib breeding designs and implementations of the "animal model" (Kruuk 2004) assume that coefficients of relatedness are the same for relatives of the same degree, irrespective of sex. Consequently, sex-linked effects influence the estimates of additive genetic variation in these models. While it is possible to simultaneously estimate autosomal and sex-linked genetic effects with the animal model (Larsen et al. 2014), few studies have attempted to do so (but see Roulin and Jensen 2015). Consequently, there are few empirical estimates of genetic variation that would allow us to answer the question of whether evolvability tends to be higher for sex-linked or autosomal loci, and for males or females.

Ignoring epistasis, the genetic variance at a sex-linked locus in the heterogametic sex is determined by the additive effects of alleles at that locus. In the homogametic sex, however, genetic variation will also depend on intralocus allelic interactions, such as dominance effects. If we assume that sex-linked alleles contribute additively to the phenotype (there is no dominance), and that they have equal hemizygous and homozygous effects on trait expression, then a locus present on the sex chromosome will contribute twice as much variance in the heterogametic compared to the homogametic sex. This has led to the hypothesis that the heterogametic sex will be more genetically more variable than the homogametic sex. Realistically, however, the average effects of alleles are unlikely to be equal in each sex because of dosage compensation, which adjusts the activity of sex-linked loci to equalize gene expression between the sexes (Muller 1932). Variation can either increase or decrease depending on the mechanism of dosage compensation (Cowley et al. 1986), the chromosomal sex-determination system, and their interaction.

There are few empirical tests of the hypothesis that the heterogametic sex has more genetic variance than the homogametic sex. Reinhold and Engqvist (2013) compared coefficients of variation for body size in four animal taxa, two where males are the heterogametic sex and two where females are heterogametic. Consistent with their prediction, the heterogametic sex had more variation in body size in all taxa. However, the authors were only able to look at patterns of phenotypic, and not genetic variation, and in a relatively small dataset. Other studies that have attempted to test this hypothesis have similarly focused on phenotypic variation (e.g., Zajitschek et al. 2020). So far, the best data come from Wyman and Rowe (2014), who compared differences in coefficients of additive genetic variation across heterogametic types. They found no evidence that the heterogametic sex tended to have more genetic variation. Future studies testing this hypothesis may want to carefully consider the mechanism of dosage compensation as well as sex chromosome system. For example, in *Drosophila,* it is the number of X chromosomes that determine whether a fly is female (XX) or male (XY, XO), and the mechanism of dosage compensation is to double the transcription rate of the X chromosome in males. This may enhance the expression of deleterious X-linked mutations in males, leading to stronger purifying selection on males. If this occurs, it could reduce genetic variation and consequently evolvability.

12.4.3 Sex-Specific Autosomal Variation

Males and females may also differ in genetic variation for traits determined by autosomal loci. The total additive genetic variance in sexual dimorphism for traits controlled by autosomal loci is:

$$\sigma_A^2(\mathrm{M}-\mathrm{F}) = \sigma_A^2(\mathrm{M}) + \sigma_A^2(\mathrm{F}) - 2\sigma_A(\mathrm{M},\mathrm{F}), \tag{12.3}$$

where σ_A^2 is the additive genetic variance, and M and F represent males and females, respectively.

Few studies have considered the relationship between sexual dimorphism in σ_A^2 and evolvability (but see Rolff et al. 2005), with most studies focusing on heritabilities (Mousseau and Roff 1989; Gershman et al. 2010; Stillwell and Davidowitz 2010), which for reasons discussed earlier may not be a good measure of evolvability in many contexts. Wyman and Rowe (2014) collated data from 279 male-female pairs of traits from 75 species and asked whether there is a systematic difference in σ_A^2, expressed as additive genetic coefficients of variation (CV_A), between males and females. The authors failed to detect any difference in the average CV_A between males and females across all trait types and species. There was, however, substantial variation in dimorphism of CV_A depending on the trait and species studied, and a difference in the skew of distributions when CV_A was higher for males versus females. When dimorphism in CV_A was male biased (male CV_A > female CV_A), the magnitude of dimorphism in trait-specific means tended to be larger than when CV_A dimorphism was female biased.

Similar results have been found for gene expression traits, which allow us to study phenome-wide differences in σ_A^2 between males and females. Allen et al. (2018) estimated sex-specific genetic variances in gene expression traits in *D. serrata* and found on average that there was no difference in the genetic variance between males and females. However, genes which had male-biased expression had disproportionately higher genetic variance than genes which had female-biased expression. Traits with male-biased expression were also more evolvable regardless of which sex was expressing the trait. In *D. melanogaster,* however, Houle and Cheng (2021) found that genetic variance of male-biased gene-expression traits was higher in males, and genetic variance of female-biased traits was higher in females. Overall, results of higher genetic variance, or faster evolutionary rates, in male-biased gene expression traits has been demonstrated in some studies (e.g., Meiklejohn et al. 2003; Zhang and Parsch 2005; Assis et al. 2012) suggesting that it may be a general pattern. Dutoit et al. (2017) tested whether nucleotide diversity, as a measure of molecular genetic variation, was sexually dimorphic in the collared flycatcher. They found that genes with sex-biased expression had more sequence variation than did unbiased genes, and had a positive relationship between genetic diversity across the genome and sex bias in gene expression. Male-biased genes expressed in the brain also had disproportionately more nucleotide diversity with increasing levels of sex-biased expression than did female-biased genes. In general, there is more evidence for rapid sequence evolution of male-biased genes than of female-biased genes (Meiklejohn et al. 2003; Ellegren and Parsch 2007; Mank et al. 2007), but this does not necessarily imply higher evolvability and could also result from differences in the strength of selection between these gene classes. Male-biased genes do tend to have more tissue-specific expression (Assis et al. 2012; Hansen and Kulathinal 2013), suggesting that they may be less constrained by pleiotropy (Rowe et al. 2018). If this is the case, males may have higher conditional evolvability than females.

The positive relationship between male sex-biased gene expression and evolvability may be explained by at least three nonexclusive mechanisms. Male-biased expression may relax selection operating through females, thus increasing evolvability. Males may be less constrained by pleiotropy than females are, or they may maintain more genetic variation in traits as a side effect of higher genetic variation in fitness. Low-fitness males have been shown to harbor more σ_A^2 for individual and multivariate sexually selected trait combinations than high-fitness males (McGuigan and Blows 2009; Delcourt et al. 2012; Sztepanacz and Rundle 2012). Whether this pattern also extends to females has not been tested. Disentangling the roles of pleiotropy and selection in determining evolvability will be key to understanding these differences.

12.4.4 Cross-Sex Genetic Covariances

Sexually homologous traits in males and females that are determined by the same genetics can be thought of as two traits that genetically covary. When the optimal trait values of males and females differ, cross-sex covariances will determine the extent to which each sex can reach their phenotypic optimum through the evolution of sexual dimorphism (Lande 1980). Consequently, cross-sex covariances hold a central role in the resolution of intralocus sexual conflict and thus in maintaining population mean fitness.

A homologous trait expressed in both males and females is conceptually equivalent to two genetically correlated traits, and we can therefore use the same tools to study the effects of cross-sex genetic covariances on evolvability that we use to study cross-trait genetic covariances. The evolvability metric e introduced above does not tell the complete story about the evolvability of a single trait when its genetic variance is correlated with other traits. To quantify the independent evolvability of correlated characters, Hansen (2003) developed the concept of conditional evolvability. Conditional evolvability (c) is the amount of genetic variation in a focal trait that is not bound up in correlations with other traits, or in other words, the response per unit directional selection in a focal trait, when a defined set of correlated traits are kept constant (as when they are under infinitely strong stabilizing selection). We may often expect directional selection on males and stabilizing selection on females over longer timescales, which makes the traditional notion of conditional evolvability relevant for assessing the potential for dimorphism evolution on longer timescales. Over shorter timescales, any shared traits may experience divergent selection due to different phenotypic optima, and we would like to know how much of the genetic variance would allow sexually antagonistic versus concordant responses to selection. Sztepanacz and Houle (2019) developed a method for partitioning genetic variation and cross-sex covariation into sexually concordant and sexually antagonistic genetic subspaces. Their partitioning characterizes genetic variation that would allow a response to selection in exactly the same direction between the sexes and in exactly opposite directions. Here, we propose that these genetic subspaces are analogous to a measure of conditional evolvability for dimorphism, and outline how Sztepanacz and Houle's method can be applied to quantify dimorphic evolvability (box 12.1).

12.4.5 Standardized Intersexual Genetic Correlations r_{mf}

Many studies have estimated standardized intersexual genetic correlations, r_{mf}, for single traits. The quantity r_{mf} is the ratio of additive genetic covariance between the sexes to the geometric mean of additive genetic variances of males and females. It can also be expressed as $r_{mf} = \sqrt{\frac{h^2_{\mathrm{FD}} \cdot h^2_{\mathrm{MS}}}{h^2_{\mathrm{MD}} \cdot h^2_{\mathrm{FS}}}}$, where h^2 represents heritability estimates based on father-daughter (FD), mother-son (MS), mother-daughter (MD), and father-son (FS) covariances (Lynch and Walsh 1998). The cross-sex genetic correlation r_{mf} is related to conditional evolvability, because $1 - r^2_{mf}$ is the autonomy of one sex conditional on the other. However, r_{mf} alone cannot tell us much about evolvability as we have defined it, because variances and covariances are what matter, not correlations. For example, high r_{mf} could reflect a high additive genetic covariance between the sexes, or just a low additive variance in one sex.

If we assume that the sexes have equal additive genetic variance for a focal trait, then the additive genetic variance (evolvability) for sexual dimorphism is $2\sigma^2_{\mathrm{A}}(1 - r_{mf})$, where σ^2_{A} is the additive genetic variance in the trait (Matthews et al. 2019). As discussed in the previous section, the assumption of equal genetic variance in males and females may be reasonable for many traits, but not for all of them, and in particular, not for sexually selected traits. Poissant et al. (2009) compiled 488 estimates of r_{mf} from more than 100

Box 12.1
Sexually Concordant and Antagonistic Evolvability

The quantitative-genetic framework to simultaneously estimate within- and cross-sex genetic covariances was put forth several decades ago (Lande 1980):

$$\mathbf{G}_{mf} = \begin{bmatrix} \mathbf{G}_m & \mathbf{B}' \\ \mathbf{B} & \mathbf{G}_f \end{bmatrix},$$

where $\mathbf{G}_m$ and $\mathbf{G}_f$ are the symmetric within-sex (co)variances among traits for males and females, respectively, and **B** (**B′**) are the covariances of homologous traits expressed in both males and females. The diagonal elements of **B** quantify the amount of genetic variation that is shared between the sexes for the same trait, whereas the off-diagonal elements quantify cross-sex cross-trait covariances. Unlike $\mathbf{G}_m$ and $\mathbf{G}_f$, **B** is not necessarily a symmetric matrix, because the genetic covariance between trait *a* in males and *b* in females may not the same as *b* in males and *a* in females.

To quantify the evolvability of males and females under the conditions of sexually concordant and sexually antagonistic selection, we can partition $\mathbf{G}_{mf}$ into sexually concordant and sexually antagonistic genetic subspaces (Sztepanacz and Houle 2019). This partitioning characterizes the genetic variation that would allow a response to selection acting in exactly the same direction between the sexes and selection acting in exactly opposite directions. Therefore, these subspaces of genetic variation reflect sexually concordant and sexually antagonistic evolvability. We illustrate this approach below with a 2-trait numerical example. For a more in-depth mathematical description, see Cheng and Houle (2020).

Our goal is to transform male and female genetic variances and covariances ($\mathbf{G}_{mf}$) into concordant and antagonistic genetic variances and covariances ($\mathbf{G}_{\mathrm{CA}}$):

$$\mathbf{G}_{mf} = \begin{bmatrix} \mathbf{G}_m & \mathbf{B}' \\ \mathbf{B} & \mathbf{G}_f \end{bmatrix} \rightarrow \mathbf{G}_{\mathrm{CA}} = \begin{bmatrix} \mathbf{G}_{\mathrm{C}} & \mathbf{B}'_{\mathrm{CA}} \\ \mathbf{B}_{\mathrm{CA}} & \mathbf{G}_{\mathrm{A}} \end{bmatrix}$$

The matrix $\mathbf{G}_{\mathrm{C}}$ is genetic variation in the concordant subspace, $\mathbf{G}_{\mathrm{A}}$ in the antagonistic subspace, and $\mathbf{B}_{\mathrm{CA}}$ is the genetic covariance that leads to indirect responses in the other space.

Let the 2×2 matrix $\mathbf{G}_{mf}$ be:

		Male		Female	
		trait 1	trait 2	trait 1	trait 2
Male	trait 1	49	5	40	10
	trait 2	5	53	3	49
Female	trait 1	40	3	40	12
	trait 2	10	49	12	70

The first step to defining concordant and antagonistic evolvability is to define a set of orthonormal vectors ($\mathbf{S}_m$) that spans the concordant and antagonistic subspaces of $\mathbf{G}_{mf}$. Here, we let this matrix be:

$$\mathbf{S}_m = \frac{1}{\sqrt{2}}\begin{bmatrix} 1 & 0 & 1 & 0 \\ 0 & 1 & 0 & 1 \\ 1 & 0 & -1 & 0 \\ 0 & 1 & 0 & -1 \end{bmatrix} = \frac{1}{\sqrt{2}}\begin{bmatrix} \mathbf{I} & \mathbf{I} \\ \mathbf{I} & -\mathbf{I} \end{bmatrix}$$

where **I** is a 2×2 identity matrix. However, any set of vectors that form the basis of a 2×2 matrix would work. The term $\frac{1}{\sqrt{2}}$ scales the concordant and antagonistic axes to unit length,

Box 12.1
(continued)

preserving the size of $\mathbf{G}_{mf}$. The next step is to project $\mathbf{G}_{mf}$ onto the concordant and antagonistic subspaces defined by $\mathbf{S}_m$:

$$\mathbf{G}_{CA} = \mathbf{S}_m \, \mathbf{G}_{mf} \, \mathbf{S}'_m$$

$\mathbf{S}_m$ is symmetric, so that $\mathbf{S}'_m = \mathbf{S}_m$ and

$$\mathbf{G}_{CA} = \frac{1}{\sqrt{4}} \mathbf{S}_m \, \mathbf{G}_{mf} \, \mathbf{S}_m.$$

Working through our example:

$$\mathbf{G}_{CA} = \frac{1}{\sqrt{4}} \begin{bmatrix} 1 & 0 & 1 & 0 \\ 0 & 1 & 0 & 1 \\ 1 & 0 & -1 & 0 \\ 0 & 1 & 0 & -1 \end{bmatrix} \begin{bmatrix} 49 & 5 & 40 & 10 \\ 5 & 53 & 3 & 49 \\ 40 & 3 & 40 & 12 \\ 10 & 49 & 12 & 70 \end{bmatrix} \begin{bmatrix} 1 & 0 & 1 & 0 \\ 0 & 1 & 0 & 1 \\ 1 & 0 & -1 & 0 \\ 0 & 1 & 0 & -1 \end{bmatrix}$$

$$\mathbf{G}_{CA} = \begin{bmatrix} 84.5 & 15 & 4.5 & -7 \\ 15 & 110.5 & 0 & -8.5 \\ 4.5 & 0 & 4.5 & -2 \\ -7 & -8.5 & -2 & 12.5 \end{bmatrix}.$$

The upper right submatrix $\mathbf{G}_C = \begin{bmatrix} 84.5 & 15 \\ 15 & 110.5 \end{bmatrix}$ is the genetic variation in the concordant subspace, which would allow a response to concordant selection. $\mathbf{G}_A = \begin{bmatrix} 4.5 & -2 \\ -2 & 12.5 \end{bmatrix}$ is the genetic variation in the antagonistic subspace, which would allow a response to sexually antagonistic selection, and $\mathbf{B}_{CA} = \begin{bmatrix} 4.5 & 0 \\ -7 & -8.5 \end{bmatrix}$ is the genetic covariance that leads to indirect responses in the other space. The total genetic variance (the trace) of $\mathbf{G}_{CA} = 212$, the same as the trace of $\mathbf{G}_{mf}$, illustrating that our transformation did not affect the overall size or total evolvability. If we compare the genetic variance in concordant and antagonistic subspaces, we find that there is 11 times more sexually concordant evolvability than there is antagonistic evolvability.

studies to test the prediction that r_{mf} and sexual dimorphism are negatively correlated. They found about half of the estimates of r_{mf} were above 0.8, translating into an autonomy of less than 36%, suggesting that males and females share much of the genetic variation that underlies homologous traits. This appeared to be general across trait types. They found little evidence that r_{mf} differed for different classes of traits, except for fitness components, which tended to have lower intersexual genetic correlations, consistent with other studies (Foerster et al. 2007). A major drawback of this study, in addition to the issues with focusing on correlations, is that only point estimates of r_{mf} were analyzed, with differences in sample size and errors of the estimates not accounted for. Quantitative-genetic estimates of variance and covariance tend to be estimated with substantial error (Sztepanacz and Blows 2017), and cross-sex covariances may be particularly affected because of the lack of residual covariances between male and females traits, which may lead to less precise estimates. Unfortunately, the historical lack of reporting of standard errors for genetic variances and covariances reduces our ability to thoroughly assess the potential issue.

Whether positive cross-sex genetic correlations increase or decrease evolvability depends on whether selection is sexually concordant or antagonistic. Positive correlations will increase evolvability under sexually concordant selection and decrease it when selection is sexually antagonistic. Morrissey's (2016) meta-analysis of 424 selection estimates for 89 traits and 34 species compiled by Cox and Calsbeek (2009) found a strong and positive correlation between male and female selection gradients: ($r(\beta_m, \beta_f) = 0.794$ [0.666, 0.928]), showing that sexually antagonistic selection was rare and not highly antagonistic. A historical emphasis on intralocus sexual conflict, and studies focusing on highly sex-dimorphic traits that typically experience sexually antagonistic selection has led to a general view that intersexual genetic correlations are an evolutionary constraint. However, the data showing many positive r_{mf} estimates and frequent sexually concordant selection suggests that intersexual genetic correlations may often increase evolvability. An analysis of published data in a model that quantified the costs of evolving sexual dimorphism also showed that only 10% of traits are associated with large costs of selection for sexual dimorphism, while the rest have modest or small costs (Matthews et al. 2019). In general, genetic correlations are often viewed from the perspective of evolutionary constraints rather than evolvability, although they may increase evolvability as often as they constrain it (Agrawal and Stinchcombe 2009). While they are two sides of the same coin, recasting our perspective from evolutionary constraints to evolvability may lead to more balanced insights.

12.4.6 Multivariate Cross-Sex Covariances

In addition to the issues of correlation-focused analyses discussed above, we also know that bivariate correlations rarely reflect the higher-dimensional distribution of genetic variation across multivariate trait combinations (Walsh and Blows 2009). There are fewer studies that quantify multivariate cross-sex covariances (i.e., the **B** matrix), and some suffer from the same issues of scaling discussed above, which complicate inferences about the relationship between evolvability and **B**. The majority of studies have tended to focus on noticeably dimorphic (Lewis et al. 2011; Gosden et al. 2012; Ingleby et al. 2014; Cox et al. 2017; Kollar et al. 2021) and sexually-selected traits (Gosden et al. 2012; Ingleby et al. 2014; Cox et al. 2017). In these cases, **B** was most often found to limit the magnitude (Lewis et al. 2011; Ingleby et al. 2014; Cox et al. 2017) and direction (Lewis et al. 2011; Gosden et al. 2012) of the predicted response to estimated selection gradients or to random skewers. The effect of **B** on evolvability under random or concordant selection has been quantified less frequently. Cox et al. (2017) reported that **B** had little effect on the predicted response to selection in random directions, while Holman and Jacomb (2017) and Sztepanacz and Houle (2019) showed that **B** facilitated the response to sexually concordant selection.

A few studies have already used the recent approach of Sztepanacz and Houle (2019) (see box 12.1) to sort genetic variation into sexually concordant and antagonistic genetic subspaces. Sztepanacz and Houle (2019) found that concordant evolvability was much higher than antagonistic evolvability for wing shape in *Drosophila melanogaster.* Kollar et al. (2021) found more antagonistic genetic variance for morphology and physiology traits in the moss *Ceratodon purpureus* and more concordant evolvability for growth and development traits. However, the confidence intervals of their estimates were large and often overlapping. Finally, Houle and Cheng (2021) reanalyzed published data on gene expression (Ayroles et al. 2009), and cuticular hydrocarbon traits (Ingleby et al. 2014) in *D. melanogaster*, and they found more concordant than antagonistic genetic variation.

Asymmetry appears to be a common feature of **B** (Steven et al. 2007; Barker et al. 2010; Lewis et al. 2011; Gosden and Chenoweth 2014; Ingleby et al. 2014; Walling et al. 2014), but its effects have been rarely quantified. Gosden and Chenoweth (2014) found that asymmetry accounted for 10–50% of the variance in **B**, and that asymmetry was positively associated with population divergence. One limitation of recasting $\mathbf{G}_{mf}$ into concordant and antagonistic genetic subspaces is that it does obviously show sex differences in multivariate evolvability that arise from asymmetry in **B**. One way to more clearly show these differences is to compare the predicted response to random (Cheverud et al. 1983) or known selection vectors using an observed versus modified $\mathbf{G}_{mf}$ (e.g., Cox et al. 2017; Holman and Jacomb 2017; Sztepanacz and Houle 2019). $\mathbf{G}_{mf}$ can be modified in a variety of ways, depending on the question being addressed. To determine the effect of asymmetries in **B** on the predicted response, **B** could be made symmetric. To quantify the effect of the modified covariances, we can use the metric $R = e_{mf}/e^{*}$, the ratio of the evolvabilities for observed $\mathbf{G}_{mf}$ and modified $\mathbf{G}^{*}$ (Agrawal and Stinchcombe 2009; Holman and Jacomb 2017). Values of $R < 1$ indicate that the covariances that were modified constrain predicted responses, whereas $R > 1$ suggests they facilitate it. Using this approach, Sztepanacz and Houle (2019) found that, on average, predicted responses to random sexually concordant and antagonistic selection vectors were biased away from the selection gradient more in females than in males. Their data suggest that females have reduced evolvability under directional selection compared to males.

12.5 The Effect of Sexual Selection on Evolvability

Males and females not only share the majority of their genome, but they also must interact during reproduction. These interactions lead to sexual selection, which is known to drive the evolution of extravagant ornamental traits, complex behaviors, and to have an important role in population persistence, divergence, and speciation. Ultimately, sexual selection arises from variation in reproductive success among individuals, which affects evolvability through its effects on the evolution and maintenance of genetic variation. There are many nonexclusive mechanisms that cause variation in reproductive success, such as intrasexual competition for access to mates or resources, pre- or post-copulatory mate choice exerted by the opposite sex, and gamete (sperm or pollen) competition. In the following sections, we first discuss how variation in reproductive success can affect evolvability by reducing the effective size of populations. We then discuss the effect of mate preferences on evolvability.

12.5.1 Reduction in Effective Population Size Due to Sexual Selection

The effective size of a population, N_e, determines the rate at which drift purges genetic variation from populations through the fixation of random alleles (see Pélabon et al., chapter 13). Sexual selection increases variation in reproductive success compared to an ideal (Wright-Fisher) population, which has the potential to reduce the effective size (N_e)

of a population below its census size (N). The effective size of a population is maximized when the effective sizes of the sexes are equal (S. Wright 1931). In most sexual systems, the reproductive success of males, including the number of males who fail to reproduce, is affected by sexual selection to a larger extent than that of females (Bateman 1948; Collet et al. 2012). This difference is expected to lead to the divergence of N_e between males and females and potentially reduce evolvability through the increasing effects of drift in the population.

The effect of sexual selection on N_e and its consequent effects on evolvability are difficult to investigate empirically, because estimating the mean and variance in reproductive success is often a challenge. Using an experiment in *D. melanogaster*, Pischedda et al. (2015) showed that sexual selection had minimal impact on the effective population size in lab-adapted populations. They found that strong sexual selection was operating in their population, with variance in reproductive success ~14 times higher in males than in females. However, inducing high rates of random offspring mortality reduced the effect of sexual selection on N_e compared to N by balancing the effective population sizes of males and females. Overall, their results showed that very strong sexual selection can have minimal effect on N_e in populations with high offspring mortality. Experimental work in *D. pseudoobscura* has also shown that sexual selection may not affect N_e. Snook et al. (2009) estimated N_e in populations undergoing experimental evolution with different intensities of sexual selection. They estimated N_e with both a census-based estimator and a genetic estimator based on molecular markers from two sampling events that took place 26 generations apart. Both genetic and census-based estimates of N_e showed that it did not differ among sexual-selection treatments, and that the ratio $\frac{N_e}{N}$ also did not differ.

Genetic analyses of other systems that experience strong sexual selection have also shown nearly equal ratios of effective and census population sizes (Broquet et al. 2009). Together these data suggest that sexual selection may not have a large effect on N_e, at least for short-lived fecund species that experience high offspring mortality. In longer-lived species or those with lower rates of offspring mortality, this may not be the case (Gagne et al. 2018).

Even if sexual selection does reduce N_e, it does not necessarily mean that evolvability will be reduced. Although none of the studies discussed above make a direct link between N_e and evolvability, there is relatively little empirical evidence, in general, to suggest that N_e has a major role in evolvability on short timescales, given a sufficient effective population size (Pélabon et al., chapter 13).

12.5.2 The Lek Paradox

The evolution of conspicuous male sexual displays generated much of the historical and current interest in sexual selection research, beginning with Darwin (1871), who struggled to reconcile how exaggerated male traits, which would otherwise reduce survival, could evolve and be maintained in populations. Darwin argued that the survival cost of bearing such traits is compensated by the increase in reproductive success that they engender, either through benefits in intrasexual competition or because they make their bearers more sexually attractive.

In many systems, females exert strong preferences to mate with certain males where there are no apparent benefits to their choice (Andersson 1994). The leading hypothesis to explain the evolution of female preferences and male displays in the absence of direct benefits is that display traits reflect the genetic quality of a male, and females are choosing to gain indirect fitness benefits for their offspring (Kirkpatrick and Ryan 1991). One of the major difficulties in understanding this "good-genes" hypothesis of sexual selection is that additive genetic variation in the sexually selected traits and their fitness benefits should be rapidly exhausted in the population through the process of sexual selection itself, negating any benefits of choice. This is the so-called lek paradox (Borgia 1979). Resolving the lek paradox, or how evolvability is maintained in the face of strong sexual selection, is a central focus of sexual-selection research.

The assumption that sexual selection depletes additive genetic variation (and therefore evolvability) appeared to be supported by the observation that life-history traits (i.e., components of fitness) had lower heritabilities than morphological traits (Mousseau and Roff 1987; Roff and Mousseau 1987). As we have emphasized throughout this chapter, however, heritability is not necessarily a good indicator of the magnitude of additive genetic variation or evolvability of a trait. Indeed, individual sexually-selected (Pomiankowski and Møller 1995) and life-history (Houle 1992) traits have been shown to have higher additive genetic variance and evolvability than do morphological traits.

12.5.3 The Maintenance of Genetic Variance under Sexual Selection

Various mechanisms have been proposed to explain how additive genetic variation is maintained under sexual selection (for a comprehensive review, see Radwan 2007). However, their relative importance remains unclear. The genic-capture hypothesis proposed by Rowe and Houle (1996) suggests that the expression of sexually-selected traits is costly and therefore depends on an individual's condition. Condition is controlled by many loci, and consequently, collects mutations that occur across a large part of the genome, allowing additive genetic variation to be maintained in the face of strong sexual selection. Although some empirical studies have found evidence for condition-dependent expression of male sexual displays (e.g., Hill 1990; Bonduriansky and Rowe 2005; Delcourt and Rundle 2011), many have not, and even fewer have investigated whether there is genetic variance in condition dependence (Cotton et al. 2004).

Genetic compatibility provides a different avenue by which evolvability could be maintained or even increased by sexual selection. It posits that offspring fitness depends on the allelic combinations of mothers and fathers (Trivers 1972) and hence on dominance and epistatic effects (Tregenza and Wedell 2000; Neff and Pitcher 2008). Interest in the genetic-compatibility hypothesis arose as a mechanism to explain polyandry and the costs associated with multiple mating. Early verbal models relied on post-copulatory genetic incompatibilities to discriminate between genetically compatible sires and provide females with fitness benefits for their offspring (Zeh and Zeh 1996), although females can also exercise precopulatory choice on the basis of genetic compatibility or dissimilarity (Mays and Hill 2004). However, one of the major theoretical difficulties of the genetic-compatibility model is how preference for nonadditive benefits can evolve when the nonadditive benefits are not heritable (Lehmann et al. 2007). Empirical studies of genetic compatibility are most extensive with respect to the major histocompatibility complex (MHC), which is a locus of polymorphic genes that control immunological function and self-recognition in vertebrates. Genetic identity of the MHC can be distinguished through odors (Penn 2002), enabling pre-copulatory choice based on this locus. Humans, mice, and fish (reviewed in Tregenza and Wedell 2000) have all been shown to prefer odor samples of males that have genetically dissimilar MHC genotypes.

Falling between the "good-genes" and "genetic compatibility" hypotheses is mate preference for heterozygosity. Females choose males that have high heterozygosity across many loci, because they will have a greater likelihood of producing genetically compatible offspring (Brown 1997), or because heterozygosity increases male fitness, contributing direct benefits to females and their offspring. Models suggest that preferences for heterozygous mates can evolve and be maintained in populations in the absence of direct benefits (Fromhage et al. 2009), and there is empirical evidence to support preference for heterozygosity in some circumstances (Landry et al. 2001; Garcia-Navas et al. 2009). For each of the three mechanisms discussed in this section, we can point to specific examples of where they appear to be operating in a population or species. However, the empirical data do not provide support for any of them as a general mechanism to maintain additive genetic variation in populations.

12.5.4 Multivariate Sexual Selection and Trade-Offs

Many, if not most, sexual displays are complex and comprise many individual traits, such as different components of wing song in crickets (Brooks et al. 2005) or *Drosophila* (Hoy et al. 1988), pheromone profiles comprising several different chemical compounds (Blows and Allan 1998), or complex color patterns (Butler et al. 2007). In these cases, additive genetic variation in the individual component traits of the display will likely tell us little about their evolvability (Blows et al. 2004). Fitness trade-offs between sexually and naturally selected traits and/or between males and females are likely to arise because of the action of sexual selection and maintain genetic variation that is beneficial for one trait type (or sex) and detrimental for the other. Radwan et al. (2016) hypothesized that stronger sexual selection is associated with increased genetic variation in sexually-selected and ecological traits as a consequence of these trade-offs. However, there is limited evidence to support this hypothesis or others that predict increased genetic variance associated with sexual selection (e.g., Petrie and Roberts 2006). Multivariate studies of sexual selection often find that multivariate sexual-selection gradients are orthogonal to the major axes of genetic variance in sexual displays (Hine et al. 2004; Van Homrigh et al. 2007). These data support the prediction that sexual selection reduces evolvability of targeted traits, and the current data also suggest that the genetic variance that is maintained is a consequence of antagonistic pleiotropy, which is another variance-maintaining mechanism (Connallon and Clark 2012). High-fitness males have less genetic variance in sexual displays than low-fitness males, suggesting that the genetic variance underlying sexual displays is under apparent stabilizing selection (McGuigan et al. 2011; Delcourt et al. 2012; Sztepanacz and Rundle 2012). Therefore, the genetic variance in multivariate sexual displays may not be relevant to evolvability under a scenario of directional sexual selection.

12.5.5 Conclusions for Sexual Selection

The effect of sexual selection on evolvability is complex. On one hand, we might expect sexual selection to reduce evolvability by reducing effective population sizes and the genetic variation among successful reproducers. On the other hand, high evolvability is observed in many individual sexually-selected traits. Despite the substantial theoretical and empirical attention given to solving this paradox of the lek, we still do not have a clear understanding of how to resolve it. The most general explanation is that antagonistic pleiotropy resulting from multivariate trade-offs maintains segregating variation in sexually-selected traits. Whether this variation is relevant to evolvability will depend on the direction and form of selection operating in the population.

12.6 Conclusion

In this chapter, we have considered how two major axes of mating-system variation affect evolvability: variation in outcrossing rate, and variation in the mating strategies and behaviors that lead to variation in reproductive success. We have shown that estimating the evolvability of quantitative traits may be difficult when properly accounting for the effect of mating system. The main challenges are the rigorous decomposition of genetic variance into additive and nonadditive components, the recasting of population-level evolvability into sexually concordant and antagonistic subspaces, and finally the consideration of linkage disequilibrium on short- and mid-term evolvability and the stochasticity associated with it. Our review of the theoretical and empirical evidence has led us to the following conclusions:

- There is no universal measure of evolvability is meaningful across mating systems.
- Selfing does not necessarily reduce evolvability, and variation in outcrossing rate has no systematic effect on the evolvability of populations.
- There is no empirical evidence that the heterogametic sex is more evolvable than the homogametic sex.
- There is no systematic difference in evolvability between males and females, but there are often differences in evolvability in particular populations.
- Within populations, when males are more evolvable, the difference in evolvability between the sexes tends to be more extreme than when females are more evolvable.
- There is more sexually concordant than antagonistic evolvability.
- In general, sexual selection does not reduce the effective size of populations and consequently may have little impact on short-term evolvability.
- Antagonistic pleiotropy may be the most important mechanism maintaining genetic variation in sexually selected traits, but this variation may not be relevant for evolvability under sexual selection.
- A multivariate view of genetic variation and selection is necessary to understand the evolvability of most traits, including those that experience sexual selection.

Acknowledgments

Thanks to the Centre for Advanced Study in Oslo; Geir Bolstad and Thomas Hansen, who made helpful comments on the manuscript; and to members of the Evolvability Journal Club for discussions and comments.

References

Abney, M., M. S. McPeek, and C. Ober. 2000. Estimation of variance components of quantitative traits in inbred populations. *American Journal of Human Genetics* 66: 629–650.

Abu Awad, D., and D. Roze. 2018. Effects of partial selfing on the equilibrium genetic variance, mutation load, and inbreeding depression under stabilizing selection. *Evolution* 72: 751–769.

Agrawal, A. F., and J. R. Stinchcombe. 2009. How much do genetic covariances alter the rate of adaptation? *Proceedings of the Royal Society B* 276: 1183–1191.

Allen, S. L., R. Bonduriansky, and S. F. Chenoweth. 2018. Genetic constraints on microevolutionary divergence of sex-biased gene expression. *Philosophical Transactions of the Royal Society B* 373: 20170427.

Andersson, M. 1994. *Sexual Selection.* Princeton, NJ: Princeton University Press.

Arnqvist, G., and L. Rowe. 2002a. Antagonistic coevolution between the sexes in a group of insects. *Nature* 415: 787–789.

Arnqvist, G., and L. Rowe. 2002b. Correlated evolution of male and female morphologies in water striders. *Evolution* 56: 936–947.

Arnqvist, G., and L. Rowe. 2005. *Sexual Conflict.* Princeton, NJ: Princeton University Press.

Ashman, T. L., and C. J. Majetic. 2006. Genetic constraints on floral evolution: A review and evaluation of patterns. *Heredity* 96: 343–352.

Assis, R., Q. Zhou, and D. Bachtrog. 2012. Sex-biased transcriptome evolution in *Drosophila. Genome Biology and Evolution* 4: 1189–1200.

Ayroles, J. F., M. A. Carbone, E. A. Stone, et al. 2009. Systems genetics of complex traits in *Drosophila melanogaster. Nature Genetics* 41: 299–307.

Barker, B. S., P. C. Phillips, and S. J. Arnold. 2010. A test of the conjecture that G-matrices are more stable than B-matrices. *Evolution* 64: 2601–2613.

Bartkowska, M. P., and M. O. Johnston. 2009. Quantitative genetic variation in populations of *Amsinckia spectabilis* that differ in rate of self-fertilization. *Evolution* 63: 1103–1117.

Bateman, A. J. 1948. Intra-sexual selection in *Drosophila. Heredity* 2: 349–368.

Blows, M. W., and R. A. Allan. 1998. Levels of mate recognition within and between two *Drosophila* species and their hybrids. *American Naturalist* 152: 826–837.

Blows, M. W., S. F. Chenoweth, and E. Hine. 2004. Orientation of the genetic variance-covariance matrix and the fitness surface for multiple male sexually selected traits. *American Naturalist* 163: 329–340.

Bonduriansky, R., and S. F. Chenoweth. 2009. Intralocus sexual conflict. *TREE* 24: 280–288.

Bonduriansky, R., and L. Rowe. 2005. Sexual selection, genetic architecture, and the condition dependence of body shape in the sexually dimorphic fly *Prochyliza xanthostoma* (Piophilidae). *Evolution* 59: 138–151.

Borgia, G. 1979. Sexual selection and the evolution of mating systems. In *Sexual Selection and Reproductive Competition in Insects,* edited by M. S. Blum and N. A. Blum, 19–80. New York: Academic Press.

Brooks, R. R., J. J. Hunt, M. W. Blows, M. J. Smith, L. F. Bussière, and M. D. Jennions. 2005. Experimental evidence for multivariate stabilizing sexual selection. *Evolution* 59: 871–880.

Broquet, T., J. Jaquiéry, and N. Perrin. 2009. Opportunity for sexual selection and effective population size in the lek-breeding European treefrog (*Hyla arborea*). *Evolution* 63: 674–683.

Brown, J. L. 1997. A theory of mate choice based on heterozygosity. *Behavioral Ecology* 8: 60–65.

Butler, M. A., S. A. Sawyer, and J. B. Losos. 2007. Sexual dimorphism and adaptive radiation in *Anolis* lizards. *Nature* 447: 202–205.

Charlesworth, D., and B. Charlesworth. 1995. Quantitative genetics in plants: The effect of the breeding system on genetic variability. *Evolution* 49: 911–920.

Charlesworth, D., and S. Mayer. 1995. Genetic variability of plant characters in the partial inbreeder *Collinsia heterophylla* (Scrophulariaceae). *American Journal of Botany* 82: 112–120.

Cheng, C., and D. Houle. 2020. Predicting multivariate responses of sexual dimorphism to direct and indirect selection. *American Naturalist* 196: 391–405.

Cheverud, J. M., J. Rutledge, and W. R. Atchley. 1983. Quantitative genetics of development: genetic correlations among age-specific trait values and the evolution of ontogeny. *Evolution* 37: 895–905.

Clo, J., and Ø. H. Opedal. 2021. Genetics of quantitative traits with dominance under stabilizing and directional selection in partially selfing species. *Evolution* 75: 1920–1935.

Clo, J., L. Gay, and J. Ronfort. 2019. How does selfing affect the genetic variance of quantitative traits? An updated meta-analysis on empirical results in angiosperm species. *Evolution* 73: 1578–1590.

Clo, J., J. Ronfort, and D. Abu Awad. 2020. Hidden genetic variance contributes to increase the short-term adaptive potential of selfing populations. *Journal of Evolutionary Biology* 33:1203–1215.

Cockerham, C. C., and B. S. Weir. 1984. Covariances of relatives stemming from a population undergoing mixed self and random mating. *Biometrics* 157–164.

Collet, J., D. S. Richardson, K. Worley, and T. Pizzari. 2012. Sexual selection and the differential effect of polyandry. *PNAS* 109: 8641–8645.

Connallon, T., and A. G. Clark. 2012. A general population genetic framework for antagonistic selection that accounts for demography and recurrent mutation. *Genetics* 190: 1477–1489.

Costa e Silva, J. C., C. Hardner, and B. M. Potts. 2010. Genetic variation and parental performance under inbreeding for growth in *Eucalyptus globulus. Annals of Forest Science* 67: 606.

Cotton, S., K. Fowler, and A. Pomiankowski. 2004. Do sexual ornaments demonstrate heightened condition-dependent expression as predicted by the handicap hypothesis? *Proceedings of the Royal Society B* 271: 771–783.

Cowley, D. E., W. R. Atchley, and J. Rutledge. 1986. Quantitative genetics of *Drosophila melanogaster.* I. Sexual dimorphism in genetic parameters for wing traits. *Genetics* 114: 549–566.

Cox, R. M., and R. Calsbeek. 2009. Sexually antagonistic selection, sexual dimorphism, and the resolution of intralocus sexual conflict. *American Naturalist* 173: 176–187.

Cox, R. M., R. A. Costello, B. E. Camber, and J. W. McGlothlin. 2017. Multivariate genetic architecture of the *Anolis* dewlap reveals both shared and sex-specific features of a sexually dimorphic ornament. *Journal of Evolutionary Biology* 30: 1262–1275.

Darwin, C. 1871. *The Descent of Man and Selection in Relation to Sex.* London: John Murray.

Darwin, C. 1876. *The Effects of Cross and Self Fertilization in the Vegetable Kingdom.* London: John Murray.

Dean, R., and J. E. Mank. 2014. The role of sex chromosomes in sexual dimorphism: Discordance between molecular and phenotypic data. *Journal of Evolutionary Biology* 27: 1443–1453.

Delcourt, M., and H. D. Rundle. 2011. Condition dependence of a multicomponent sexual display trait in *Drosophila serrata. American Naturalist* 177: 812–823.

Delcourt, M., M. W. Blows, J. D. Aguirre, and H. D. Rundle. 2012. Evolutionary optimum for male sexual traits characterized using the multivariate Robertson-Price Identity. *PNAS* 109: 10414–10419.

Dutoit, L., R. Burri, A. Nater, C. F. Mugal, and H. Ellegren. 2017. Genomic distribution and estimation of nucleotide diversity in natural populations: perspectives from the collared flycatcher (*Ficedula albicollis*) genome. *Molecular Ecology Resources* 17: 586–597.

Edwards, J. W., and K. R. Lamkey. 2002. Quantitative genetics of inbreeding in a synthetic maize population. *Crop Science* 42: 1094–1104.

Ellegren, H., and J. Parsch. 2007. The evolution of sex-biased genes and sex-biased gene expression. *Nature Reviews Genetics* 8: 689–698.

Falconer, D. S., and T. F. C. Mackay. 1996. *Introduction to Quantitative Genetics.* London: Longman.

Foerster, K., T. Coulson, B. C. Sheldon, J. M. Pemberton, T. H. Clutton-Brock, and L. E. B. Kruuk. 2007. Sexually antagonistic genetic variation for fitness in red deer. *Nature* 447: 1107–U1109.

Fowler, K., and M. C. Whitlock. 1999. The distribution of phenotypic variance with inbreeding. *Evolution* 53: 1143–1156.

Fromhage, L., H. Kokko, and J. M. Reid. 2009. Evolution of mate choice for genome-wide heterozygosity. *Evolution* 63: 684–694.

Fry, J. D. 2010. The genomic location of sexually antagonistic variation: Some cautionary comments. *Evolution* 64: 1510–1516.

Gagne, R. B., M. T. Tinker, K. D. Gustafson, K. Ralls, S. Larson, L. M. Tarjan, M. A. Miller, and H. B. Ernest. 2018. Measures of effective population size in sea otters reveal special considerations for wide-ranging species. *Evolutionary Applications* 11: 1779–1790.

Garcia-Navas, V., J. Ortego, and J. J. Sanz. 2009. Heterozygosity-based assortative mating in blue tits (*Cyanistes caeruleus*): Implications for the evolution of mate choice. *Proceedings of the Royal Society B* 276: 2931–2940.

Geber, M. A., and L. R. Griffen. 2003. Inheritance and natural selection on functional traits. *International Journal of Plant Sciences* 164: S21–S42.

Gershman, S. N., C. A. Barnett, A. M. Pettinger, C. B. Weddle, J. Hunt, and S. K. Sakaluk. 2010. Inbred decorated crickets exhibit higher measures of macroparasitic immunity than outbred individuals. *Heredity* 105: 282–289.

Godfrey, R. M., and M. T. J. Johnson. 2014. Effects of functionally asexual reproduction on quantitative genetic variation in the evening primroses (*Oenothera,* Onagraceae). *American Journal of Botany* 101: 1906–1914.

Goodwillie, C., S. Kalisz, and C. G. Eckert. 2005. The evolutionary enigma of mixed mating systems in plants: Occurrence, theoretical explanations, and empirical evidence. *AREES* 36: 47–79.

Gosden, T. P., and S. F. Chenoweth. 2014. The evolutionary stability of cross-sex, cross-trait genetic covariances. *Evolution* 68: 1687–1697.

Gosden, T. P., K.-L. Shastri, P. Innocenti, and S. F. Chenoweth. 2012. The B-matrix harbors significant and sex-specific constraints on the evolution of multicharacter sexual dimorphism. *Evolution* 66: 2106–2116.

Grieshop, K., and G. Arnqvist. 2018. Sex-specific dominance reversal of genetic variation for fitness. *PLOS Biology* 16: e2006810.

Hansen, M. E., and R. J. Kulathinal. 2013. Sex-biased networks and nodes of sexually antagonistic conflict in *Drosophila. International Journal of Evolutionary Biology* 2013: 545392.

Hansen, T. F. 2003. Is modularity necessary for evolvability? Remarks on the relationship between pleiotropy and evolvability. *Biosystems* 69: 83–94.

Hansen, T. F., and C. Pélabon. 2021. Evolvability: A quantitative-genetics perspective. *AREES* 52: 153–175.

Hansen, T. F., C. Pélabon, W. S. Armbruster, and M. L. Carlson. 2003. Evolvability and genetic constraint in *Dalechampia* blossoms: Components of variance and measures of evolvability. *Journal of Evolutionary Biology* 16: 754–766.

Hansen, T. F., C. Pélabon, and D. Houle. 2011. Heritability is not evolvability. *Evolutionary Biology* 38: 258–277.

Herlihy, C. R., and C. G. Eckert. 2007. Evolutionary analysis of a key floral trait in *Aquilegia canadensis* (Ranunculaceae): Genetic variation in herkogamy and its effect on the mating system. *Evolution* 61: 1661–1674.

Hill, G. E. 1990. Female house finches prefer colourful males: Sexual selection for a condition-dependent trait. *Animal Behaviour* 40: 563–572.

Hill, W. G. 1996. Sewall Wright's "systems of mating." *Genetics* 143: 1499.

Hine, E., S. F. Chenoweth, and M. W. Blows. 2004. Multivariate quantitative genetics and the lek paradox: Genetic variance in male sexually selected traits of *Drosophila serrata* under field conditions. *Evolution* 58: 2754–2762.

Hoeschele, I., and A. Vollema. 1993. Estimation of variance components with dominance and inbreeding in dairy cattle. *Journal of Animal Breeding and Genetics* 110: 93–104.

Hoffmann, A. A., J. Merilä, and T. N. Kristensen. 2016. Heritability and evolvability of fitness and nonfitness traits: Lessons from livestock. *Evolution* 70: 1770–1779.

Holeski, L. M., and J. K. Kelly. 2006. Mating system and the evolution of quantitative traits: An experimental study of *Mimulus guttatus. Evolution* 60: 711–723.

Holman, L., and F. Jacomb. 2017. The effects of stress and sex on selection, genetic covariance, and the evolutionary response. *Journal of Evolutionary Biology* 30: 1898–1909.

Houle, D. 1992. Comparing evolvability and variability of quantitative traits. *Genetics* 130: 195–204.

Houle, D., and C. Cheng. 2021. Predicting the evolution of sexual dimorphism in gene expression. *Molecular Biology and Evolution* 38: 1847–1859.

Hoy, R. R., A. Hoikkala, and K. Kaneshiro. 1988. Hawaiian courtship songs: Evolutionary innovation in communication signals of *Drosophila. Science* 240: 217–219.

Husby, A., H. Schielzeth, W. Forstmeier, L. Gustafsson, and A. Qvarnström. 2013. Sex chromosome linked genetic variance and the evolution of sexual dimorphism of quantitative traits. *Evolution* 67: 609–619.

Ingleby, F. C., P. Innocenti, H. D. Rundle, and E. H. Morrow. 2014. Between-sex genetic covariance constrains the evolution of sexual dimorphism in *Drosophila melanogaster. Journal of Evolutionary Biology* 27: 1721–1732.

Jarne, P., and D. Charlesworth. 1993. The evolution of the selfing rate in functionally hermaphrodite plants and animals. *AREES* 24: 441–466.

Jullien, M., M. Navascués, J. Ronfort, K. Loridon, and L. Gay. 2019. Structure of multilocus genetic diversity in predominantly selfing populations. *Heredity* 123: 176–191.

Kamran-Disfani, A., and A. F. Agrawal. 2014. Selfing, adaptation and background selection in finite populations. *Journal of Evolutionary Biology* 27: 1360–1371.

Kelly, J. K. 1999a. Response to selection in partially self-fertilizing populations. I. Selection on a single trait. *Evolution* 53: 336–349.

Kelly, J. K. 1999b. Response to selection in partially self-fertilizing populations. II. Selection on multiple traits. *Evolution* 53: 350–357.

Kelly, J. K. 2003. Deleterious mutations and the genetic variance of male fitness components in *Mimulus guttatus. Genetics* 164: 1071–1085.

Kelly, J. K., and H. S. Arathi. 2003. Inbreeding and the genetic variance in floral traits of *Mimulus guttatus. Heredity* 90: 77–83.

Kelly, J. K., and S. Williamson. 2000. Predicting response to selection on a quantitative trait: A comparison between models for mixed-mating populations. *Journal of Theoretical Biology* 207: 37–56.

Kirkpatrick, M., and M. J. Ryan. 1991. The evolution of mating preferencces and the paradox of the lek. *Nature* 350: 33–38.

Kollar, L. M., S. Kiel, A. J. James, et al. 2021. The genetic architecture of sexual dimorphism in the moss *Ceratodon purpureus. Proceedings of the Royal Society B* 288: 20202908.

Kruuk, L. E. B. 2004. Estimating genetic parameters in natural populations using the "animal model." *Philosophical Transactions of the Royal Society B* 359: 873–890.

Lande, R. 1977. The influence of the mating system on the maintenance of genetic variability in polygenic characters. *Genetics* 86: 485–498.

Lande, R. 1979. Quantitative genetic analysis of multivariate evolution, applied to brain: body size allometry. *Evolution* 33: 402–416.

Lande, R. 1980. Sexual dimorphism, sexual selection, and adaptation in polygenic characters. *Evolution* 34: 292–305.

Lande, R., and E. Porcher. 2015. Maintenance of quantitative genetic variance under partial self-fertilization, with implications for evolution of selfing. *Genetics* 200: 891–906.

Landry, C., D. Garant, P. Duchesne, and L. Bernatchez. 2001. "Good genes as heterozygosity": The major histocompatibility complex and mate choice in Atlantic salmon (*Salmo salar*). *Proceedings of the Royal Society B* 268: 1279–1285.

Larsen, C. T., A. M. Holand, H. Jensen, I. Steinsland, and A. Roulin. 2014. On estimation and identifiability issues of sex-linked inheritance with a case study of pigmentation in Swiss barn owl (*Tyto alba*). *Ecology and Evolution* 4: 1555–1566.

Lehmann, L., L. F. Keller, and H. Kokko. 2007. Mate choice evolution, dominance effects, and the maintenance of genetic variation. *Journal of Theoretical Biology* 244: 282–295.

Lewis, Z., N. Wedell, and J. Hunt. 2011. Evidence for strong intralocus sexual conflict in the Indian meal moth, *Plodia interpunctella. Evolution* 65: 2085–2097.

Lynch, M., and B. Walsh. 1998. *Genetics and Analysis of Quantitative Traits.* Sunderland, MA: Sinauer Associates.

Mallet, M. A., C. M. Kimber, and A. K. Chippindale. 2012. Susceptibility of the male fitness phenotype to spontaneous mutation. *Biology Letters* 8: 426–429.

Mank, J. E., L. Hultin-Rosenberg, E. Axelsson, and H. Ellegren. 2007. Rapid evolution of female-biased, but not male-biased, genes expressed in the avian brain. *Molecular Biology and Evolution* 24: 2698–2706.

Manna, F., G. Martin, and T. Lenormand. 2011. Fitness landscapes: An alternative theory for the dominance of mutation. *Genetics* 189: 923–937.

Marriage, T. N., and J. K. Kelly. 2009. Inbreeding depression in an asexual population of *Mimulus guttatus. Journal of Evolutionary Biology* 22: 2320–2331.

Massouras, A., S. M. Waszak, M. Albarca-Aguilera, et al. 2012. Genomic variation and its impact on gene expression in *Drosophila melanogaster. PLOS Genetics* 8: e1003055.

Matthews, G., S. Hangartner, D. G. Chapple, and T. Connallon. 2019. Quantifying maladaptation during the evolution of sexual dimorphism. *Proceedings of the Royal Society B* 286: 20191372.

Mays Jr., H. L., and G. E. Hill. 2004. Choosing mates: Good genes versus genes that are a good fit. *TREE* 19: 554–559.

McGuigan, K., and M. W. Blows. 2009. Asymmetry of genetic variation in fitness-related traits: Apparent stabilizing selection on g(max). *Evolution* 63: 2838–2847.

McGuigan, K., L. Rowe, and M. W. Blows. 2011. Pleiotropy, apparent stabilizing selection and uncovering fitness optima. *TREE* 26: 22–29.

Meiklejohn, C. D., J. Parsch, J. M. Ranz, and D. L. Hartl. 2003. Rapid evolution of male-biased gene expression in *Drosophila. PNAS* 100: 9894–9899.

Moeller, D. A., R. D. Briscoe Runquist, A. M. Moe, et al. 2017. Global biogeography of mating system variation in seed plants. *Ecology Letters* 20: 375–384.

Morrissey, M. B. 2016. Meta-analysis of magnitudes, differences and variation in evolutionary parameters. *Journal of Evolutionary Biology* 29: 1882–1904.

Mousseau, T. A., and D. A. Roff. 1987. Natural selection and the heritability of fitness components. *Heredity* 59: 181–197.

Mousseau, T. A., and D. A. Roff. 1989. Geographic variability in the incidence and heritability of wing dimorphism in the striped ground cricket, *Allonemobius fasciatus. Heredity* 62: 315–318.

Muller, H. J. 1932. Some genetic aspects of sex. *American Naturalist* 66: 118–138.

Neff, B. D., and T. E. Pitcher. 2008. Mate choice for non-additive genetic benefits: A resolution to the lek paradox. *Journal of Theoretical Biology* 254: 147–155.

Noel, E., P. Jarne, S. Glemin, A. MacKenzie, A. Segard, V. Sarda, and P. David. 2017. Experimental evidence for the negative effects of self-fertilization on the adaptive potential of populations. *Current Biology* 27: 237–242.

Opedal, Ø. H. 2018. Herkogamy, a principal functional trait of plant reproductive biology. *International Journal of Plant Sciences* 179: 677–687.

Opedal, Ø. H., G. H. Bolstad, T. F. Hansen, W. S. Armbruster, and C. Pélabon. 2017. The evolvability of herkogamy: Quantifying the evolutionary potential of a composite trait. *Evolution* 71: 1572–1586.

Penn, D. J. 2002. The scent of genetic compatibility: Sexual selection and the major histocompatibility complex. *Ethology* 108: 1–21.

Petrie, M., and G. Roberts. 2006. Sexual selection and the evolution of evolvability. *Heredity* 98: 198–205.

Picard, M. A. L., B. Vicoso, S. Bertrand, and H. Escriva. 2021. Diversity of modes of reproduction and sex determination systems in invertebrates, and the putative contribution of genetic conflict. *Genes* 12: 1136.

Pischedda, A., U. Friberg, A. D. Stewart, P. M. Miller, and W. R. Rice. 2015. Sexual selection has minimal impact on effective population sizes in species with high rates of random offspring mortality: An empirical demonstration using fitness distributions. *Evolution* 69: 2638–2647.

Poissant, J., A. J. Wilson, and D. W. Coltman. 2009. Sex-specific genetic variance and the evolution of sexual dimorphism: A systematic review of cross-sex genetic correlations. *Evolution* 64: 97–107.

Pomiankowski, A., and A. P. Møller. 1995. A resolution of the lek paradox. *Proceedings of the Royal Society B* 260: 21–29.

Radwan, J. 2007. Maintenance of genetic variation in sexual ornaments: A review of the mechanisms. *Genetica* 134: 113–127.

Radwan, J., L. Engqvist, and K. Reinhold. 2016. A paradox of genetic variance in epigamic traits: Beyond "good genes" view of sexual selection. *Evolutionary Biology* 43: 267–275.

Reinhold, K., and L. Engqvist. 2013. The variability is in the sex chromosomes. *Evolution* 67: 3662–3668.

Rieseberg, L. H., M. A. Archer, and R. K. Wayne. 1999. Transgressive segregation, adaptation and speciation. *Heredity* 83: 363–372.

Rieseberg, L. H., A. Widmer, A. M. Arntz, and B. Burke. 2003. The genetic architecture necessary for transgressive segregation is common in both natural and domesticated populations. *Philosophical Transactions of the Royal Society B* 358: 1141–1147.

Roff, D. A., and T. A. Mousseau. 1987. Quantitative genetics and fitness: Lessons from *Drosophila*. *Heredity* 58: 103–118.

Rolff, J., S. A. Armitage, and D. W. Coltman. 2005. Genetic constraints and sexual dimorphism in immune defense. *Evolution* 59: 1844–1850.

Roulin, A., and H. Jensen. 2015. Sex-linked inheritance, genetic correlations and sexual dimorphism in three melanin-based colour traits in the barn owl. *Journal of Evolutionary Biology* 28: 655–666.

Rowe, L., and D. Houle. 1996. The lek paradox and the capture of genetic variance by condition dependent traits. *Proceedings of the Royal Society B* 263: 1415–1421.

Rowe, L., S. F. Chenoweth, and A. F. Agrawal. 2018. The genomics of sexual conflict. *American Naturalist* 192: 274–286.

Sharp, N. P., and A. F. Agrawal. 2012. Male-biased fitness effects of spontaneous mutations in *Drosophila melanogaster*. *Evolution* 67: 1189–1195.

Shaw, F. H., and J. A. Woolliams. 1999. Variance component analysis of skin and weight data for sheep subjected to rapid inbreeding. *Genetics Selection Evolution* 31: 43–59.

Shaw, R. G., D. L. Byers, and F. H. Shaw. 1998. Genetic components of variation in *Nemophila menziesii* undergoing inbreeding: Morphology and flowering time. *Genetics* 150: 1649–1661.

Siol, M., J.-M. Prosperi, I. Bonnin, and J. Ronfort. 2008. How multilocus genotypic pattern helps to understand the history of selfing populations: A case study in *Medicago truncatula*. *Heredity* 100: 517–525.

Snook, R. R., L. Brüstle, and J. Slate. 2009. A test and review of the role of effective population size on experimental sexual selection patterns. *Evolution* 63: 1923–1933.

Stebbins, G. L. 1974. *Flowering Plants: Evolution above the Species Level*. London: Edward Arnold.

Steven, J. C., L. F. Delph, and E. D. Brodie. 2007. Sexual dimorphism in the quantitative-genetic architecture of floral, leaf, and allocation traits in *Silene latifolia*. *Evolution* 61: 42–57.

Stillwell, R. C., and G. Davidowitz. 2010. A developmental perspective on the evolution of sexual size dimorphism of a moth. *Proceedings of the Royal Society B* 277: 2069–2074.

Sztepanacz, J. L., and M. W. Blows. 2017. Accounting for sampling error in genetic eigenvalues using random matrix theory. *Genetics* 206: 1271–1284.

Sztepanacz, J. L., and D. Houle. 2019. Cross-sex genetic covariances limit the evolvability of wing-shape within and among species of *Drosophila. Evolution* 73: 1617–1633.

Sztepanacz, J. L., and H. D. Rundle. 2012. Reduced genetic variance among high fitness individuals: Inferring stabilizing selection on male sexual displays in *Drosophila serrata. Evolution* 66: 3101–3110.

Tregenza, T., and N. Wedell. 2000. Genetic compatibility, mate choice and patterns of parentage: Invited review. *Molecular Ecology* 9: 1013–1027.

Trivers, R. L. 1972. Parental investment and sexual selection. In *Sexual Selection and the Descent of Man,* edited by B. Campbell, pp. 136–179. Chicago: Aldine.

Van Homrigh, A., M. Higgie, K. McGuigan, and M. W. Blows. 2007. The depletion of genetic variance by sexual selection. *Current Biology* 17: 528–532.

Walling, C. A., M. B. Morrissey, K. Foerster, T. H. Clutton-Brock, J. M. Pemberton, and L. E. B. Kruuk. 2014. A multivariate analysis of genetic constraints to life history evolution in a wild population of red deer. *Genetics* 198: 1735–1749.

Walsh, B., and M. W. Blows. 2009. Abundant genetic variation + strong selection = multivariate genetic constraints: A geometric view of adaptation. *AREES* 40: 41–59.

Wardyn, B. M., J. W. Edwards, and K. R. Lamkey. 2007. The genetic structure of a maize population: The role of dominance. *Crop Science* 47: 467–474.

Whitehead, M. R., R. Lanfear, R. J. Mitchell, and J. D. Karron. 2018. Plant mating systems often vary widely among populations. *Frontiers in Ecology and Evolution* 6: 38.

Wolak, M. E., and L. F. Keller. 2014. Dominance genetic variance and inbreeding in natural populations. In *Quantitative Genetics in the Wild,* edited by A. Charmantier, D. Garant, and L. E. B. Kruuk, 104–127. Oxford: Oxford University Press.

Wolak, M. E., P. Arcese, L. F. Keller, P. Nietlisbach, and J. M. Reid. 2018. Sex-specific additive genetic variances and correlations for fitness in a song sparrow (*Melospiza melodia*) population subject to natural immigration and inbreeding. *Evolution* 72: 2057–2075.

Wright, A., and C. C. Cockerham. 1985. Selection with partial selfing. I. Mass selection. *Genetics* 109: 585–597.

Wright, S. 1921a. Systems of mating. I. The biometric relations between parent and offspring. *Genetics* 6: 111–123.

Wright, S. 1921b. Systems of mating. II. The effects of inbreeding on the genetic composition of a population. *Genetics* 6: 124–143.

Wright, S. 1921c. Systems of mating. III. Assortative mating based on somatic resemblance. *Genetics* 6: 144–161.

Wright, S. 1921d. Systems of mating. IV. The effects of selection. *Genetics* 6: 162–166.

Wright, S. 1931. Evolution in Mendelian populations. *Genetics* 16: 97.

Wright, S. 1952. The theoretical variance within and among subdivisions of a population that is in a steady state. *Genetics* 37: 312.

Wyman, M. J., and L. Rowe. 2014. Male bias in distributions of additive genetic, residual, and phenotypic variances of shared traits. *American Naturalist* 184: 326–337.

Zajitschek, S. R., F. Zajitschek, R. Bonduriansky, et al. 2020. Sexual dimorphism in traitvariability and its eco-evolutionary and statistical implications. *eLife* 9: e63170.

Zeh, J. A., and D. W. Zeh. 1996. The evolution of polyandry I: Intragenomic conflict and genetic incompatibility. *Proceedings of Royal Society B* 263: 1711–1717.

Zhang, Z., and J. Parsch. 2005. Positive correlation between evolutionary rate and recombination rate in *Drosophila* genes with male-biased expression. *Molecular Biology and Evolution* 22: 1945–1947.

13 Can We Explain Variation in Evolvability on Ecological Timescales?

Christophe Pélabon, Michael B. Morrissey, Jane M. Reid, and Jacqueline L. Sztepanacz

The ability of populations to respond to selection, their evolvability, depends on the amount of additive genetic variance, V_A, they harbor. Estimates of V_A are particularly variable among populations and studies, however. Population size, gene flow, selection, and environmental variation are expected to change evolvability on short timescales, but the magnitude and the adaptive significance of those changes remain unclear. In this chapter, we summarize theoretical expectations before reviewing empirical evidence of changes in evolvability on short timescales due to these factors. Experiments confirm that rapid changes in evolvability occur, but these changes are often idiosyncratic, making it difficult to predict how ecological factors such as habitat fragmentation, specialist versus generalist lifestyle, or stressful environments, affect short-term evolvability. In a few cases, however, changes in evolvability seem predictable but not necessarily adaptive.

13.1 Short-Term Evolvability

Understanding and predicting the rate at which selection can drive phenotypic changes over short timeframes is a central aim in evolutionary ecology. This makes evolvability, the ability of phenotypes to respond to selection, a key parameter of both theoretical and empirical focus. The Lande equation, $\Delta\bar{z} = V_A \beta$ (Lande 1979; Lande and Arnold 1983) predicts that the response of a trait mean to one generation of selection, $\Delta\bar{z}$, is the product of the selection gradient, β, and the additive genetic variance of the trait, V_A. From this prediction, it follows that the evolvability of a trait can be quantified by its additive genetic variance (Houle 1992; Hansen et al. 2003; Hansen, chapter 5; Houle and Pélabon, chapter 6).[1] Over the past 50 years, many studies have estimated V_A in captive and wild populations. This work has shown that V_A, and thus evolvability, is particularly variable among traits, populations, and species (Hansen and Pélabon 2021). More surprising, however, is the variation in V_A commonly observed for single traits among closely related species, among populations, or even sometimes within population when repeatedly estimated at different times. For example, if we consider V_A in tarsus length measured in 11 species of small passerine birds with a mean tarsus length ranging from 15 to 25 mm (43 estimates), the among-species coefficient of variation (CV) of the additive genetic standard

1. References to chapter numbers in the text are to chapters in this volume.

deviations ($\sqrt{V_A}$) is 0.41, with values ranging from 0 in several populations to 1.01 in *Geospiza conirostris* (Grant 1983), while the CV of the total phenotypic standard deviations is 0.33, and the CV of the means is only 0.12. A considerable part of this variation results from the difficulty of estimating V_A and the large error variance generally associated with these estimates (Sztepanacz and Blows 2017; Pélabon et al. 2021), but we also expect selection, drift, gene flow, and environmental variation to affect V_A. The effects of these factors on V_A are not always understood, however, and it is rarely clear when they are necessary or whether they can explain differences between estimates given the estimation error. It also remains controversial whether, as suggested by some authors (e.g., Hoffmann and Parson 1997), rapid changes in evolvability due to some of these different factors could facilitate adaptation.

Under some assumptions, V_A of polygenic traits controlled by n diallelic loci with additive effects equals $n\overline{H}\,\overline{a^2}$, where $\overline{H}$ and $\overline{a^2}$ are respectively the average over all loci of the heterozygosity and the squared allelic effect size expressed as the deviance from the trait mean (Hansen and Pélabon 2021). Thus, V_A is primarily affected at short timescales by changes in allele frequency and/or changes in allelic effect size, whereas changes in the number of loci are happening at longer timescales. Because allelic variation is maintained by mutation-selection balance, fluctuating selection, and/or local adaptation and migration (Lande 1975, 1992; Sasaki and Ellner 1997; Bürger 2000), we expect changes in patterns of selection, gene flow, and genetic drift (the random sampling of alleles at each generation), to affect allele frequency and thus V_A. Additionally, V_A will be affected by changes in linkage disequilibrium, the nonrandom association of alleles among loci due to selection and gene flow (Bulmer 1971, 1980). Environmental changes can also modify allelic effects when there are gene-by-environment (G×E) interactions (DeWitt and Scheiner 2004). Similarly, changes in the genetic background due to changes in allele frequency can also alter allelic effects in the presence of epistatic interactions among loci. In particular, directional epistasis (i.e., when allelic effect sizes are consistently correlated with the trait mean) can generate systematic changes in V_A when allele frequencies are changing (figure 13.1A; Carter et al. 2005, Hansen 2013; Hansen and Wagner, chapter 7). Although directional epistasis has received relatively little attention outside theoretical studies, it may be common and biologically important.

In this chapter, we consider how, and to what extent, population size, gene flow, selection, and environmental variation affect evolvability on short ecological timescales, that is, over 1 to 100 generations. First, we summarize theoretical predictions concerning the effects that each of these factors can produce on evolvability via changes in allele frequency, allelic effect size, or linkage disequilibrium. We then present empirical evidence of rapid changes in evolvability drawn from experimental work and studies of natural populations. In natural populations, changes in evolvability are often studied with respect to specific ecological contexts that combine different mechanisms (see table 13.1). We thus discuss whether changes in evolvability observed under these circumstances can be explained by the effect of our four factors.

Whether an increase in V_A facilitates adaptation depends on the correspondence between the direction of selection and patterns of univariate and multivariate genetic variance summarized by the G-matrix (figure 13.1B). Additionally, asymmetry in the distribution of the

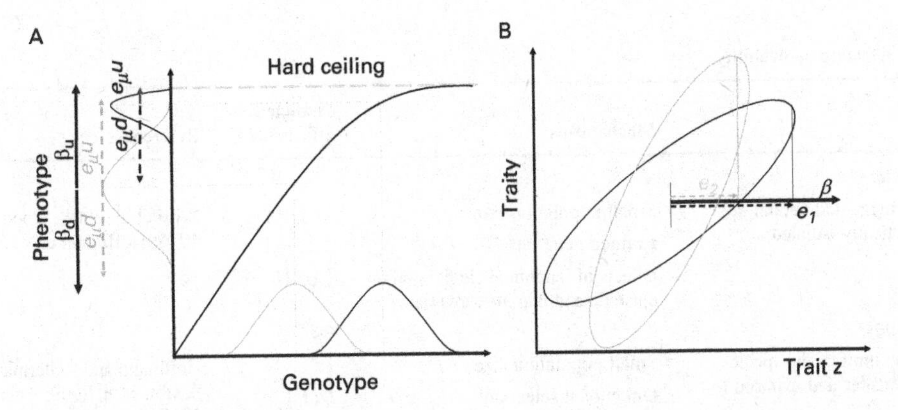

Figure 13.1
Changes in evolvability due to (A) directional epistasis and (B) changes in the genetic variance-covariance matrix, **G**. In both panels, solid arrows (β) represent selection and dashed arrows (e_μ) represent mean-scaled evolvability. (A) A nonlinear genotype-phenotype map generated by the presence of a hard ceiling causes changes in allelic effect size when the genetic background changes. Consequently, a similar genotypic variation translates into different phenotypic variations as the trait mean changes. This generates negative epistasis, where V_A decreases when the trait mean increases (from the gray to the black distribution). It also generates asymmetry in evolvability. When the trait mean approaches the hard ceiling, evolvabilities to increase ($e_\mu u$) or decrease ($e_\mu d$) the mean differ. (B) Changes in the orientation of **G** due to changes in the relative variance and covariance in traits z and y affects evolvability in the direction of the selection gradient β (for details, see Hansen, chapter 5). When the alignment between selection gradient and the direction of high multivariate genetic variance decreases (from the black to the gray ellipse), evolvability is reduced from e_1 to e_2. Evolvability will also change when the total size of **G**, that is, the total additive variance, changes. Finally, a change in the correlation between the 2 traits will affect the shape of **G**; an increase in the correlation will increase the ellipse eccentricity, also referred to as the integration of **G**. When integration increases, the ability of trait z to respond to selection when trait y is maintained constant, that is, the conditional evolvability of z, decreases (see figure 5.4 in Hansen, chapter 5).

breeding values generated, for example by a nonlinear genotype-phenotype (GP) map, could generate asymmetry in response to selection in opposite directions (figure 13.1A). While presenting empirical results, we thus discuss whether changes in evolvability solely affect the variance on which selection can act (i.e., *selectability*) or facilitate adaptation in the direction of selection (i.e., increase *adaptability*; see Hansen, chapter 5, this volume).

Many studies have tested the effects of ecological factors on the evolutionary potential. Their results, however, often have been difficult to interpret, because changes were expressed as changes in heritability that confound changes in other sources of phenotypic variance (e.g., nonadditive genetic variance, environmental variance, or developmental instability). Here, we provide new insights into these questions by focusing on mean-scaled evolvability ($e_\mu = V_A/\mu^2$). Mean-scaled evolvability offers a useful metric to compare changes in V_A when other components of the phenotypic variance are changing simultaneously or when trait means strongly differ (Hansen et al. 2011; Houle and Pélabon, chapter 6). In this chapter, we use the term *evolvability* as a synonym for additive genetic variance, while the term *mean-scaled evolvability* designates the mean-standardized measure of V_A.

Table 13.1
Ecological factors affecting evolvability

Ecological factor	Mechanisms	Predicted effects on V_A	References
Habitat fragmentation*			
– Populations in fragmented habitat are smaller and genetically isolated	Small population size	↓	Willi et al. 2006; Wood et al. 2016; Hoffmann et al. 2017
	Limited gene flow	↓	
	Effects of random allele fixation on epistatic and dominance variance	↑ or ↓	
Marginal populations*			
– Populations at the limit of the species distribution are smaller and exposed to harsher environments	Small population size	↓	Hoffmann and Kellermann 2006; Sexton et al. 2009; Angert et al. 2020; Pennington et al. 2021
– The lack of genetic variation limits population expansion	Directional selection†	↑ or ↓	
→ marginal populations have less genetic variation	Limited gene flow	↓	
	Unidirectional net gene flow	↑	
Extreme, novel or stressful environments*			
– Extreme environments generate a release in cryptic genetic variation	Release of cryptic genetic variation	↑	Hoffmann and Parsons 1997; Hoffmann and Merilä 1999; Rowiński and Rogell 2017
– Directional selection is stronger in extreme environment	G×E interaction‡	↓ or ↑	
	Directional selection†	↓ or ↑	
Specialist versus generalist*			
– Specialists are exposed to stabilizing selection while generalists are exposed to fluctuating selection	Stabilizing selection in specialists	↓	Kassen 2002; Martinossi-Allibert et al. 2017
	Fluctuating selection in generalists	↑	
Niche construction			
– Selection pressures decrease with niche construction	Relaxed selection	↑	Donohue 2005; Laland and Sterelny 2006
– Niche construction generates environmental changes	G×E interaction‡	↓ or ↑	
Pace of life			
– Species with fast life-histories are exposed to more fluctuating selection from one generation to the next	Fluctuating selection	↓ in slow sp.	Wright et al. 2019
– Overlapping generations combined with fluctuating selection maintain more V_A	Overlapping generation + Fluctuating selection	↑ in slow sp.	Ellner and Hairston 1994
Invasive species*			
– Invasive species have small population size when introduced into new environments	Small population size	↓	Ellstrand and Schierenbeck. 2000; Lee 2002; Lee and Gelembiuk 2008
– Introgression among divergent populations after introduction increases genetic variation	Genetic introgression	↑	
Adaptive introgression*			
– Hybridization increases genetic variation or generates new phenotypes not observed in the parental populations	Transgressive segregation	↑	Anderson and Stebbins 1954; Hamilton and Miller 2016; Pfenning et al. 2016; Arnold and Kunte 2017
– Hybridization provides new beneficial alleles or genes	Decreasing genetic integration	↑ or ↓	
	Capture of key beneficial alleles	↑	

Note: For each factor, the table lists the different mechanisms suggested to affect V_A and the predicted direction of these effects (arrows).

* Ecological factors considered in this chapter.

† The direction of the effect on V_A depends on the direction of epistasis (see text).

‡ The direction of the effect on V_A depends on the presence of directional epistasis and the effect of plasticity on the trait mean.

13.2 Summary of Theoretical Expectations

13.2.1 Population Size

All else being equal, larger non-inbred populations facing novel selection are expected to be more evolvable than smaller ones for two reasons. First, the probability of occurrence of newly beneficial alleles that are normally present at low frequencies or appear via mutation increases with population size. Second, larger populations are less affected by genetic drift, which decreases V_A and hence the size of **G** at a rate of $1/(2N_e)$ per generation, where N_e is the effective population size (Lande 1976, 1979).

Several authors also argued that random fixation of alleles with dominance and epistatic effects during bottleneck events (i.e., rapid reduction of the population size for a few generations) could increase V_A (Willis and Orr 1993; Cheverud and Routman 1996; Barton and Turelli 2004; Turelli and Barton 2006). Exact predictions, however, rely on details of the genetic architecture, such as the number of genes affecting the traits, the distribution of allelic effects, or patterns of pleiotropy, dominance, and epistasis (Barton and Turelli 2004; Turelli and Barton 2006), and Hansen and Wagner (2001) showed that V_A could either increase or decrease following changes in allele frequencies during bottlenecks, depending on the directionality of epistasis.

Drift-induced fixation of alleles with pleiotropic or epistatic effects, or alleles with large effects, may also change the orientation or integration of **G**. However, these changes are not expected to systematically modify the relative proportion of additive variance and covariance. Consequently, drift should not change the shape of **G** on average (Jones et al. 2003; Lopez-Fanjul et al. 2004, 2006).

13.2.2 Gene Flow

Gene flow counteracts the effect of genetic drift by reintroducing lost alleles or introducing new ones in the recipient populations (Lande 1992; Whitlock 1999). Additionally, gene flow between genetically differentiated populations will increase V_A by causing linkage disequilibrium and skewed distribution of genotypic values in the recipient population (Bulmer 1980; Tufto 2000). Recombination decreases linkage disequilibrium, and after several generations, the degree to which V_A in the recipient population exceeds its value before introgression of immigrant alleles depends on the among-individual variance in the proportion of the genome inherited from the original immigrants and native populations (V_q), and on the difference in mean breeding value between the two parental populations (*g*). Assuming similar V_A in the different populations, the total additive genetic variance for the trait of interest in the recipient population after introgression is $V_{At} = V_{An} + g^2V_q$, where V_{An} is the additive genetic variance before introgression (Reid and Arcese 2020; see also Muff et al. 2019). Consequently, V_A should rapidly increase following the establishment of new gene flow between populations adapted to different optima. Constant gene flow should maintain V_A at a high level with a skewed distribution of genotypic values. If gene flow ceases, recombination will remove linkage disequilibrium, and V_A will return to its initial level.

By increasing genetic variance in the multivariate direction of divergence between populations, gene flow can also affect evolvability and adaptability by changing the orientation of **G** (Nei and Li 1973; Guillaume and Whitlock 2007). Depending on the alignment between **G** of the recipient population and the direction of population divergence, gene flow may reinforce or weaken existing genetic correlations. If the off-diagonal elements of **G** (i.e., nonzero genetic covariances) mostly result from current patterns of selection and therefore linkage disequilibrium, new gene flow will rapidly affect those covariances, and the changes may be observed for many generations if the recombination rate is low (e.g., in the presence of physical linkage). With higher recombination rates, new covariances should rapidly come to an equilibrium between gene flow (building of linkage disequilibrium) and recombination, or they may rapidly return to their original value if gene flow stops (Nei and Li 1973). The simulation study from Guillaume and Whitlock (2007) suggests, however, that genetic correlation due to physical linkage or pleiotropy should be much less affected by gene flow.

When genetic differentiation between populations increases, larger changes in V_A should follow from genetic introgression, but we should also expect maladaptation to increase due to the shift in the phenotypic mean of the recipient population toward the mean of the immigrants, a phenomenon referred to as migration load (Kirkpatrick and Barton 1997; Lenormand 2002). Additionally, nonadditive genetic effects—such as transgressive segregation, the appearance of novel phenotypes in hybrids not observed in the parental species (Rieseberg et al. 1999), or outbreeding depression due to epistasis (Dobzhansky-Muller incompatibilities)—may strongly affect genetic variation of the hybrids and their descendants, as well as their fitness (Orr and Turelli 2001).

13.2.3 Selection

For traits controlled by few loci with large effects, changes in V_A under directional selection mostly results from changes in allele frequency, but the direction of the changes depends on the initial allele frequency. Indeed, the increasing frequency or fixation of beneficial alleles already present in the population at high frequency will only generate a small reduction in V_A. In contrast, increasing frequency of beneficial alleles that are initially rare can dramatically increase V_A (Sorensen and Hill 1982), whereas changes toward low or high frequency of alleles normally present at intermediate frequency will decrease V_A.

For traits controlled by many loci with small effect size spread across the genome, changes in allele frequency due to selection are expected to be small, and changes in V_A should mostly result from changes in linkage disequilibrium (Bulmer 1971, 1980; Lande 1975). These changes are expected to be limited, however, because they accumulate at a rate proportional to the effect of selection on the phenotypic variance and to the heritability squared, h^4. Additionally, at each generation, recombination destroys half of whatever disequilibrium contribution to V_A has accumulated (Bulmer 1971). Even under strong selection, the decrease in V_A due to linkage disequilibrium is limited and only occurs when moving from one equilibrium (e.g., no selection) to the new equilibrium value under selection. Thus, if one studies a system under reasonably consistent directional selection, no ongoing reduction in V_A is expected to occur due to linkage disequilibrium. Even with fluctuating or intermittent selection, the effects of linkage disequilibrium on V_A will typically not be dramatic. Although such effects could be stronger with physical linkage, recent genomic analyses increasingly suggest that causal loci of quantitative traits are typically spread throughout the genome (Pitchers et al. 2019).

Selection on variance (stabilizing and disruptive selection) may generate larger changes in V_A due to linkage disequilibrium (a decrease with stabilizing selection and an increase with disruptive selection), provided that any concurrent directional selection is relatively weak (Bulmer 1971, 1980; Sorensen and Hill 1983). For disruptive selection, however, these effects strongly depend on the type of mating (random, assortative, or disassortative; Thoday 1972).

The shape of **G** depends on both pleiotropic mutations and correlational selection (selection on trait combination) that generates linkage disequilibrium (Lande 1980; Turelli 1985; Arnold et al. 2008; Chantepie and Chevin 2020). On short timescales, changes in the direction of selection should affect the shape and the orientation of **G** mostly via changes in linkage disequilibrium, although changes in the frequency of alleles with large effects may also affect the shape of **G** (Agrawal et al. 2001). If covariance among traits essentially results from linkage disequilibrium, a decrease in evolvability due to changes in the fitness landscape will be transient, and evolvability should progressively be restored as recombination breaks down non-adaptive linkages while selection builds new adaptive ones. In contrast, genetic correlations resulting from pleiotropy or physical linkage may generate long-term changes in evolvability when the direction of selection changes.

The above results derive from models that assume additive allelic effects. With additivity, an allele substitution always has the same effect, regardless of the genetic background, and changes in the trait mean are not accompanied by changes in V_A. With directional epistasis, however, selection on the trait mean will increase V_A when it is in the direction of positive epistasis and decrease V_A in the opposite direction (figure 13.1A; Carter et al. 2005; Hansen et al. 2006; Hansen 2013).

The occurrence of directional epistasis also affects the predictions concerning the evolution of genetic canalization (i.e., the reduced sensitivity of a genotype to allelic changes; Flatt 2005). The original theory suggests that genetic canalization, and thus a reduction of V_A, should result from stabilizing selection (Waddington 1957; Scharloo 1991). However, models have shown instead that only stabilizing selection of intermediate strength should favor genetic canalization, because the epistatic interactions necessary for canalization require genetic variation at loci controlling the trait, and because reduced allelic effects during canalization decrease the strength of stabilizing selection (Wagner et al. 1997; Le Rouzic et al. 2013). This further suggests that selection on the trait mean in direction of negative epistasis may be more efficient than stabilizing selection to generate genetic canalization. However, if genetic canalization can benefit a population under stabilizing selection by increasing its adaptive precision (sensu Pélabon et al. 2012), genetic canalization resulting from selection in the direction of negative epistasis may decrease the population adaptability and compromise its fitness if selection persists.

Finally, epistasis also offers a mechanism for pleiotropic effects to evolve under selection, when changes in allelic effect on one trait modify effects of alleles controlling other traits (Wagner and Mezey 2000; see Pavličev and Cheverud 2015, for a review).

13.2.4 Environmental Variation

Populations exposed to extreme environments often show novel phenotypes whose frequency can be increased via artificial selection (Waddington 1953). These observations have been interpreted as the release of cryptic genetic variation from environmental canalization under extreme environmental changes (Scharloo 1991; Flatt 2005; Paaby and Rockman 2014). Therefore, the classical prediction has been that genetic variation should increase in novel or stressful environments.

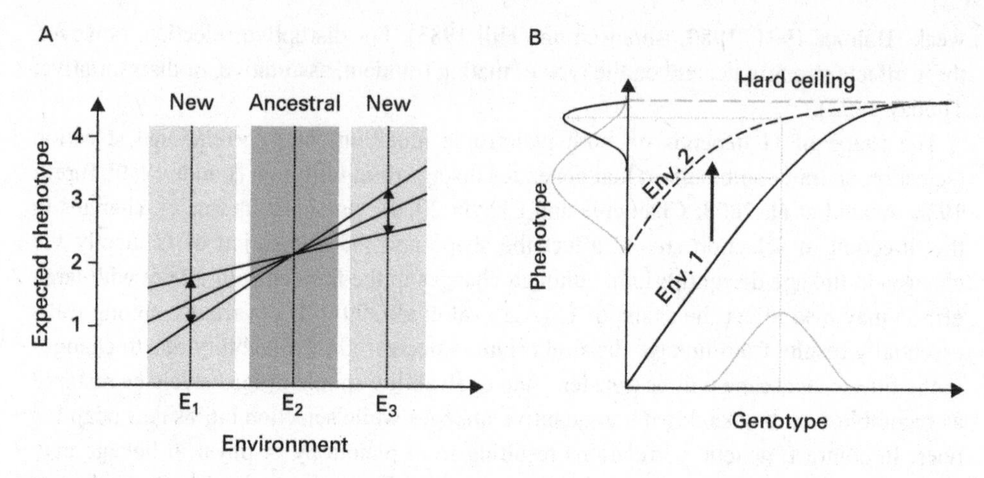

Figure 13.2
Changes in evolvability with environmental variation due to G×E interaction (in panel A) and a nonlinear GP map (in panel B). (A) The G×E interaction entails genetic differences in the slope of the reaction norms (3 different genotypes are represented here). Stabilizing selection in the ancestral environment E_2 generates convergence of the reaction norms and a reduction of V_A. In novel environments E_1 or E_3, V_A increases (black double-head arrows) due to differences in the slope of the reaction norms (Lande 2009). (B) Phenotypic plasticity (back arrow) associated with a nonlinear GP map causes a change in the trait mean from environment 1 to environment 2 that affects V_A (distributions along the y axis).

Accordingly, Hermisson and Wagner (2004) showed that G×E interactions could modify allelic effects in different environments, generating differences in short-term evolvability similar to those generated by epistasis. Lande's (2009) model of genetic accommodation also predicts an increase in genetic variance and evolvability when populations are exposed to novel environments (figure 13.2A).

Similar models extended to two traits suggest that environmental variation can affect genetic covariance when the environment of maximum canalization differs among traits (De Jong 1990; Stearns et al. 1991). Thus, environmental changes are expected to affect the shape of **G**, but contrary to the positive effect on V_A expected for univariate traits, effects on **G** are unpredictable. Furthermore, even when **G** remains constant, environmental changes generating a reorientation of the fitness landscape could affect multivariate evolvability.

13.3 Empirical Evidence

13.3.1 Population Size

Many experiments confirm that population size positively affects the response to artificial selection (Frankham et al. 1968, 1999; Jones et al. 1968; Eisen 1975; Weber 1990). For example, in the study by Weber (1990), the mean-scaled selection response after 55 generations of selection was 2.3 times larger in populations with 1,000 parents than in populations with 40 parents (21% versus 9% change in the trait mean). Similarly, evolutionary rescue studies generally show that the probability for populations to evolve and persist in new environments depends on their initial size (Bell and Gonzales 2009; Ramsayer et al. 2013).

In contrast, experimental evidence for a negative effect of drift on V_A has been less conclusive, possibly because the populations used were rarely at equilibrium and produced noisy results (Houle 1989). Still, in a large-scale experiment on *Drosophila,* where 52 lines experienced one generation of full-sib mating followed by two generations at large population size (inbreeding coefficient $F = 0.25 - 0.32$), the V_A values of six wing-size and shape traits decreased on average by 32%, and the environmental variance increased by 11% compared to three large populations ($N > 1{,}000$, $F \approx 0.001$; Whitlock and Fowler 1999).

With habitat fragmentation, the decrease in evolutionary potential of populations has been a reason for concern, but studies of natural populations have provided little support for a decrease in evolvability in small populations (reviewed in Willi et al. 2006; Wood et al. 2016). Inferences, however, are mostly based on correlations between population size and heritabilities, which do not reliably reflect V_A (Houle 1992; Hansen et al. 2011). Still, the few studies reporting the necessary statistics to estimate e_μ (Widen and Andersson 1993; Podolsky 2001; Oakley 2015 on plants; Wood et al. 2015 on fish) confirm the absence of marked effects of population size on evolvability in natural populations. Willi et al. (2006) argued that the small number of generations between estimates in bottlenecked populations, together with gene flow, may explain the general absence of effects. Indeed, the negative effect of drift on V_A is a slow process, except for very small populations, and a decrease in V_A of 10% due solely to drift still takes approximately 20 generations when $N_e = 100$.

Predictions of an increase in V_A following bottleneck events have received inconsistent support (Bryant and Meffert 1988, 1995; Cheverud et al. 1999; Whitlock and Fowler 1999; Phillips et al. 2001). Furthermore, in a large-scale experiment on *Drosophila bunnanda,* van Heerwaarden et al. (2008) showed that despite an average increase in e_μ of desiccation resistance from 0.2% in the control lines to 0.7% in the lines that experienced bottleneck, the response to selection to increase desiccation resistance was not enhanced, suggesting that the increase in V_A did not increase adaptability.

Finally, experimental studies on *Drosophila* indicate that changes in the shape of **G** following bottleneck events are sometimes important and long-lasting, but they do not show systematic patterns (Phillips et al. 2001; Whitlock et al. 2002), as expected from the theory.

13.3.2 Gene Flow

The positive effect of gene flow on evolvability has been experimentally confirmed by Swindell and Bouzat (2006), who compared selection responses of *Drosophila* populations with an $N_e \approx 14$ flies either maintained in isolation for 40 generations before selection or reconnected by gene flow of 1 male or 1 female per generation, for 3 generations before selection. Gene flow increased V_A by about 37%, and the response to selection on the trait by 29%.

In natural populations, the absence of a negative relationship between V_A and population size suggests that gene flow often counteracts the effects of drift (Willi et al. 2006), but difficulties in estimating gene flow have limited our ability to quantify this effect. In a recent study on the song sparrows (*Melospiza melodia*) of Mandarte Island, Reid et al. (2021) estimated the different parameters necessary to quantify the effects of gene flow on V_A. They showed that gene flow from immigrants increases V_A in juvenile survival by 10–40%. However, they also showed that juvenile survival decreases with an increasing proportion of the genome inherited from immigrant ancestors. Gene flow also increases evolvability of the liability for extra-pair paternity in the island population by increasing V_A for female extra-pair reproduction by 10% and V_A for male paternity loss by 40%. The evolvability of the mating system is further affected by a decrease of the cross-sex genetic correlation between the two traits (Reid and Arcese 2020).

In the walking stick (*Timema cristinae*), gene flow between populations inhabiting different environments generates new genetic covariances between host preference and color pattern that compromise adaptation by rendering individuals less cryptic (Nosil et al. 2006; Bolnick and Nosil 2007). These results illustrate how changes in **G** resulting from gene flow can be detrimental for local adaptation. In some cases, however, gene flow can favor adaptation to a new environment, for example, when trees in northern latitudes exposed to warmer environments due to climate change receive advantageous alleles via long-distance pollination from southern populations (Kremer et al. 2012). Overall, the effect of gene flow on adaptability will depend on the orientation and the steepness of the environmental gradient relative to the net gene flow between divergent populations.

Several authors have suggested that genetic introgression between diversified genotypes during multiple introductions improves evolvability of invasive species (Kolbe et al. 2004; Lee and Gelembiuk 2008). This effect may also result from the release of cryptic genetic variation generated by admixture of alleles with nonadditive effects (Dlugosh et al. 2015). This hypothesis is partly supported by the study of the grass *Phalaris arundinacea,* where introgression between distantly related genotypes introduced to North America increased broad-sense heritability (total genetic variance/phenotypic variance) in two out of the eight traits studied (Lavergne and Molofsky 2007). It remains unclear, however, how much the success of this invasive species is due to an increase in genetic variance and adaptability.

By introducing potentially beneficial alleles in the genome of a focal species, introgressive species hybridization may increase evolvability and adaptability (Anderson and Stebbins 1954; Hamilton and Miller 2016). For example, populations of the Gulf killifish (*Fundulus grandis*) in Galveston Bay recently acquired resistance to hydrocarbon pollution by introgression with the Atlantic killifish (*F. heteroclitus;* Oziolor et al. 2019). Similarly, in the *Anopheles gambiae* complex, molecular analyses revealed that the presence in several species of the chromosomal inversion 2La associated with survival in arid environments is possibly due to recent introgression (Sharakhov et al. 2006).

Interspecific hybridization may also allow species to expand their range or invade new habitats when there is transgressive segregation (Rieseberg et al. 1999; Pfenning et al. 2016), or when hybridization affects selectability by increasing phenotypic variation or decreasing phenotypic covariance among traits (Parsons et al. 2011; Selz et al. 2014; Lucek et al. 2017). Quantifying and interpreting these effects in terms of evolvability may prove difficult, however. First, it is challenging to quantify changes in evolvability resulting from the appearance of novel traits that require new measurement methods. In such cases, only qualitative assessment may be possible. Second, analyses are often restricted to the F1 and F2 hybrids, thus limiting our ability to infer how much of the variation mostly due to nonadditive effects in early generations eventually translates into "potentially adaptive" genetic variation. Finally, changes in the phenotypic variance are often idiosyncratic and eliminated from the data by variance standardization, rendering it difficult to assess whether variances generally increase during hybridization.

13.3.3 Selection

Artificial-selection experiments comparing genetic parameters before and after directional selection have provided inconsistent results concerning the effects of selection on V_A. Some studies suggest that neither linkage disequilibrium nor changes in allele frequency strongly affect V_A during the first ~10 generations (e.g., Atkins and Thompson 1986), while others report a rapid decrease in V_A. For example, in fewer than 5 generations of artificial selection, e_μ of bristle number in *Drosophila* decreased from 0.48% in the base line to 0.33% and 0.25% in the up- and down-selected lines, respectively (population size of the selected lines $N = 100$; Clayton et al. 1957). Unfortunately, imprecision of the estimates and stochasticity of the responses often limit inferences about changes in V_A that can be drawn from many of these short-term experiments.

Artificial-selection experiments longer than 10 generations more consistently report a decrease in evolvability. For example, 85 generations of selection to increase bristle number in *Drosophila* reduced e_μ from 0.91% in the base population to 0.22% and 0.52% in the 2 selected lines (Yoo 1980). The cause of these changes is unclear, however, because drift alone could have generated a 40% reduction in V_A if N_e was 100, which was the census size of the selected lines.

At much longer timescales, it has been suggested that sustained directional selection should deplete genetic variation due to allele fixation, and that evolutionary potential should vary according to the mode and strength of selection acting on different categories of traits. Contrary to the initial expectation, studies comparing mean-scaled evolvability across trait categories showed that life-history traits, supposedly under constant directional selection, harbor higher evolutionary potential than do morphological traits (Houle 1992; Wheelwright et al. 2014; Hansen and Pélabon 2021). These results do not directly inform about the effect of sustained directional selection on V_A, but they suggest, together with the correlation observed between additive and mutational variance across traits (Houle 1998; Houle et al. 2017), that mutational target size (the number of loci affecting a trait) is an important factor controlling V_A in a mutation-selection balance regime. More relevant here would be data about changes in V_A in traits that have been under stabilizing selection and then are suddenly subjected to directional selection for many generations. Unfortunately, we are not aware of such data for natural populations.

Kelly (2008) experimentally tested whether an increase in frequency of rare alleles with large effect increased V_A during artificial selection of corolla size in Monkeyflowers (*Mimulus*). Although V_A increased in the line selected to increase corolla size and decreased in the lines selected in the opposite direction, those changes are most likely explained by directional epistasis on the measurement scale. Indeed, although V_A differed by a factor of two between the up- and down-selected lines, this difference nearly vanished when expressed as mean-scaled evolvability (e_μ down-selected = 2.26%; e_μ up-selected = 2.46%).

More generally, when the variance does not increase proportionally to the mean squared on the measurement scale, we should expect directional epistasis and the evolution of evolvability under directional selection (Hansen and Wagner, chapter 7; G. Wagner chapter 10). An artificial-selection experiment on critical maximal thermal tolerance, CT_{max}, in zebra fish (*Danio rerio*) illustrates this (Morgan et al. 2020). In this experiment, the response of the up-selected lines was much weaker than that of the down-selected lines. The authors speculated that denaturation of proteins with temperature generates a hard ceiling for CT_{max} and consequently a nonlinear GP map, as illustrated in figure 13.1A. This hypothesized GP map represents a case of directional epistasis where an increase in mean CT_{max} systematically decreases allelic effect sizes and V_A, while a decrease in mean CT_{max} has the opposite effect. Accordingly, phenotypic variance strongly increased in the down-selected lines, while it decreased in the up-selected lines (from 0.2°C² in the parental generation to 0.12°C² in the up-selected lines and 4.6°C² in the down-selected lines). A similar process could explain why the increase in e_μ observed after bottleneck events in the study by van Heerwaarden et al. (2008) did not enhance the response to selection for increasing desiccation resistance.

Few experiments have tested the effect of stabilizing selection on V_A. Pélabon et al. (2010) applied different types of artificial selection to two shape traits of the *Drosophila* wing and showed that stabilizing and fluctuating selection had very little effect on the phenotypic variance, possibly because the selected traits were already under stabilizing selection before the start of the experiment. In contrast, disruptive selection increased phenotypic variance, doubling the coefficient of phenotypic variation of the selected traits. Although this increase in variance could have resulted from the accumulation of linkage disequilibrium, a more likely explanation is that under strong disruptive selection, only offspring of mated pairs of low × low and high × high phenotype parents had a substantial chance of producing selected offspring, similar to a "kill-the-hybrids" speciation experiment (Rice and Hostert 1993), resulting in two populations with substantial differences in allele frequency. However, the effect of linkage disequilibrium on V_A during disruptive selection is suggested in an experiment by Sorensen and Hill (1983), who showed a temporal increase in heritability during three generations of disruptive selection followed by a return to the original value when selection was relaxed. Sztepanacz and Blows (2017) imposed disruptive selection on two multivariate phenotypes with high and low additive genetic variation; they found that phenotypic variance decreased in the trait with high V_A and increased in the trait with low V_A, possibly because of differences in genetic architecture between these multivariate phenotypes. These results suggest that changes in V_A under disruptive selection are unpredictable, and it also remains unclear how much the changes in phenotypic variance and heritability observed in some of these experiments reflect changes in V_A, because disruptive selection can also affect environmental and developmental variance (Halliburton and Gall 1981; Pélabon et al. 2010).

In contrast, the prediction of higher evolvability maintained by fluctuating selection has been repeatedly confirmed by experimental evolution studies where microorganisms grown in spatially or temporally heterogenous environments maintained higher genetic variation than those grown in homogenous environments (see Kassen 2002 for a review). Martinossi-Allibert et al. (2017) tested this hypothesis in natural populations by comparing evolvability between generalist and specialist bird species. Using long-term studies of passerine birds that provided data on environmental variation and quantitative genetic parameters, they tested whether specialist species experiencing supposedly more homogenous habitat were less evolvable than generalist species exposed to more fluctuating environments. Although generalists tended to have higher levels of heterozygosity at microsatellite loci, no effect of habitat specialization on evolvability was detected.

Several artificial-selection or experimental evolution studies show that **G** can evolve rapidly, but the observed changes are often idiosyncratic and not correlated with putative or experimental patterns of selection. Furthermore, the outcomes of these experiments often depend on the type of traits selected. For example, genetic (or phenotypic) correlations among life-history traits are generally altered by artificial selection perpendicular to the direction of correlation (Stanton and Young 1994; Delph et al. 2011; Steven et al. 2020). In contrast, artificial selection on morphological allometry showed that some correlations are particularly difficult to alter (Egset et al. 2012; Bolstad et al. 2015). Consequently, whether and how pleiotropy can evolve in response to selection for particular functional association among traits is still uncertain. Pavličev et al. (2008, 2011) identified regulatory quantitative trait loci (rQTL) that modify pleiotropic effects of other loci and change genetic covariance depending on the direction of selection. The effects of these changes of covariance on evolvability are unclear, however, because the effects of rQTL or other genetic mechanisms changing pleiotropic interactions on the total genetic variance are generally not considered.

Empirical evidence thus confirms the wide range of effects that selection can have on evolvability, depending on the genetic architecture of the traits. By fitting phenomenological models with different genetic architectures to time-series from an artificial selection on wing shape in *Drosophila*, Le Rouzic et al. (2011) showed that the best model included up to nine parameters (e.g., drift, segregation of large effect alleles, and epistasis). Unfortunately, the limited knowledge we have of the genetic architecture of most quantitative traits often limits our interpretation of observed changes in V_A. Still, observations of asymmetrical response to bidirectional selection for life-history traits (Frankham 1990) or key physiological traits (e.g., maximal thermal tolerance; Gerken et al. 2016; Morgan et al. 2020) suggest that directional epistasis may be common and can affect evolvability on short timescales under directional selection. Accordingly, negative directional epistasis could explain the decrease in V_A sometimes observed for key ecological traits in marginal populations exposed to extreme environments (Pujol and Pannell 2008; Kellermann et al. 2009; van Heerwaarden et al. 2008). This explanation contrasts with the classical one involving changes in allele frequency due to factors such as small population size, limited gene flow, and constant directional selection (reviewed in Hoffmann and Kellermann 2006; Angert et al. 2020). Still, the recent study by Pennington et al. (2021) comparing changes in mean-scaled evolvability along geographic or niche gradients for 38 species showed that evolvability tends to decrease in marginal (isolated) populations but slightly increases in more extreme environments (marginal niche). Although these results weakly support the gene-flow-population-size hypothesis, they illustrate the difficulties of predicting the effects of ecological factors on evolvability when these factors involve different mechanisms with conflicting effects on evolvability.

13.3.4 Environmental Variation

Many experimental studies have documented the effects of extreme environments on qualitative genetic variation, such as missing wing veins, deformed or absent eyes in *Drosophila*, or changes in symmetry patterns in *Arabidopsis* (Waddington 1953; Scharloo 1991; Rutherford and Lindquist 1998; Queitsch et al. 2002). Although the increase in genetic variation observed in these experiments is beyond doubt, the extent to which this new variation could help future adaptation has been questioned (Wagner et al. 1999), and it is difficult to interpret the consequences of these qualitative changes on evolvability and adaptability. However, quantitative genetic studies that have tested the effects of stressful environments on evolutionary potential (for a review, see Hoffmann and Merilä 1999), have provided ambiguous results because their analyses were of heritabilities rather than evolvabilities.

We compiled data from experimental studies that specifically tested the effect of environmental stress on V_A and for which e_μ could be computed. Despite considerable scatter (figure 13.3), evolvability estimates tend to be larger in more stressful environments. On average, e_μ increases under stressful conditions by 53%±16% for morphological traits ($n=72$) and by 12%±32% for nonmorphological traits ($n=32$). Contrary to Rowiński and Rogell (2017), we did not observe a larger change in life history traits. The percentage difference in e_μ was calculated as log(e_μ *stress*/e_μ *control*), and standard errors were estimated by nonparametric bootstrap. Removing outliers with differences above 400% in absolute value yielded an increase of e_μ in stressful environment by 37%±12% for morphological traits and 35%±21% for nonmorphological traits. Although these results partly support the hypothesis of an increase in evolvability under stressful conditions, they should be tempered by the fact that imprecise estimates render most direct comparisons statistically nonsignificant, as exemplified by the recent review of the effect of thermal stress on genetic variation (Fischer et al. 2021). Furthermore, evidence for an increase in V_A in stressful environments is often based on traits with little relevance for population persistence in those stressful or novel environments, leaving open the question concerning the adaptive significance of those changes in evolvability.

Studies of thermal tolerance offer possible examples of more predictable changes in evolvability induced by environmental variation. In the artificial selection experiment on CT_{max} in zebra fish by Morgan et al. (2020), the response of the up-selected lines vanished when fish were acclimated to high temperature prior to selection. These results were interpreted as an additional consequence of the hard ceiling in CT_{max} that generated a change in phenotypic variance when mean CT_{max} increased toward the maximum value due to acclimation to higher temperature (figure 13.2B). The decrease in evolvability with acclimation to higher temperature was confirmed by the lack of response to selection toward higher CT_{max} in the acclimated lines. These results parallel those by Mitchell and Hoffmann (2010), who observed a decrease in e_μ in CT_{max} when two populations of *D. melanogaster* were exposed to ramping temperature (28 to 38°C at 0.06°C min^{-1}) instead of static increase (10°C min^{-1}). Mean-scaled evolvability decreased from 0.14% to 0.02% in the first replicate, and it entirely vanished (from 1% to 4×10^{-6}%) in the second replicate.

Charmantier and Garant (2005) reviewed studies that compared evolutionary potential between natural populations inhabiting favorable and unfavorable environments or between favorable and unfavorable periods for a single population. Based on the lack of consistent changes in heritability, they concluded that there was no clear evidence for an increase in evolutionary potential with a decrease in environmental quality. Martínez-Padilla et al. (2017) tested this hypothesis anew using mean-scaled evolvability and a quantification of habitat favorability based on species distribution and habitat characteristics of European wild bird populations. They showed that evolvability decreases in both highly favorable and highly non-favorable habitats. These results are questionable, however, because they

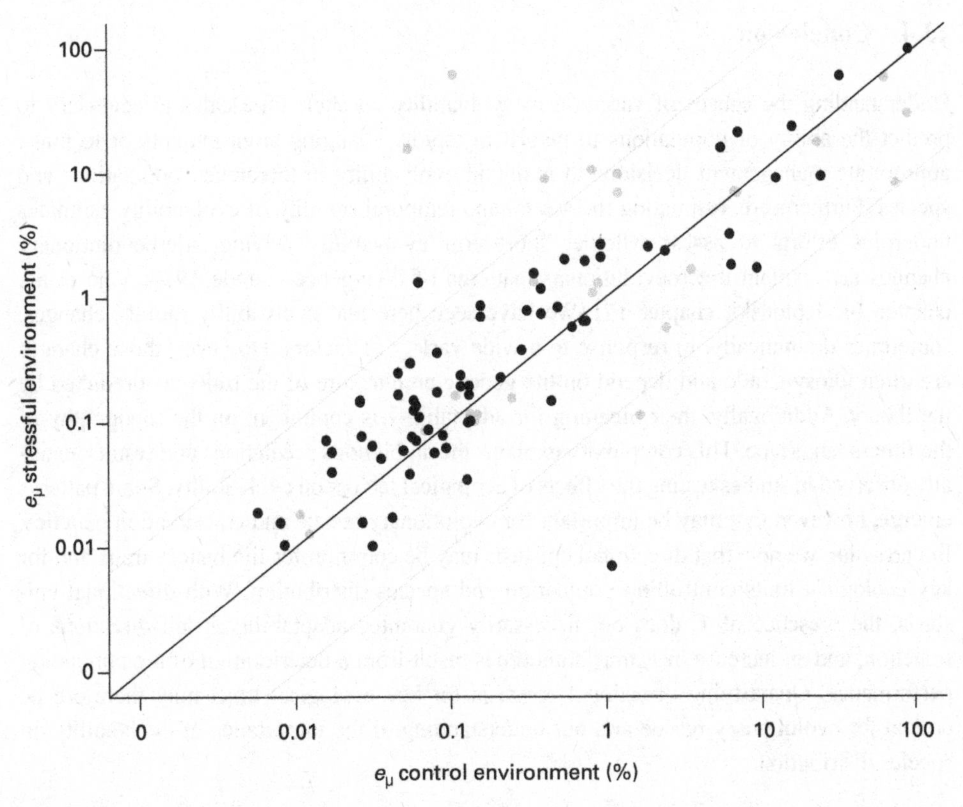

Figure 13.3
Effect of stress on mean-scaled evolvability (e_μ). Data are from experiments estimating V_A for individuals of the same population exposed to benign (control) or stressful environments (black dots: morphological traits $N = 72$, gray dots: nonmorphological traits $N = 32$). The solid line represents a 1:1 relationship and the 2 gray lines represent a 10-fold increase (up) or decrease (down) of e_μ under stress (supplementary material is available online for details about the studies: https://mitpress.mit.edu/9780262545624/evolvability/). Evolvability less than 0.001% in one of the environments is not presented.

depend on the uneven distribution along the favorability gradient of 17 estimates of body mass with high evolvability, while the other 110 estimates are linear morphological traits (for the effect of trait dimensionality on mean-scaled variance, see Lande 1977; Houle 1992; Pélabon et al. 2020). A reanalysis of the data including only linear traits shows no relationship between evolvability and environmental favorability. Furthermore, a recent analysis of the adaptive potential across species' geographic and niche ranges also reports a weak positive relationship ($r^2 < 0.1\%$) between mean-scaled evolvability and the distance from the niche center (Pennington et al. 2021).

Comparing the sizes and shapes of 61 pairs of G matrices estimated in different environments, Wood and Brodie (2015) showed that environment weakly affected integration, but it strongly affected the orientation of **G**, as well as the size, which changed by up to 200%. These differences, however, were not more pronounced when one of the environments was novel or when the effect of the environment on trait means increased. Although these results confirm the unpredictability of environmentally induced changes in **G**, they remain difficult to interpret in terms of evolvability due to the lack of standardization of the genetic variance and the absence of standard errors associated with the estimates.

13.4 Conclusion

Understanding the causes of variation in evolvability on short timescales is necessary to predict the ability of populations to persist in rapidly changing environments or to make appropriate management decisions to maintain evolvability in threatened populations and species. Furthermore, estimating the spatial and temporal stability of evolvability estimates underpins efforts to assess whether short-term evolvability driving microevolutionary changes can explain macroevolutionary patterns of divergence (Lande 1979; Voje et al., chapter 14; Jablonski, chapter 17). We have seen here that evolvability rapidly changes, sometimes dramatically, in response to a wide variety of factors. However, those changes are often idiosyncratic and depend on the genetic architecture of the traits, as predicted by the theory. Additionally, their meaning for adaptability is contingent on the topography of the fitness landscape. This complexity explains the ambiguous predictions and results generally observed in studies testing the effects of ecological factors on evolvability. Some patterns emerge, however, that may be important for evolutionary rescue and conservation genetics. In particular, we note that directional epistasis may be common for life history traits and for key ecological traits controlling population and species distribution. With directional epistasis, the presence of V_A does not necessarily guarantee adaptability in all directions of selection, and an increase in V_A may sometimes result from a deterioration of the population performance. Quantifying directional epistasis for key ecological traits may therefore be crucial for evolutionary rescue and our understanding of the importance of evolvability on species distribution.

Acknowledgments

We thank the Centre for Advanced Study (CAS) in Oslo for supporting us during the preparation of this manuscript. We also thank T. F. Hansen, D. Houle, G. H. Bolstad, S. Orzak, F. Galis, and the participants of the Evolvability Journal Club for discussions and comments on the manuscript. CP is supported by Norwegian Research Council grant 287214, and JMR is supported by the Research Council of Norway through its Centre of Excellence funding scheme, project 223257.

References

Agrawal, A. F., E. D. Brodie, and L. H. Rieseberg. 2001. Possible consequences of genes of major effect: Transient changes in the G-matrix. *Genetica* 112: 33–43.

Anderson, E., and G. L. Stebbins. 1954. Hybridization as an evolutionary stimulus. *Evolution* 8: 378–388.

Angert, A. L., M. G. Bontrager, and J. Ågren. 2020. What do we really know about adaptation at range edges? *AREES* 51: 341–361.

Arnold, M. L., and K. Kunte. 2017. Adaptive genetic exchange: A tangled history of admixture and evolutionary innovation. *TREE* 32: 601–611.

Arnold, S. J., R. Bürger, P. A. Hohenlohe, B. C. Ajie, and A. G. Jones. 2008. Understanding the evolution and stability of the G-matrix. *Evolution* 62: 2451–2461.

Atkins, K. D., and R. Thompson. 1986. Predicted and realized responses to selection for an index of bone length and body weight in Scottish Blackface sheep 1. Responses in the index and component traits. *Animal Science* 43: 421–435.

Barton, N. H., and M. Turelli. 2004. Effects of genetic drift on variance components under a general model of epistasis. *Evolution* 58: 2111–2132.

Bell, G., and A. Gonzalez. 2009. Evolutionary rescue can prevent extinction following environmental change. *Ecology letters* 12: 942–948.

Bolnick, D. I., and P. Nosil. 2007. Natural selection in populations subject to a migration load. *Evolution* 61: 2229–2243.

Bolstad, G. H., J. A. Cassara, E. Márquez, T. F. Hansen, K. van der Linde, D. Houle, and C. Pélabon. 2015. Complex constraints on allometry revealed by artificial selection on the wing of *Drosophila melanogaster. PNAS* 112: 13284–13289.

Bulmer, M. G. 1971. The effect of selection on genetic variability. *American Naturalist* 105: 201–211.

Bulmer, M. G. 1980. *The Mathematical Theory of Quantitative Genetics.* Oxford: Clarendon Press.

Bürger, R. 2000. *The Mathematical Theory of Selection, Recombination, and Mutation.* New York: Wiley.

Bryant, E. H., and L. M. Meffert. 1988. Effect of an experimental bottleneck on morphological integration in the housefly. *Evolution* 42: 698–707.

Bryant, E. H., and L. M. Meffert. 1995. An analysis of selectional response in relation to a population bottleneck. *Evolution* 49: 626–634.

Carter, A. J. R., J. Hermisson, and T. F. Hansen. 2005. The role of epistatic gene interactions in the response to selection and the evolution of evolvability. *Theoretical Population Biology* 68: 179–196.

Chantepie, S., and L. M. Chevin. 2020. How does the strength of selection influence genetic correlations? *Evolution Letters* 4: 468–478.

Charmantier, A., and D. Garant. 2005. Environmental quality and evolutionary potential: Lessons from wild populations. *Proceeding of the Royal Society B* 272: 1415–1425.

Cheverud, J. M., and E. J. Routman. 1996. Epistasis as a source of increased additive genetic variance at population bottlenecks. *Evolution* 50: 1042–1051.

Cheverud, J. M., T. Ty, L. Vaughn, et al. 1999. Epistasis and the evolution of additive genetic variance in populations that pass through a bottleneck. *Evolution* 53: 1009–1018.

Clayton, G. A., J. A. Morris, and A. Robertson. 1957. An experimental check on quantitative genetical theory I. Short-term responses to selection. *Journal of Genetics* 55: 131–151.

De Jong, G. 1990. Quantitative genetics of reaction norms. *Journal of Evolutionary Biology* 3: 447–468.

Delph, L. F., J. C. Steven, I. A. Anderson, C. R. Herlihy, and E. D. Brodie III. 2011. Elimination of a genetic correlation between the sexes via artificial correlational selection. *Evolution* 65: 2872–2880.

DeWitt, T. J., and S. M. Scheiner, eds. 2004. *Phenotypic Plasticity: Functional and Conceptual Approaches.* Oxford: Oxford University Press.

Dlugosch, K. M., S. R. Anderson, J. Braasch, F. A. Cang, and H. D. Gillette. 2015. The devil is in the details: Genetic variation in introduced populations and its contributions to invasion. *Molecular Ecology* 24: 2095–2111.

Donohue, K. 2005. Niche construction through phenological plasticity: Life history dynamics and ecological consequences. *New Phytologist* 166: 83–92.

Egset, C. K., T. F. Hansen, A. Le Rouzic, G. H. Bolstad, G. Rosenqvist, and C. Pélabon. 2012. Artificial selection on allometry: Change in elevation but not slope. *Journal of Evolutionary Biology* 25: 938–948.

Eisen, E. J. 1975. Population size and selection intensity effects on long-term selection response in mice. *Genetics* 79: 305–323.

Ellner, S., and N. G. Hairston. 1994. Role of overlapping generations in maintaining genetic variation in a fluctuating environment. *American Naturalist* 143: 403–417.

Ellstrand, N. C., and K. A. Schierenbeck. 2000. Hybridization as a stimulus for the evolution of invasiveness in plants? *PNAS* 97: 7043–7050.

Fischer, K., J. Kreyling, M. Beaulieu, et al. 2021. Species-specific effects of thermal stress on the expression of genetic variation across a diverse group of plant and animal taxa under experimental conditions. *Heredity* 126: 23–37.

Flatt, T. 2005. The evolutionary genetics of canalization. *Quarterly Review of Biology* 80: 287–316.

Frankham, R. 1990. Are responses to artificial selection for reproductive fitness characters consistently asymmetrical? *Genetics Research* 56: 35–42.

Frankham, R., L. P. Jones, and J. S. F. Barker. 1968. The effects of population size and selection intensity in selection for a quantitative character in *Drosophila:* I. Short-term response to selection. *Genetics Research* 12: 237–248.

Frankham, R., K. Lees, M. E. Montgomery, P. R. England, E. H. Lowe, and D. A. Briscoe. 1999. Do population size bottlenecks reduce evolutionary potential? *Animal Conservation* 2: 255–260.

Gerken, A. R., T. F. Mackay, and T. J. Morgan. 2016. Artificial selection on chill-coma recovery time in *Drosophila melanogaster:* Direct and correlated responses to selection. *Journal of Thermal Biology* 59: 77–85.

Grant, P. R. 1983. Inheritance of size and shape in a population of Darwin's finches, *Geospiza conirostris. Proceedings of the Royal Society B* 220: 219–236.

Guillaume, F., and M. C. Whitlock. 2007. Effects of migration on the genetic covariance matrix. *Evolution* 61: 2398–2409.

Halliburton, R., and G. A. E. Gall. 1981. Disruptive selection and assortative mating in *Tribolium castaneum*. *Evolution* 35: 829–843.

Hamilton, J. A., and J. M. Miller. 2016. Adaptive introgression as a resource for management and genetic conservation in a changing climate. *Conservation Biology* 30: 33–41.

Hansen, T. F. 2013. Why epistasis is important for selection and adaptation. *Evolution* 67: 3501–3511.

Hansen, T. F., and C. Pélabon. 2021. Evolvability: A quantitative genetics perspective. *AREES* 52: 153–175.

Hansen, T. F., and G. P. Wagner. 2001. Modeling genetic architecture: A multilinear theory of gene interaction. *Theoretical Population Biology* 59: 61–86.

Hansen, T. F., C. Pélabon, W. S. Armbruster, and M. L. Carlson. 2003. Evolvability and genetic constraint in *Dalechampia* blossoms: Components of variance and measures of evolvability. *Journal of Evolutionary Biology* 16: 754–766.

Hansen, T. F., J. M. Álvarez-Castro, A. J. Carter, J. Hermisson, and G. P. Wagner. 2006. Evolution of genetic architecture under directional selection. *Evolution* 60: 1523–1536.

Hansen, T. F., C. Pélabon, and D. Houle. 2011. Heritability is not evolvability. *Evolutionary Biology* 38: 258–277.

Hermisson, J. and G. P. Wagner. 2004. The population genetic theory of hidden variation and genetic robustness. *Genetics* 168: 2271–2284.

Hoffmann, A. A., and V. Kellermann. 2006. Revisiting heritable variation and limits to species distribution: Recent developments. *Israel Journal of Ecology and Evolution* 52: 247–261.

Hoffmann, A. A., and J. Merilä. 1999. Heritable variation and evolution under favourable and unfavourable conditions. *TREE* 14: 96–101.

Hoffmann, A. A., and P. A. Parsons. 1997. *Extreme Environmental Change and Evolution.* Cambridge: Cambridge University Press.

Hoffmann, A. A., C. M. Sgrò, and T. N. Kristensen. 2017. Revisiting adaptive potential, population size, and conservation. *TREE* 32: 506–517.

Houle, D. 1989. The maintenance of polygenic variation in finite populations. *Evolution* 43: 1767–1780.

Houle, D. 1992. Comparing evolvability and variability of quantitative traits. *Genetics* 130: 195–204.

Houle, D. 1998. How should we explain variation in the genetic variance of traits? *Genetica* 102: 241–253.

Houle, D., C. Pélabon, G. P. Wagner, and T. F. Hansen. 2011. Measurement and meaning in biology. *Quarterly Review of Biology* 86: 3–34.

Houle, D., G. H. Bolstad, K. van der Linde, and T. F. Hansen. 2017. Mutation predicts 40 million years of fly wing evolution. *Nature* 548: 447–450.

Jones, A. G., S. J. Arnold, and R. Bürger. 2003. Stability of the G-matrix in a population experiencing pleiotropic mutation, stabilizing selection, and genetic drift. *Evolution* 57: 1747–1760.

Jones, L. P., R. Frankham, and J. S. F. Barker. 1968. The effects of population size and selection intensity in selection for a quantitative character in *Drosophila:* II. Long-term response to selection. *Genetics Research* 12: 249–266.

Kassen, R. 2002. The experimental evolution of specialists, generalists, and the maintenance of diversity. *Journal of Evolutionary Biology* 15: 173–190.

Kelly, J. K. 2008. Testing the rare-alleles model of quantitative variation by artificial selection. *Genetica* 132: 187–198.

Kellermann, V., B. Van Heerwaarden, C. M. Sgrò, and A. A. Hoffmann. 2009. Fundamental evolutionary limits in ecological traits drive *Drosophila* species distributions. *Science* 325: 1244–1246.

Kirkpatrick, M., and N. H. Barton. 1997. Evolution of a species' range. *American Naturalist* 150: 1–23.

Kolbe, J. J., R. E. Glor, Schettino, A. C. Lara, A. Larson, and J. B. Losos. 2004. Genetic variation increases during biological invasion by a Cuban lizard. *Nature* 431: 177–181.

Kremer, A., O. Ronce, J. J. Robledo-Arnuncio, et al. 2012. Long-distance gene flow and adaptation of forest trees to rapid climate change. *Ecology Letters* 15: 378–392.

Laland, K. N., and K. Sterelny. 2006. Perspective: Seven reasons (not) to neglect niche construction. *Evolution* 60: 1751–1762.

Lande, R. 1975. The maintenance of genetic variability by mutation in a polygenic character with linked loci. *Genetics Research* 26: 221–235.

Lande, R. 1976. Natural selection and random genetic drift in phenotypic evolution. *Evolution* 30: 314–334.

Lande, R. 1979. Quantitative genetic analysis of multivariate evolution, applied to brain-body size allometry. *Evolution* 33: 402–416.

Lande, R. 1980. The genetic covariance between characters maintained by pleiotropic mutations. *Genetics* 94: 203–215.

Lande, R. 1992. Neutral theory of quantitative genetic variance in an island model with local extinction and colonization. *Evolution* 46: 381–389.

Lande, R. 2009. Adaptation to an extraordinary environment by evolution of phenotypic plasticity and genetic assimilation. *Journal of Evolutionary Biology* 22: 1435–1446.

Lande, R., and S. J. Arnold. 1983. The measurement of selection on correlated characters. *Evolution* 37: 1210–1226.

Lavergne, S., and J. Molofsky. 2007. Increased genetic variation and evolutionary potential drive the success of an invasive grass. *PNAS* 104: 3883–3888.

Lee, C. E. 2002. Evolutionary genetics of invasive species. *TREE* 17: 386–391.

Lee, C. E., and G. W. Gelembiuk. 2008. Evolutionary origins of invasive populations. *Evolutionary Applications* 1: 427–448.

Lenormand, T. 2002. Gene flow and the limits to natural selection. *TREE* 17: 183–189.

Le Rouzic, A., D. Houle, and T. F. Hansen. 2011. A modelling framework for the analysis of artificial-selection time series. *Genetics Research* 93: 155–173.

Le Rouzic, A., J. M. Álvarez-Castro, and T. F. Hansen. 2013. The evolution of canalization and evolvability in stable and fluctuating environments. *Evolutionary Biology* 40: 317–340.

López-Fanjul, C., A. Fernández, and M. A. Toro. 2004. Epistasis and the temporal change in the additive variance-covariance matrix induced by drift. *Evolution* 58: 1655–1663.

López-Fanjul, C., A. Fernández, and M. A. Toro. 2006. The effect of genetic drift on the variance/covariance components generated by multilocus additive×additive epistatic systems. *Journal of Theoretical Biology* 239: 161–171.

Lucek, K., L. Greuter, O. M. Selz, and O. Seehausen. 2017. Effects of interspecific gene flow on the phenotypic variance–covariance matrix in Lake Victoria Cichlids. *Hydrobiologia* 791: 145–154.

Martínez-Padilla, J., A. Estrada, R. Early, and F. García-González. 2017. Evolvability meets biogeography: evolutionary potential decreases at high and low environmental favourability. *Proceeding of the Royal Society B* 284: 20170516.

Martinossi-Allibert, I., J. Clavel, S. Ducatez, I. L. Viol, and C. Teplitsky. 2017. Does habitat specialization shape the evolutionary potential of wild bird populations? *Journal of Avian Biology* 48: 1158–1165.

Mitchell, K. A., and A. A. Hoffmann. 2010. Thermal ramping rate influences evolutionary potential and species differences for upper thermal limits in *Drosophila. Functional Ecology* 24: 694–700.

Morgan, R., M. H. Finnøen, H. Jensen, C. Pélabon, and F. Jutfelt. 2020. Low potential for evolutionary rescue from climate change in a tropical fish. *PNAS* 117: 33365–33372.

Muff, S., A. K. Niskanen, D. Saatoglu, L. F. Keller, and H. Jensen. 2019. Animal models with group-specific additive genetic variances: Extending genetic group models. *Genetics Selection Evolution* 51: 1–16.

Nei, M., and W. H. Li. 1973. Linkage disequilibrium in subdivided populations. *Genetics* 75: 213–219.

Nosil, P., B. J. Crespi, C. P. Sandoval, and M. Kirkpatrick. 2006. Migration and the genetic covariance between habitat preference and performance. *American Naturalist* 167: E66–E78.

Oakley, C. G. 2015. The influence of natural variation in population size on ecological and quantitative genetics of the endangered endemic plant *Hypericum cumulicola. International Journal of Plant Sciences* 176: 11–19.

Orr, H. A., and M. Turelli. 2001. The evolution of postzygotic isolation: Accumulating Dobzhansky-Muller incompatibilities. *Evolution* 55: 1085–1094.

Oziolor, E. M., N. M. Reid, S. Yair, et al. 2019. Adaptive introgression enables evolutionary rescue from extreme environmental pollution. *Science* 364: 455–457.

Paaby, A. B., and M. V. Rockman, 2014. Cryptic genetic variation: Evolution's hidden substrate. *Nature Reviews Genetics* 15: 247–258.

Parsons, K. J., Y. H. Son, and R. C. Albertson. 2011. Hybridization promotes evolvability in African cichlids: Connections between transgressive segregation and phenotypic integration. *Evolutionary Biology* 38: 306–315.

Pavličev, M., and J. M. Cheverud. 2015. Constraints evolve: Context dependency of gene effects allows evolution of pleiotropy. *AREES* 46: 413–434.

Pavličev, M., J. P. Kenney-Hunt, E. A. Norgard, C. C. Roseman, et al. 2008. Genetic variation in pleiotropy: Differential epistasis as a source of variation in the allometric relationship between long bone lengths and body weight. *Evolution* 62: 199–213.

Pavličev, M., J. M. Cheverud, and G. P. Wagner. 2011. Evolution of adaptive phenotypic variation patterns by direct selection for evolvability. *Proceedings of the Royal Society B* 278: 1903–1912.

Pélabon, C., T. F. Hansen, A. J. Carter, and D. Houle. 2010. Evolution of variation and variability under fluctuating, stabilizing, and disruptive selection. *Evolution* 64: 1912–1925.

Pélabon, C., W. S. Armbruster, T. F. Hansen, G. H. Bolstad, and R. Pérez-Barrales. 2012. Adaptive accuracy and adaptive landscapes. In *The Adaptive Landscape in Evolutionary Biology*, edited by E. I. Svensson and R. Calsbeek, 150–168. Oxford: Oxford University Press.

Pélabon, C., C. H. Hilde, S. Einum, and M. Gamelon. 2020. On the use of the coefficient of variation to quantify and compare trait variation. *Evolution Letters* 4: 180–188.

Pélabon, C., E. Albertsen, A. Le Rouzic, C. Firmat, et al. 2021. Quantitative assessment of observed vs. predicted responses to selection. *Evolution* 75: 2217–2236.

Pennington, L. K., R. A. Slatyer, D. V. Ruiz-Ramos, S. D. Veloz, and J. P. Sexton. 2021. How is adaptive potential distributed within species ranges? *Evolution* 75: 2152–2166.

Pfennig, K. S., A. L. Kelly, and A. A. Pierce. 2016. Hybridization as a facilitator of species range expansion. *Proceedings of the Royal Society B* 283: 20161329.

Phillips, P. C., M. C. Whitlock, and K. Fowler. 2001. Inbreeding changes the shape of the genetic covariance matrix in *Drosophila melanogaster*. *Genetics* 158: 1137–1145.

Pitchers, W., J. Nye, E. J. Márquez, A. Kowalski, I. Dworkin, and D. Houle. 2019. A multivariate genome-wide association study of wing shape in *Drosophila melanogaster*. *Genetics* 211: 1429–1447.

Podolsky, R. H. 2001. Genetic variation for morphological and allozyme variation in relation to population size in *Clarkia dudleyana*, an endemic annual. *Conservation Biology* 15: 412–423.

Pujol, B., and J. R. Pannell. 2008. Reduced responses to selection after species range expansion. *Science* 321: 96.

Queitsch, C., T. A. Sangster, and S. Lindquist. 2002. Hsp90 as a capacitor of phenotypic variation. *Nature* 417: 618–624.

Ramsayer, J., O. Kaltz, and M. E. Hochberg. 2013. Evolutionary rescue in populations of *Pseudomonas fluorescens* across an antibiotic gradient. *Evolutionary Applications* 6: 608–616.

Reid, J. M., and P. Arcese. 2020. Recent immigrants alter the quantitative genetic architecture of paternity in song sparrows. *Evolution Letters* 4: 124–136.

Reid, J. M., P. Arcese, P. Nietlisbach, M. E. Wolak, et al. 2021. Immigration counter-acts local micro-evolution of a major fitness component: Migration-selection balance in free-living song sparrows. *Evolution Letters* 5: 48–60.

Rice, W. R., and E. E. Hostert. 1993. Perspective: Laboratory experiments on speciation: What have we learned in forty years? *Evolution* 47: 1637–1653.

Rieseberg, L. H., M. A. Archer, and R. K. Wayne. 1999. Transgressive segregation, adaptation and speciation. *Heredity* 83: 363–372.

Rowiński, P. K., and B. Rogell. 2017. Environmental stress correlates with increases in both genetic and residual variances: A meta-analysis of animal studies. *Evolution* 71: 1339–1351.

Rutherford, S. L., and S. Lindquist. 1998. Hsp90 as a capacitor for morphological evolution. *Nature* 396: 336–342.

Sasaki, A., and S. Ellner. 1997. Quantitative genetic variance maintained by fluctuating selection with overlapping generations: Variance components and covariances. *Evolution* 51: 682–696.

Scharloo, W. 1991. Canalization: Genetic and developmental aspects. *AREES* 22: 65–93.

Selz, O. M., K. Lucek, K. A. Young, and O. Seehausen. 2014. Relaxed trait covariance in interspecific cichlid hybrids predicts morphological diversity in adaptive radiations. *Journal of Evolutionary Biology* 27: 11–24.

Sexton, J. P., P. J. McIntyre, A. L. Angert, and K. J. Rice. 2009. Evolution and ecology of species range limits. *AREES* 40: 415–436.

Sharakhov, I. V., B. J. White, M. V. Sharakhova, et al. 2006. Breakpoint structure reveals the unique origin of an interspecific chromosomal inversion (2La) in the *Anopheles gambiae* complex. *PNAS* 103: 6258–6262.

Sorensen D. A., and W. G. Hill. 1982. Effect of short-term directional selection on genetic-variability–experiments with *Drosophila Melanogaster*. *Heredity* 48: 27–33

Sorensen, D. A., and W. G. Hill. 1983. Effects of disruptive selection on genetic variance. *Theoretical and Applied Genetics* 65: 173–180.

Stanton, M., and H. J. Young. 1994. Selecting for floral character associations in wild radish, *Raphanus sativus* L. *Journal of Evolutionary Biology* 7: 271–285.

Stearns, S., G. de Jong, and B. Newman. 1991. The effects of phenotypic plasticity on genetic correlations. *TREE* 6: 122–126.

Steven, J. C., I. A. Anderson, E. D. Brodie III, and L. F. Delph. 2020. Rapid reversal of a potentially constraining genetic covariance between leaf and flower traits in *Silene latifolia*. *Ecology and Evolution* 10: 569–578.

Swindell, W. R., and J. L. Bouzat. 2006. Reduced inbreeding depression due to historical inbreeding in *Drosophila melanogaster:* Evidence for purging. *Journal of Evolutionary Biology* 19: 1257–1264.

Sztepanacz, J. L., and M. W. Blows. 2017. Artificial selection to increase the phenotypic variance in g max fails. *American Naturalist* 190: 707–723.

Thoday, J. M. 1972. Disruptive selection. *Proceedings of the Royal Society B* 182: 109–143.

Tufto, J. 2000. Quantitative genetic models for the balance between migration and stabilizing selection. *Genetics Research* 76: 285–293.

Turelli, M. 1985. Effects of pleiotropy on predictions concerning mutation-selection balance for polygenic traits. *Genetics* 111: 165–195.

Turelli, M., and N. H. Barton. 2006. Will population bottlenecks and multilocus epistasis increase additive genetic variance? *Evolution* 60: 1763–1776.

Van Heerwaarden, B., Y. Willi, T. N. Kristensen, and A. A. Hoffmann. 2008. Population bottlenecks increase additive genetic variance but do not break a selection limit in rain forest *Drosophila. Genetics* 179: 2135–2146.

Waddington, C. H. 1953. Genetic assimilation of an acquired character. *Evolution* 7: 118–126.

Waddington, C.H. 1957. *The Strategy of the Genes.* London: Allen & Unwin.

Wagner, G. P., G. Booth, and H. Bagheri-Chaichian. 1997. A population genetic theory of canalization. *Evolution* 51: 329–347.

Wagner, G. P., C. H. Chiu, and T. F. Hansen. 1999. Is Hsp90 a regulator of evolvability? *Journal of Experimental Zoology B* 285: 116–118.

Wagner, G. P., and J. Mezey. 2000. Modeling the evolution of genetic architecture: A continuum of alleles model with pairwise A×A epistasis. *Journal of Theoretical Biology* 203: 163–175.

Walsh, B., and M. W. Blows. 2009. Abundant genetic variation + strong selection = multivariate genetic constraints: A geometric view of adaptation. *AREES* 40: 41–59.

Weber, K. E. 1990. Increased selection response in larger populations. I. Selection for wing-tip height in *Drosophila melanogaster* at three population sizes. *Genetics* 125: 579–584.

Wheelwright, N. T., L. F. Keller, and E. Postma. 2014. The effect of trait type and strength of selection on heritability and evolvability in an island bird population. *Evolution* 68: 3325–3336.

Whitlock, M. C. 1999. Neutral additive genetic variance in a metapopulation. *Genetics Research* 74: 215–221.

Whitlock, M. C., and K. Fowler. 1999. The changes in genetic and environmental variance with inbreeding in *Drosophila melanogaster. Genetics* 152: 345–353.

Whitlock, M.C., P. C. Phillips, and K. Fowler. 2002. Persistence of changes in the genetic covariance matrix after a bottleneck. *Evolution* 56: 1968–1975.

Widén, B., and S. Andersson. 1993. Quantitative genetics of life-history and morphology in a rare plant, *Senecio integrifolius. Heredity* 70: 503–514.

Willi, Y., J. Van Buskirk, and A. A. Hoffmann. 2006. Limits to the adaptive potential of small populations. *AREES* 37: 433–458.

Willis, J. H., and H. A. Orr. 1993. Increased heritable variation following population bottlenecks: the role of dominance. *Evolution* 47: 949–957.

Wood, C. W., and E. D. Brodie III. 2015. Environmental effects on the structure of the G-matrix. *Evolution* 69: 2927–2940.

Wood, J. L. A., D. Tezel, D. Joyal, and D. J. Fraser. 2015. Population size is weakly related to quantitative genetic variation and trait differentiation in a stream fish. *Evolution* 69: 2303–2318.

Wood, J. L., M. C. Yates, and D. J. Fraser. 2016. Are heritability and selection related to population size in nature? Meta-analysis and conservation implications. *Evolutionary Applications* 9: 640–657.

Wright, J., G. H. Bolstad, Y. G. Araya-Ajoy, and N. J. Dingemanse. 2019. Life-history evolution under fluctuating density-dependent selection and the adaptive alignment of pace-of-life syndromes. *Biological Reviews* 94: 230–247.

Yoo, B. H. 1980. Long-term selection for a quantitative character in large replicate populations of *Drosophila melanogaster. Theoretical and Applied Genetics* 57: 25–32.

14 Does Lack of Evolvability Constrain Adaptation? If So, on What Timescales?

Kjetil L. Voje, Mark Grabowski, Agnes Holstad, Arthur Porto, Masahito Tsuboi, and Geir H. Bolstad

The relevance of genetic constraints for evolutionary change beyond microevolutionary timescales is debated. The high evolvability of natural populations predicts rapid adaptation, but evolvability is often found to correlate with phenotypic divergence on longer timescales, which makes sense if evolvability constrains divergence. This chapter attempts to reconcile the observation of high evolvability of populations with the idea that genetic constraints may still be relevant on long timescales. We first establish that a relationship between evolvability and divergence is a common empirical phenomenon both among populations within species (microevolution) and among species (macroevolution). We then argue that a satisfactory model for the prevalence of this empirical relationship is lacking. Linking microevolutionary theory with the dynamics of the adaptive landscape across time—moving toward a proper quantitative theory of phenotypic change on macroevolution timescales—is key to better understanding the relative importance of genetic constraints on phenotypic evolution beyond a handful of generations.

14.1 Introduction

The study of adaptation—how natural selection improves organisms' fit to their environment—is central to evolutionary biology. Adaptations enable lineages to survive and thrive in vastly different habitats, or they may represent fine-tuned differences among populations, like the relationship between pericarp thickness in the fruits of populations of *Camellia japonica* and the length of the rostrum of the seed-predatory weevil *Curculio camelliae* (Toju and Sota 2006). But not all populations are well adapted. For example, *Crescentia alata* and several other plant species in Central America have large fruits that do not get dispersed due to the late-Pleistocene extinction of the many large herbivores that acted as their agents for seed dispersal (Janzen and Martin 1982). Why is fruit size evolving fast in populations of *Camellia japonica* in Japan while the large and energy-expensive fruit of *Crescentia alata* is not? In this chapter, we ask whether lack of evolvability—the potential (or disposition) of a population to evolve—may be an explanation for why "evolutionary failure is commonplace" (Bradshaw 1991, 289). We find that evolvability and phenotypic divergence are often positively correlated, both on short and on longer timescales, an intriguing result, given the lack of models that readily predict this correlation.

To say something meaningful about a potential relationship between adaptation and evolvability, we first clarify what we mean by adaptation, as the term has accumulated numerous definitions (e.g., Reeve and Sherman 1993). In the context of evolvability in quantitative genetics (see Hansen and Houle 2008; Hansen and Pélabon 2021), adaptation can be understood and defined in relation to an adaptive landscape. Simpson (1944) outlined the concept of the adaptive landscape as a representation of possible combinations of phenotypic traits where elevations in the landscape represent higher population fitness. Adaptation can be both a process and an outcome. In the context of an adaptive landscape, the process of adaptation is about climbing peaks, and selection will always push the population up along the steepest slope of a fitness surface it resides on (Lande 1979; 2007). The outcome of this climbing process is increased adaptation (and a reduced maladaptiveness); a well-adapted population will be at or close to a peak in the landscape. Because elevation on this landscape reflects the fitness of the population, the degree of maladaptation increases with the vertical distance to the closest peak. The different populations of *Camellia japonica* in Japan probably reside at or close to local peaks in the adaptive landscape for pericarp thickness. The South American plants lacking large-bodied agents for seed-dispersal are probably closer to the foot than the top of a mountain in the adaptive landscape or are trapped on a local peak that has been reduced from a high summit to a small hill.

Changes in the environment experienced by a population can affect the adaptive landscape and thus decrease adaptiveness (i.e., cause maladaptation). The extinction of a seed disperser is an obvious example. But several other processes can also displace a population from a peak or hinder it from efficiently ascending peaks in the adaptive landscape. Gene flow among populations may hinder local adaptation (Savolainen et al. 2007), small population size will increase the prevalence of mildly deleterious alleles (Ohta 1992) and enable genetic drift to play an increasing role on the evolutionary dynamics (Walsh and Lynch 2018). Genetic architecture (e.g., pleiotropy) may generate a deviation in the response to selection, causing the evolving population to take a curved path toward the peak (Lande 1979). Different degrees of maladaptation may therefore be a common state in nature (Crespi 2000), even for apparently well-adapted populations. Indeed, a large-scale analysis of selection gradients indicated that most of the populations studied (64%) had a trait mean that deviated more than 1 standard deviation from the estimated optimum and about one third had a mismatch between trait mean and optimum of more than 2 standard deviations (Estes and Arnold 2007).

The ample evidence of maladaptation in natural populations suggests that the ability to evolve—and potentially a lack thereof—matters on short timescales. When the position of the optimum changes, a highly evolvable population will track and re-ascend the peak, while less evolvable populations will remain displaced from the peak. Lineage extinction is the ultimate failure of adapting sufficiently rapidly to changes in the environment (Gomulkiewicz and Houle 2009), a fate common to the great majority of all lineages that have ever existed (Jablonski 2004).

Are constraints imposed by the lack of evolvability relevant on timescales beyond microevolution? This question has a long and controversial history in evolutionary biology (e.g., Simpson 1944; Kluge and Kerfoot 1973; Schluter 1996). Low genetic variation in the direction of selection is commonly assumed to be a soft constraint, because it can be overcome given enough time (Maynard Smith et al. 1985). Therefore, as long as a sustainable population size is maintained during the time interval in which the population reclimbs the peak, extinction will be avoided. Indeed, currently living species must have been able to surmount changes in the adaptive landscape in their past, which suggests little relevance of evolvability on macroevolutionary timescales. A growing body of empirical work suggests otherwise.

Genetic constraints are influencing evolution if the closest adaptive peak has not been reached by the population due to lack of available genetic variation (Arnold 1992). Schluter (1996) was the first to detect that phenotypic differentiation between populations and species tended to be biased in the multivariate direction containing the greatest additive genetic variance (i.e., the direction with highest evolvability). Later studies have found a similar pattern between evolvability and divergence, sometimes across macroevolutionary timescales. For example, Houle et al. (2017) showed that the evolvability of a population of the fruit fly *Drosophila melanogaster* strongly correlated with trait divergence among Drosophilid species that shared a common ancestor 40 million years ago. Empirical evidence in favor of evolvability constraining the process of adaptation on both long and short timescales is paradoxical, given the apparent high evolvability of natural populations (Bolstad et al. 2014). On short timescales, evolvability depends on the amount of additive genetic variation, and most quantitative traits seem to contain enough variation to quickly respond to directional selection (Hansen and Pélabon 2021). Directional selection on traits is also common in nature (Hereford et al. 2004), and populations typically respond rapidly—just as predicted by theory (Hendry and Kinnison 1999; Kinnison and Hendry 2001). Many populations are therefore seemingly sufficiently evolvable to readily overcome even serious cases of maladaptation and to rapidly ascend peaks in the adaptive landscape. But why then are the large fruits of *Crescentia alata* rotting close to the individual producing them?

This chapter discusses how to reconcile the apparent high evolvability of natural populations with the hypothesis that a population's ability to evolve might act as a constraint on the process of adaptation. After introducing the quantitative genetic concept of evolvability, we discuss methodological issues when investigating correlations between evolvability and divergence. Reviewing published studies, we show that a positive correlation between evolvability and phenotypic divergence is a common empirical pattern. We then briefly discuss trait evolution models and conclude that we currently lack a satisfactory model that fully explains the commonness of the relationship between evolvability and phenotypic divergence. Because the realism of the different models depends on the dynamical nature of the adaptive landscape, we discuss new developments in our understanding of how adaptive landscapes change on different time intervals. We end by pointing to future directions of research that will help us further assess the relevance of evolvability for adaptation and phenotypic divergence.

14.2 General Introduction to Evolvability

To understand the relationship between evolvability and constraint, we need first to understand the measurement of evolvability. Quantitative genetic theory posits that short-term evolvability can be quantified using a metric reflecting standing genetic variation. Houle (1992; see also Hansen, chapter 5; Houle and Pélabon, chapter 6)[1] proposed that evolvability, e, can be operationalized using the mean-scaled additive genetic variance:

1. References to chapter numbers in the text are to chapters in this volume.

$$e = \frac{V_A}{\bar{z}^2},$$

where V_A and $\bar{z}$ are respectively the trait's additive genetic variance and mean before selection. Hansen et al. (2003a) showed that e can be interpreted as the proportional evolutionary response of a trait to 1 unit strength of directional selection, where the unit is defined as the strength of selection on fitness itself. This definition of evolvability serves as a metric, allowing us to assess and compare the ability of different types of traits to evolve.

Reported estimates of univariate evolvabilities suggest abundant additive genetic variation for virtually any trait of interest (Hansen et al. 2011). On a trait-by-trait basis, that would suggest a sufficiently large supply of "fuel" for the evolutionary process to cast doubt on any hypotheses claiming evolvability could act as an evolutionary constraint. Still, observed evolutionary rates are often orders of magnitude smaller than predicted from univariate evolvabilities. For example, Lande (1976, 333) found that only about 1 selective death per million individuals per generation is needed to explain the observed evolution in tooth characters of Tertiary mammals in the fossil record (see also Lynch 1990). One possible explanation is that univariate evolvability estimates are not representative of the true capacity for traits to evolve. Empirical studies indicate that variation in single traits is often bound to variation in other traits of the same organism due to genetic correlations (e.g., Walsh and Blows 2009). The immediate implication is that evolutionary change for any one trait is often not possible without substantial changes in other traits. Strong stabilizing selection on pleiotropically linked traits may therefore severely reduce the amount of "free" additive genetic variance available for a given trait to evolve (Hansen and Houle 2004).

Suggestions of multivariate constraint as an essential component of adaptation have been made for decades (e.g., Dickerson 1955), and evolutionary biology has witnessed an increasing use of quantitative genetic approaches aimed at understanding evolution in multivariate morpho-space. Most of these approaches rely on the genetic variance-covariance matrix, **G**, as the central entity with which to study evolvability. For example, several studies have attempted to find dimensions of **G** with little to no additive genetic variance and have framed issues surrounding evolvability in terms of "nearly null spaces" (e.g., Gomulkiewicz and Houle 2009), that is, subspaces of **G** with very low evolvability. These studies argue that finding such dimensions is essential to understanding evolvability, as they would represent multivariate constraints due to diminished evolutionary potential in these directions. However, studying these dimensions is complicated, because estimating variance in nearly null spaces may be confounded with measurement error. It may also be that the absence of genetic variance in short time spans is not representative of long-term evolvability, as both new mutations or changes in allele frequencies (because of dominance or epistasis) may lead to increased additive variance.

Another popular approach to studying multivariate evolvability is framed in terms of lines of least evolutionary resistance (sensu Schluter 1996). The term "lines of least evolutionary resistance" refers to dimensions of multivariate space with a larger-than-average amount of the total additive genetic variance along which evolution could proceed at a fast pace (Hansen and Houle 2008). Although lines of least resistance are often much easier to estimate and study than are nearly null spaces, they also have shortcomings. Most notably, there are usually multiple dimensions with abundant additive genetic variance in a population, so lack of population divergence along the primary axis of genetic variance is not an indication that those populations did not diverge along an axis associated with greater-than-average additive variance (Hansen and Voje 2011).

Hansen and Houle (2008) proposed an approach to unify these perspectives on multivariate evolution into a single framework, suggesting multiple direct measurements of evolvability that take into account the extent to which variation in individual traits are bound to other traits during adaptation. These are defined as unconditional and conditional evolvabilities and depend on assumptions about the adaptive landscape. Unconditional evolvability is measured as the magnitude of the projection of the response on the selection vector; it represents the magnitude of the evolutionary response in the direction of selection. Conditional evolvability is measured as the response along the selection vector when no other directions (with measurements) of response are allowed (Hansen et al. 2003b). This represents a situation where evolvability is the genetic variation available for selection in one direction when other multivariate directions are under strong stabilizing selection. The importance of such operational definitions of evolvability is that they provide a truly multivariate view of evolution.

Although some researchers have argued that explanations for stasis are "far outside the domain of genetic constraints" (Arnold 2014, 743), others have argued that the multivariate nature of evolution may provide a partial resolution to the problem of stasis (Hansen and Houle 2004; Walsh and Blows 2009). Indeed, most conditional evolvabilities can be much smaller than unconditional evolvabilities, highlighting once again that most individual trait variance is bound to other traits (Hansen 2012). One explanation for a lack of adaptation despite abundant variation may therefore be that we simply do not have a good understanding of all the relevant traits that make up **G**, or how a high-dimensional **G** impacts and is impacted by natural selection. To complicate the matters further, studies of multivariate evolvability and divergence are also plagued with methodological issues.

14.3 Methods Matter!

Analyzing the relationship between evolvability and divergence is not straightforward. A first challenge is that **G** is hard to measure with high accuracy (Cheverud 1988), making the comparison to divergence imprecise. A second methodological issue is the use of correlation matrices. In a genetic correlation matrix, elements are standardized by the trait variances, removing the magnitude of variation and, therefore, obscuring the relationship between the genetic variance and divergence. A third methodological issue is the tendency to solely assess the angle between the divergence vector and the dominant eigenvector of **G** (g_{max}) when investigating for a relationship between evolvability and divergence, as there may be many directions in phenotype space with high evolvability (Hansen and Voje 2011). There are additional issues with interpreting several of the matrix comparison methods (see discussion in Bolstad et al. 2014), and their power to detect a true evolvability-divergence relationship might be weak (e.g., see the reanalysis of Lofsvold's data later in this section).

To analyze the relationship between evolvability and divergence, we advocate using mean standardization or natural log transformation before employing the framework suggested by Hansen and Houle (2008). These two methods are interchangeable for small variances, as mean standardization is the first order (local) approximation of the natural log (see Grabowski and Roseman 2015). Not all traits can be meaningfully log-transformed or mean standardized, however (see Houle et al. 2011; Pélabon et al. 2020). After such standardization, the estimated evolvabilities in a direction of divergence can be compared with the average evolvability of all traits (Hansen and Houle 2008; Hansen and Voje 2011), or evolvabilities can be compared to divergence variance or rates across traits (e.g., as in Bolstad et al. 2014). For the latter approach, one would typically do a regression with log divergence variance or rate as response and log evolvability as predictor, to estimate the scaling relationship between the two.

The approach we advocate also has methodological issues. A first issue is that traits of different dimensionality will have systematically different evolvabilities and divergence rates (Gingerich 1993; Hansen et al. 2011). Note, however, that these differences are not statistical artifacts but should be interpreted as a dimensionality-scaling effect rather than a potentially constraining effect of evolvability. Therefore, to test for a relationship between evolvability and divergence, it is advisable to include only traits measured in the same physical dimension in the same analysis. A second issue is the choice of how to linearly transform the traits before fitting the regression between evolvability and divergence (for more on this point, see Houle et al. 2020; Jiang and Zhang 2020).

We illustrate the impact that different methodologies can have on the conclusions regarding the relationship between **G** and the among-population variance-covariance matrix, **D**, by reanalyzing the data on different subspecies of the genus *Peromyscus* presented in Lofsvold (1986, 1988). Lofsvold (1988) concluded that, overall, there is no significant similarity between **G** and **D** (**L** in Lofsvold 1988). His analysis was based on comparing angles of the first 5 eigenvectors between matrices, computing matrix correlations, and performing Mantel tests. Conveniently, the variance-covariance matrices presented by Lofsvold are based on natural log transformed traits, and hence the genetic variances (V_A) can be interpreted as evolvabilities and the among-population variances (V_D) are on the same scale. We analyzed the scaling relationship between **D** and **G** by using a simple least squares regression with log V_D as response and log V_A as predictor. We detected moderate to strong relationships between the **D**s and **G**s, with scaling exponents (b) in the range 0.70–0.93, and R^2 in the range 29–89% (figure 14.1a). In two of the subspecies, the among-population divergence was best explained by the **G** of the same subspecies, indicating that constraints break down over time, whereas in the other subspecies (*P. maniculatus bairdii*), this was not the case. Interestingly, the relationships are generally steeper and stronger when using **P**, the phenotypic variance-covariance matrix, in place of **G** (figure 14.1b). This may be because **G** is poorly estimated compared to **P**, and therefore the shape of **P** is a better representation of the shape of the true **G** (see Cheverud 1988). Alternatively, it can be caused by a component of plasticity shared by **P** and **D**. In any case, our analysis reaches the opposite conclusion of Lofsvold (1988).

With the data of Lofsvold (1988), we can also test whether there is a relationship between **G** and divergence among species and subspecies. Because there are only 3 species, calculating **D** at this level is not informative. However, we can quantify whether the divergence vectors

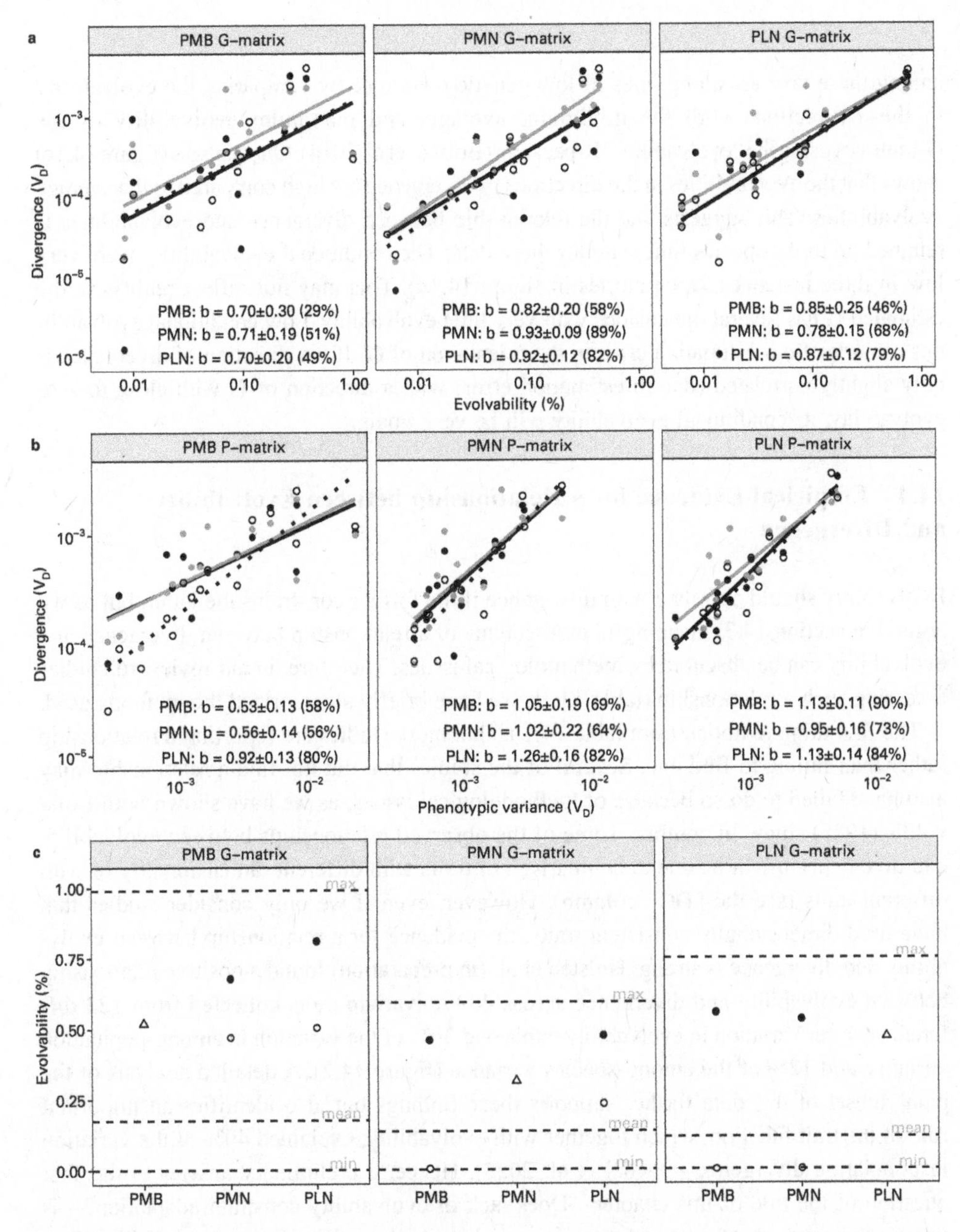

Figure 14.1
Analysis of scaling relationship between divergence and (a) evolvability and (b) phenotypic variance in different subspecies of the genus *Peromyscus* ("deer mouse," PMB = *P. maniculatus bairdii*, PMN = *P. maniculatus nebrascensis*, PLN = *P. leucopus noveborascensis*). Divergence (V_D) and phenotypic variance (V_P) are in units of $\ln^2$(mm), and evolvability is in units of $100 \times \ln^2$(mm) (i.e., $100 \times V_A$), which can be interpreted as percentage change in the trait mean under unit selection. The scaling exponents $b \pm$ SE (R^2) were estimated from the slope of least squares regression on log transformed variances of the 15 traits at the 2 levels. The traits used in the analysis were defined by the eigenvectors of the corresponding **P** when **G** was used as the explanatory variable and by the eigenvectors of the corresponding **G** when **P** was used as the explanatory variable (data are from Lofsvold 1986, 1988). We used the original **G**-matrices presented in Lofsvold (1986) and not the bent **G**-matrices presented in Lofsvold (1988). One obvious sign error was corrected. (c) Evolvability in the direction of divergence from the focal subspecies, for which **G** was estimated, to the subspecies indicated on the x-axis. For comparison, the open triangles show the average evolvability of the 2-dimensional plane with highest divergence in each subspecies (this plane accounted for approximately 70% of the divergence). The vertical lines show (from top to bottom) maximum evolvability, average evolvability, and minimum evolvability, respectively, of the **G**-matrix. Filled circles show evolvability, and open circles show conditional evolvability. The average conditional evolvabilities are not shown as they were visually indistinguishable from the minimum evolvabilities. The figure is based on the bent **G**-matrices published in Lofsvold (1988), to avoid negative minimum evolvabilities.

among these taxa are along lines of low genetic resistance by comparing the evolvability in these directions with the minimum, average, and maximum evolvability of the **G**-matrices using the "evolvability" R-package (Bolstad et al. 2014). Our analysis (figure 14.1c) shows that the evolvabilities in the directions of divergence are high compared to the average evolvabilities. This suggests that the relationship between divergence and evolvabilities is retained up to the species timescale for these data. The conditional evolvabilities were very low in three instances (open circles in figure 14.1c). This may not reflect reality, as the estimated **G** has several dimensions with very little evolvability. Low conditional evolvabilities can arise from estimation error in the orientation of **G**. If the direction of divergence is only slightly correlated (due to estimation error) with a direction of **G** with close to zero evolvability, its conditional evolvability will be very small.

14.4 Empirical Evidence for a Relationship between Evolvability and Divergence

Evolvability should correlate with divergence if the former constrains the latter, but as we argued in section 14.3, meaningful assessments of a relationship between divergence and evolvability can be obscured by methodological issues. Therefore, in our review of studies assessing such a relationship (table 14.1), we have briefly summarized the methods used.

The first thing to notice from table 14.1 is that more studies are reporting a relationship rather than failing to find one. Several of the studies that did not find a relationship may also have failed to do so because of methodological issues, as we have shown with Lofsvold's (1988) study. In contrast, some of the observed relationships between evolvability and divergence might be due to comparison of traits with different dimensionality or with different units (see the "DC" column). However, even if we only consider studies that have used dimensionally consistent traits, the evidence for a relationship between evolvability and divergence is strong. Holstad et al. (in preparation) found a positive relationship between evolvability and divergence across 409 univariate traits collected from 123 different species. Variation in evolvability explained 30% of the variation in among-population variance and 12% of the among-species variance (figure 14.2). A detailed analysis of the plant subset of this data further supports these findings but also identifies an important role of the trait function, which together with evolvability, explained 40% of the variation in population divergence (Opedal et al. 2023). Hence, a preliminary answer to the first question of the title of this chapter—Does lack of evolvability constrain adaptation?—is yes, in the sense that plenty of circumstantial evidence indicates that evolvability does constrain evolution and therefore also adaptation. This result aligns well with the many studies reporting a relationship between within-population phenotypic variation and covariation (i.e., the **P** matrix) and divergence (e.g., Hunt 2007b; Grabowski et al. 2011; Baab 2018; Tsuboi et al. 2018).

The evolvability-divergence relationship is commonly observed both on the population and the species timescales (table 14.1). Hence, the answer to the second question of the title is that constraints appear to be common even on a macroevolutionary timescale, where divergence times are often on the order of millions of years. Holstad et al. (in preparation) observed a weakening in the evolvability-divergence relationship at the species timescale compared to the relationship observed at the timescale of population divergence. Other

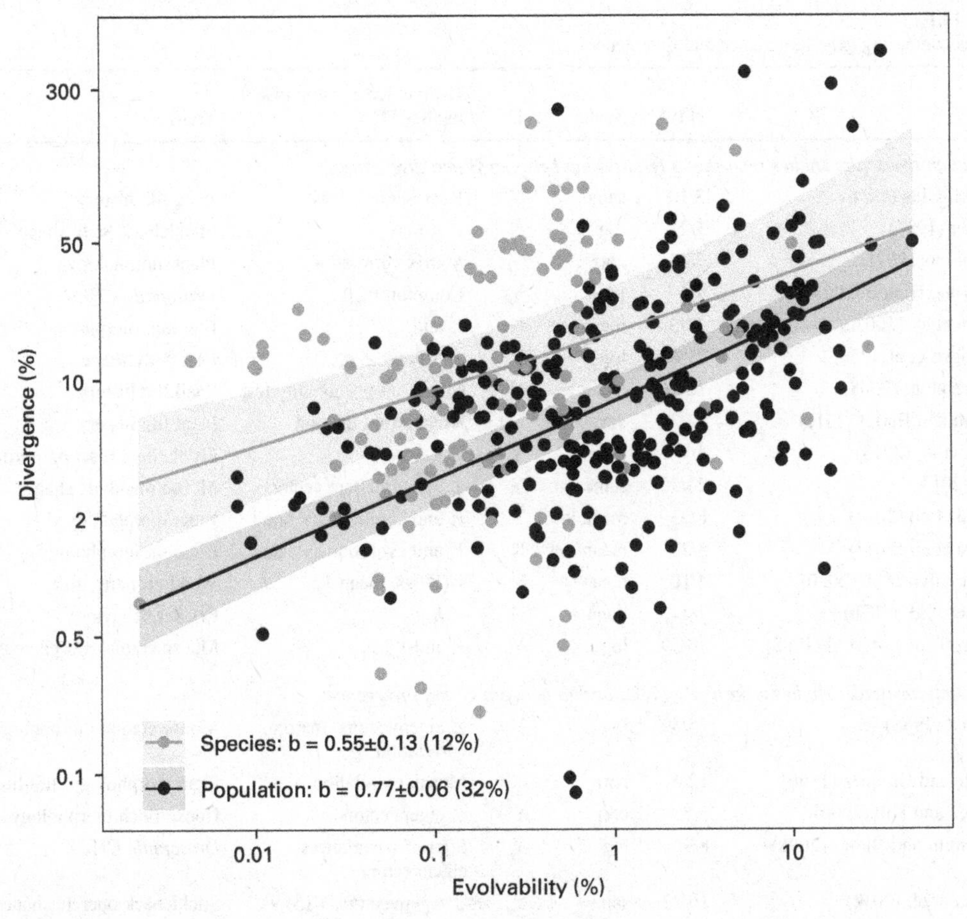

Figure 14.2
Divergence among populations and species predicted by evolvability. Divergence is expressed as expected proportional divergence in percentage change from the mean of the measured populations per trait. Evolvability is expressed as the mean percentage potential evolutionary change. The scaling exponents $b \pm \mathrm{SE}$ (and marginal R^2) are obtained from mixed-effect models on natural log-transformed variables (divergence and evolvability) with closest shared taxa as random effect. The figure is rendered with permission from Holstad et al. (in preparation).

studies likewise report a weakening relationship with divergence time (Schluter 1996; Berger et al. 2014; Chakrabarty and Schielzeth 2020; but see Innocenti and Chenoweth 2013), supporting the idea that constraints break down over time.

The studies listed in table 14.1 cover a variety of traits, including thermal reaction norms, cuticular hydrocarbons (CHCs), morphological shape, and gene expression, as well as a wide variety of taxa. Hence, the positive relationship between **G** and divergence seems to be very general, at least within each trait group.

The positive relationship between evolvability and divergence is not a given, considering that most quantitative traits seem to harbor levels of additive genetic variance that could generate rates of evolution that far exceed those we observe. Furthermore, both evolvability and divergence are estimates of variance at particular levels of biological organization, which require substantial amounts of data to be estimated with high accuracy. The estimates reported in the studies listed in table 14.1 thus all come with rather large errors, which will

Table 14.1
Studies comparing genetic variance and divergence

Study	N/tx*	Scale‡	DC§	G-divergence comparison method**	Traits
Population timescale: studies reporting a relationship between G and divergence					
Mitchell-Olds (1996)	3/10	same	Y	Regression slopes	Plant life history
Schluter (1996)	5/21	log	N	∠ g-max	Stickleback body shape
Andersson (1997)	7/12	corr	Y	Matrix correlation	Plant morphology
Blows and Higgie (2003)	4/6	log	Y	Common PCA	*Drosophila* CHCs
Hansen et al. (2003a)	24/5	mean	N	V_A and V_D	Blossom morphology
McGuigan et al. (2005)	21/8	log	Y	∠ g-max; ∠ $\boldsymbol{p}_i$	Fish body shape
Chapuis et al. (2008)	12/16	mean	N	Matrix proportionality test	Snail life history
Colautti and Barret (2011)	12/20	var	Y§§	Krzanowski method	Plant life history
Berger et al. (2013)	5/7	mean	Y	∠ g-max	Fly thermal reaction norms
Boell (2013)	24/50†	same	Y	∠ genetic effect vectors	Mouse mandible shape
Bolstad et al. (2014)	6/23	mean/log	Y	V_A and evolutionary rate	Bract morphology
Bolstad et al. (2014)	5/23	mean/log	N	V_A and evolutionary rate	Blossom morphology
Costa e Silva et al. (2020)	4/10	mean	N	$V_A(\boldsymbol{\beta})$ vs. mean V_A	Wood property traits
Royauté et al. (2020)	7/4	none	N	∠ $\boldsymbol{h}_i$	Cricket behavior
Reanalysis of Lofsvold (1988)	15/59	log	Y	V_A and V_D	Mouse cranial morphology
Population timescale: studies reporting no relationship between G and divergence:					
Lofsvold (1988)	15/59	log	Y	∠ eigenvectors; matrix correlation	Mouse cranial morphology
Venable and Búrquez (1990)	12/6	corr	Y§§	Matrix correlation	Plant morphology /life-history
Badyaev and Hill (2000)	5/7	corr	Y§§	∠ eigenvectors	House finch morphology
Chenoweth and Blows (2008)	8/9	log	Y	Sign of covariances; eigenvectors	*Drosophila* CHCs
Kimmel et al. (2012)	10/22	same	Y	∠ eigenvectors; $V_A(\boldsymbol{\beta})$ vs. mean V_A	Stickleback opercle shape
Species timescale: studies reporting a relationship between G and divergence					
Schluter (1996)	5/26	log	Y	∠ g-max	Bird and mouse morphology
Baker and Wilkinson (2003)	9/15	corr	Y	Matrix correlation	Stalk-eyed fly morphology
Bégin and Roff (2003)	5/3	log	Y	∠ g-max	Cricket morphology
Bégin and Roff (2004)	5/7	log	Y	∠ eigenvectors	Cricket morphology
Marroig and Cheverud (2005)	39/16	same	Y	∠ g-max	Monkey cranial morphology
Hansen and Houle (2008)	8/20	same	Y	$V_A(\boldsymbol{\beta})$ vs. mean V_A	*Drosophila* wing shape
Boell (2013)	24/50†	same	Y	∠ genetic effect vectors	Mouse mandible shape
Innocenti and Chenoweth (2013)	36/7	same	Y	$V_A(\boldsymbol{\beta})$ vs. mean V_A	*Drosophila* gene expression
Porto et al. (2015)	30/6	same	Y	V_A and V_D	Marsupial cranial morphology
Houle et al. (2017)	17/117	same	Y	V_A and evolutionary rate	*Drosophila* wing shape
Lucas et al. (2018)	69/8	corr	Y	PCA similarity index	Butterfly wing pattern
McGlothlin et al. (2018)	8/7	log	Y	V_A and V_D; ∠ $\boldsymbol{h}_i$	*Anolis* lizard skeletal shape
Polly and Mock (2018)	14/13	same	Y	∠ eigenvectors; matrix correlation	Shrew molar shape
Chakrabarty and Schielzeth (2020)	10/3	same	Y	V_A and V_D	Grasshopper morphology
Reanalysis of Lofsvold (1988)	15/3	log	Y	$V_A(\boldsymbol{\beta})$ vs. mean V_A	Mouse cranial morphology

Table 14.1
(continued)

Study	N/tx*	Scale‡	DC§	G-divergence comparison method**	Traits
Species timescale: studies reporting no relationship between G and divergence					
Hohenlohe and Arnold (2008)	2/39	same	Y	Matrix size, shape and orientation	Snake vertebral number

Note: The studies are categorized by the timescale of divergence (population or species) and whether they report a relationship between the two levels of variation.

* N = number of traits; tx = number of taxa.

† Total number of taxa (mix of species, subspecies, and populations within subspecies).

‡ Same = measured in same units; log = naturally log transformed; corr = correlation matrices; mean = mean scaling; var = phenotypic variance scaling; mean/log = variances mean scaled, evolutionary rates log transformed; none = no standardization.

§ Dimensional consistency.

§§ The traits have different dimensions, but their correlations are comparable.

** V_A = genetic variance, V_D = among taxa variance, ∠ = angle between divergence vector(s) and PCA = Principal Component Analysis, $\boldsymbol{p}_i$ = ith resultant projection of genetic variance closest to the direction of phenotypic divergence, $V_A(\boldsymbol{\beta})$ = genetic variance along a vector of species divergence, $\boldsymbol{h}_i$ = ith eigenvector from Krzanowski's common subspace analysis of several **G**s.

tend to obscure a potential relationship between evolvability and divergence. One interpretation of the data is that the underlying relationship is so strong that even rather poor estimates are sufficient to detect the signal. If a strong signal between divergence and evolvability is the norm, this can inform us about the likely historical trait dynamics, as different models make different predictions regarding a relationship between divergence and evolvability. Section 14.5 therefore reviews various theoretical models of phenotypic divergence and the relationships between evolvability and divergence that they predict.

14.5 Predicted Relationships between Evolvability and Divergence

In this section, we present a sample of models predicting scaling relationships between evolvability and divergence. Some models of trait evolution predict a relationship, while others do not (Hansen and Martins 1996). The models differ primarily in their assumptions about the adaptive landscape and how it changes over time. Contrasting data with theoretical predictions is a fruitful approach to better understand correlations between evolvability and divergence.

14.5.1 Neutral and Linear Selection

Models of neutral evolution (flat adaptive landscape) or constant or fluctuating linear selection (tilted adaptive surface) predict a positive, linear relationship between evolvability and divergence. Predicted levels of trait divergence, however, are far larger than empirical observations (e.g., Lynch 1990; Estes and Arnold 2007; Houle et al. 2017).

14.5.2 Fixed Optimum

Lande (1976) developed a model with a single optimum, where the variance among taxa is given by a balance between selection and genetic drift. The stationary variance of the trait mean under this model (assuming weak selection) is $\mathrm{Var}(\bar{z}) = 1/(4sN_e)$, where N_e is the effective population size, and s is the curvature of the quadratic fitness function (i.e., the selection gradient $\beta = -2s(\bar{z} - \theta)$, where θ is the optimum; see also Hansen and Martins 1996). Hence, at equilibrium, this model does not predict any relationship between evolvability and divergence. The initial approach to the optimum generates a positive relationship between evolvability and divergence, but it requires an assumption of short timescale, very weak stabilizing selection, and/or low evolvability.

14.5.3 Moving Optimum (Ignoring Genetic Drift)

Bolstad et al. (2014) analyzed an evolutionary model in which the optimum moved according to an Ornstein-Uhlenbeck (OU) process (figure 14.3). The OU process of the optimum is given by $d\theta = -\alpha(\theta - \bar{\theta})dt + \sigma dB$, where α describes the "pull" of the trait toward the primary optimum $\bar{\theta}$, and σ is a parameter scaling the white noise (dB) process. Under this model, the stationary variance in the species means is given by $\text{Var}(\bar{z}) = 2Ves/(2es + \alpha)$, where $V = \sigma^2/(2\alpha)$ is the stationary variance of the OU-process, and e is the evolvability. If the movement of the optimum is much faster than the response to selection, then the population cannot track the optimum and the variance of the trait mean goes toward 0. If adaptation is much faster than the movement of the optimum, the populations would track it perfectly, and variance of the trait mean would converge on the variance of the optimum V. Between these two extremes, the relationship between evolvability and among population variance is concave (i.e., negative second derivative), and we therefore expect a scaling relationship between evolvability and divergence between 0 and 1. The value of the relationship depends on the value of α relative to the product $2es$. If $\alpha \approx 2es$, populations lag far behind their optimum, and the scaling becomes close to isometry. When α is smaller than $2es$, populations will track the optimum faster, and the scaling coefficient will decrease.

If trait means evolve according to a stationary OU-process, the phylogenetic signal decreases over time. Therefore, if we replace α with $2es$ and use reasonable values of e

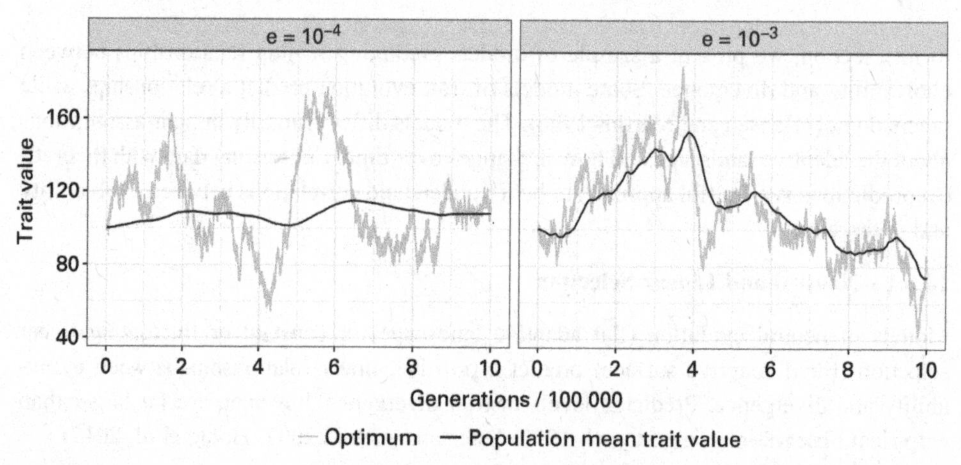

Figure 14.3
Tracking a moving optimum. Shown are the dynamics of two traits differing in evolvability (10^{-4} and 10^{-3}), both tracking a moving optimum following an OU-process with parameters $\alpha = 10^{-5}$, $\bar{\theta} = 100$, and $\sigma = 0.1$, with weak stabilizing selection ($s = 0.01$). The trait with the highest evolvability tracks the optimum much better than the trait with low evolvability. Consequently, the evolvability will be positively related to population divergence in this scenario (given that trait optima move independently among populations).

and s, we can evaluate at what timescales we would expect to observe both a nearly isometric scaling relationship and a phylogenetic signal in the traits. Mean-scaled evolvability is often around 10^{-3} (Hansen et al. 2011), while moderately strong stabilizing selection would be given by $s=1$. These values give a half-life ($\ln(2)/\alpha$) of about 350 generations, showing that this model is only consistent with observing a phylogenetic signal on very short timescales. For traits varying around a low level of evolvability, say $e=10^{-4}$, and experiencing very weak stabilizing selection, say $s=0.01$, the half-life would be about 350,000 generations, which would be consistent with observing a phylogenetic signal on the population timescale but not the species timescale. The latter would require even weaker selection or lower evolvability. This model can explain a relationship between evolvability and divergence but only in a very restricted part of parameter space.

The above OU-model converges on a Brownian motion when $\alpha \rightarrow 0$, and σ is finite. In this situation, the variance in the trait means settles on the same rate of increase as the variance in the optimum, but with a constant lag that is inversely proportional to the evolvability, resulting in a weak relationship between evolvability and divergence.

14.5.4 Natural Selection Shaping within and among Species Variances

A relationship between divergence and evolvability may result from selection shaping evolvability to align with the adaptive landscape (e.g., Pavličev et al. 2011; Jones et al. 2014), which in turn may align with directions of divergence among populations. Following Arnold et al. (2001), this alignment can happen if peak movement follows directions of "selective lines of least resistance." In this model, the adaptive landscape is Gaussian in all trait dimensions, and directions with weaker stabilizing selection (wider bell curves) are assumed to be more prone to peak movement, and hence, divergence. In addition, the strength of stabilizing selection must be negatively related to evolvability, but this is not necessarily the case (Hermisson, et al. 2003; Le Rouzic et al. 2013).

14.5.5 Local Adaptation with Gene Flow

In a system with gene flow between populations, among-population variance in a trait will be determined by the balance between gene flow reducing variation and local adaptation to different optima increasing variation. Because the response to natural selection depends on the evolvability, we would expect traits with high evolvability to reside closer to their optima compared to traits with low evolvability, and therefore a positive relationship between evolvability and among-population variance. In addition, we would expect an increase in the evolvability due to the build-up of linkage disequilibrium (Bulmer 1980; Tufto 2000; Pélabon et al., chapter 13). The increase in evolvability due to linkage disequilibrium would depend on the among-population variance, which would further strengthen the relationship between evolvability and divergence. However, this model cannot explain the observed relationship at the species level.

14.6 Dynamics of the Adaptive Landscape across Time

Understanding the nature of how the adaptive landscape changes across time is key to assessing whether evolvability is likely to constrain adaptation. Evolvability as a constraint should be common if peak movements generally outpace the ability of populations to track the topological changes; in contrast, it should not be important if landscape changes are slow or rare relative to the evolvability (see section 14.5.3 on moving-optimum models). The observation that populations generally are displaced from their optimum (Estes and Arnold 2007) might indicate that the adaptive landscape is in constant flux (see also Chevin et al. 2015 and Gamelon et al. 2018). Studies of the fossil record on the sub-million-year timescale support this view. Changes in trait means within a limited range, which we term stationary trait dynamics, are a common mode of evolution in lineages on this timescale (e.g., Gingerich 2001; Hunt 2007a; Uyeda et al. 2011; Voje 2016). The magnitudes of trait change during such a stationary phase are frequently too large for a fixed optimum model to explain (e.g., Arnold 2014; Voje et al. 2018).

If the adaptive landscape changes on short timescales, optima must be able to show larger changes on macroevolutionary timescales. Despite many verbal models of macroevolution—for example, adaptive radiation (Schluter 2000), punctuated equilibrium (Eldredge and Gould 1972), and Red Queen (Van Valen 1973)—there are currently few formal models of the dynamics of the adaptive landscape on macroevolutionary timescales. Existing models are phenomenological in the sense that they are derived solely from the fit of stochastic models, such as Brownian motion or Ornstein-Uhlenbeck processes, to empirical data (e.g., Hansen 2012; Uyeda and Harmon 2014). For example, several studies have explored shifts in the adaptive landscape along branches of a phylogeny using Ornstein-Uhlenbeck models (e.g., Mahler et al. 2013). Whether these estimated shifts represent cumulative changes in the position of adaptive peaks across time or they represent sudden large-scale changes in the adaptive landscape is currently hard to disentangle (e.g., Uyeda and Harmon 2014). Unifying analyses of microevolutionary, fossil, and phylogenetic data is one way forward to improve our understanding of adaptive landscape dynamics. For example, analyses of evolutionary sequences describing how single linages evolve on a sub-million-year timescale (e.g., Hunt et al. 2008; Reitan et al. 2012; Voje 2020) could assess whether large-scale shifts in adaptive optima happen more frequently than predicted based on phylogenetic comparative data. Incorporating measurements of evolvability into comparative methods is also likely to better our understanding of the relationship between evolvability and divergence along the timescale continuum (for a statistical framework, see Hansen et al. 2021).

14.7 Conclusion

The predicted effectiveness of adaptation suggested by univariate estimates of evolvability strongly indicates that maladaptation should be a transient phenomenon in natural populations. Still, maladaptation seems to be a common state in nature. The large body of work showing a correlation between phenotypic divergence and evolvability may suggest that genetic constraints are important, but we lack evolutionary models adequately explaining how constraints can be so pervasive. Contrasting data with clear theoretical predictions on the role of evolvability in phenotypic divergence can help answer a range of currently unanswered questions:

- Does the relationship between divergence and evolvability weaken with time?
- What is the relative explanatory power of genetic constraints and selection on observed correlations between divergence and evolvability?

- How much is evolvability reduced when conditioning on traits known to be under stabilizing selection?
- How similar are the inferred dynamics of the adaptive landscape when analyses are based on different types of data spanning different time intervals?

Acknowledgments

This chapter emerged from our participation in the project "Evolvability: A New and Unifying Concept for Evolutionary Biology?" (2019–2020), funded by the Norwegian Academy of Science and Letters and hosted by the Centre for Advanced Study (Oslo) in 2019–2020. We thank all involved in the project for feedback on an earlier version of this chapter. We thank Thomas F. Hansen, Christophe Pélabon, Mihaela Pavličev, David Houle, Laura Nuño de la Rosa, and an anonymous reviewer for thoughtful and thorough comments on the manuscript. KLV was supported by an ERC–2020–STG (Grant agreement ID: 948465), MG was supported by the Fulbright U.S. Scholars Program, AH was supported by the Norwegian Research Council (Grant #287214), MT was funded by the Swedish Research Council (2016-06635), and GHB was supported by the Norwegian Research Council (Grants # 275862 and 287214).

References

Andersson, S. 1997. Genetic constraints on phenotypic evolution in *Nigella* (Ranunculaceae). *Biological Journal of the Linnean Society* 62: 519–532.

Arnold, S. J. 1992. Constraints on phenotypic evolution. *American Naturalist* 140: S85–S107.

Arnold, S. J. 2014. Phenotypic evolution: The ongoing synthesis. *American Naturalist* 183: 729–746.

Arnold, S. J., M. E. Pfrender, and A. G. Jones. 2001. The adaptive landscape as a conceptual bridge between micro- and macroevolution. *Genetica* 112–113: 932.

Baab, K. L. 2018. Evolvability and craniofacial diversification in genus *Homo*. *Evolution* 72: 2781–2791.

Badyaev, A. V., and G. E. Hill. 2000. The evolution of sexual dimorphism in the house finch. I. Population divergence in morphological covariance structure. *Evolution* 54: 1784–1794.

Baker, R. H., and G. S. Wilkinson. 2003. Phylogenetic analysis of correlation structure in stalk-eyed flies (*Diasemopsis*, Diopsidae). *Evolution* 57: 87–103.

Bégin, M., and D. A. Roff. 2003. The constancy of the G matrix through species divergence and the effects of quantitative genetic constraints on phenotypic evolution: A case study in crickets. *Evolution* 57: 1107–1120.

Bégin, M., and D. A. Roff. 2004. From micro- to macroevolution through quantitative genetic variation: Positive evidence from field crickets. *Evolution* 58: 2287–2304.

Berger, D., E. Postma, W. U. Blanckenhorn, and R. J. Walters. 2013. Quantitative genetic divergence and standing genetic (co)variance in thermal reaction norms along latitude. *Evolution* 67: 2385–2399.

Berger, D., R. J. Walters, and W. U. Blanckenhorn. 2014. Experimental evolution for generalists and specialists reveals multivariate genetic constraints on thermal reaction norms. *Journal of Evolutionary Biology* 27: 1975–1989.

Blows, M. W., and M. Higgie. 2003. Genetic constraints on the evolution of mate recognition under natural selection. *American Naturalist* 161: 240–253.

Boell, L. 2013. Lines of least resistance and genetic architecture of house mouse (*Mus musculus*) mandible shape. *Evolution and Development* 15: 197–204.

Bolstad, G. H., T. F. Hansen, C. Pélabon, M. Falahati-Anbaran, R. Pérez-Barrales, and W. S. Armbruster. 2014. Genetic constraints predict evolutionary divergence in *Dalechampia* blossoms. *Philosophical Transactions of the Royal Society B* 369: 20130255.

Bradshaw, A. D. 1991. The Croonian Lecture, 1991. Genostasis and the limits to evolution. *Philosophical Transactions of the Royal Society* B 333: 289–305.

Bulmer, M. G. 1980. *The Mathematical Theory of Quantitative Genetics.* London: Clarendon Press.

Chakrabarty, A., and H. Schielzeth. 2020. Comparative analysis of the multivariate genetic architecture of morphological traits in three species of Gomphocerine grasshoppers. *Heredity* 124: 367–382.

Chapuis, E., G. Martin, and J. Goudet. 2008. Effects of selection and drift on G matrix evolution in a heterogeneous environment: A multivariate Q_{st}-F_{st} test with the freshwater snail *Galba truncatula. Genetics* 180: 2151–2161.

Chenoweth, S. F., and M. W. Blows. 2008. Q_{st} meets the G matrix: The dimensionality of adaptive divergence in multiple correlated quantitative traits. *Evolution* 62: 1437–1449.

Cheverud, J. M. 1988. A comparison of genetic and phenotypic correlations. *Evolution* 42: 958–968.

Chevin, L., M. E. Visser, and J. Tufto. 2015. Estimating the variation, autocorrelation, and environmental sensitivity of phenotypic selection. *Evolution* 69: 2319–2332.

Colautti, R. I., and S. C. H. Barrett. 2011. Population divergence along lines of genetic variance and covariance in the invasive plant *Lythrum salicaria* in eastern North America. *Evolution* 65: 2514–2529.

Costa e Silva, J., B. M. Potts, and P. A. Harrison. 2020. Population divergence along a genetic line of least resistance in the tree species *Eucalyptus globulus. Genes* 11: 1095.

Crespi, B. J. 2000. The evolution of maladaptation. *Heredity* 84: 623–629.

Dickerson, G. E. 1955. Genetic slippage in response to selection for multiple objectives. *Cold Spring Harbor Symposia on Quantitative Biology* 20: 213–224.

Eldredge, N., and S. J. Gould. 1972. Punctuated equilibria: An alternative to phyletic gradualism. In *Models in Paleobiology,* edited by T. Schopf, 82–115. San Francisco: Freeman Cooper.

Estes, S., and S. J. Arnold. 2007. Resolving the paradox of stasis: Models with stabilizing selection explain evolutionary divergence on all timescales. *American Naturalist* 169: 227–244.

Gamelon, M., J. Tufto, A. L. K. Nilsson, K. Jerstad, O. W. Røstad, N. C. Stenseth, and B.-E. Sæther. 2018. Environmental drivers of varying selective optima in a small passerine: A multivariate, multiepisodic approach. *Evolution* 72: 2325–2342.

Gingerich, P. D. 1993. Quantification and comparison of evolutionary rates. *American Journal of Science* 293-A: 453–478.

Gingerich, P. D. 2001. Rates of evolution on the time scale of the evolutionary process. *Genetica* 112–113: 127–144.

Gomulkiewicz, R., and D. Houle. 2009. Demographic and genetic constraints on evolution. *American Naturalist* 174: E218–E229.

Grabowski, M., and C. C. Roseman. 2015. Complex and changing patterns of natural selection explain the evolution of the human hip. *Journal of Human Evolution* 85: 94–110.

Grabowski, M. W., J. D. Polk, and C. C. Roseman. 2011. Divergent patterns of integration and reduced constraint in the human hip and the origins of bipedalism. *Evolution* 65: 1336–1356.

Hansen, T. F. 2012. Adaptive landscapes and macroevolutionary dynamics. In *The Adaptive Landscape in Evolutionary Biology,* edited by E. I. Svensson and R. Calsbeek, 205–226. Oxford: Oxford University Press.

Hansen, T. F., and D. Houle. 2004. Evolvability, stabilizing selection, and the problem of stasis. In *Phenotypic Integration: Studying the Ecology and Evolution of Complex Phenotypes,* edited by M. Pigliucci and K. Preston, 130–150. Oxford: Oxford University Press.

Hansen, T. F., and D. Houle. 2008. Measuring and comparing evolvability and constraint in multivariate characters. *Journal of Evolutionary Biology* 21: 1201–1219.

Hansen, T. F., and E. P. Martins. 1996. Translating between microevolutionary process and macroevolutionary patterns: The correlation structure of interspecific data. *Evolution* 50: 1404–1417.

Hansen, T. F., and C. P. Pélabon. 2021. Evolvability: A quantitative-genetics perspective. *AREES* 52: 153–175.

Hansen, T. F., and K. L. Voje. 2011. Deviation from the line of least resistance does not exclude genetic constraints: A comment on Berner et al. (2010). *Evolution* 65: 1821–1822.

Hansen, T. F., C. Pélabon, W. S. Armbruster, and M. L. Carlson. 2003a. Evolvability and genetic constraint in *Dalechampia* blossoms: Components of variance and measures of evolvability. *Journal of Evolutionary Biology* 16: 754–766.

Hansen, T. F., W. S. Armbruster, M. L. Carlson, and C. Pélabon. 2003b. Evolvability and genetic constraint in *Dalechampia* blossoms: Genetic correlations and conditional evolvability. *Journal of Experimental Zoology B* 296: 23–39.

Hansen, T. F., C. Pélabon, and D. Houle. 2011. Heritability is not evolvability. *Evolutionary Biology* 38: 258–277.

Hansen, T. F., G. H. Bolstad, and M. Tsuboi. 2021. Analyzing disparity and rates of morphological evolution with model-based phylogenetic comparative methods. *Systematic Biology* 71: 1054–1072.

Hendry, A. P., and M. T. Kinnison. 1999. The pace of modern life: Measuring rates of contemporary microevolution. *Evolution* 53: 1637–1653.

Hereford, J., T. F. Hansen, and D. Houle. 2004. Comparing strengths of directional selection: How strong is strong? *Evolution* 58: 2133–2143.

Hermisson, J., T. F. Hansen, and G. P. Wagner. 2003. Epistasis in polygenic traits and the evolution of genetic architecture under stabilizing selection. *American Naturalist* 161: 708–734.

Hohenlohe, P. A., and S. J. Arnold. 2008. MIPoD: A hypothesis-testing framework for microevolutionary inference from patterns of divergence. *American Naturalist* 171: 366–385.

Holstad, A., K. L. Voje, Ø. H. Opedal, G. H. Bolstad, S. Bourg, T. F. Hansen, and C. Pélabon. (in preparation). Evolvability explains divergence among populations of extant and extinct species.

Houle, D. 1992. Comparing evolvability and variability of quantitative traits. *Genetics* 130: 195–204.

Houle, D., C. Pélabon, G. P. Wagner, and T. F. Hansen. 2011. Measurement and meaning in biology. *Quarterly Review of Biology* 86: 3–34.

Houle, D., G. H. Bolstad, K. van der Linde, and T. F. Hansen. 2017. Mutation predicts 40 million years of fly wing evolution. *Nature* 548: 447–450.

Houle, D., G. H. Bolstad, and T. F. Hansen. 2020. Fly wing evolutionary rate is a near-isometric function of mutational variation. BioRxiv, https://doi.org/10.1101/2020.08.27.268938.

Hunt, G. 2007a. The relative importance of directional change, random walks, and stasis in the evolution of fossil lineages. *PNAS* 104: 18404–18408.

Hunt, G. 2007b. Evolutionary divergence in directions of high phenotypic variance in the ostracode genus *Poseidonamicus. Evolution* 61: 1560–1576.

Hunt, G., M. A. Bell, and M. P. Travis. 2008. Evolution toward a new adaptive optimum: Phenotypic evolution in a fossil stickleback lineage. *Evolution* 62: 700–710.

Innocenti, P., and S. F. Chenoweth. 2013. Interspecific divergence of transcription networks along lines of genetic variance in *Drosophila:* Dimensionality, evolvability, and constraint. *Molecular Biology and Evolution* 30:1358–1367.

Jablonski, D. 2004. Extinction: Past and present. *Nature* 427: 589.

Janzen, D. H., and P. S. Martin. 1982. Neotropical anachronisms: The fruits the gomphotheres ate. *Science* 215: 19–27.

Jiang, D., and J. Zhang. 2020. Fly wing evolution explained by a neutral model with mutational pleiotropy. *Evolution* 74: 2158–2167.

Jones, A. G., R. Bürger, and S. J. Arnold. 2014. Epistasis and natural selection shape the mutational architecture of complex traits. *Nature Communications* 5: 1–10.

Kimmel, C. B., W. A. Cresko, P. C. Phillips et al. 2012. Independent axes of genetic variation and parallel evolutionary divergence of opercle bone shape in threespine stickleback. *Evolution* 66: 419–434.

Kinnison, M. T., and A. P. Hendry. 2001. The pace of modern life II: From rates of contemporary microevolution to pattern and process. *Genetica* 112–113: 145–164.

Kluge, A. G., and W. C. Kerfoot. 1973. The predictability and regularity of character divergence. *American Naturalist* 107: 426–442.

Lande, R. 1976. Natural selection and random genetic drift in phenotypic evolution. *Evolution* 30: 314–334.

Lande, R. 1979. Quantitative genetic analysis of multivariate evolution, applied to brain: body size allometry. *Evolution* 33: 402–416.

Lande, R. 2007. Expected relative fitness and the adaptive topography of fluctuating selection. *Evolution* 61: 1835–1846.

Le Rouzic, A., J. M. Álvarez-Castro, and T. F. Hansen. 2013. The evolution of canalization and evolvability in stable and fluctuating environments. *Evolutionary Biology* 40: 317–340.

Lofsvold, D. 1986. Quantitative genetics of morphological differentiation in *Peromyscus.* I. Tests of the homogeneity of genetic covariance structure among species and subspecies. *Evolution* 40: 559–573.

Lofsvold, D. 1988. Quantitative genetics of morphological differentiation in *Peromyscus.* II. Analysis of selection and drift. *Evolution* 42: 54–67.

Lucas, L. K., C. C. Nice, and Z. Gompert. 2018. Genetic constraints on wing pattern variation in *Lycaeides* butterflies: A case study on mapping complex, multifaceted traits in structured populations. *Molecular Ecology Resources* 18: 892–907.

Lynch, M. 1990. The rate of morphological evolution in mammals from the standpoint of the neutral expectation. *American Naturalist* 136: 727–741.

Mahler, D. L., T. Ingram, L. J. Revell, and J. B. Losos. 2013. Exceptional convergence on the macroevolutionary landscape in island lizard radiations. *Science* 341: 292–295.

Marroig, G., and J. M. Cheverud. 2005. Size as a line of least evolutionary resistance: Diet and adaptive morphological radiation in New World monkeys. *Evolution* 59: 1128–1142.

Maynard Smith, J., R. Burian, S. Kauffman et al. 1985. Developmental constraints and evolution. *Quarterly Review of Biology* 60: 265–287.

McGlothlin, J. W., M. E. Kobiela, H. V. Wright, D. L. Mahler, J. J. Kolbe, J. B. Losos, and E. D. Brodie. 2018. Adaptive radiation along a deeply conserved genetic line of least resistance in *Anolis* lizards. *Evolution Letters* 2: 310–322.

McGuigan, K., S. F. Chenoweth, and M. W. Blows. 2005. Phenotypic divergence along lines of genetic variance. *American Naturalist* 165: 32–43.

Mitchell-Olds, T. 1996. Pleiotropy causes long-term genetic constraints on life-history evolution in *Brassica rapa*. *Evolution* 50: 1849–1858.

Ohta, T. 1992. The nearly neutral theory of molecular evolution. *AREES* 23: 263–286.

Opedal, Ø. H., W. S. Armbruster, T. F. Hansen, et al. 2023. Trait function and evolvability predict phenotypic divergence of plant populations. *PNAS* 120 (1) e2203228120.

Pavličev, M., J. M. Cheverud, and G. P. Wagner. 2011. Evolution of adaptive phenotypic variation patterns by direct selection for evolvability. *Proceedings of the Royal Society B* 278: 1903–1912.

Pélabon, C., C. H. Hilde, S. Einum, and M. Gamelon. 2020. On the use of the coefficient of variation to quantify and compare trait variation. *Evolution Letters* 4: 180–188.

Polly, P. D., and O. B. Mock. 2018. Heritability: the link between development and the microevolution of molar tooth form. *Historical Biology* 30: 53–63.

Porto, A., H. Sebastião, S. E. Pavan, J. L. Vandeberg, G. Marroig, and J. M. Cheverud. 2015. Rate of evolutionary change in cranial morphology of the marsupial genus *Monodelphis* is constrained by the availability of additive genetic variation. *Journal of Evolutionary Biology* 28: 973–985.

Reeve, H. K., and P. W. Sherman. 1993. Adaptation and the goals of evolutionary research. *Quarterly Review of Biology* 68: 1–32.

Reitan, T., T. Schweder, and J. Henderiks. 2012. Phenotypic evolution studied by layered stochastic differential equations. *Annals of Applied Statistics* 6: 1531–1551.

Royauté, R., A. Hedrick, and N. A. Dochtermann. 2020. Behavioural syndromes shape evolutionary trajectories via conserved genetic architecture. *Proceedings of the Royal Society B* 287: 20200183.

Savolainen, O., T. Pyhäjärvi, and T. Knürr. 2007. Gene flow and local adaptation in trees. *AREES* 38: 595–619.

Schluter, D. 1996. Adaptive radiation along genetic lines of least resistance. *Evolution* 50: 1766–1774.

Schluter, D. 2000. *The Ecology of Adaptive Radiation.* New York: Oxford University Press.

Simpson, G. G. 1944. *Tempo and Mode in Evolution.* New York: Columbia University Press.

Toju, H., and T. Sota. 2006. Adaptive divergence of scaling relationships mediates the arms race between a weevil and its host plant. *Biology Letters* 2: 539–542.

Tsuboi, M., W. van der Bijl, B. T. Kopperud et al. 2018. Breakdown of brain–body allometry and the encephalization of birds and mammals. *Nature Ecology & Evolution* 2: 1492–1500.

Tufto, J. 2000. The evolution of plasticity and non-plastic spatial and temporal adaptations in the presence of imperfect environmental cues. *American Naturalist* 156: 121–130.

Uyeda, J. C., and L. J. Harmon. 2014. A novel Bayesian method for inferring and interpreting the dynamics of adaptive landscapes from phylogenetic comparative data. *Systematic Biology* 63: 902–918.

Uyeda, J. C., T. F. Hansen, S. J. Arnold, and J. Pienaar. 2011. The million-year wait for macroevolutionary bursts. *PNAS* 108: 15908–15913.

Van Valen, L. 1973. A new evolutionary law. *Evolutionary Theory* 1: 1–30.

Venable, D. L., and A. Búrquez. 1990. Quantitative genetics of size, shape, life-history, and fruit characteristics of the seed heteromorphic composite *Heterosperma pinnatum.* II. Correlation structure. *Evolution* 44: 1748–1763.

Voje, K. L. 2016. Tempo does not correlate with mode in the fossil record. *Evolution* 70: 2678–2689.

Voje, K. L. 2020. Testing eco-evolutionary predictions using fossil data: Phyletic evolution following ecological opportunity. *Evolution* 74: 188–200.

Voje, K. L., J. Starrfelt, and L. H. Liow. 2018. Model adequacy and microevolutionary explanations for stasis in the fossil record. *American Naturalist* 191: 509–523.

Walsh, B., and M. W. Blows. 2009. Abundant genetic variation + strong selection = multivariate genetic constraints: A geometric view of adaptation. *AREES* 40: 41–59.

Walsh, B., and M. Lynch. 2018. *Evolution and Selection of Quantitative Traits.* Oxford: Oxford University Press.

15 Evolvability of Flowers: Macroevolutionary Indicators of Adaptive Paths of Least Resistance

W. Scott Armbruster

Flowering plants are good organisms for analysis of the macroevolutionary signals of differential evolvability. The repeated evolutionary paths followed by flowering plants during their 150+ Myr history provide clues about genetic and developmental biases that yield high evolvability. Analysis of heterochronic variation suggests that evolutionary paths of low resistance (high evolvability) and much of the diversification of floral structure and function have been facilitated by heterochrony. There also appear to be links between the development and evolution of floral orientation, a feature surprisingly important in pollination. The modular independence of flowers relative to vegetative traits may enhance their evolvability, as probably do patterns of intra-floral modularity. Another good indicator of evolutionary paths of low resistance and high evolvability is the high levels of homoplasy (parallel evolution and reversals) of some traits. Parallel and convergent evolution is clearly facilitated by effective preaptations being in place.

15.1 Introduction: Looking for Signals of Differential Evolvability in Flowering Plants

Since their invasion of land some 500 million years ago (Morris et al. 2018), plants have undergone two major pulses of increasing complexity, disparity, and diversity. The first was associated with the origin of seeds in the late Devonian (ca. 350 Mybp), and the second, even larger pulse was subsequent to the origin of flowers in the early Cretaceous (or possibly late Jurassic). Thus, unlike in most animal lineages, major increases in complexity and disparity in plants occurred relatively late in their evolutionary history (Leslie et al. 2021). Flowering plants (angiosperms), the subject of this chapter, are thus characterized by this key innovation: flowers. The evolvability of floral traits will be the focus of this chapter.

My goal is to assess patterns and mechanisms of evolutionary divergence of flowering-plant populations and species in relation to potential evolutionary biases (differential evolvabilities) detected (or hinted at) within populations. Although plant evolution is not fundamentally different from animal evolution, plants provide experimental and analytical opportunities not available to most animal biologists. Plants' bodies are highly modular, with many iterative (replicated) structures on the same individual (e.g., leaves, flowers, fruits), which are basically genetically identical and usually also functionally identical. Variation among repeated units reflects ontogenetic variation, positional effects, and adaptive (e.g., sun versus shade leaves) or nonadaptive plastic responses to microenvironmental variation (Diggle 2014). This nongenetic variation can be ecologically important, yielding direct insights into form-function relationships, as well as patterns of stabilizing and canalizing selection. Iterative plant organs also promote direct investigation into the evolutionary significance of phenotypic and genetic integration and modularity (e.g., Berg 1960; Hansen et al. 2007; Pélabon et al. 2011; see reviews in Armbruster et al. 2014; Conner and Lande 2014). Finally, most plants are easy to clone, allowing investigators to address directly environmental sources of phenotypic (co)variation (but see the cautionary notes in Schwaegerle et al. 2000; Schwaegerle 2005).

Because flowering plants are largely immobile, the majority of species employ animals to solve one or more life-history tasks, such as pollen or seed transport. Although plants have evolved a diversity of solutions to their reproductive tasks, there are recurrent themes in their evolutionary diversification, themes that suggest evolutionary "paths of least resistance" (Stebbins 1950, 497; Schluter 1996, 2000). These repeated evolutionary paths presumably reflect routes of elevated evolvability, interacting with differential fitness advantages.

Most early investigation into the evolutionary paths of least resistance in plants have focused on flower structure and related aspects of reproductive systems, a bias I continue in this chapter. However, recent molecular-genetic studies of plant evolution have explored other aspects of plant metabolism, morphology, and function (see Jaramillo and Kramer 2007; Wessinger and Heileman 2020; Julca et al. 2021; Sengupta and Heileman 2022). I will not attempt to review this research, except to acknowledge that it is beginning to yield important insights into the role of evolvability in determining divergence and convergence of organ development, structure, and function between species.

Here, I review patterns of developmental and morphological variation of flowers within and among plant populations and species. I focus on the role of developmentally based shifts and other "exaptive" transitions (sensu Gould and Vrba 1982; ≈ "preadaptations"; i.e., co-option of preexisting features for new functions; see Arnold 1994) and their roles in the origin of phenotypic and ecological novelty associated with the divergence of populations and species. Exaptive transitions suggest evolutionary lines of least resistance, because preaptations often precede repeated, parallel origins of the same or similar novel feature across related lineages ("homoplasy"). Darwin (1872, 175), Simpson (1944), and Mayr (1963) suggested that when a "preadaptive" (= "preaptive," sensu Gould and Vrba 1982) trait is in place, subsequent evolutionary change can happen rapidly (and presumably easily), because the basis for the change is already there (see Arnold 1994; Armbruster 1997; McLennan 2008). Thus exaptation, as a process or result, can yield important clues about evolutionary paths of low resistance.

For the purposes of this chapter, I will assume that repeated transitions in character states reflect population-level evolvability as it interacts with the transformative efficacy of selection (taken together, "evolutionary lability"), where the signal is examined at the among-population level (and above). Thus, I use variation in trait evolutionary lability in plants as a signal hinting at differential evolvability ("ease" of genetically based phenotypic change within populations), while hoping that selection is reasonably constant. This assumption seems reasonable when studying most floral evolution, because selection usually "compares" the pollination consequences of floral changes rather than the transition processes themselves. For example, selection for a new pollinator due to local extinction of an old one will be equally strong whether the needed evolutionary response involves sepals or petals, colors or fragrances. The most likely response will be the most evolvable floral change that attracts a new pollinator. To this end, I discuss both rampant parallel evolution (homoplasy) and the mechanistic bases of such repeated transitions, as ways to gain insights into evolutionary routes of high evolvability.

Within populations, multivariate phenotypic covariation reflects the trajectories of ontogenetic and genetic variation. Lande (1979), Schluter (1996, 2000), Hansen and Houle (2008), and others since have argued cogently that the multivariate directions of greater genetic variation constitute the trajectories of highest evolvability, at least over the short term (see cautionary note in Hansen and Voje 2011). In turn, much of the heritable phenotypic covariation expressed in plants has an ontogenetic basis (i.e., subtle variation in the ontogenetic stage at which certain functional events occur, e.g., in flowers, the opening of petals or dehiscence of stamens). Such variation in the functional chronology in relation to the ontogenetic chronology ("heterochrony"; Gould 1977) creates heritable patterns of phenotypic covariance. We can expect, therefore, that populations, and perhaps species, will tend to diverge along these trajectories of high genetic and phenotypic variation (e.g., Haber 2016; see also Kluge and Kerfoot 1973; Johnson and Mickevitch 1977; Pierce and Mitton 1979; but cf. Sokal 1976 and Riska 1979). This view, although still contentious, is supported by some studies in plants that show population and species divergence to occur largely along the with-population phenotypic or quantitative-genetic trajectories (Armbruster 1991; Andersson 1991; Bolstad et al. 2014). This leads to the suggestion that the study of ontogenetic trajectories of covariation in flowers can reveal evolvability biases that "predict" the divergence of populations and species. The basic take-home message from this review is that plant evolution appears to proceed largely by building on preexisting states or structures or by simple changes in genetically controlled developmental mechanisms (i.e., via exaptation in the broadest sense). Preaptations define the phenotypic starting points for subsequent evolutionary paths of low genetic/developmental resistance.

15.2 Some Methodological Assumptions

The present review assumes that trait evolutionary lability provides some insight into trait evolvability. This relationship is not necessarily a very tight one, however, because evolutionary lability of a trait is also influenced by the efficacy of divergent or diversifying natural selection acting on the trait. Additional factors may further obscure the relationship between evolutionary lability and evolvability (e.g., variation in effective population size, influencing drift; and proximity of divergent populations, influencing gene flow). However, it is probably safe to conclude that the dominant relationship looks something like:

$$\textit{Evolutionary Lability} = \textit{Evolvability} * \textit{Selection} \tag{15.1}$$

where * is some interactive function (e.g., multiplicative if properly scaled), and "selection" refers to transformational efficacy of divergent or diversifying selection (including effects of directionality, consistency, strength, and duration). Note that "lability" can be interpreted as a disposition (i.e., propensity or capacity) for evolutionary change in phenotype, as in the discussions of evolutionary developmental mechanisms in sections 15.3 and 15.4.

However, trait evolutionary lability can also be an observation, as when the evidence for the capacity of evolutionary change is purely macroevolutionary (i.e., drawn from the observation of phenotypic variation among relatives and measured as phenotypic disparity). I use the term also in this sense. Homoplasy (convergence, parallelisms, or reversals in trait evolution) is a useful indicator of trait evolutionary lability in the empirical sense. Given an accurately estimated phylogeny, a high degree of homoplasy in a trait indicates high evolutionary lability, although the reverse is certainly not true (see Wake 1991; Wake et al. 2011).

15.3 Heterochrony: A Repeated Path of Low Evolutionary Resistance

Heterochronic variation within and across species reflects the effect on phenotype of differences in timing of various developmental events, given a sufficient degree of developmental modularity. The two main manifestations of heterochrony that emerge from species comparisons are paedomorphosis (retention of juvenile traits into sexual maturity via truncated ontogeny) and peramorphosis (exaggeration of adult traits at sexual maturity via extended ontogeny; table 15.1). Because such transitions are the result of simple changes in speed or timing of ontogenetic sequences relative to maturation, heterochronic change is a likely route of low evolutionary resistance (i.e., high evolvability), in response to selection for novel morphology and function, at least when the favored phenotype is within the domain of heterochronic possibilities.

An overlapping area of research concerns allometric/isometric transitions between species ("evolutionary allometry"). As an example, consider Gould's (1974, 1977) presentation of allometry in and among cervine species (deer, sensu lato): log (mature-male antler mass) scales closely and positively (slope > 1) with log (body mass). Thus the Irish elk (*Megaloceros giganteus*) could be expected to have exceptionally large antlers by virtue of its large body size alone, although probably both have evolved in concert along a path of low resistance in response to sexual selection (Gould 1974). Evolutionary allometry usually has a heterochronic developmental basis, at least in part (see Gould 1977). The

Table 15.1
Dictionary of heterochrony terms, as applied to plants (including flowers)

Phenotype term	Meaning	Process term	Process description
Paedomorphosis/ Paedomorphy	Retention of juvenile characteristics into sexual maturity via truncated ontogeny	Progenesis	Period of growth of the descendant form is stopped prematurely; advancement of sexual maturation relative to ontogeny of nonsexual structures
		Neoteny	Rate of growth is less in the descendant than in the ancestor; retardation of ontogeny of nonsexual structures relative to sexual maturation
		Post displacement	Delayed onset of growth of nonsexual structures is delayed
Peramorphosis	Exaggeration of adult traits at sexual maturity via extension of ontogenetic trajectory	Hypermorphosis	Extended ontogeny relative to timing of maturity via delayed sexual maturity (delayed offset)
		Acceleration	Extended ontogeny relative to timing of maturity by accelerated ontogeny; growth rate is increased [relative to sexual maturity] (increase in rate)
		Predisplacement	Onset of growth occurs earlier in the descendant than in the ancestor (earlier onset)

Note: Both paedomorphosis and peramorphosis can be produced by one or more of three processes: variation in time of termination of ontogenetic growth, variation in time of initiation of ontogenetic growth, and change in rate of ontogenetic growth relative to sexual maturity (or other temporal landmark; Alberch et al. 1979).

developmental factors that contribute to evolutionary allometry can sometimes be examined by assessing within-population allometric variation ("ontogenetic" and "static" allometries; Pélabon et al. 2013, 2014; see Armbruster 1991 for a plant example). Because both heterochronic and allometric differences between species have their origins in differing developmental trajectories, I include examples of both in this section without distinguishing between them.

15.3.1 Heterochronic Changes in Flowers Can Lead to Pollinator Shifts

One of the earliest well documented examples of ecologically important heterochronic change in flower morphology was presented by Guerrant (1982). He noticed that the shape of the flowers of hummingbird-pollinated *Delphinium nudicaule* in California closely resembled the buds of several bee-pollinated species of *Delphinium*. Elegant formal analyses presented a convincing case of a transition to hummingbird pollination through retention of bud-like floral shape into anthesis (i.e., floral "maturity"), which he recognized as neoteny (figure 15.1).

Figure 15.1
Adaptation to hummingbird pollination in *Delphinium* flowers (see Guerrant 1982 for details). (A) Mature, orange-red flower of hummingbird-pollinated *Delphinium nudicaule*. (B) Floral bud of bee-pollinated *Delphinium glaucum*. (C) Three mature, receptive flowers of *D. glaucum*, one with pollinating bumble bee obtaining nectar. Note the striking shape similarity between the mature *D. nudicaule* flower and the bud of *D. glaucum*. Photos by W. S. Armbruster.

Heterochronic shifts in floral development have led to changes in the length of nectar spurs (outgrowth of the perianth, with nectar at the distal end) in other taxa. These morphological changes have occurred in concert with ecological changes in principal pollinators or evolutionary changes in pollinator morphology (usually proboscis length). The ecological and evolutionary significance of changes in nectar-spur length was examined in detail by Darwin (1862) and has been investigated extensively in subsequent years (e.g., Nilsson 1998; Maad 2000; Whittall and Hodges 2007; Sletvold and Ågren 2010, Boberg et al. 2014). Changes in spur length may reflect plant-pollinator coevolution (Darwin 1862; Wallace 1867), adaptive responses to shifts in pollinator species mediating selection (Whittall and Hodges 2007), or a combination of the two (Boberg et al. 2014). The development of spurs and increases in their lengths reflect localized cell proliferation and/or cell elongation over developmental time (see Wessinger and Hileman 2020, fig. 3). Thus, evolutionary increases in spur length can easily occur via peramorphosis. The ease of the transition from spurless flowers to shallowly and deeply spurred flowers is demonstrated by the large number of independent origins of spurred flowers in multiple families (e.g., Balsaminaceae, Geraniaceae, Orchidaceae, Plantaginaceae, Ranunculaceae, Scrophulariaceae, Tropaeolaceae).

In the case of spurs in *Aquilegia* (Ranunculaceae), there appears to be a trend toward increasing spur length in North American species, mediated by cell elongation late in development (Puzey et al. 2012); this is associated with sequential shifts to pollinators with longer tongues (Whittall and Hodges 2007). However, in tropical *Angraecum* and related angrecoid orchids, there is molecular-phylogenetic evidence suggesting evolution of shorter spurs in some lineages as well as longer spurs in other lineages (Andriananjamanantsoa et al. 2016). Similarly, in many temperate terrestrial orchids, the evolutionary trend seems to be from longer to shorter spurs via paedomorphosis (Box et al. 2008; Box and Glover 2010). Differences in spur length in *Diascia* spp. (Scrophulariaceae) also appear to reflect adaptation to different *Rediviva* bee pollinators (Melittidae) of different leg lengths (Steiner and Whitehead 1990, 1991; Melin et al. 2021), but increasing spur length appears not to have influenced the degree of specialization (Hollens et al. 2017). Thus, changes in spur length leading to shifts in pollinators at the species level appear to be both evolutionarily labile and reversible, as would be expected for a highly evolvable trait, with change mediated by heterochrony.

Much like spurs, floral tubes (elongated bases of fused corollas and/or calyces, usually with nectar secreted at the bottom) range from short to long, and, when narrow, limiting reward access to only those animals with long-enough tongues. Transitions between tube lengths probably also have a heterochronic basis, with the evolution to shorter tubes usually reflecting paedomorphosis and evolution to longer tubes usually reflecting peramorphosis. For example, Ezcurra and de Azkue (1989) suggested that peramorphosis via accelerated corolla development is the best explanation for the evolutionary transition to elongated floral tubes associated with a shift from bee pollination to sphingid moth pollination in *Ruellia* (Acanthaceae).

An example of paedomorphosis leading to specialization in which animals can access a pollen reward is seen in a clade of *Dalechampia* vines (Euphorbiaceae) in Madagascar. All *Dalechampia* species have unisexual flowers united into hermaphroditic inflorescences, which function, in nearly all cases, as single blossoms (i.e., pollination units, or "pseudanthia"). In the basal-most species in Madgascar (i.e., resembling African ancestors), the staminate flowers open fully, and pollen is eaten or collected by pollinating beetles, flies, or bees. However, in one or more clades of derived species, the staminate flowers retain two "juvenile" characteristics into anthesis (time of flower maturation): (1) the sepals fail to split and reflex, so that the flower remains in the spherical shape of a bud, although of "adult" size and with mature pollen; and (2) the receptacle to which the stamens are attached fails to elongate because of "suppressed" cell elongation, and thus the stamens remain enclosed by the sepals. At the time of stamen dehiscence, the margins of the sepals are separated by narrow cracks near the calyx apex; otherwise, the flowers look like enlarged floral buds (figure 15.2). A consequence of this arrangement is that the pollen is largely protected from being eaten by pollinivores (e.g., flies and beetles) or collected by bees unable to buzz their thoracic muscles (e.g., honey bees). Thus, paedomorphic members of this clade of *Dalechampia* have shifted from generalist pollination by beetles, flies, and/or pollen-collecting bees (Armbruster and Baldwin 1998) to more specialized pollination by only those bees, including *Xylocopa, Amegilla,* and *Nomia,* that can "buzz-pollinate" by vibrating their thoracic muscles at high frequencies (Armbruster et al. 2013; Plebani et al. 2015; see review in Vallejo-Marín 2019).

Another common pollinator shift seen in plants is attraction and utilization of pollinators of different body sizes. In most cases, this involves developmentally based allometric/isometric shifts in flower size, with smaller, scaled-down flowers adapted to smaller pollinators and larger, scaled-up flowers adapted to larger pollinators (Armbruster 1990, 1991, 1993; cf. Marroig and Cheverud 2010). Surprisingly little work has been done on this kind pollinator transition, perhaps because it is obvious, or perhaps because flower-size evolution is so often assumed to be correlated with changes in mating system (e.g., the "selfing syndrome;" see section 15.3.2 and Armbruster et al. 2002; Sicard and Lenhard 2011; Cutter 2019; Mazer et al. 2020). Such size shifts can be explained by changes in growth rates or duration, involving cell enlargement, proliferation, or both, and affecting some or all floral parts (Wessinger and Hileman 2020).

15.3.2 Heterochronic Changes Can Lead to Higher Self-Pollination Rates

The above examples notwithstanding, the vast majority of the literature on paedomorphic shifts associated with species divergence and changes in pollination systems concerns shifts from plants with large, cross-pollinating flowers to plants with smaller, self-pollinating flowers (e.g., Hill et al. 1992; Gallardo et al. 1993; Stewart and Canne-Hilliker 1998; Ehlers and Pedersen 2000; Sherry and Lord 2000; Box and Glover 2010; Li and Johnston 2010). Stebbins (1950, 1970, 1974) recognized this as the commonest evolutionary transition seen across flowering plants. There are reputed to be many hundreds of independent transitions from outcrossing to selfing and possibly dozens in the other direction (see Igic and Busch 2013; Whitehead et al. 2018). Why is this transition so common? Although recurrent strong selection for mating-system shifts certainly cannot be ruled out as a factor, several aspects argue that the transition is highly evolvable. Response to selection for self-pollination under pollen limitation (selection for reproductive assurance) may be particularly easy, at least in species that are self-compatible, because allometric miniaturization can occur easily via early sexual maturation relative to flower-size growth. An automatic correlate of allometric floral miniaturization is reduced herkogamy (i.e., reduced *absolute* distance between anthers and stigmas in the same flower or blossom), which is associated with higher rates

Figure 15.2
Blossom inflorescences (pseudanthia) of two species of *Dalechampia* (Euphorbiaceae), illustrating heterochronic transition in development of the staminate (male) flowers (see Armbruster et al. 2013 for details). (A) *Dalechampia tamifolia* with "normal" development, where sepals reflex in a few seconds, and staminal column elongates in less than an hour, just prior to anther dehiscence. The pollen reward is open and available to many kinds of pollinators, including beetles. Note the vertical blossom orientation, which creates a large landing platform. (B) *Dalechampia* aff. *bernieri* with paedomorphic staminate flowers, having sepals that do not reflex and staminal columns that do not elongate. This is an adaptation for buzz pollination, where pollen is available only to those species of bees that can vibrate their thoracic muscles at the right frequency and intensity to sonicate pollen out of the cracks at the tips of the nearly closed, bud-like male flowers. Note the lateral blossom orientation. Symbols: b = staminate flower in bud; m = mature staminate flower at anthesis. Photos by W. S. Armbruster.

of self-pollination (e.g., Armbruster 1988a; Motten and Stone 2000; Armbruster et al. 2002; Opedal et al. 2017). Another consequence of floral miniaturization is, however, reduced investment in advertisements (e.g., smaller petals) and/or rewards (components of the "selfing syndrome," as noted above), with the ecological consequence likely being reduced attraction of pollinators and thus greater dependence on autofertility. This, combined with loss of genetic diversity, has led to the view that self-fertilization in flowering plants and other organisms may be an evolutionary "dead end" or "blind alley" (Stebbins 1950, 1957; Takebayashi and Morrell 2001; Busch and Delph 2017), although there is not unanimity in this conclusion (Takebayashi and Morrell 2001; Igic and Busch 2013). The dead-end hypothesis implies a directionality to evolvability: ease of transition in one direction, but not the other (see discussion in Igic and Busch 2013).

A special case of heterochronic transitions promoting self-pollination is seen in the origin of cleistogamous flowers, which undergo self-pollination and seed maturation without the perianth (petals and/or sepals) opening. This trait is scattered across the angiosperm phylogeny, reflecting dozens of independent origins. The bud-like nature of the perianth of sexually mature cleistogamous flowers indicates that heterochrony probably played a role in each of the independent origins (Lord 1981, 1982; Minter and Lord 1983; Ezcurra and de Azkue 1989; Ezcurra 1993; Porras and Munoz 2000).

15.4 Evolution and Development of Floral and Fruit Orientation

The proper functioning of most animal-pollinated flowers in pollination and seed production depends on the "correct" orientation of flowers and floral parts relative to gravity. This is because pollinator flight performance is directly constrained by the pull of gravity, which thus influences animal orientation on landing or hovering. In addition, rain usually falls from above (although not always), which might influence optimal floral orientation in wet weather (pollen viability and dispersal are generally compromised by immersion in water; Huang et al. 2002; Mao and Huang 2009). Floral orientation can also play a role in filtering pollinators, allowing specialization on the most effective pollinators (i.e., those whose visits provide the greatest increase in fitness; Fenster et al. 2009; Armbruster 2017) or better fit to those pollinators increasing fitness marginally without trade-offs (Aigner 2001). For example, pendent or semi-pendent flowers favor hummingbird pollinators by making bee visitation difficult (Gegear et al. 2017). Pendent flowers combined with viscid corolla secretions in *Proboscidea* and related Martyniaceae prevent or reduce floral entry by small nectar- or pollen-seeking insects, restricting access to large bees that are good pollinators (Armbruster, unpublished observations). While vertical (upward) floral orientation attracts sphingid moth pollination, lateral floral orientation discourages their visitation in *Zaluzianskya* (Scrophulariaceae; Campbell et al. 2016). "Correct" floral orientation is achieved developmentally by the operation of one or more of at least three distinct processes: (1) bending of the peduncle (main floral stem), (2) bending of the pedicel (secondary floral stem), or (3) twisting of the pedicel.

Below I provide examples that support the view that small developmental changes can lead to major transitions in floral orientation and pollination ecology, and hence that there is high evolvability of flower function via such simple changes. First, I build the case that floral orientation is often evolutionarily labile at the species level. I then show that this lability can be accounted for by simple changes in development, as is reflected in the widespread occurrence of serial developmental changes in flower orientation within the "lifetime" of a flower. A third line of evidence supporting the high evolvability and ease of evolutionary change in floral orientation derives from the fact that many plants exhibit "behavioral" plasticity in floral attachment angle.

15.4.1 Floral Orientation Is Evolutionarily Labile at the Species Level

The evolution and functional significance of flower attachment angle (via peduncle or pedicel bending or twisting) has been studied at the species level only to a limited extent. Most research has been on rotation of flowers through twisting of pedicels, known as resupination, particularly in orchids. A paper on the heterochronic evolution of temperate orchids showed that the 180° shift in floral orientation of *Gymnadenia austriaca* flowers, compared to relatives, was likely the result of the suppression of resupination, the last event in the developmental sequence prior to flower opening (Box et al. 2008). Similarly, *Angraecum* orchids show considerable evolutionary lability in resupination (where resupinate is the basal condition; Andriananjamanantsoa et al. 2016). The authors' phylogenetic reconstruction shows at least 5 losses of resupination. These transitions all presumably reflect curtailment of the final stage of floral development. Adaptive loss of resupination has also been reported in lobelioids (Campanulaceae) by Ayers (1994, 1997), who noted that losses were easier than gains and that there was some degree of adaptive plasticity in degree of resupination (see below).

Variation in inflorescence and flower orientation is a dominant theme in the evolution of the tropical giant herb *Heliconia* s.l. (Heliconiaceae). The neotropical group contains 200–250 species, most of which are pollinated primarily by hummingbirds (Stiles 1975). Flowers are resupinate or non-resupinate, and inflorescences are erect or pendent. The evolution of each of these two traits is contingent on the state of the other (yielding correlated evolution; Iles et al. 2017). Resupinate flowers in erect inflorescences is the inferred basal condition; this leads to sternotribic (in this case, "under-the-chin") pollen placement and stigma contact. The likelihood of resupination loss depends on whether the inflorescence is pendent. Loss of resupination, apparently by early termination of pedicel development, has occurred only a few times in erect-inflorescence lineages. In contrast, suppression of resupination has evolved independently in pendant-inflorescence lineages about 13 times (Iles et al. 2017). This latter combination of traits results in nototribic (in this case, "on-the-forehead") pollen placement and stigma contact. These various developmental transitions potentially allow sympatric species of *Heliconia* to partition locations of pollen placement on pollinators, reducing loss of pollen during heterospecific visits and stigmatic clogging by heterospecific pollen.

Species-level evolution of floral orientation caused by pedicel or peduncle (inflorescence stem) bending has received less attention that that by pedicel rotation. As just described in *Heliconia* spp., bending of the peduncle during development, in combination with flower resupination or lack thereof, can result in changes in pollen placement and stigma contact across species. Evolutionary lability in floral orientation via pedicel bending was recently described in *Lonicera* by Xiang et al. (2021). Floral orientation evolved apparently in response to both the seasonal trends in the physical environment and in the main pollinator species. The authors detected 3 or 4 cases of upward-facing flowers evolving from downward-facing ancestors, and at least one reversal, in their species sample.

Species blooming early in the season tended to have downward-facing flowers, possibly facilitating heat retention and reducing pollen damage and nectar dilution by rain, while those blooming later in the season tended to have upward-facing flowers. All moth-pollinated species had upward-facing flowers, whereas the one hummingbird-pollinated species had downward-facing flowers (see also Sapir and Dudley 2013); the bee-pollinated species had either flower orientation (Xiang et al. 2021).

Dalechampia vines exhibit 1 or 2 origins of lateral-facing, 2 origins of pendent, and 1 origin of upward-facing inflorescences as a result of peduncle bending. In this group, lateral blossom orientation is associated with pollination by female bees collecting floral resin or pollen, pendent orientation is associated with pollination by fragrance-collecting male euglossine bees, and upward orientation is associated with pollination by beetles (see figure 15.2; Armbruster 1993; Armbruster et al. 1993; Plebani et al. 2015).

15.4.2 Changes in Floral Orientation during Normal Development Suggest Floral Orientation Can Evolve Easily through Heterochrony

In many plants, the orientation of floral buds, flowers, and fruits changes during development due to changes in the angle of the stem (pedicel or peduncle). Heterochronic changes in such developmental sequences could be an evolutionary route of low resistance, leading to evolution of novel floral orientation angle in response to changes in the pollinator environment or selection to optimize seed-dispersal mechanics during fruiting.

Other than the resupination literature noted earlier in this chapter, the literature on other developmental changes in floral orientation is surprisingly scarce. In a small survey, I combined what little I could find in the literature with original observations (table 15.2). The survey shows a remarkable amount of change in floral orientation associated with development from bud, to anthesis, to fruiting. This is not surprising, because the optimal orientations of buds, receptive flowers, developing fruits, and mature or dehiscing fruits are likely to differ. Niu et al. (2016) showed, for example, that, although the optimal orientation of flowers of many plants is lateral or pendent, the optimal orientation of the fruits of those species is upright when the fruits are capsular and split open at the top. Indeed, experiments demonstrated better seed dispersal from the fruits of *Silene chungtienensis* that were in the upright position. A published survey showed that most plants with dry, partially dehiscent, "container-like" capsules had vertically oriented fruits, allowing gradual release of seeds with wind or other disturbances. A substantial proportion of these had to reorient floral/fruit structures to achieve upward-facing fruits (Niu et al. 2016; cf. table 15.2).

15.4.3 "Behavioral" Plasticity of Floral Orientation Also Suggests Orientation Is Highly Evolvable

That flowers can change their orientation "behaviorally" (i.e., exhibit rapid phenotypic plasticity in response to an external or internal stimuli) has been known at least since Darwin's descriptions of the phenomenon (e.g., petal closure in response to cold or nightfall; Darwin, 1862, 1880). Probably the best-known floral behavior is floral heliotropism: movement in response to the sun's position (see review in van der Kooi et al. 2019). Recent work by Yon et al. (2017) and Haverkamp et al. (2019) shows daily shifts in the orientation of flowers of *Nicotiana attenuata:* from upward by night (promoting moth pollination) to downward by day (keeping the interior of flowers cooler). In contrast, the flowers of *Eriocapitella* sp.

Table 15.2
Developmental changes in orientation of floral and fruit structures in a sample of flowers from field and garden surveys and the literature

Floral Orientation: Taxon	In Bud	In Flower (Anthesis)	In Fruit	Source
Aquilegia vulgaris L (Ranunculaceae)	lateral?	pendent	erect	original
Anisodus luridus (Solanaceae)	pendent	pendent	erect	Wang et al. 2010
Collinsia spp. (Plantaginaceae)	lateral	lateral	erect	original
Dactylorhiza fuchsii, Phalaenopsis amabilis, P. equestris, (Orchidaceae) and most other orchids with lateral floral orientation and enlarged labellum	unrotated-not resupinate	rotated 90–180° to place labellum in lowermost position (resupinate)	rotated 180° (still resupinate)	original
Dalechampia spathulata (Scheidw.) Baill. (Euphorbiaceae)	lateral	lateral	pendent, then half-erect at dehiscence	original
Diascia personata Hilliard & Burtt (Scrophulariaceaae)	erect	lateral	erect	original
Digitalis purpurea L (Plantaginaceae)	erect to lateral	half-pendent	lateral, then erect at dehiscence	original
Fuchsia sp. (Onagraceae)	erect to lateral	pendent	pendent	original
Gymnadenia spp. (Orchidaceae)	unrotated	rotated 180° (resupinate)	resupinate	Box et al. (2008)
Gymnadenia austriaca (Orchidaceae): apomictic	unrotated	unrotated	unrotated?	Box et al. (2008)
Nemesia sp. (Scrophulariaceaae)	erect	lateral	erect	original
Oenothera glazioviana Micheli (Onagraceae)	erect	lateral	erect	original
Papaver rhoeas L (papaveraceae)	pendent	generally upwards facing	erect	original
Pulsatilla cernua	erect	pendent	erect	Huang et al. (2002), fig. 2
Salvia spp. (Lamiaceae)	lateral	lateral	lateral	original
Silene chungtienensis (Caryophyllaceae)	—	pendent	erect	Niu et al. (2016)
Stylidium spp. (Stylidiaceae)	most erect	most lateral, some species facing upwards	most erect	original

Note: "Erect" is facing vertically upward; "half-erect" is facing upward at an angle between 45° and 90° (90° is vertical); "lateral" is facing sideways, horizontally (=0°±45°); "pendent" is hanging or facing downward; "half-pendent" is facing downward at an angle between −45° and −90° (−90° is fully pendent).

(Ranunculaceae) and many other plant species close partially and nod (face downward) at night and in inclement weather (Armbruster, unpublished observations).

Perhaps the most dramatic example of behavioral change in floral orientation is corrective floral reorientation after damage to flower-supporting structures. Many species of flowering plants have the ability to restore optimal floral orientation after mechanical injury caused by inclement weather, falling debris, or animals (Armbruster and Muchhala 2020). These corrective changes can occur through bending of peduncles (e.g., *Dactylorhiza fuchsii*), bending of pedicels (e.g., *Aconitum* and *Delphinium*), or rotation of flowers by twisting of pedicels (e.g., *Stylidium* spp.; Armbruster and Muchhala 2020).

Together these three observations (evolutionary lability, developmental sequences in orientation, and floral behavior) suggest high evolvability of floral orientation via heterochronic changes in bending or twisting of floral stems. Future research could test the prediction that lineages whose members have obvious developmental or behavioral sequences of floral reorientation also exhibit high evolutionary lability in floral orientation at the among-species level.

15.5 Preaptations Suggest Paths of Low Evolutionary Resistance

Another clue about evolutionary paths of low resistance comes from the study of the process of exaptation, where the existence of a preaptation appears to greatly increases the likelihood of an adaptive shift (Gould and Vrba 1982). In this view, preaptations are complex traits that coincidentally take on a second adaptive function. Thus preaptations are expected to be rare, but, arguably, evolutionarily important. As noted in section 15.1, the existence of a preaptation and its subsequent co-option into a novel adaptive function (the exaptation process) is often associated with parallel evolution at the macroevolutionary scale. In this section, I focus on several examples of apparent exaptation in plants.

15.5.1 Biosynthetic Paths of Least Resistance (Biochemical Exaptation)

Biosynthetic preaptations are probably common in plants, and they most likely indicate evolutionary paths of low resistance. If a secondary compound is synthesized by a plant in a particular place for a particular purpose, it may be easily co-opted elsewhere for other purposes, given that all cells in a plant have the same genetic machinery for biosynthesis. One of the earliest suggestions of the exaptive origins of pollinator attractants was that floral-scent compounds may have originated for defense and secondarily taken on signaling and attraction roles (Pellmyr and Thien 1986). In a series of papers, my collaborators and I tested a similar idea with respect to floral resin produced by *Dalechampia* vines, finding that the resin, which today is involved in attracting pollinators in most species, retains the putative ancestral feature of being defensive against insect herbivores. Curiously, the same oxygenated triterpenoid compounds have later (in the evolution of the genus) been co-opted to again play a defensive role for both floral and nonfloral tissues (Armbruster 1997; Armbruster et al. 1997, 2009).

Another example is correlated evolution (across related species) of blossom color (petals or bracts); leaf or stem color; and in some cases, fruit color. Most of the data are anecdotal, although at least one formal analysis has been conducted. I found that the presence of anthocyanins (red, purple, or deep orange in color) in autumn leaves was a good predictor that maple (*Acer* spp.) lineages would later evolve flowers with anthocyanin-rich (red) petals. In the large sample of species studied, lineages not having red or deep orange autumn leaves never evolved red petals, and instead exhibit yellow or greenish petals, as is basal in the genus. Thus, possessing certain protective pigments in the leaves constitutes a biochemical preaptation for production and use of the same pigments in flowers to attract pollinators (Armbruster 2002). Similar vegetative-flower color correlations were seen in *Dalechampia, Solanum,* and *Syringa* (Armbruster 2002; see also Sobel and Streisfeld 2013; Renoult et al. 2014; Larter et al. 2018). Thus, in lineages that already produce anthocyanin pigments for protection of leaves and buds, evolutionary shifts to pollination by butterflies (e.g., pink flowers) or birds (e.g., red flowers, at least in temperate North America) may be particularly easy.

Similar evolvability arguments may hold for evolution of pigments attracting seed dispersers. For example, some strawberry lines (*Fragaria* × *ananassa*) have both bright-red autumn foliage and red berries, although flower petals are white. In contrast, in the hummingbird-pollinated lineage, *Fuchsia* sect. *Quelusia,* the stems and sepals are red, but the ripe fruit is blue-purple like the petals (see Berry 1989). The leathery capsules of *Euonymus europea* are a distinctive deep-pink when ripe (attracting birds to the arillate seeds). The autumn leaves of some varieties are the same distinctive color (at least to human eyes).

15.5.2 Morphological Exaptation: Leaves to Bracts

Leaves and floral bracts are serially homologous, and the homology is sometimes very close and hints at an evolutionary path of low resistance for achieving pollinator attraction or floral protection. For example, in the Southeast Asian plant *Saururus chinensis* (Saururaceae), white bracts subtending inflorescences are involved in attracting pollinators, although they are morphologically identical to leaves (Song et al. 2018). In fact, prior to floral anthesis, they are green and indistinguishable from the lower leaves (except by position). After fruit set, the bracts turn green again and look and function like leaves (Song et al. 2018). It is easy to interpret this as an evolutionary path of low resistance in the evolution of pollinator-attraction structures.

The involucral (subtending) bracts of *Dalechampia* blossoms are a step further along a least-resistance line of evolutionary differentiation. In addition to diverging from leaves in color and function (at least during flowering), they also diverge from leaves in shape. Apparently to more tightly enclose the flowers when closed, the bracts of most species lack petioles. That this morphological transition is "easy" is suggested by occasional developmental errors, when bracts are replaced by leaf-shaped structures, or nearby leaves are replaced with bract-like structures. Another developmental link between leaves and bracts in *Dalechampia* is the correlated shape evolution of the two. Most species with 3-lobed or 3-leafleted leaves have 3-lobed involucral bracts. Species with unlobed leaves usually have unlobed involucral bracts. This correlation in also seen in populations of some polymorphic species (e.g., *D. heteromorpha;* Armbruster, unpublished observations).

Once *Dalechampia* bracts were evolutionarily "in place," serial modifications down various lineages have led to greater divergence from the function of the ancestral leaf. In several independent lineages, bracts evolved the ability to open by day and close protectively by night, perhaps by modification of "sleep movements" seen in many leaves (Darwin 1880; Armbruster 1997). In many *Dalechampia,* the bract movements have evolved to synchronize blossom opening with peak activity periods of pollinators; for example, opening in late morning in *D. brownsbergensis* (Armbruster and McCormick 1990) and in late afternoon in *D. magnistipulata* (Armbruster and Webster 1979). In two independent lineages, bracts have evolved to persist into fruiting and close protectively, enveloping the developing fruits (Armbruster 1997). In both lineages they re-suffuse with chlorophyll, become cryptic, and probably contribute to the photosynthetic budget (Pélabon et al. 2015).

15.5.3 Morphological Exaptation: Transitions to Pollen Rewards

A common transition in flowering plants is the transition from rewarding pollinators with nectar to rewarding them with only pollen. A very conservative estimate of the number of independent transitions from nectar to pollen rewards can be obtained by counting the number of plant families identified as containing lineages with species exhibiting buzz pollination (deduced from direct observations and characteristic floral morphology). A survey in 2013 counted 65 families that probably used buzz pollination (De Luca and Vallejo-Marín 2013). This number is a gross underestimate of the total number of transitions to pollen rewards, however, because it omits taxa that have switched to pollen rewards but not buzz pollination. Phylogenetic analysis of *Dalechampia* indicated 4 or 5 independent origins of pollen rewards in this one genus alone (Armbruster 1993; Armbruster and Baldwin 1998). At least one of these shifts, by *Dalechampia shankii,* involved so little genetic or morphological change that it and its resin-reward sister species were considered conspecific until 1988 (Armbruster 1988b; Armbruster et al. 2009).

Why is this such a common transition and a line of easy response to selection? All plants with hermaphroditic flowers have stamens (the basal condition in angiosperms), and thus have an obvious preaptation in place for using pollen as a reward. Even when such plants have another reward, such as nectar, there may be some pollinators collecting pollen, setting the stage for a shift in importance of the pollen collectors as pollinators (i.e., exploiting a pre-existing ecological opportunity). Further response to selection for shifting to a pollen reward (e.g., dispersal beyond range of usual pollinators, as for *Dalechampia* colonizing Madagascar; see Armbruster and Baldwin 1998) is then reflected in reduction of nectar production (or, for *Dalechampia,* resin production), with concomitant energy and mineral-nutrient savings.

15.5.4 Morphological Exaptation: Repeated Evolution of Unisexual Flowers Based on a Simple Preaptation

Another extremely common transition seen across the evolutionary history of flowering plants is a change in the sexual system from hermaphroditic flowers to unisexual flowers (Stebbins 1974; Thompson 1986). Transitions are estimated as, minimally, 100 (Charlesworth and Guttman 1999) and probably many more than this just for dioecy (male and female flowers on different plants) alone (Mitchell and Diggle 2005). To this we can add the many origins of monoecy (male and female flowers on the same plant). As Mitchell and Diggle (2005) point out, this is an exceptionally high level of homoplasy, through both parallelisms and convergent evolution. Evidence of convergence comes from the diversity of developmental-genetic mechanism by which stamens or styles are lost (Mitchell and Diggle 2005, e.g., their figure 7). There are, however, very few, if any, examples of the reverse transition (unisexual to hermaphroditic flowers).

A simplistic explanation for this bias (i.e., the ease of transition to unisexuality) is that it is easier to lose structures possessed than to gain structure not possessed. Thus, it seems more likely that the evolvability side of my equation (15.1) (in section 15.2) is responsible for this bias rather than the selection side. (There is no obvious selective reason to expect loss to be favored over gain.) If this evolvability bias holds true for stamens and pistils, then the presence of both sex parts in hermaphroditic flowers can be viewed as a preaptation for the evolution of unisexual flowers, creating an evolutionary line of low resistance, as would be consistent with the large number of independent transformations from hermaphroditic to unisexual flowers.

15.6 Floral Modularity May Increase Evolvability

Flowers are modular units whose variation is often quasi-independent of variation in vegetative structures (Berg 1960, Armbruster et al. 1999, 2014; Hansen et al. 2007; Pélabon et al. 2011; Conner and Lande 2014;). This floral-vegetative modularity should enhance evolvability of both sets of traits in the face of conflicting selection on floral and vegetative traits. Modular separation reduces or eliminates trade-offs otherwise manifested through pleiotropy and other integrating processes (see Hansen 2003).

Intra-floral modularity (i.e., when a flower comprises multiple variational modules) may also increase flower evolvability. This modularity may be exhibited at the between-whorl level (e.g., petals versus stamens; Armbruster and Wege 2019; Dellinger et al. 2019b), the within-whorl level (e.g., among stamens in heterantherous flowers), or at the within-organ level (e.g., anther versus anther appendage; Dellinger et al. 2019b). This level of modularity is the subject of increasing research interest, because selection on floral structures diverges if their functions differ (e.g., Ordano et al. 2008; Diggle 2014). For example, response to directional selection for larger petals better to attract the pollinators may be constrained by stabilizing selection acting on floral sexual parts so that they continue to "fit" the pollinator. This generates selection for intra-floral modularity, with variation in pollination-efficiency (sexual) structures decoupled from the variation in attraction structures (such as petals; Rosas-Guerrero et al. 2011; Armbruster and Wege 2019; Dellinger et al. 2019a, 2019b). In some cases, ovaries are decoupled from both of the above modules (Armbruster et al. 1999; Armbruster and Wege 2019).

15.7 Discussion and Conclusions

This review identifies properties and trends in flowering-plant evolution that indicate evolutionary lability and allow inference of elevated evolvability. Heterochrony and ontogenetic and static allometry hint at evolutionary paths of low resistance. Another clue comes from the link between developmental and behavioral variation in floral orientation and the divergence in floral orientation among species. The modularity of flowers may enhance their evolvability, both in their variational independence from vegetative traits and in the variational independence of different floral parts with different functions (see Opedal 2019). Another possible indicator of elevated trait evolvability is the rampant homoplasy such traits exhibit. Parallel and convergent evolution seems often to have been facilitated by preaptations.

This investigation into floral-trait evolvability is based on the assumption that evolutionary lability of a trait reflects, to some extent, its evolvability. Of course, another major factor controlling evolutionary lability is the efficacy of divergent natural selection. Selection and evolvability interact along with other intrinsic and extrinsic factors, respectively, to determine evolutionary lability (see Jablonski 2017a,b, chapter 17).[1] Ideally, one should compare the evolutionary labilities of traits evolving under the same or similar selection regimes, so that differences in lability are more directly attributable to differences in evolvability. This goal is not easily achieved, however, and the attempts employed have often fallen short in this respect. It would have been desirable to contrast the evolutionary labilities of traits likely to exhibit heterochrony or preaptation, for example, with those unlikely to exhibit such characteristics. The above caveat notwithstanding, it seems fair to argue that selection usually "compares" the pollination and seed-set consequences of floral changes rather than the transition processes themselves, at least if mineral-nutrient and energy demands are fairly similar.

1. References to chapter numbers in the text are to chapters in this volume.

Although a few studies show that population evolvability influences the course of population and species divergence (e.g., Bolstad et al. 2014; Holstad 2020; Opedal et al. 2023), other studies suggest instead that selection does or should overwhelm such evolvability biases at the macroevolutionary scale, at least in plants (see discussion in Bolstad et al. 2014). This may be the case in the evolution of leaf stomatal traits, where among-taxon correlations were apparently the result of selection for adaptive combinations of genetically/developmentally independent traits and not the result of any genetic or developmental factors creating an evolvability bias (Muir et al. 2021). In *Dalechampia,* some traits show among-taxon correlations that are consistent with genetic and developmental biases (e.g., involucral bract size and shape; Armbruster 1991; Hansen et al. 2003; Bolstad et al. 2014). However, other traits, such as gland-anther distance and gland-stigma distance, which influence floral fit with pollinators, show strong among-taxon correlations consistent with their functional interaction (Armbruster 1991; Armbruster et al. 2009), but they lack strong intrinsic (genetic) integration ($r_A = 0.27 - 0.33$; *D. scandens,* Tulum and Tovar populations, respectively; Bolstad et al. 2014) or extrinsic (selective) integration at the within-population level (Armbruster et al., in prep.). These examples underscore the need for caution in interpreting macroevolutionary trends as always reflecting paths of high evolvability. These caveats notwithstanding, examination of macroevolutionary trends can sometimes yield important insights into patterns of differential evolutionary lability and their intrinsic (population evolvability) and extrinsic (selective) causes.

Acknowledgments

I thank the Centre for Advanced Study (Oslo) for travel support and hospitality, and Thomas Hansen, Geir Bolstad, Frietson Galis, David Jablonski, Øystein Opedal, Mihaela Pavličev, Christophe Pélabon, and an anonymous reviewer for discussions and/or comments on earlier versions of this contribution.

References

Aigner, P. A. 2001. Optimality modeling and fitness trade-offs: when should plants become pollinator specialists? *Oikos* 95: 177–184.

Alberch, P., S. J. Gould, G. F. Oster, and D. B. Wake. 1979. Size and shape in ontogeny and phylogeny. *Paleobiology* 5: 296–317.

Andersson, S. 1991. Quantitative genetic variation in a population of *Crepis tectorum* subsp. *pumila* (Asteraceae). *Biological Journal of the Linnean Society* 44: 381–393.

Andriananjamanantsoa, H. N., S. Engberg, E. E. Louis, L. Brouilletli. 2016. Diversification of *Angraecum* (Orchidaceae, Vandeae) in Madagascar: Revised phylogeny reveals species accumulation through time rather than rapid radiation. *PLOS One* 11: e0163194.

Armbruster, W. S. 1988a. Multilevel comparative analysis of morphology, function, and evolution of *Dalechampia* blossoms. *Ecology* 69: 1746–1761.

Armbruster, W. S. 1988b. A new species, section, and synopsis of *Dalechampia* (Euphorbiaceae) from Costa Rica. *Systematic Botany* 13: 303–312.

Armbruster, W. S. 1990. Estimating and testing the shapes of adaptive surfaces: The morphology and pollination of *Dalechampia* blossoms. *American Naturalist* 135: 14–31.

Armbruster, W. S. 1991. Multilevel analyses of morphometric data from natural plant populations: Insights into ontogenetic, genetic, and selective correlations in *Dalechampia scandens. Evolution* 45: 1229–1244.

Armbruster, W. S. 1993. Evolution of plant pollination systems: Hypotheses and tests with the neotropical vine *Dalechampia. Evolution* 47: 1480–1505.

Armbruster, W. S. 1997. Exaptations link the evolution of plant-herbivore and plant-pollinator interactions: A phylogenetic inquiry. *Ecology* 78: 1661–1674.

Armbruster, W. S. 2017. The specialization continuum in pollination systems: Diversity of concepts and implications for ecology, evolution and conservation. *Functional Ecology* 31: 88–100.

Armbruster, W. S. 2002. Can indirect selection and genetic context contribute to trait diversification? A transition-probability study of blossom-color evolution in two genera. *Journal of Evolutionary Biology* 15: 468–486.

Armbruster, W. S., and B. G. Baldwin. 1998. Switch from specialized to generalized pollination. *Nature* 394: 632.

Armbruster, W. S., and K. D. McCormick. 1990. Diel foraging patterns of male euglossine bees: Ecological causes and evolutionary response by plants. *Biotropica* 22: 160–171.

Armbruster, W. S., and N. Muchhala. 2020. Floral reorientation: The restoration of pollination accuracy after accidents. *New Phytologist* 227: 232–243.

Armbruster, W. S., and G. L. Webster. 1979. Pollination of two species of *Dalechampia* (Euphorbiaceae) in Mexico by euglossine bees. *Biotropica* 11: 278–283.

Armbruster W. S., and J. A. Wege. 2019. Detecting canalization and intra-floral modularity in *Stylidium* flowers: Correlations alone do not suffice. *Annals of Botany* 123: 355–372.

Armbruster, W. S., M. E. Edwards, J. F. Hines, R. L. A. Mahunnah, and P. Munyenyembe. 1993. Evolution and pollination of Madagascan and African *Dalechampia* (Euphorbiaceae). *National Geographic Research and Exploration* 9: 430–444.

Armbruster, W. S., J. J. Howard, T. P. Clausen, et al. 1997. Do biochemical exaptations link evolution of plant defense and pollination systems? Historical hypotheses and experimental tests with *Dalechampia* vines. *American Naturalist* 149: 461–484.

Armbruster, W. S., V. S. Di Stilio, J. D. Tuxill, et al. 1999. Covariance and decoupling of floral and vegetative traits in nine neotropical plants: A reevaluation of Berg's correlation-pleiades concept. *American Journal of Botany* 86: 39–55.

Armbruster, W. S., C. P. H. Mulder, B. G. Baldwin, S. Kalisz, B. Wessa, and H. Nute. 2002. Comparative analysis of late floral development and mating-system evolution in tribe Collinsieae (Scrophulariaceae, s.l.). *American. Journal of Botany* 89: 37–49.

Armbruster, W. S., J. Lee, and B. G. Baldwin. 2009. Macroevolutionary patterns of defense and pollination in *Dalechampia* vines: Adaptation, exaptation, and evolutionary novelty. *PNAS* 106: 18085–18090.

Armbruster, W. S., J. Lee, M. E. Edwards, and B. G. Baldwin. 2013. Floral paedomorphy leads to secondary specialization in pollination of Madagascar *Dalechampia* (Euphorbiaceae). *Evolution* 67: 1196–1203.

Armbruster, W. S., C. Pélabon, G. H. Bolstad, and T. F. Hansen. 2014. Integrated phenotypes: Understanding trait covariation in plants and animals. *Philosophical Transactions of the Royal Society B* 369: 20130245.

Arnold, E. N. 1994. Investigating the origin of performance advantage: Adaptation, exaptation and lineage effects. In *Phylogenetics and Ecology*, edited by P. Eggleton and R. Vane-Wright, 123–168. London: Academic Press.

Ayers, T. J. 1994. Floral resupination in the Lobeliaceae: A twist on a twist. *American Journal of Botany* 81 (6, Suppl.): 140.

Ayers, T. J. 1997. Three new species of *Lysipomia* (Lobeliaceae) endemic to the paramos of southern Ecuador. *Brittonia* 49: 433–440.

Berg, R. L. 1960. The ecological significance of correlation pleiades. *Evolution* 14: 171–180.

Berry, P. E. 1989 A Systematic revision of *Fuchsia* Sect. *Quelusia* (Onagraceae). *Annals of the Missouri Botanical Garden* 76: 532–584.

Boberg, E., R. Alexandersson, M. Jonsson, J. Maad, J. Ågren, and L. A. Nilsson. 2014. Pollinator shifts and the evolution of spur length in the moth-pollinated orchid *Platanthera bifolia. Annals of Botany* 113: 267–275.

Bolstad G. H., T. F. Hansen, C. Pélabon, et al. 2014. Genetic constraints predict evolutionary divergence in *Dalechampia* blossoms. *Philosophical Transactions of the Royal Society B* 369: 20130255.

Box, M. S., and B. J. Glover. 2010. A plant developmentalist's guide to paedomorphosis: Reintroducing a classic concept to a new generation. *Trends in Plant Science* 15: 241–246.

Box, M. S., R. M. Bateman, B. J. Glover, and P. J. Rudall. 2008. Floral ontogenetic evidence of repeated speciation via paedomorphosis in subtribe Orchidinae (Orchidaceae). *Botanical Journal of the Linnean Society* 157: 429–454.

Busch, J. W., and L. F. Delph. 2017. Evolution: Selfing takes species down Stebbins's blind alley. *Current Biology* 27: R61–R63.

Campbell, D. R., A. Jurgens, and S. D. Johnson. 2016. Reproductive isolation between *Zaluzianskya* species: The influence of volatiles and flower orientation on hawkmoth foraging choices. *New Phytologist* 210: 333–342.

Charlesworth, D., and D. S. Guttman. 1999. The evolution of dioecy and plant sex chromosome systems. In *Sex Determination in Plants,* edited by C. C. Ainsworth, 25–49. Oxford: BIOS.

Conner, J. K., and R. Lande. 2014. Raissa L. Berg's contributions to the study of phenotypic integration, with a professional biographical sketch. *Philosophical Transactions of the Royal Society B* 369: 20130250.

Cutter, A. D. 2019. Reproductive transitions in plants and animals: Selfing syndrome, sexual selection and speciation. *New Phytologist* 224: 1080–1094.

Darwin, C. R. 1862. *The Various Contrivances by Which Orchids Are Fertilized by Insects.* London: John Murray.

Darwin, C. R. 1872. *The Origin of Species,* 6th ed. London: John Murray.

Darwin C. R. 1880. *The Power of Movement in Plants.* London: John Murray.

Dellinger, A. S., M. Chartier, D. Fernández-Fernández, et al. 2019a. Beyond buzz-pollination—Departures from an adaptive plateau lead to new pollination syndromes. *New Phytologist* 221: 1136–1149.

Dellinger, A. S., S. Artuso, S. Pamperl, et al. 2019b. Modularity increases rate of floral evolution and adaptive success for functionally specialized pollination systems. *Communications Biology* 2: 453.

De Luca, P. A., and M. Vallejo-Marín. 2013. What's the "buzz" about? The ecology and evolutionary significance of buzz-pollination. *Current Opinion in Plant Biology* 16: 429–435.

Diggle, P. K. 2014. Modularity and intra-floral integration in metameric organisms: Plants are more than the sum of their parts. *Philosophical Transactions of the Royal Society B* 369: 20130253.

Ehlers, B. K., and H. Æ. Pedersen. 2000. Genetic variation in three species of Epipactis (Orchidaceae): Geographic scale and evolutionary inferences. *Botanical Journal of the Linnean Society* 69: 411–430.

Ezcurra, C. 1993. Systematics of *Ruellia* (Acanthaceae) in southern South America. *Annals of the Missouri Botanical Garden* 80: 787–845.

Ezcurra, C., and D. de Azkue. 1989. Validation and genetic and morphological relationships of *Ruellia macrosolen* (Acanthaceae) from southern South America. *Systematic Botany* 14: 297–303.

Fenster, C. B., W. S. Armbruster, M. R. Dudash. 2009. Specialization of flowers: Is floral orientation an overlooked first step? *New Phytologist* 183: 502–506.

Gallardo, R., E. Dominguez, and J. M. Muñoz. 1993. The heterochronic origin of the cleistogamous flower in *Astragalus cymbicarpos* (Fabaceae). *American Journal of Botany* 80: 814–823.

Gegear, R. J., R. Burns, and K. A. Swoboda-Bhattarai. 2017. "Hummingbird" floral traits interact synergistically to discourage visitation by bumble bee foragers. *Ecology* 98: 489–499.

Gould, S. J. 1974. The origin and function of "bizarre" structures: Antler size and skull size in the "Irish elk," *Megaloceros giganteus. Evolution* 28: 191–220.

Gould, S. J. 1977. *Ontogeny and Phylogeny.* Cambridge, MA: Belknap Press.

Gould, S. J., and E. S. Vrba. 1982. Exaptation—a missing term in the science of form. *Paleobiology* 8: 4–15.

Guerrant, E. O., Jr. 1982. Neotenic evolution of *Delphinium nudicaule* (Ranunculaceae): A hummingbird-pollinated larkspur. *Evolution* 36: 699–712.

Haber, A. 2016. Phenotypic covariation and morphological diversification in the ruminant skull. *American Naturalist* 187: 576–591.

Hansen, T. F. 2003. Is modularity necessary for evolvability? Remarks on the relationship between pleiotropy and evolvability. *Biosystems* 69: 83–94.

Hansen, T. F., and D. Houle. 2008. Measuring and comparing evolvability and constraint in multivariate characters. *Journal of Evolutionary Biology* 21: 1201–1219.

Hansen, T. F., and K. L. Voje. 2011. Deviation from the line of least resistance does not exclude genetic constraints: A comment on Berner et al. (2010). *Evolution* 65: 1821–1822.

Hansen, T. F., W. S. Armbruster, M. L. Carlson, and C. Pélabon. 2003. Evolvability and genetic constraint in *Dalechampia* blossoms: Genetic correlations and conditional evolvability. *Journal of Experimental Zoology B* 296: 23–39.

Hansen, T. F., C. Pélabon, and W. S. Armbruster. 2007. Comparing variational properties of homologous floral and vegetative characters in *Dalechampia scandens:* Testing the Berg hypothesis. *Evolutionary Biology* 34: 86–98.

Haverkamp A., X. Li, B. S. Hansson, I. T. Baldwin, M. Knaden, and F. Yon. 2019. Flower movement balances pollinator needs and pollen protection. *Ecology* 100: e02553.

Hill, J. P., E. M. Lord, and R. G. Shaw. 1992. Morphological and growth-rate differences among outcrossing and self-pollinating races of *Arenaria uniflora* (Caryophyllaceae). *Journal of Evolutionary Biology* 5: 559–573.

Hollens, H., T. Van der Niet, R. Cozien, and M. Kuhlmann. 2017. A spur-ious inference: Pollination is not more specialized in long-spurred than in spurless species in *Diascia-Rediviva* mutualisms. *Flora* 232: 73–82.

Holstad, A. 2020. Does evolvability predict evolutionary divergence? M.Sc. thesis. Trondheim, Norway: Norwegian University of Science & Technology.

Huang, S.-Q., Y. Takahashi, and A. Dafni. 2002. Why does the flower stalk of *Pulsatilla cernua* (Ranunculaceae) bend during anthesis? *American Journal of Botany* 89: 1599–1603.

Igic, B., and J. W. Busch. 2013. Is self-fertilization an evolutionary dead end? *New Phytologist* 198: 386–397.

Iles, W. J. D., C. Sass, L. Lagomarsino, G. Benson-Martin, H. Driscoll, and C. D. Specht. 2017. The phylogeny of *Heliconia* (Heliconiaceae) and the evolution of floral presentation. *Molecular Phylogenetics and Evolution* 117: 150–167.

Jablonski, D. 2017a. Approaches to macroevolution: 1. General concepts and origin of variation. *Evolutionary Biology* 44: 427–450.

Jablonski, D. 2017b. Approaches to macroevolution: 2. Sorting of variation, some overarching issues, and general conclusions. *Evolutionary Biology* 44: 451–475.

Jaramillo, M. A., and E. M. Kramer. 2007. The role of developmental genetics in understanding homology and morphological evolution in plants. *International Journal of Plant Sciences* 168: 61–72.

Johnson, M. S., and M. F. Mickevich. 1977. Variability and evolutionary rates of characters. *Evolution* 31: 642–648.

Julca, I., C. Ferrari, M. Flores-Tornero, et al. 2021. Comparative transcriptomic analysis reveals conserved programmes underpinning organogenesis and reproduction in land plants. *Nature Plants* 7: 1143–1159.

Kluge, A. G., and W. C. Kerfoot. 1973. Predictability and regularity of character divergence. *American Naturalist* 107: 426–442

Lande, R. 1979. Quantitative genetic-analysis of multivariate evolution, applied to brain–body size allometry. *Evolution* 33: 402–416.

Larter, M., A. Dunbar-Wallis, A. E. Berardi, and S. D. Smith. 2018. Convergent evolution at the pathway level: Predictable regulatory changes during flower color transitions. *Molecular Biology and Evolution* 35: 2159–2169.

Leslie, A. B., C. Simpson, and L. Mander. 2021. Reproductive innovations and pulsed rise in plant complexity. *Science* 373: 1368–1372.

Li, P., and M. O. Johnston. 2010. Flower development and the evolution of self-fertilization in *Amsinckia:* The role of heterochrony. *Evolutionary Biology* 37: 143–168.

Lord, E. M. 1981. Cleistogamy: A tool for the study of floral morphogenesis, function and evolution. *Botanical Review (Lancaster)* 47: 421–449.

Lord, E. M. 1982. Floral morphogenesis in *Lamium amplexicaule* (Labiatae) with a model for the evolution of the cleistogamous flower. *Botanical Gazette* 143: 63–72.

Maad J. 2000. Phenotypic selection in hawkmoth-pollinated *Platanthera bifolia:* Targets and fitness surfaces. *Evolution* 54: 112–123.

Mao, Y. Y., and S. Q. Huang. 2009. Pollen resistance to water in 80 angiosperm species: Flower structures protect rain-susceptible pollen. *New Phytologist* 183: 892–899.

Marroig, G., and J. Cheverud. 2010. Size as a line of least resistance ii: Direct selection on size or correlated response due to constraints? *Evolution* 64: 1470–1488.

Mayr, E. 1963. *Animal Species and Evolution.* Cambridge, MA: Belknap Press.

Mazer, S. J., I. M. Park, M. Kimura, E. M. Maul, A. M. Yim, and K. Peach. 2020. Mating system and historical climate conditions affect population mean seed mass: Evidence for adaptation and a new component of the selfing syndrome in *Clarkia. Journal of Ecology* 108: 1523–1539.

McLennan, D. 2008. The concept of co-option: Why evolution often looks miraculous. *Evolution: Education and Outreach* 1: 246–258.

Melin, A., R. Altwegg, J. C. Manning, and J. F. Colville. 2021. Allometric relationships shape foreleg evolution of long-legged oil bees (Melittidae: *Rediviva*). *Evolution* 75: 437–449.

Mintner, T. C., and E. M. Lord. 1983. A comparison of cleistogamous and chasmogamous floral development in *Collomia grandiflora* Dougl. ex Lindl. (Polemoniaceae). *American Journal of Botany* 70: 1499–1508.

Mitchell, C. H., and P. K. Diggle. 2005. The evolution of unisexual flowers: Morphological and functional convergence results from diverse developmental transitions. *American Journal of Botany* 92: 1068–1076.

Morris J. L., M. N. Puttick, J. W. Clark, et al. 2018. The timescale of early land plant evolution. *PNAS* 115: E2274–E2283.

Motten, A. F., and J. L. Stone. 2000. Heritability of stigma position and the effect of stigma-anther separation on outcrossing in a predominantly self-fertilizing weed, *Datura stramonium* (Solanaceae). *American Journal of Botany* 87: 339–347.

Muir, C. D., M. À. Conesa, J. Galmés, et al. 2021. How important are functional and developmental constraints on phenotypic evolution? An empirical test with the stomatal anatomy of flowering plants. BioRxiv, https://www.biorxiv.org/content/10.1101/2021.09.02.457988v3.

Nilsson, N. 1998. Deep flowers for long tongues. *TREE* 13: 259–260.

Niu, Y., Z. Zhou, W. Sha, and H. Sun. 2016. Post-floral erection of stalks provides insight into the evolution of fruit orientation and its effects on seed dispersal. *Scientific Reports* 6: 20146.

Opedal, Ø. H. 2019. The evolvability of animal-pollinated flowers: towards predicting adaptation to novel pollinator communities. *New Phytologist* 221: 1128–1135.

Opedal Ø. H., G. H. Bolstad, T. F. Hansen, W. S. Armbruster, and C. Pélabon. 2017. The evolvability of herkogamy: Quantifying the evolutionary potential of a composite trait. *Evolution* 71: 1572–1586.

Opedal, Ø. H., W. S. Armbruster, T. F. Hansen, et al. 2023. Trait function and evolvability predict phenotypic divergence of plant populations. *PNAS*, 120 (1) e2203228120.

Ordano, M., J. Fornoni, K. Boege, and C. A. Domínguez. 2008. The adaptive value of phenotypic floral integration. *New Phytologist* 179: 1183–1192.

Pélabon, C., W. S. Armbruster, and T. F. Hansen. 2011. Experimental evidence for the Berg hypothesis: Vegetative traits are more sensitive than pollination traits to environmental variation. *Functional Ecology* 25: 247–257.

Pélabon, C., G. H. Bolstad, C. Egset, J. M. Cheverud, M. Pavličev, and G. Rosenqvist. 2013. On the relationship between ontogenetic and static allometry. *American Naturalist* 181: 195–212.

Pélabon, C., C. Firmat, G. H. Bolstad, et al. 2014. Evolution of morphological allometry. *Annals of the New York Academy of Sciences* 1320: 58–75.

Pélabon C., L. Hennet, R. Strimbeck, H. Johnson, and W. S. Armbruster. 2015. Blossom colour change after pollination provides carbon for developing seeds. *Functional Ecology* 9: 1137–1143.

Pellmyr, O., and L. B. Thien. 1986. Insect reproduction and floral fragrances—keys to the evolution of the angiosperms. *Taxon* 35: 76–85.

Pierce, B. A., and J. B. Mitton. 1979. Relationship of genetic-variation within and among populations—extension of the Kluge-Kerfoot phenomenon. *Systematic Zoology* 28: 63–70.

Plebani, M., O. Imanizabayo, D. M. Hansen, and W. S. Armbruster. 2015. Pollination ecology and circadian patterns of inflorescence opening of the Madagascan climber *Dalechampia* aff. *bernieri* (Euphorbiaceae). *Journal of Tropical Ecology* 31: 99–101.

Porras, R., and J. M. Munoz. 2000. Cleistogamous capitulum in *Centaurea melitensis* (Asteraceae): Heterochronic origin. *American Journal of Botany* 87: 925–933.

Puzey, J. R., S. J. Gerbode, S. A. Hodges, E. M. Kramer, and L. Mahadevan. 2012. Evolution of spur-length diversity in *Aquilegia* petals is achieved solely through cell-shape anisotropy. *Proceedings of the Royal Society B* 279: 1640–1645.

Renoult, J. P., A. Valido, P. Jordano, and H. M. Schaefer. 2014. Adaptation of flower and fruit colours to multiple, distinct mutualists. *New Phytologist* 201: 678–686.

Riska, B. 1979. Character variability and evolutionary rate in *Menidia. Evolution* 33: 1001–1004.

Rosas-Guerrero, V., M. Quesada, W. S. Armbruster, R. Pérez-Barrales, and S. D. Smith. 2011. Influence of pollination specialization and breeding system on floral integration and phenotypic variation in *Ipomoea. Evolution* 65: 350–364.

Rümpler, F., and G. Theißen. 2019. Reconstructing the ancestral flower of extant angiosperms: The 'war of the whorls' is heating up. *Journal of Experimental Botany* 70: 2615–2622.

Sapir, N., and R. Dudley. 2013. Implications of floral orientation for flight kinematics and metabolic expenditure of hover-feeding hummingbirds. *Functional Ecology* 27: 227–235.

Schluter, D. 1996. Adaptive radiation along genetic lines of least resistence. *Evolution* 50: 1766–1774.

Schluter, D. 2000. *The Ecology of Adaptive Raduation.* Oxford: Oxford University Press.

Schwaegerle, K. E. 2005. Quantitative genetic analysis of plant growth: Biases arising from vegetative propagation. *Evolution* 59: 1259–1267.

Schwaegerle, K. E., H. McIintyre, and C. Swingley. C. 2000. Quantitative genetics and the persistence of environmental effects in clonally propagated organisms. *Evolution* 54: 452–461.

Sengupta, A., and L. C. Hileman. 2022. A CYC–RAD–DIV–DRIF interaction likely pre-dates the origin of floral monosymmetry in Lamiales. *EvoDevo* 13: 3.

Sherry, R. A., and E. M. Lord. 2000. A comparative developmental study of the selfing and outcrossing flowers of *Clarkia tembloriensis* (Onagraceae). *International Journal of Plant Sciences* 161: 563–574.

Sicard, A., and M. Lenhard. 2011. The selfing syndrome: A model for studying the genetic and evolutionary basis of morphological adaptation in plants. *Annals of Botany* 107: 1433–1443.

Simpson, G. G. 1944. *Tempo and Mode in Evolution.* New York: Columbia University Press.

Sletvold, N., and J. Ågren. 2010. Pollinator-mediated selection on floral display and spur length in the orchid *Gymnadenia conopsea. International Journal of Plant Sciences* 171: 999–1009.

Sobel, J. M., and M. A. Streisfeld. 2013. Flower color as a model system for studies of plant evo-devo. *Frontiers in Plant Science* 4: 321.

Sokal, R. R. 1976. Kluge-Kerfoot phenomenon reexamined. *American Naturalist* 110: 1077–1091.

Song, B., J. Stöcklin, W. S. Armbruster, et al. 2018. Reversible color change in leaves enhances pollinator attraction and reproductive success in *Saururus chinensis* (Saururaceae). *Annals of Botany* 121: 641–650.

Stebbins, G. L. 1950. *Variation and Evolution in Plants.* New York: Columbia University Press.

Stebbins, G. L. 1957. Self-fertilization and population variability in the higher plants. *American Naturalist* 91: 337–354.

Stebbins, G. L. 1970. Adaptive radiations in Angiosperms: Pollination mechanisms. *AREES* 1: 307–326.

Stebbins, G. L. 1974. *Flowering Plants. Evolution above the Species Level.* Cambridge, MA: Belknap Press.

Steiner, K. E., and V. B. Whitehead. 1990. Pollinator adaptation to oil-secreting flowers—*Rediviva* and *Diascia. Evolution* 44: 1701–1707.

Steiner, K. E., and V. B. Whitehead. 1991. Oil flowers and oil bees—further evidence for pollinator adaptation. *Evolution* 45: 1493–1501.

Stewart, H. M., and J. M. Canne-Hilliker. 1998. Floral development of *Agalinis neoscotica, Agalinis paupercula* var. *borealis,* and *Agalinis purpurea* (Scrophulariaceae): Implications for taxonomy and mating system. *International Journal of Plant Sciences* 159: 418–439.

Stiles, F. G. 1975. Ecology, flowering phenology, and hummingbird pollination of some Costa Rican *Heliconia* species. *Ecology* 56: 285–301.

Takebayashi, N., and P. L Morrell. 2001. Is self-fertilization an evolutionary dead end? Revisiting an old hypothesis with genetic theories and a macroevolutionary approach. *American Journal of Botany* 88: 1143–1150.

Thompson, K. 1986. Are unisexual flowers primitive? *New Phytologist* 103: 597–601.

Vallejo-Marín, M. 2019. Buzz pollination: Studying bee vibrations on flowers. *New Phytologist* 224: 1068–1074.

van der Kooi, C. J., P. G. Kevan, and M. H. Koski. 2019. The thermal ecology of flowers. *Annals of Botany* 124: 343–353.

Wake, D. B. 1991. Homoplasy: The result of natural selection, or evidence of design limitations? *American Naturalist* 138: 543–567.

Wake, D. B., M. H. Wake, and C. D. Specht. 2011. Homoplasy: From detecting pattern to determining process and mechanism of evolution. *Science* 331: 1032–1035.

Wallace, A. R. 1867. Creation by law. *Quarterly Journal of Science* 4: 471–488.

Wang, Y., L. L. Meng, Y. P. Yang, and Y. W. Duan. 2010. Change in floral orientation in *Anisodus luridus* (Solanaceae) protects pollen grains and facilitates development of fertilized ovules. *American Journal of Botany* 97: 1618–1624.

Wessinger, C. A., and L. C. Hileman. 2020. Parallelism in flower evolution and development. *AREES* 51: 387–408.

Whittall, J. B., and S. A. Hodges. 2007. Pollinator shifts drive increasingly long nectar spurs in columbine flowers. *Nature* 447: 706–712.

Whitehead, M. R., R. Lanfear, R. J. Mitchell, and J. D. Karron. 2018. Plant mating systems often vary widely among populations. *Frontiers in Ecology and Evolution* 6: 38.

Yon, F., D. Kessler, Y. Joo, L. C. Llorca, S. G. Kim, and I. T. Baldwin. 2017. Fitness consequences of altering floral circadian oscillations for *Nicotiana attenuata. Journal of Integrative Plant Biology* 59: 180–189.

Xiang, G. J., Y. H. Guo, and C. F. Yang. 2021. Diversification of floral orientation in *Lonicera* is associated with pollinator shift and flowering phenology. *Journal of Systematics and Evolution* 59: 557–566.

16 Evolvability of Body Plans: On Phylotypic Stages, Developmental Modularity, and an Ancient Metazoan Constraint

Frietson Galis

The evolvability of animal body plans is limited. For instance, strong constraints exist against evolutionary change of early organogenesis, also called the phylotypic stage. Most of the body plan is usually laid out during this stage, and as a consequence, its conservation is implicated in the conservation of body plans. Two hypotheses have been proposed to explain the strong conservation of the phylotypic stage. One states that the conservation reflects a strong interactivity between developmental modules, so that mutations would have many pleiotropic effects, resulting in stabilizing selection against the mutations. The other states that, at least in insects, the conservation is caused by the robustness of a centrally important organizer gene network against mutational changes. I describe how the empirical and theoretical support for the robustness hypothesis is weak, but it is strong for the pleiotropy hypothesis. This highlights the importance of developmental modularity for evolvability. Finally, I discuss how an ancient metazoan constraint on the division of differentiated cells causes the early loss of pluripotentiality of cells. Consequently, the layout of the body plan occurs early, when the embryo is small, and the number of inductive interactions is too limited to allow for effective developmental modularity. Hence, this constraint on simultaneous cell division and differentiation causes another constraint: The one against changes of the phylotypic stage and of the body plan traits that are determined at these stages.

16.1 Introduction: Limited Evolvability of Phylotypic Stages and Body Plans

Evolution has produced an astonishing array of organisms, yet early stages of organogenesis in animals have been remarkably conserved across many higher taxa (figures 16.1 and 16.2; Medawar 1954; Seidel 1960; Ballard 1981; Sander 1983; Raff 1994; Gilbert 1997; Hall 1997). Furthermore, most of the body plan traits that are determined during these stages have been conserved. Evolution, thus, appears to be subject to constraints and there are limits to the evolvability of body plans. These conserved early organogenesis stages, at which the morphological and genetic similarity appears to be greater than at earlier or later stages, are usually called phylotypic stages. During these phylotypic stages most of the body plan is laid out. Sander (1983) introduced the term phylotypic stage as an alternative to the terms Körpergrundgestalt of Seidel (1960) and phyletic stage of Cohen (1977). Since then, the term has not only been applied to phyla, but also to other higher taxa, for example, to the class of insects (Sander 1983; Sander and Schmidt-Ott 2004).

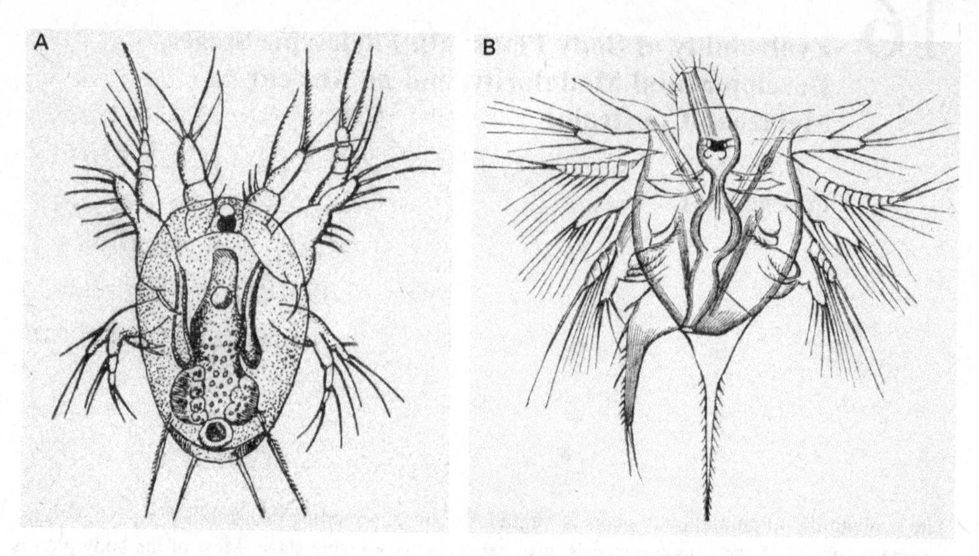

Figure 16.1
Early organogenesis stages are very similar in crustaceans. Nauplius stages in (A) *Cyclops*, (B) Cirripedia species (from Claus and Grobben 1917).

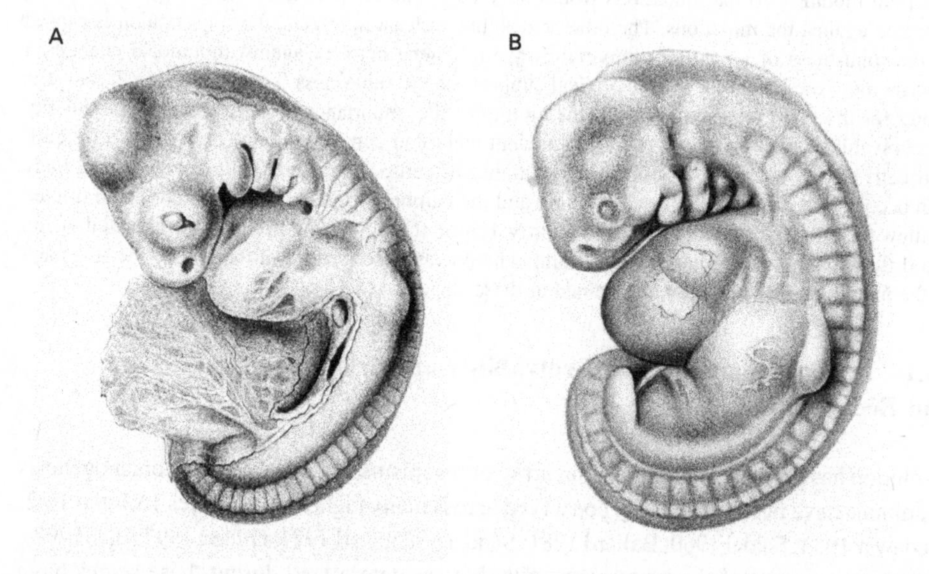

Figure 16.2
Early organogenesis stages in vertebrates are more similar than earlier or later stages. Especially in amniotes, the similarity is striking. Pharyngula stage in (A) *Lacerta* and (B) human (from Keibel 1904 and 1908).

Von Baer (1828) was the first to hypothesize that the constraint against evolutionary changes of these stages might be caused by the negative cascading consequences of early changes (pleiotropic effects), with later stages having fewer cascading consequences. Although this proposed constraint is undoubtedly real (Buss 1987), its importance is challenged by the existence of considerable variation in the embryonic stages before early organogenesis, cleavage, and gastrulation (Seidel 1960; Sander 1983; Gilbert and Raunio 1997; Galis and Sinervo 2002). Sander (1983) already pointed out that the stages preceding the phylotypic stage are highly variable, but that thereafter, the developmental pathways converge (see also Seidel 1960). The larger variability of the earlier stages is not always immediately apparent, as there are striking cases of morphological similarity that result from convergent or parallel evolution (Buss 1987; Gilbert and Raunio 1997; Hall 1999; Galis and Sinervo 2002). The morphological similarity is reflected in similarity of gene expression patterns (Levin et al. 2016). Similarity is almost unavoidable in the early developmental stages, because of the complete reset of the organism at the initial single-celled stage (Galis and Sinervo 2002, Galis et al. 2018). Only a limited number of permutations is possible in embryos with a few, not yet differentiated cells. Further reasons for similarity of cleavage and gastrulation are caused by convergent locomotory and nutritional adaptations plus maternal attempts at dictating offspring features (reviewed in Buss 1987). A good example of remarkable convergence is cleavage and gastrulation in the yolk-rich embryos of cephalopods, fishes, reptiles, and birds, because yolk impedes cleavage and as a result, the embryo develops as a disk on top of the yolk (figure 16.3). Yet, despite often remarkable similarity, within phyla and classes the processes of cleavage and gastrulation are far more diverse than is the end product of gastrulation: the beginning of the phylotypic stage. For instance, cnidarians have seven different types of gastrulation (Gilbert and Raunio 1997). In insects, there are drastic differences in gastrulation between short, intermediate, and long-germ-band insects, poly-embryonic wasps being even more derived (Grbič 2000). Within teleosts and mammals, gastrulation is also highly variable (Collazo et al. 1994; Viebahn 1999). Yet, at the end of gastrulation, the developmental pathways converge,

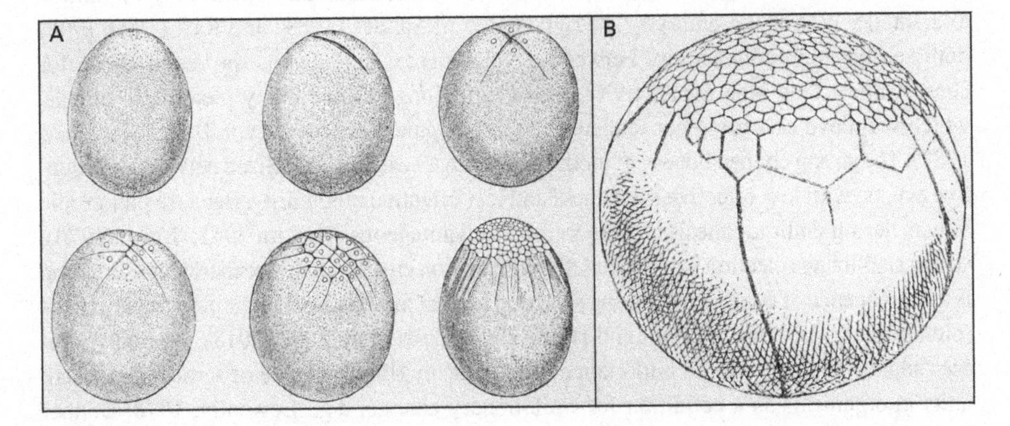

Figure 16.3
Convergence of cleavage stages. Cleavage of the embryo on top of the yolk, as yolk impedes cell division (meroblastic cleavage), in (A) the cephalopod *Loligo pelalei* (after Claus and Grobben 1917) and (B) in the longnose gar *Lepisosteus osseus* (after Balfour 1881).

gastrulae invariably have no more than two or three germ layers, and the organ systems emerging from the germ layers are similarly conserved (e.g., skin and nervous system arise from ectoderm and the digestive system from endoderm; Hall 1999). A key outcome of the process of gastrulation is that sheets of cells come into contact with each other, enabling the conserved embryonic inductions that are essential for the organization of the body plan at the phylotypic stage. These inductions between adjacent cell populations appear to form a severe spatiotemporal constraint on the outcome of gastrulation, which is the starting point of the conserved phylotypic stage (Galis and Sinervo 2002; Zalts and Yanai 2017).

In contrast to the strong conservation of the phylotypic stages within phyla or classes, these stages differ dramatically between phyla and classes (see below for a discussion). The segmented germ-band stage in insects, the nauplius stage of crustaceans, and the neurula/pharyngula stage in vertebrates are examples of this diversification (cf. figures 16.1 and 16.2). Nonetheless, what all these stages have in common is that they start at the end of gastrulation and that most of the body plan is being laid out during this part of development.

16.2 Pleiotropy Proposed as the Cause of the Conservation of Phylotypic Stages

Sander (1983) hypothesized that the evolutionary conservation of the phylotypic stages in animals is caused by pleiotropic effects resulting from strong interactivity between developmental modules. Raff (1994, 1996) proposed a similar hypothesis to explain the presumed strong stabilizing selection against evolutionary changes influencing the phylotypic stage: The web of intense interactions among organ primordia of the embryo at this stage causes any small mutational change to lead to many pleiotropic effects elsewhere in the embryo, thus reducing the chance of a favorable mutation. Raff further argued that at earlier stages, fewer inductive interactions occur, as there are not yet organ primordia, and thus fewer pleiotropic effects (although changes may still affect the entire embryo). At later, larger, stages there are many more inductive interactions, but they take place within semi-independent modules (e.g., limbs, the heart); hence changes will usually be limited to a smaller part of the embryo. The hypotheses of Sander (1983) and Raff (1994, 1996) both see the high connectivity between developmental modules as the major cause for conservation. This high connectivity causes mutations to have many pleiotropic effects, which will have cascading consequences as development proceeds (von Baer 1828; Buss 1987). The strong connectedness of modules implies an easily destabilized network of inductive events, with low effective robustness and low effective modularity. Because pleiotropic effects during embryogenesis are generally disadvantageous (Hadorn 1961; Wright 1970), strong stabilizing selection against mutational variation ensues. In this scenario, conservation is a consequence of consistently strong selection against mutations via their pleiotropic effects (pleiotropic constraints; Hansen and Houle 2004; Galis et al. 2002, 2018). These hypotheses support ideas about the importance of modularity (the existence of semi-independent units in organisms as a condition for evolutionary change; e.g., Lewontin, 1978; Bonner 1988; Galis 1996; G. Wagner and Altenberg 1996; Galis and Metz 2001; Galis et al. 2002; Schlosser 2002; Mitteröcker 2009; Armbruster et al. 2014). Developmental modularity limits the effects of mutational changes to only part of the organism, thereby greatly reducing the probability that advantageous changes are associated with adverse pleiotropic effects elsewhere. It is thus the lack of developmental modularity that is hypothesized to be the cause of the limited evolvability of phylotypic stages.

16.3 Robustness as an Alternative Cause for the Conservation of the Phylotypic Stage

Von Dassow and Munro (1999) proposed an alternative hypothesis to explain the conservation of the phylotypic stage in insects (the segmented/extended germ-band stage) that is diametrically opposite to the pleiotropy hypothesis from Sander and Raff (Galis et al. 2002). Here, the robustness of a centrally important organizer gene network and module, the segment polarity gene network, is hypothesized to be causally involved in the conservation (figure 16.4). Hu et al (2017) also mention the robustness of gene networks as a possible cause for the conservation of phylotypic stages in vertebrates, although their major conclusion is that pleiotropic constraints appear to be involved.

This robustness hypothesis implies strikingly different roles for modularity in evolution than does the pleiotropy hypothesis of Sander and Raff (see table 16.1), that is, constraining rather than facilitating evolutionary change (Galis et al. 2002).

Von Dassow and Munro (1999) proposed that conservation of the segment polarity network occurs despite accumulation of genetic changes, as these changes have little phenotypic effect and mainly lead to hidden variation (von Dassow et al. 2000; von Dassow and Odell 2002). Hence, the robustness of the segment polarity network in each segment is proposed to provide a buffer against phenotypic effects of mutational changes of the segmented germ-band stage. In robust gene networks, by definition, developmental noise and mutations do not lead to clear phenotypic effects, because gene interactions neutralize perturbations and in particular make mutations recessive (Gibson and G. Wagner 2000; A. Wagner 2000).

16.4 Evaluating Pleiotropic Constraints and Robustness

The modeling of the robustness of the network by von Dassow and colleagues is valuable, but what matters here is whether robustness can cause long-term conservation. Although it is not possible to directly test these hypotheses about the evolutionary past, they can be tested indirectly, as they lead to very different predictions for mutations affecting the phylotypic stage (table 16.1). The robustness hypothesis proposes that mutations will have minimal phenotypic effects and mainly produce hidden, or cryptic, variation (i.e., phenotypically similar, but genotypically different), whereas the pleiotropy hypothesis proposes that the

Table 16.1
Predictions of the robustness and pleiotropy hypotheses

	Pleiotropy hypothesis	Robustness hypothesis
Genetic mutational variation	Visible at the phenotypic level	Hidden
Direct phenotypic effects	Potentially large	Small
Dominance of direct effects	Haplo-insufficiency possible	Recessivity, or near recessivity
Pleiotropic effects	Many	Few

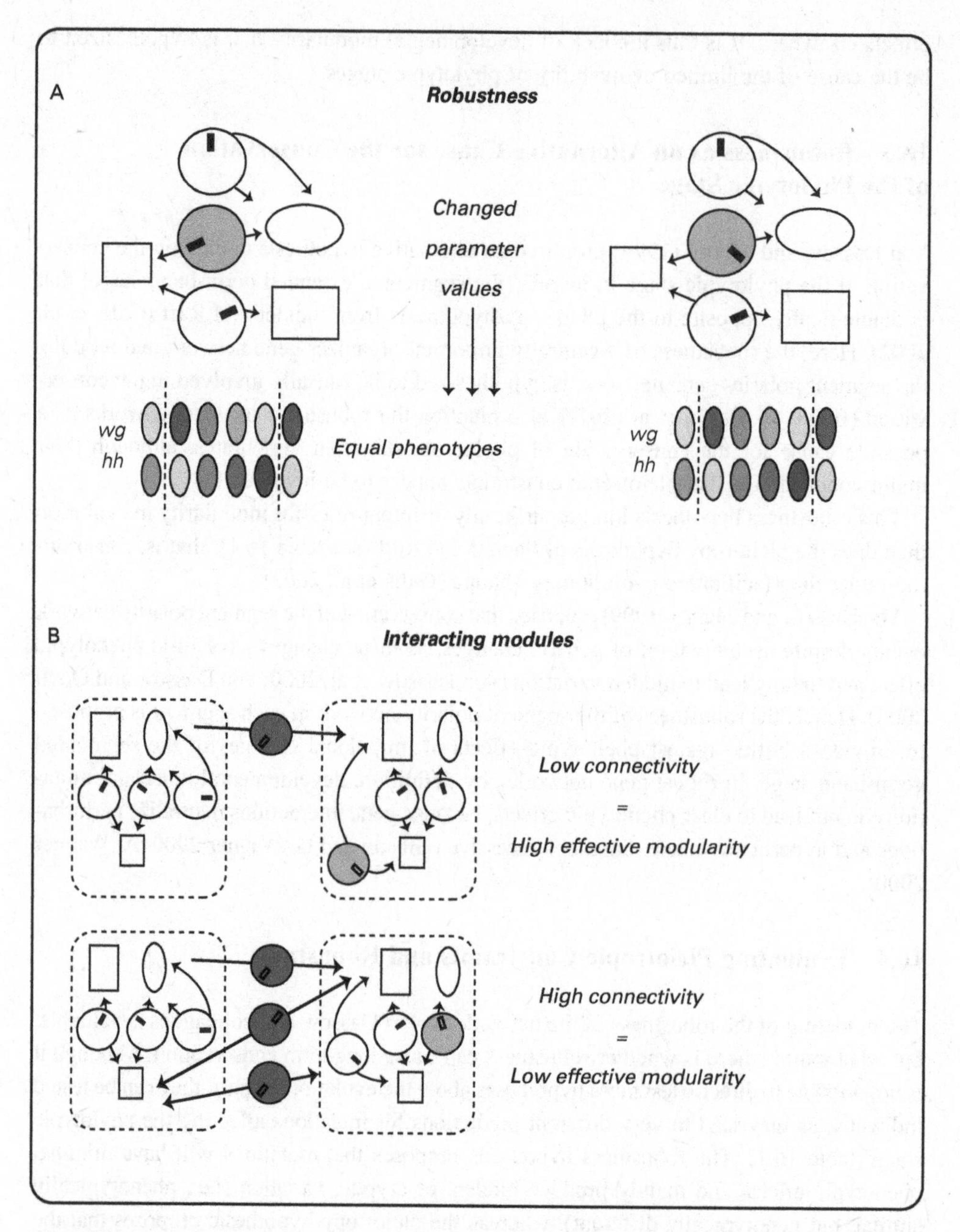

Figure 16.4
Robustness and effective modularity. (A) Robustness. When a parameter is changed in a robust genetic network, the resulting phenotype does not change (in this case illustrated with the concentration of the organizing morphogens *Wingless* (WG) and *Hedgehog* (HH) in the cells of the ectoderm of *Drosophila* during the segmented germ-band stage.) (B) Effective modularity. Modules are discernible and discrete units in large genetic networks that have some autonomy and a clear physical location (Raff 1996). These modules can differ in the amount of connectedness. Low connectivity (i.e., few connections having small effects) implies high effective modularity. High connectivity implies low effective modularity (from Galis et al. 2002).

phylotypic stage is vulnerable to mutational change and that mutations will have large deleterious effects.

16.4.1 Phylotypic Stage in *Drosophila:* Importance of Pleiotropy

When evaluating the empirical evidence for these hypotheses in *Drosophila,* Galis et al. (2002) found little support for effective robustness of the segment polarity gene network (or other gene networks active during the stage) against mutational change acting on the phylotypic stage. Small changes in the position, shape, and intensity of segment polarity stripes lead to dramatic phenotypic effects, so that the system does not appear robust. The organizer function of segment polarity and other regulatory genes causes mutations in these genes generally to result in a cascade of pleiotropic effects. This pleiotropy is not surprising, as during this stage, the segment polarity genes are, together with *Hox* and other genes, involved in the specification and early differentiation of virtually all organ primordia and the patterning of drastic morphogenetic events (e.g., germband retraction, dorsal closure, and head involution). Many auto-regulatory and cross-regulatory interactions provide feedback on the input of the segment polarity gene network, further lowering the robustness and modularity of this gene network. Consequently, phenotypic effects of mutations with an effect on the segment polarity network (and more generally on the phylotypic stage) are severe and include many cascading pleiotropic effects. The severe phenotypic effects of even weakly hypo-morphic mutations illustrate the observation of Lande et al. (1994) that mutations of small, nearly additive effects are usually expressed relatively late in development, whereas lethal mutations are usually expressed early (see also Hadorn 1961; Wright 1970).

Although extremely strong selection for the robustness of early organogenesis undoubtedly occurs, given the deleteriousness of most mutations affecting that stage, the number of interactions involved in the morphogenetic patterning is probably too limited to prevent substantial and global interactivity between developmental modules (Galis et al. 2002).

16.4.2 Phylotypic Stage in Mammals: Importance of Pleiotropy

Teratological data on rodents strongly support the pleiotropy hypothesis for the vertebrate phylotypic stage: Early organogenesis was found to be more vulnerable to disturbances than earlier or later stages, as disturbances caused more abnormalities and lethality (Galis and Metz 2001). The interdependent pattern of the numerous induced abnormalities (pleiotropic effects) indicates that the high interactivity and low effective modularity is the root cause of the vulnerability of this stage. The vulnerability, thus, is not due to a single vulnerable process (e.g., neural tube closure), as is the case for the vulnerability of cell divisions during the cleavage stage (Galis et al. 2018); instead, it is due to the global interactivity during the stage. Indeed, in rodents and humans, almost all major congenital abnormalities find their origin in disturbances of the phylotypic stage (Russell 1950; Shenefelt 1972; Sadler 2010). The global interactivity implies that a particular, potentially useful change of this stage (e.g., a change in the number of limbs and antennae in insects or the number of kidneys, lungs, spleens, eyes, ears, long bones, and digits in vertebrates) will nearly always be accompanied by other abnormalities and by early lethality.

For example, in humans, changes to the number of (normally) 7 cervical vertebrae are induced at the phylotypic stage and are associated with a wide variety of abnormalities and early lethality (Galis et al. 2006; Furtado et al. 2011; Varela-Lasheras et al. 2011; Ten

Broek et al. 2012; Schut et al. 2020a–c; Galis et al. 2021). Changes of the cervical number are usually manifested as cervical ribs (rudimentary or full ribs on the seventh vertebra) or rudimentary or absent first ribs. Approximately 90% of individuals with such changes are dead at birth, and a strong association is seen with cardiovascular, nervous, urogenital, and other congenital abnormalities (op. cit.). After birth, pleiotropic effects increase the risk for embryonal tumors (Schumacher and Gutjahr 1992; Galis 1999; Galis and Metz 2003; Merks et al. 2005) and miscarriages (Schut et al. 2020b). The increased risk for miscarriages in humans is in agreement with fertility problems in thoroughbred horses either with rudimentary and absent first ribs, or cervical ribs (termed flared ribs and bifid first rib; May-Davis 2017).

Although the deleterious pleiotropic effects and the resulting strong selection against changes of the number of cervical vertebrae have been best investigated in humans, further support comes from studies on a wide variety of mammals, including afrotherians and xenarthrans (Varela-Lasheras et al. 2011), thoroughbred horses (May-Davis 2017), extinct woolly mammoths and rhinoceroses (Reumer et al. 2014; van der Geer and Galis 2017), and domesticated dogs (Brocal et al. 2018).

Further support for the importance of the disturbance of global interactivity for the induction of cervical ribs comes from the large heterogeneity of genetic and environmental causes of cervical ribs. Many different genetic abnormalities can disrupt the patterning of early organogenesis and lead to the induction of cervical ribs, including single gene disorders, large copy number variations and most aneuploidies (Keeling and Kjaer 1999; Galis et al. 2006; Furtado et al. 2011; Schut et al. 2019, 2020a). Moreover, a wide variety of teratogenic disruptions can lead to the development of cervical ribs in rodents (Li and Shiota 1999; Kawanishi et al. 2003; Wéry et al. 2003). The timing and duration of the disruption of the patterning matters more than the specific nature of the disruption. The many different possible disruptions leading to cervical ribs probably explain their extraordinarily frequent occurrence in humans. Approximately half of all deceased fetuses and infants have cervical ribs (McNally et al. 1990; Furtado et al. 2011; Ten Broek et al. 2012; Schut et al 2019). Assuming that ~15% of clinically recognized pregnancies end in a miscarriage (Forbes 1997), and approximately half of these fetuses have cervical ribs, it follows that almost 8% of all human conceptions experience a disturbed early phylotypic stage and develop cervical ribs. This makes cervical ribs one of the most common congenital abnormalities and emphasizes the vulnerability of the phylotypic stage.

As with changes in the number of cervical vertebrae, ~90% of human individuals with an extra digit are dead at birth (Opitz et al. 1987, Galis et al. 2010), and at least 290 syndromes are associated with extra digits (Biesecker 2011).

The medical and veterinary literature shows that increases in the number of replicated organs, which is determined during the phylotypic stage, are observed very rarely—for example, spleens, kidneys, ureters, vaginas, penises, testicles, long bones, and (even if extremely rarely) additional arms and legs (e.g., J. Uchida et al. 2006; Galis and Metz 2007, Lilje et al. 2007). These duplications appear to be strongly selected against, owing to associated deleterious pleiotropic effects (Grüneberg 1963; Lande 1978; Galis et al. 2001, 2010; Lilje et al. 2007; Biesecker 2011). Without doubt, extra organs, including digits, can have strong selective advantages. For instance, extra digits are advantageous in digging and swimming. Yet, despite the potential functional advantages and the extremely frequent occurrence of mutations for polydactyly, extra digits in amniotes are extremely rare, whereas extra digit-like structures (from modified carpal or tarsal bones, or sesamoid bones) are common across a range of animals (e.g., the panda's thumb, the mole's thumb, and an extra digit-like structure in the sea turtle *Chelone;* Galis et al. 2001).

Similarly, a higher number of cervical vertebrae would most likely be advantageous in long-necked mammals. Even giraffes have only 7 cervical vertebrae, whereas swans have 23–25 (Woolfenden 1961). A giraffe's neck is quite stiff, requires substantial force to lift it, and it is too short to reach the ground unless the front legs are spread wide apart (figure 16.5). This awkward position renders giraffes vulnerable to predators while drinking (Valeix et al. 2009). Hence, absence of directional selection for change is often not a plausible alternative explanation of the strong conservation.

Correspondingly, Grüneberg found that loss of digits (oligodactyly) is associated with a multitude of pleiotropic effects in both the appendicular and axial skeleton in mice. Presumably, due to this pleiotropic constraint, loss of organs typically occurs via the slow and continued evolution of an earlier stop of development, followed by partial or complete degeneration of the organ (Lande 1978; Raynaud and Brabet 1994; Galis et al. 2001, 2002; Bejder and Hall 2002). Hence, construction is followed by destruction. For instance, in horses and cows, 5-digit condensations initially still develop. The developmental interactions with these digit condensations apparently cannot be easily discarded evolutionarily. Similarly, in blind cave fishes and salamanders, eye development always proceeds until the lenses have been formed, after which degeneration follows (e.g., Dufton et al. 2012). Other examples of vestigial organs that have evolutionarily lost functionality but not yet fully disappeared during early development are teeth in baleen whales; the clavicle in canids, felids, and lagomorphs; wings in emus and kiwis; and pelvises in whales (Glover 1916; Klima 1990; Bejder and Hall 2002; Senter and Moch 2015). Generally, more of these structures are seen during early development than later on, due to their subsequent degeneration. As a result of the slow accumulation of mutations during the loss of complex organs, re-evolution is virtually impossible, in agreement with Dollo's law (Goldberg and Igić 2008; Galis et al. 2010).

16.4.3 Transcriptomic Data: Importance of Pleiotropy

Further support for the pleiotropy hypothesis comes from transcriptomic studies on insects and vertebrates, which show conserved expression of regulatory genes during the phylotypic stage, including expression of microRNA genes (Kalinka et al. 2010; De Mendoza et al. 2013; Stergachis et al. 2013; Ninova et al. 2014; Levin et al. 2016), genes with pleiotropic activity in other parts of the embryo (Cheng et al. 2014; Hu et al. 2017), and those with pleiotropic activity at other stages during development (Levin et al. 2012; Hu et al. 2017; see also Fish et al. 2017).

16.4.4 No Theoretical Support for Robustness as a Cause of Long-Term Conservation

Theory supports robustness as a cause of short-term—but not long-term—conservation. Stabilizing selection is expected to lead to robustness to protect optimized traits against developmental noise and mutations, potentially leading to short-term conservation (A. Wagner 2000; Metz 2011; Papakostas et al. 2014; Austin 2016; Melzer and Theißen

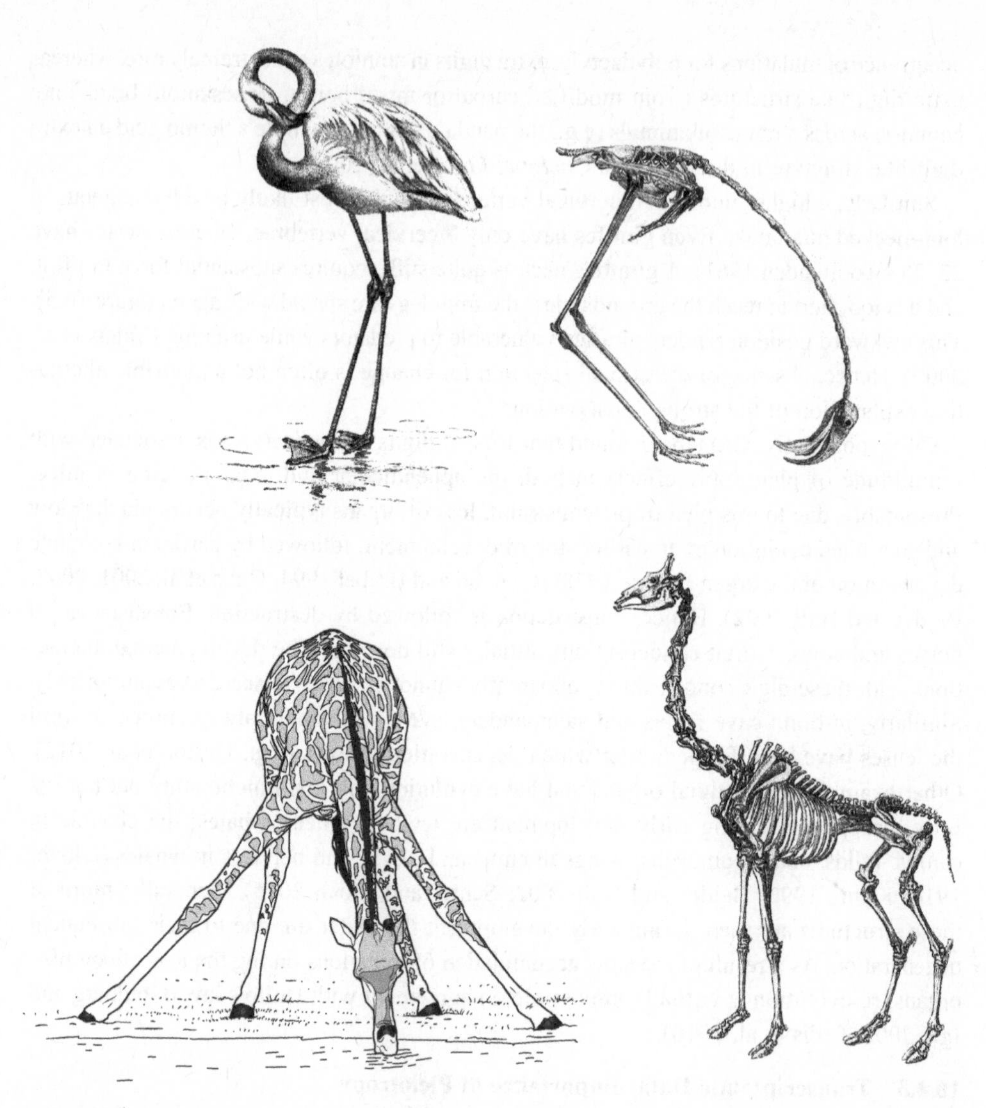

Figure 16.5
The long neck of the giraffe has only 7 vertebrae, which makes it rather stiff, so that lifting it costs considerable force. Despite the length of each cervical vertebra, the neck is not long enough to reach the ground unless the front legs are spread wide apart, which is an awkward position that makes the giraffe vulnerable to predators. Many vertebrae make a long and flexible neck in flamingoes. Top-left and bottom-right figures reproduced from Owen (1866), top right from Evans (1900), bottom left drawing by Erik-Jan Bosch (Naturalis).

2016). However, over long evolutionary scales, hidden variations will continue to accumulate, leading to a diversification of genetic backgrounds for new mutations (Gibson and G. Wagner 2000; Metz 2011; A. Wagner 2012; Siegal and Leu 2014). Hence, during periods with drastic environmental changes that lead to strong directional selection, the robustness of development alone cannot be sufficient to prevent change, because of the accumulated cryptic variation. In fact, in the long-term, cryptic variation is expected to lead to increased evolvability (A. Wagner 2012; Siegal and Leu 2014; Melzer and Theißen 2016).

In conclusion, empirical and theoretical support for the robustness hypothesis as an explanation for the conservation of the phylotypic stages is weak, and for the pleiotropy hypothesis it is strong, emphasizing the importance of modularity of developmental pathways for evolvability. The high interactivity between the developmental pathways not only constrains the evolvability of the phylotypic stages but also the evolvability of body plans, as the layout of most of the body plans occurs during the phylotypic stages.

16.5 Increased Modularity and Evolvability of Later Stages

When the final number of organs is determined after the vulnerable phylotypic stage, and development becomes more modular, the constraint on changes is considerably weaker. In arthropods with sequential production of segments continuing past the conserved segmented germ band stage, segment numbers vary strongly. Another instructive case is the probable polydactyly in frogs (Galis et al. 2001; Hayashi et al. 2015). Anterior to the first digit, a prehallux or prepollux is present in various species of frogs, most likely representing a rudimentary sixth digit. In amphibians with aquatic larvae, limb development occurs later than in amniotes and generally occurs after the phylotypic stage (Galis et al. 2001; Galis et al. 2003). Limb development is especially late in anurans with an extreme mode of metamorphosis. Thus almost all limb development is unaffected by the interactivity of the phylotypic stage, a drastic difference compared with amniotes. This result is in agreement with the extremely high self-organizing capacity of the limb buds in many amphibians compared with those in amniotes (Balinsky 1970). Amphibian limb buds can be grafted to very different places, such as the head, and still successfully develop into limbs, which are even capable of movement (Detwiler 1930). After the initiation of the limb field, limb development can proceed almost as an independent module, with few interactions with other parts of the body (Galis et al. 2003). This independence reduces the number of pleiotropic effects of mutations that affect limb development. Hence, the probable polydactyly in frogs represents another example of relaxed selection against evolutionary changes due to the reduced pleiotropy of later developmental stages.

In agreement with the weaker conservation of later-determined numbers, changes of the number of thoracic vertebrae are not significantly associated with congenital abnormalities in humans (weaker pleiotropy), and stabilizing selection is considerably weaker compared to that of cervical vertebrae changes (Galis et al. 2006).

The late determination of the number of thoracic vertebrae is comparable to that of cervical vertebrae in birds and many reptiles. The number of cervical vertebrae is highly variable across bird species, with 12 in pigeons and up to 25 in swans (Woolfenden 1961). Even in the shortest necks, the number of vertebrae is considerably larger than in mammals.

Also in reptiles, the number of vertebrae can be highly variable; however, this is mainly the case in reptiles with long necks (e.g., many plesiosaurs and dinosaurs), whereas the number is quite constant in families with 9 or fewer cervical vertebrae, such as pterosaurs, crocodiles, turtles, geckos, and many other lizards (Hofstetter and Gasc 1969; Bennett 2014). The later determination of cervical vertebrae when there are many of them, as in birds and reptiles, should weaken the constraint, because it is expected that number changes will be associated with fewer pleiotropic effects, as is the case for thoracic and lumbar vertebrae in humans (Galis et al. 2006; Ten Broek et al. 2012). In agreement with this, the largest intraspecific variability in number is found in birds and reptiles with the longest necks (e.g., swans can have 21–25 vertebrae; Woolfenden 1961). Hence, the constraint on changes in the number of cervical vertebrae in mammals is less exceptional than it seems, it is just somewhat stronger than in short-necked reptiles with limbs, presumably because of a combination of the generally greater mobility and higher metabolic rates (Galis et al. 2021).

Other examples of late-determined structures are the number of segments in insects, carpal and tarsal elements, phalanges, teeth, trunk, and caudal vertebrae in amniotes, and nipples in mammals, which are all highly evolvable.

16.6 Relaxation of Stabilizing Selection Increases Evolvability

The conservation of the phylotypic stages should not be taken overly strictly. What matters is that in many higher taxa, development is remarkably conserved during phylotypic stage, more strongly than during earlier or later stages (Sander and Schmidt-Ott 2004). As argued above, stabilizing selection against the pleiotropic effects of mutations appears to be the reason that mutations affecting the phylotypic stage hardly ever persist in populations. Thus, when external or internal conditions relax the normally strong stabilizing selection, evolutionary changes may occur on rare occasions. In the wild, such breaking of constraints is often associated with the start of adaptive radiations and the emergence of key innovations, when for instance many empty niches suddenly become available, or a massive extinction of predators has taken place (Galis 2001; Galis and Metz 2021). Arguably, relaxed selection allows novelties to persist for some time, despite associated pleiotropic effects. This longer persistence in the population most likely leads to selection against the most deleterious pleiotropic effects, such that when stabilizing selection increases again, the chance for persistence of the novelties is increased (McCune 1990).

The effect of relaxation of stabilizing selection is most clearly illustrated in domesticated animals, where human care keeps those individuals alive that would likely not survive in nature. Extra digits in chickens, cats, and dogs are good examples of persistence of traits that are strongly selected against in nature (Galis et al. 2001). Dog and horse breeds with unusually high frequencies of cervical ribs and rudimentary first ribs provide further examples, where human care allows the survival of traits despite their frequent co-occurrence with congenital abnormalities or fertility problems (May-Davis 2017; Brocal et al. 2018). Sloths and manatees, carrying exceptional numbers of cervical vertebrae, provide good natural examples of the effect of relaxed stabilizing selection. Sloths and manatees commonly have skeletal and other congenital abnormalities that are associated with cervical ribs in deceased human fetuses (Varela-Lasheras et al. 2011). These abnormalities would probably be fatal in most more active mammals. Their extremely slow metabolic rates, combined with weak locomotory constraints, appear to provide a relaxation of stabilizing selection against some of the associated skeletal and congenital abnormalities (Varela-Lasheras et al. 2011). Interestingly, sloths and manatees also have extremely low cancer rates (Galis and Metz 2003; Tollis et al 2020). Several other species show the association between cervical ribs and low metabolic and activity rates, providing support that these reduced rates may be involved in breaking the constraint. Slow loris and pottos (primates with extremely low metabolic and activity rates) often have cervical ribs (Galis et al. 2022). Whales and dolphins are also exceptional in regularly having cervical ribs. They also often have skeletal abnormalities. Relaxed selection against skeletal abnormalities is probably also involved; in this case, the relaxation is thought to be caused by the supporting effect of water (Galis et al. 2021). Furthermore, whales and dolphins also have low cancer rates, which probably also weakens the stabilizing selection against cervical ribs.

Thus, the difficulty of breaking specific constraints varies among taxa, due to differences in the selection regimes experienced and in the specific pleiotropic effects associated with trait changes (Galis and Metz 2018).

16.7 Evolvability of Vulnerable Early Cleavage and Gastrulation Stages

Raff (1994, 1996) proposed that the larger evolvability of the earlier developmental stages of cleavage and gastrulation may be due to the lower number of inductive interactions in the embryo, which should lead to fewer pleiotropic effects. Yet effective modularity is low, and cleavage is a vulnerable stage, particularly with respect to high doses of toxicants and radiation (e.g., Russell 1950; Shenefelt 1972; Galis and Metz 2001; Jacquet 2004). This vulnerability of cleavage has been used to argue that the vulnerability of organogenesis can, therefore, not be involved in the conservation of the phylotypic stage, as cleavage is evolvable, despite its vulnerability (Y. Uchida et al. 2018). However, in contrast to the phylotypic stage, the vulnerability in cleavage is mainly due to one vulnerable process: cell division. Furthermore, dividing cleavage cells are greatly similar (not yet differentiated) and are capable of self-renewal. The high capacity for self-renewal of the cleavage cells implies that, either too many cells are killed and the embryo dies, or the damage is reversible and development proceeds largely normally without adverse embryonic outcome (Russell 1950; Shenefelt 1972; Jacquet 2004; Adam 2012). In medicine, this is known as the all-or-none phenomenon, which has been extensively used in genetic counseling of pregnant women, who have inadvertently undergone an exposure to teratogenic substances in the early stages of pregnancy, frequently before the pregnancy has been recognized (Jacquet 2004; Adam 2012). The vulnerability of the subsequent phylotypic stage differs critically in that the strong global interactivity restricts the potential for reversal of damage. And mutations that affect cleavage and gastrulation presumably have a greater probability of being successful, as it is more difficult to destabilize a simple pattern than a more complicated one (Galis and Sinervo 2002; Galis et al. 2018). Hence, the weaker conservation of cleavage and gastrulation may well be due to the greater simplicity of the early forms.

16.8 Diversity of Phylotypic Stages among Metazoans

The remarkable diversity of phylotypic stages among metazoans (e.g., the Nauplius stage in crustaceans and the neurula in vertebrates; figures 16.1 and 16.2) seemingly contradicts the explanation of conservation caused by interactivity of the phylotypic stage. But the pattern of divergence suggests an early rapid phase of diversification in the evolution of metazoans during the Ediacaran Cambrian times, followed by strong conservation of discrete taxon-specific phylotypic stages and body plans (Buss 1987). The cause for this pattern of early rapid diversification of body plans followed by stasis is not well understood. Davidson and Erwin (2006) proposed that the evolution of more hierarchical and interconnected gene regulatory networks and their increased cooption for other developmental functions may be involved.

One hypothesis that did not receive much attention was by Buss (1987), who suggested that the initial diversification occurred during the early, chaotic phase in the evolution from unicellular to multicellular individuals (presumably during the Cambrian explosion). During this process, the level of selection shifted from individual cells to individual organisms. Early during this transition, somatic mutations in cells that could gain access to reproduction had a chance to be maintained in future generations (as in plants). Later, when selection was firmly established at the level of the individual, heritable mutations became limited to those that occur in the germ line or in the short period before germ line sequestration. This hypothesis thus assumes that during the early chaotic transition, when control was not yet at the level of the individual, the lack of integration increased evolvability. This early diversification scenario is intuitively appealing, but it has received surprisingly little attention, and hardly any research has been carried out to investigate this important question in evolutionary biology. Mutagenesis experiments with simple colonial organisms and theoretical modeling could probably contribute to a better understanding of this possibility (Galis and Sinervo 2002).

16.9 An Ancient Metazoan Constraint Causes the Early Layout of the Body Plan

In animals, most of the body plan traits are initiated early, during a highly interactive stage limiting evolvability of the body plans. In complex animals, most flexibility is provided by the changes in the number of segments, which allow the multiplication of legs and other organs of the segment. Another solution to the problem of evolvability is the vegetative production of modules that are morphological repeats of the body plan—for instance, those found in cnidarians, bryozoans, and colonial ascidians (Bell 1982). In contrast, the body plan in plants is not laid out early, and new organs can be initiated throughout life (Heidstra and Sabatini 2014; Cridge et al. 2016). Why does the layout of body plans occur so early in animals? Even in animals with metamorphosis, the organ primordia emerge early, with the adult fate already determined, like imaginal discs in insects (Held 2005). A major difference between plants and animals concerns an ancient animal constraint: Differentiated cells cannot divide by mitosis unless they first dedifferentiate (which does occur in regeneration and in cancer; Buss 1987; Galis et al. 2018). The conflict between differentiation and mitosis stems from the presence in cells of a single centrosome, which is necessary for both processes in animals. Plants do not have centrosomes and use other structures to organize their microtubuli during cell division, and cilia (discussed later in this chapter) only occur in sperm cells of certain taxa (Schmit 2002). Even the cells outside the stem cell niches are able to return to a proliferative pluripotent state (Heidstra and Sabatini 2014).

16.9.1 Single Centrosome Precludes Simultaneous Cell Division and Differentiation

Centrosomes change dynamically during the cell cycle. During mitosis in animals, centrosomes are duplicated, precisely once per cell cycle. The duplicated centrosomes form the bipolar spindles that precisely segregate the duplicated chromosomes, producing an equal distribution of chromosomes between daughter cells (Sir et al. 2013; Meraldi 2016). Although centrosomes are not absolutely required for mitotic spindle formation and division in many cells, mitosis in the absence of centrosomes is an error-prone process that leads to chromosomal instability (Bonaccorsi et al. 2000; Basto et al. 2008; Sir et al. 2013; Meraldi 2016). When cells stop dividing, they form a primary cilium, for which one of the centrioles of the centrosome (the mother centriole) is necessary. This centriole (the mother centriole) converts into a basal body and migrates to the cell surface, where it organizes the primary cilium. Primary cilia were long thought to be vestigial organelles, not present in many cells. It was not until the 1990s, owing to technical improvements of visualization techniques, that it was discovered that primary cilia are present as antennae on almost all metazoan cells, including quiescent stem cells (Wheatley 1995). Furthermore, they do not function only as sensory organelles but also have a key function in intercellular signaling (Dawe et al. 2007; Walz 2017). Signaling in the cilium is involved in the organization of most (if not all) developmental processes, including left-right patterning, cell migration, reentry of cells into the cell cycle (proliferation), cell size, cell shape, specification of the plane of cell division, apoptosis, and cell fate decisions.

When cell division resumes, re-entry of the cell cycle begins with the resorption of the primary cilium, detachment of the basal body (the mother-centriole) from the cell surface, and migration of the centrosome to near the nucleus. Recent studies have shown that the cell cycle is not so much regulating centrosome and cilium dynamics; instead, the dynamics of the centrosome and primary cilium actively regulate cell cycle progression and arrest or exit followed by differentiation (Walz 2017). For example, the physical presence of the primary cilium appears to block cell division, whereas primary ciliary resorption is thought to unblock cell division, and the length of the cilium influences cell cycle duration, which in turn influences cell-fate decisions (Walz 2017). Hence, these discoveries, which began in the 1990s and continue today, allow us to understand the metazoan constraint on simultaneous cell division and differentiation.

16.9.2 Metazoan Constraint Already Proposed in 1898

At the end of the 19th century, the constraint on cell division by ciliated cells had already been independently proposed by Henneguy (1898) and Lenhossék (1898). Buss could not have known in 1987 that virtually all differentiated metazoan cells have primary cilia and thus, the constraint that he proposed on the incompatibility of cell division and differentiation, equates in essence to the constraint proposed by Henneguy (1898) and Lenhossék (1898). Incidentally, Buss (1987) erroneously attributed the proposal of the constraint on cell division by ciliated cells to Margulis (1981), instead of to Henneguy and Lenhossék. The hypothesis of Henneguy and Lenhossék remains uncontested for metazoans (for a review of the constraint, including rare exceptions, see Galis et al. 2018). Even lymphocytes, which were long thought to be exceptional in not having primary cilia, are now thought to have a modified primary cilium (Finetti et al. 2009; Dustin 2014). Furthermore, in the differentiation of some cells, the cilium or centrosome is discarded, which also prevents further cell division (Bornens 2012; Das and Storey 2014). The few claims that differentiated cells can divide during cell renewal and regeneration are controversial (Dor et al. 2004; Brennand et al. 2007; Afelik and Rovira 2017). For a detailed review of the constraint, including rare exceptions, see Galis et al. (2018).

16.9.3 Extra Centrosomes Cannot Break the Evolutionary Constraint

Exceptionally, de novo generation of extra centrosomes occurs. But it is not a viable evolutionary road to the breaking of the constraint. Extra centrosomes pose a grave risk and may lead to the formation of multiple spindle poles, aneuploidy, cell cycle arrest, apoptosis, genomic instability, cell migration (e.g., in metastasis of cancer cells; Basto et al. 2008; Godinho and Pelman 2014; Gönczy 2015), and perhaps cancer, as already proposed by Theodor Boveri (1902). Furthermore, in cells with supernumerary centrosomes, extra cilia are often formed and compromise the functioning of primary cilium signaling, which may lead to cancer and other diseases (Mahjoub and Stearns 2012). The importance of having only one centrosome per cell is also supported by the elimination of the centrioles from animal egg cells before fertilization, such that the zygote receives centrioles only from the sperm cell and does not end up with two centrosomes instead of one (Boveri 1901; Bell 1989; Manandhar et al. 2005). The cost of centriole elimination is that meiotic divisions in egg cells are less reliable. We conclude that supernumerary centrosomes are usually seriously disadvantageous for the individual and will be strongly selected against. Therefore, it is unlikely that the inability of ciliated cells to form proper mitotic spindles could be compensated for by the evolution of extra centrosomes.

16.9.4 Multiciliated Cells

Some cells in animals can have hundreds of cilia, each requiring its own basal body to be assembled de novo (Dawe et al. 2007). However, the pathways used to produce these centrioles are different from those involved in the duplication of centrioles during the cell division cycle (op. cit.). Furthermore, multiciliated cells are terminally differentiated, and for cell renewal, unciliated progenitor cells are employed (Bird et al. 2014). Hence, multiciliated cells are not an exception to the rule that differentiated cells cannot divide.

16.9.5 Pluripotent Stem Cells Only Present During Early Development

Without a doubt, the incompatibility of simultaneous cell division and ciliation has crucially shaped development and evolution of metazoan body plans. The body plan is mostly defined during early embryonic development, when there are still zones that produce pluripotent stem cell colonies that subsequently migrate to other places in the embryo to start their paths of differentiation. Thus, the ancient metazoan constraint generally causes most of the layout of the body plans to occur early, when the number of inductive interactions does not yet provide sufficient effective modularity to prevent major pleiotropy. Later in life, pluripotent cells are absent, which prevents the initiation of organ primordia. As already mentioned, also in animals with metamorphosis, organ primordia emerge early during embryogenesis (Held 2005). Adults generally only have multipotent, tissue-specific, stem cells that function in cell renewal, wound healing, and regeneration (Tanaka and Reddien 2011).

16.9.6 Constraint Inherited from Unicellular Metazoan Ancestors

Buss (1987) argued that metazoans inherited the possession of only one organizing center for microtubules, the centrosome, from their unicellular protist ancestors and that, in contrast, other unicellular groups (e.g., euglenophytes, cryptophytes, and chlorophytes) with multiple of such organizing centers do not have this constraint. These groups are capable of accomplishing simultaneous cell movement and mitotic cell division by using some centers exclusively as organizers for undulipodia (cilia, flagella) and others for cell division.

16.10 Conclusion

The ancient metazoan constraint on simultaneous cell division and differentiation (Buss 1987), thus causes the layout of body plans to occur early, when the embryo is small and there are still sufficient pluripotential cells that can differentiate into a large number of different types of cells. The limited number of inductive interactions in the small embryo does not yet allow developmental modularity, and so the interactivity is global. As a result, mutations affecting this stage will almost always have many pleiotropic effects and will, therefore, be selected against. Thus, the ancient metazoan constraint causes another developmental constraint, the one on changes at the early organogenesis stage: the conserved phylotypic stage. Furthermore, as most of the body plan is laid out during the phylotypic stage, the minimal early developmental modularity not only constrains the evolvability of the stage itself but also of animal body plans in general. Together, these appear to be two of the rare hard developmental constraints that prevent evolvability at macro-evolutionary scales and have had a major influence on animal evolution (Galis et al 2018; Galis and Metz 2021).

Acknowledgments

I thank Scott Armbruster, Christophe Pélabon, Thomas Hansen, Jacques van Alphen, Gene Hunt, and Mihaela Pavličev for useful comments and discussion on the manuscript.

References

Adam, M. P. 2012. The all-or-none phenomenon revisited. *Birth Defects Research* (A) 94: 664–69.

Afelik, S., and M. Rovira. 2017. Pancreatic β-cell regeneration: facultative or dedicated progenitors? *Molecular and Cellular Endocrinology* 445: 85–94.

Armbruster, W., C. Pélabon, G. H. Bolstad, and T. F. Hansen 2014. Integrated phenotypes: Understanding trait covariation in plants and animals. *Philosophical Transactions of the Royal Society B* 369: 20130245.

Austin, C. J. 2016. The ontology of organisms: Mechanistic modules or patterned processes? *Biology and Philosophy* 31: 639–662.

Balfour, F. M. 1881. *Comparative Embryology.* London: Macmillan and Co.

Balinsky, B. I. 1970. *An Introduction to Embryology*, 3rd ed. Philadelphia: W. B. Saunders.

Ballard, W. W. 1981. Morphogenetic movements and fate maps of vertebrates. *American Zoologist* 21: 391–399.

Basto, R., K. Brunk, T. Vinogradova, N. Peel, A. Franz, A. Khodjakov and J. W. Raff. 2008. Centrosome amplification can initiate tumorigenesis in flies. *Cell* 133: 1032–1042.

Bejder, L., and B. K. Hall. 2002. Limbs in whales and limblessness in other vertebrates: Mechanisms of evolutionary and developmental transformation and loss. *Evolution and Development* 4: 445–458.

Bell G. 1982. *The Masterpiece of Nature: The Evolution and Genetics of Sexuality*. London: Croomhelm.

Bell, G. 1989. Darwin and biology. *Journal of Heredity* 80: 417–421.

Bennett, S. C. 2014. A new specimen of the pterosaur *Scaphogranthus crassirostris*, with comments on constraint of cervical vertebrae number in pterosaurs. *Neues Jahrbuch für Geologie und Paläontologischen Abhandlungen* 27: 327–348.

Biesecker, L. G. 2011. Polydactyly: How many disorders and how many genes? 2010 update. *Developmental Dynamics* 240: 931–942.

Bird, A. M., G. von Dassow, and S. A. Maslakova. 2014. How the pilidum larva grows. *EvoDevo* 5: 13.

Bonaccorsi, S., M. G. Giansanti, and M. Gatti. 2000. Spindle assembly in *Drosophila* neuroblasts and ganglion mother cells. *Nature Cell Biology* 2: 54–56.

Bonner, J. T. 1988. *The Evolution of Complexity by Means of Natural Selection*. Princeton, NJ: Princeton University Press.

Bornens, M. 2012. The centrosome in cells and organisms. *Science* 335: 422–426.

Boveri, T. 1901. *Zellenstudien: Über die Natur der Centrosomen. 4*. Jena, Germany: Fischer.

Boveri, T. 1902. Ueber mehrpolige Mitosen als Mittel zur Analyse des Zellkerns *Verhandlungen der Physikalisch-Medizinischen Gesellschaft zu Würzburg* 35: 67–90.

Brennand, K., D. Huangfu, and D. Melton. 2007. All β cells contribute equally to islet growth and maintenance. *PLOS Biology* 5: e163.

Brocal, J., S. De Decker, R. José-López et al. 2018. C7 vertebra homeotic transformation in domestic dogs—are Pug dogs breaking mammalian evolutionary constraints? *Journal of Anatomy* 233: 255–265.

Buss, L. W. 1987. *The Evolution of Individuality*. Princeton, NJ: Princeton University Press.

Cheng, Y., Z. Ma, B.-H. Kim, W. Wu, P. Cayting et al. 2014. Principles of regulatory information conservation between mouse and human. *Nature* 515: 371–375.

Claus, C., and K. Grobben. 1917. *Lehrbuch der Zoologie*. Marburg: Elwert.

Cohen, J. 1977. *Reproduction*. London: Butterworth.

Collazo, A., J. A. Bolker, and R. Keller. 1994. A phylogenetic perspective on teleost gastrulation. *American Naturalist* 144: 133–152.

Cridge A. G., P. K. Dearden, and L. R. Brownfield. 2016. The mid-developmental transition and the evolution of animal body plans. *Annals of Botany* 117:833–843.

Das, R. M., and K. G. Storey. 2014. Apical abscission alters cell polarity and dismantles the primary cilium during neurogenesis. *Science* 343: 200–204.

Davidson, E. H., and D. H. Erwin. 2006. Gene regulatory networks and the evolution of animal body plans. *Science* 311: 796–800.

Dawe, H. R., H. Farr, and K. Gull. 2007. Centriole/basal body morphogenesis and migration during ciliogenesis in animal cells. *Journal of Cell Science* 120: 7–15.

DeMendoza, A., A. Sebé-Pedrós, M. S. Šestak, M. Matejcic, G. Torruella, T. Domazet-Loso, and I. Ruiz-Trillo. 2013. Transcription factor evolution in eukaryotes and the assembly of the regulatory toolkit in multicellular lineages. *PNAS* 110: E4858–4866.

Detwiler, S. R. 1930. Observations upon the growth, function, and nerve supply of limbs when grafted to the head of salamander embryos. *Journal of Experimental Zoology B* 55: 319–370.

Dor, Y., J. Brown, O. I. Martinez, and D. A. Melton. 2004. Adult pancreatic beta-cells are formed by self-duplication rather than stem-cell differentiation. *Nature* 420: 41–46.

Dufton, M., B. K. Hall, and T. A. Franz-Odendaal. 2012. Early lens ablation causes dramatic long-term effects on the shape of bones in the craniofacial skeleton of *Astyanax mexicanus*. *PLOS One* 7: e50308.

Dustin, M. L. 2014. T cells play the classics with a different spin. *Molecular Biology of the Cell* 25: 1699–1670.

Evans, T. H. 1900. *Birds. The Cambridge Natural History*, vol. 3. London: Macmillan.

Finetti, F., S. Rossi Paccani, M. G. Riparbelli et al. 2009. Intraflagellar transport is required for polarized recycling of the TCR/CD3 complex to the immune synapse. *Nature Cell Biology* 11: 1332–1339.

Fish, A., L. Chen, and L. A. Capra. 2017. Gene regulatory enhancers with evolutionarily conserved activity are more pleiotropic than those with species-specific activity. *Genome Biology and Evolution* 9: 2615–2625.

Forbes, L. S. 1997. The evolutionary biology of spontaneous abortion in humans. *TREE* 12: 446–450.

Furtado, L. V., H. M. Thaker, L. K. Erickson, B. H. Shirts, and J. M. Opitz. 2011. Cervical ribs are more prevalent in stillborn fetuses than in live-born infants and are strongly associated with fetal aneuploidy. *Pediatric and Developmental Pathology* 14: 431–437.

Galis, F. 1996. The application of functional morphology to evolutionary studies. *TREE* 11: 124–129.

Galis, F. 1999. Why do almost all animals have seven cervical vertebrae? Developmental constraints, Hox genes, and cancer. *Journal of Experimental Zoology B* 285: 19–26.

Galis, F. 2001. Key innovations and radiations. In *The Character Concept in Evolutionary Biology,* edited by G. P. Wagner. London: Academic Press.

Galis, F., and J. A. J. Metz. 2001. Testing the vulnerability of the phylotypic stage: On modularity and evolutionary conservation. *Journal of Experimental Zoology B* 291: 195–274.

Galis, F., and J. A. J. Metz. 2003. Anti-cancer selection as a source of developmental and evolutionary constraints. *BioEssays* 25: 1035–1039.

Galis, F., and J. A. J. Metz. 2007. Evolutionary novelties: The making and breaking of pleiotropic constraints. *Integrative and Comparative Biology* 47: 409–419.

Galis, F., and J. A. J. Metz. 2021. A macroevolutionary approach on developmental constraints in animals. In *Evolutionary Developmental Biology,* edited by L. Nuño de la Rosa and G. B. Müller. Cham, Switzerland: Springer.

Galis, F., and B. Sinervo. 2002. Divergence and convergence in early embryonic stages of metazoans. *Contributions to Zoology* 71: 101–113.

Galis, F., J. J. M. van Alphen, and J. A. J. Metz. 2001. Why five fingers? Evolutionary constraints on digit numbers. *TREE* 16: 637–646.

Galis, F., T. J. M. Van Dooren and J. A. J. Metz. 2002. Conservation of the segmented germband stage: Robustness or pleiotropy? *Trends in Genetics* 18: 504–509.

Galis, F., T. J. M. Van Dooren, H. Feuth et al. 2006. Extreme selection against homeotic transformations of cervical vertebrae in humans. *Evolution* 60: 2643–2654.

Galis, F., J. W. Arntzen, and R. Lande. 2010. Dollo's law and the irreversibility of digit loss in Bachia. *Evolution* 64: 2466–2476.

Galis, F., J. A. J. Metz, and J. J. M. van Alphen. 2018. Development and evolutionary constraints in animals. *AREES* 49: 499–522.

Galis, F., P. C. Schut, T. E. Cohen-Overbeek, and C. M. A. ten Broek. 2021. Evolutionary and developmental issues of cervical ribs. In *Thoracic Outlet Syndrome,* edited by K. A. Illig et al., 23–35. Cham, Switzerland: Springer.

Galis, F., T. J. M. Van Dooren, A. A. E. van der Geer. 2022. Breaking the constraint on the number of cervical vertebrae in mammals: on homeotic transformations in lorises and pottos. *Evolution & Development* 24: 196–210.

Gibson, G., and G. P. Wagner 2000. Canalization in evolutionary genetics: A stabilizing theory? *BioEssays* 22: 372–380.

Gilbert, S. F. 1997. *Developmental Biology,* 5th ed. Sunderland, MA: Sinauer.

Gilbert, S. F. and A. M. Raunio. 1997. *Embryology. Constructing the Organism.* Sunderland, MA: Sinauer.

Glover, A. M. 1916. *The Whalebone Whales of New England.* Boston: Society of Natural History.

Goldberg, E. E., and B. Igić. 2008. On phylogenetic tests of irreversible evolution. *Evolution* 62: 2727–2741.

Gönczy, P. 2015. Centrosomes and cancer: Revisiting a long-standing relationship. *Nature Reviews Cancer* 15: 639–652.

Grbič, M. A., 2000. Alien wasps and evolution of development. *Bioessays* 22: 920–932.

Grüneberg, H. 1963. *The Pathology of Development: A Study of Inherited Skeletal Disorders in Animals.* Oxford: Blackwell Scientific.

Hadorn, E. 1961. *Developmental Genetics and Lethal Factors.* London: Methuen and Co.

Hall, B. K. 1997. Phylotypic stage or phantom: Is there a highly conserved embryonic stage in vertebrates? *TREE* 12: 461–463.

Hall, B. K. 1999. *Evolutionary Developmental Biology,* 2nd ed. Dordrecht: Kluwer Academic.

Hansen, T. F., and D. Houle. 2004. Evolvability, stabilizing selection, and the problem of stasis. In *Phenotypic Integration: Studying the Ecology and Evolution of Complex Phenotypes,* edited by M. Pigliucci and K. Preston, 130–150. Oxford: Oxford University Press.

Hayashi, S., T. Kobayashi, T. Yano et al. 2015. Evidence for an amphibian sixth digit. *Zoological Letters* 1:17.

Heidstra, R., and S. Sabatini. 2014. Plant and animal stem cells: Similar yet different. *Nature Reviews in Molecular Cell Biology* 15: 301–312.

Held, L. J. 2005. *Imaginal Discs*. Cambridge: Cambridge University Press.

Henneguy, L. F. 1898. Sur les rapports des cils vibratiles avec les centrosomes. *Archives d'Anatomie Microscopique* 1: 481–496.

Hofstetter, R., and J.-P. Gasc. 1969. Vertebrae and ribs of modern reptiles. In *Biology of Reptiles*, edited by C. Gans, 201–310. London: Academic Press.

Hu, H., M. Uesaka, S. Guo et al. 2017. Constrained vertebrate evolution by pleiotropic genes. *Nature Ecology & Evolution* 1:1722–1730.

Jacquet, P. 2004. Sensitivity of germ cells and embryos to ionizing radiation. *Journal of Biological Regulators and Homeostatic Agents* 18: 106–114.

Kalinka, A. T., K. M. Varga, D. T. Gerrard et al. 2010. Gene expression divergence recapitulates the developmental hourglass model. *Nature* 468: 811–816.

Kawanishi, C. Y., P. Hartig, K. L. Bobseine, J. Schmid, M. Cardon, G. Massenburg, and N. Chernoff. 2003. Axial skeletal and *Hox* expression domain alterations induced by retinoic acid, valproic acid, and bromoxynil during murine development. *Journal Biochemical Molecular Toxicology* 17: 346–356.

Keeling, J. W., and I. Kjaer. 1999. Cervical ribs: Useful marker of Monosomy X in fetal hydrops. *Pediatric and Developmental Pathology* 2: 119–123.

Keibel, F. 1904. *Normentafeln zur Entwicklungsgeschichte der Wirbeltiere*, Heft IV, Jena: Gustav Fisher.

Keibel, F. 1908. *Normentafeln zur Entwicklungsgeschichte der Wirbeltiere*, Heft VIII, Jena: Gustav Fisher.

Klima, M. 1990. Rudiments of the clavicle in the embryos of whales. *Zeitschrift für Saugetierkunde* 55: 202–212.

Lande, R. 1978. Mechanisms of limb loss in tetrapods. *Evolution* 32: 73–92.

Lande, R., D. W. Schemske, and S. T. Schultz. 1994. High inbreeding depression, selective interference among loci, and the threshold selfing rate for purging recessive lethal mutations. *Evolution* 48: 965–978.

Lenhossék, M. V. 1898. Ueber Flimmerzellen. *Verhandlungen der Anatomischen Gesellschaft, Kiel* 12: 106–128.

Levin, M., Hashimshony, F. Wanger, and I. Yanai. 2012. Developmental milestones punctuate gene expression in the *Caenorhabditis* embryo. *Developmental Cell* 22:1101–1108.

Levin, M., L. Anavy, A. G. Cole et al. 2016. The mid-developmental transition and the evolution of animal body plans. *Nature* 531: 637–641.

Lewontin, R. C. 1978. Adaptation. *Scientific American* 239: 212–231.

Li, Z.-L., and K. Shiota. 1999. Stage-specific homeotic vertebral transformations in mouse fetuses induced by maternal hyptertheria during somitogenesis. *Developmental Dynamics* 216: 336–348.

Lilje, C., L. J. Finger, and R. J. Ascuitto. 2007. Complete unilateral leg duplication with ipsilateral renal agenesis. *Acta Paediatrica* 96: 461–471.

Mahjoub, M., and T. Stearns. 2012. Supernumerary centrosomes nucleate extra cilia and compromise primary cilium signaling. *Current Biology* 22: 1628–1634.

Manandhar, G., H. Schatten, and P. Sutovsky. 2005. Centrosome reduction during gametogenesis and its significance. *Biological Reproduction* 72: 2–13.

Margulis, L. 1981. *Symbiosis and Cell Evolution*. San Francisco: Freeman

May-Davis, S. 2017. Congenital malformations of the first sternal rib. *Journal of Equine Veterinary Science* 49: 92–100.

McCune, A. R. 1990. Morphological anomalies in the *Semionotus* complex: Relaxed selection during colonization of an expanding lake. *Evolution* 44: 71–85.

McNally, E., B. Sandin, and R. A. Wilkins. 1990. The ossification of the costal element of the seventh cervical vertebra with particular reference to cervical ribs. *Journal of Anatomy* 170: 125–129.

Medawar, P. B. 1954. The significance of inductive relationships in the development of vertebrates. *Journal of Embryology and Experimental Morphology* 2: 172–174.

Melzer, R., and G. Theißen. 2016. The significance of developmental robustness for species diversity. *Annals of Botany* 117: 725–732.

Meraldi, P. 2016. Centrosomes in spindle organization and chromosome segregation: A mechanistic view. *Chromosome Research* 24: 19–34.

Merks, J. H. M., A. M. Smets, R. R. van Rijn, J. Kobes, H. N. Caron, M. Maas, and R. C. Hennekam. 2005. Prevalence of rib anomalies in normal Caucasian children and childhood cancer patients. *European Journal of Medical Genetics* 48: 113–129.

Metz J. A. J. 2011. Thoughts on the geometry of meso-evolution: collecting mathematical elements for a post-modern synthesis. In *The Mathematics of Darwin's Legacy,* edited by F. A. C. C. Chalub and J. F. Rodrigues, 197–234. Basel: Birkhauser.

Mitteröcker, P. 2009. The developmental basis of variational modularity: Insights from quantitative genetics, morphometrics, and developmental biology. *Evolutionary Biology* 36: 377–385.

Ninova, M., M. Ronshaugen, and S. Griffiths-Jones. 2014. Conserved temporal patterns of microRNA expression in *Drosophila* support a developmental hourglass model. *Genome Biology and Evolution* 6: 2459–2467.

Opitz, J. M., J. M. FitzGerald, J. F. Reynolds, S. O. Lewin, A. Daniel, L. S. Ekblom, and S. Phillips. 1987. The Montana Fetal Genetic Pathology Program and a Review of Prenatal Death in Humans. *American Journal of Medical Genetics* Supplement 3: 93–112.

Owen, R. 1866. *On the Anatomy of Vertebrates.* London: Longmans, Green, and Co.

Papakostas S., L. A. Vøllestad, M. Bruneaux et al. 2014. Gene pleiotropy constrains gene expression changes in fish adapted to different thermal conditions. *Nature Communications* 5: 4071.

Raff, R. A. 1994. Developmental mechanisms in the evolution of animal form: Origins and evolvability of body plans. In *Early Life on Earth,* edited by S. Bengston, 489–500. New York: Columbia University Press.

Raff, R. A. 1996. *The Shape of Life.* Chicago: University of Chicago Press.

Raynaud, A., and J. Brabet. 1994. New data on embryonic development of the limbs in the slow-worm, *Anguis fragilis* (Linné 1758). *Annales des Sciences Naturelles- Zoologie et Biologie Animale* 15: 97–113.

Reumer, J. W. F., C. M. A. ten Broek, and F. Galis. 2014. Extraordinary incidence of cervical ribs indicates vulnerable condition in Late Pleistocene mammoths. *PeerJ* 2: e318.

Russell, L. B. 1950. X-ray induced developmental abnormalities in the mouse and their use in the analysis of embryological patterns. *Journal of Experimental Zoology B* 114: 545–602.

Sadler, T. W. 2010. Birth defects and prenatal diagnosis. In *Langman's Medical Embryology,* 11th edition, edited by T. W. Sadler, 113–115. Baltimore: Lippincott.

Sander, K. 1983. The evolution of patterning mechanisms: Gleanings from insect embryogenesis and spermatogenesis. In *Development and Evolution,* edited by B. C. Goodwin, N. Holder, and C. C. Wylie, 137–154. Cambridge: Cambridge University Press.

Sander, K., and U. Schmidt-Ott. 2004. Evo-devo aspects of classical and molecular data in a historical perspective. *Journal of Experimental Zoology B* 302: 69–91.

Schlosser, G. 2002. Modularity and the units of evolution. *Theory in Biosciences* 121: 1–80.

Schmit, A.-C. 2002. Acentrosomal microtubule nucleation in higher plants. *International Review of Cytology* 220: 257–289.

Schumacher, R., and P. Gutjahr. 1992. Association of rib anomalies and malignancy in childhood. *European Journal of Pediatrics* 151: 432–434.

Schut, P. C., E. Brosens, A. J. Eggink et al. 2020a. Exploring copy number variants in deceased fetuses and neonates with abnormal vertebral patterns and cervical ribs. *Birth Defects Research* 112: 1513–1525.

Schut, P. C., A. J. Eggink, T. E. Cohen-Overbeek, T. J. M. Van Dooren, G. J. de Borst, and F. Galis. 2020b. Miscarriage is associated with cervical ribs in thoracic outlet syndrome. *Early Human Development* 144: 105027.

Schut, P. C., A. J. Eggink, M. Boersma et al. 2020c. Cervical ribs and other abnormalities of the vertebral pattern in children with esophageal atresia and anorectal malformations. *Pediatric Research* 87: 773–778.

Schut, P. C., C. M. A. ten Broek, T. E. Cohen-Overbeek, M. Bugiani, E. A. P. Steegers, A. J. Eggink, and F. Galis. 2019. Increased prevalence of abnormal vertebral patterning in fetuses and neonates with trisomy 21. *Journal of Maternal-Fetal and Neonatal Medicine* 32: 2280–2286.

Seidel, F. 1960. Körpergrundgestalt und Keimstruktur. Eine Erörterung über die Grundlagen der vergleichenden und experimentellen Embryologie und deren Gültigkeit by phylogenetischen Überlegungen. *Zoologisch Anzeiger* 164: 245–305.

Senter, P., and J. G. Moch. 2015. A critical survey of vestigial structures in the postcranial skeletons of extant mammals. *PeerJ* 3: e1439.

Shenefelt, R. E. 1972. Morphogenesis of malformations in hamsters caused by retinoic acid: relation to dose and stage at treatment. *Teratology* 5: 103–118.

Siegal, M. L., and J.-Y. Leu. 2014. On the nature and evolutionary impact of phenotypic robustness mechanisms. *AREES* 45: 496–517.

Sir, J.-H., M. Pütz, O. Daly, G.G. Morrison, M. Dunning, J. V. Klimartin, and F. Gergely. 2013. Loss of centrioles causes chromosomal instability in vertebrate somatic cells. *Journal of Cell Biology* 203: 747–756.

Stergachis, A., S. Neph, A. Reynolds, R. Humbert, and B. Miller B. 2013. Developmental fate and cellular maturity encoded in human regulatory DNA landscapes. *Cell* 154: 888–903.

Tanaka, E. M., and P. W. Reddien. 2011. The cellular basis for animal regeneration. *Developmental Cell* 21: 172–185.

Ten Broek, C. M. A., A. J. Bakker, I. Varela-Lasheras, M. Bugiani, S. Van Dongen, and F. Galis. 2012. Evo-devo of the human vertebral column: On homeotic transformations, pathologies and prenatal selection. *Evolutionary Biology* 39: 456–471.

Tollis, M., A. K. Schnieder-Utaka, and C. C. Maley. 2020. The evolution of human cancer gene duplications across mammals. *Molecular Biology and Evolution* 37: 2875–2886.

Uchida, J. T. Naganuma, Y, Machida, K. Kitamoto, T. Yamazaki, T. Iwai and T. Nakatani. 2006. Modified extra-vesical ureteroneocystostomy for completely duplicated ureters in renal transplantation. *Urologia Internationalis* 77: 104–106.

Uchida, Y., M. Uesaka, T. Yamamoto, H. Takeda, and N. Irie. 2018. Embryonic lethality is not sufficient to explain hourglass-like conservation of vertebrate embryos. *Evodevo* 9: 7.

Valeix, M., H. Fritz, A. J. Loveridge, Z. Davidson, I. E. Hunt, F. Murindagomo, and D. W. Macdonald. 2009. Does the risk of encountering lions influence African herbivore behaviour at waterholes? *Behavioural Ecology and Sociobiology* 63: 1483–1494.

Van der Geer, A. A. E., and F. Galis. 2017. High incidence of cervical ribs indicates vulnerable condition in Late Pleistocene woolly rhinoceroses. *PeerJ* 5: e3684.

Varela-Lasheras, I., A. J. Bakker, S. van der Mije, J. van Alphen, and F. Galis. 2011. Breaking evolutionary and pleiotropic constraints in mammals: on sloths, manatees and homeotic mutations. *EvoDevo* 2:11.

Viebahn, C. 1999. The anterior margin of the mammalian gastrula: Comparative and phylogenetic aspects of its role in axis formation and head induction. *Current Topics in Developmental Biology* 46: 64–103.

Von Baer, K. E. 1828. *Entwicklungsgeschichte der Tiere: Beobachtung und Reflexion*. Königsberg: Bornträger.

Von Dassow, G., and E. M. Munro. 1999. Modularity in animal development and evolution: Elements of a conceptual framework for EvoDevo. *Journal of Experimental Zoology B* 285: 307–325.

Von Dassow, G., E. M. Munro, and G. M. Odell. 2000. The segment polarity network is a robust development module. *Nature* 406: 188–192.

Von Dassow, G., and G. M. Odell. 2002, Design and constraints of the *Drosophila* segment polarity module: Robust spatial patterning emerges from intertwined cell state switches. *Journal of Experimental Zoology B* 294: 179–215.

Wagner, A. 2000. Robustness against mutations in genetic networks of yeast. *Nature Genetics* 24: 355–361.

Wagner, A. 2012. The role of robustness in phenotypic adaptation and innovation. *Proceedings of the Royal Society B* 278: 1249–1258.

Wagner, G. P., and L. Altenberg. 1996. Complex adaptations and the evolution of evolvability. *Evolution* 50: 967–976.

Walz, G. 2017. Role of primary cilia in non-dividing and post-mitotic cells. *Cell and Tissue Research* 369: 11–25.

Wéry, N., M. G. Narotsky, N. Pcico, R. J. Kavlock, J. J. Picard, and F. Gofflot. 2003. Defects in cervical vertebrae in boric acid-exposed rat embryos are associated with anterior shifts of Hox gene expression domains. *Birth Defects Research (A)* 66: 59–67.

Wheatly, D. N. 1995. Primary cilia in normal and pathological tissues. *Pathobiology* 63: 222–238.

Woolfenden, G. E. 1961. Postcranial osteology of the waterfowl. *Biological Sciences* 6: 1–129.

Wright, T. R. 1970. The genetics of embryogenesis in *Drosophila*. *Advances in Genetics* 15: 261–395.

Zalts, H., and I. Yanai. 2017. Developmental constraints shape the evolution of the nematode mid-developmental transition. *Nature Ecology & Evolution* 1: 0113.

17 Evolvability and Macroevolution

David Jablonski

Evaluating evolvability from a multilevel, macroevolutionary perspective is difficult, and integration of paleobiological and neontological data is essential for a deeper understanding. The operational approach proposed here tests for among-clade differences in phenotypic diversification in response to an opportunity, such as that encountered after a mass extinction, entering a new adaptive zone, or entering a new geographic area. By analyzing the dynamics of clades under similar environmental conditions, the aim is to approximate a macroevolutionary common-garden experiment that factors out shared external drivers to recognize intrinsic differences in evolvability. Diversity-disparity plots can track clades to determine when their phenotypic productivity exceeds stochastic expectation from their taxonomic diversification. Factors that evidently can promote evolvability include modularity (albeit contingent on alignment of selection with modular structure or with morphological integration), pronounced ontogenetic changes in morphology, genome size, and a variety of evolutionary novelties, which might be evaluated using macroevolutionary lags and dead-clade-walking patterns. High speciation rates may indirectly foster phenotypic evolvability. Although mechanisms are controversial, clade evolvability may be higher in the Cambrian, and possibly early in the history of clades at other times; in the tropics; and, for marine organisms, in shallow-water disturbed habitats. An expanded version of this chapter has been published in the journal *Evolutionary Biology* (Jablonski 2022).

17.1 Introduction

As Jane Austen might have said with a little biological training, it is a truth universally acknowledged that not all traits, populations, species, or clades have been equally labile or productive over their evolutionary lifetimes. A fundamental challenge in addressing such contrasts lies in distinguishing the role of intrinsic factors at various levels (from the configuration of gene-regulatory networks in an organism to the geographic extent of a clade) and extrinsic factors (from local competition to global climatic upheavals) in determining such differences. (See Jablonski 2017a,b for a general discussion of intrinsic and extrinsic factors in macroevolution, i.e., evolution above the species level.)

One potential intrinsic factor is evolvability. Evolvability has been defined in many ways (see Brown 2014; Nuño de la Rosa, chapter 2),[1] but when treated in general terms—the disposition or propensity to evolve, often referring specifically to adaptive evolution—

1. References to chapter numbers in the text are to chapters in this volume.

it can reside at any level within the biological hierarchy. In the macroevolutionary perspective adopted here, the focus will be on species (i.e., reproductively isolated, genealogical units) and clades (i.e., sets of species that comprise all, and only, descendants of a single ancestral species). To understand macroevolutionary dynamics, we need to determine whether species and clades differ in their intrinsic evolvability, and if so, why—and whether those differences are stable over a clade's history. Conversely, we need to determine whether the genetic and developmental mechanisms thought to promote evolvability in the short term have predictable long-term, large-scale evolutionary consequences. This is a challenging agenda, because inferences at the requisite scale and hierarchical level almost always rely on indirect evidence. This chapter cannot provide definitive answers, but in it, I attempt to outline macroevolutionary approaches to evolvability, first among clades regarding intrinsic traits that may promote or reduce evolvability, and then addressing variation in evolvability across time and space. The aim is to present an operational macroevolutionary approach, and to organize questions and potential examples to stimulate further theoretical and empirical research.

17.2 Operationalizing Evolvability in a Historical Context: Testing Macroevolutionary Hypotheses

The term *evolvability* might apply to any macroevolutionary currency, such as taxonomic diversity, functional variety, or morphological disparity; indeed, a long-standing question has been the degree of covariation among those currencies in different situations (Jablonski 2017a,b; Folk et al. 2019; Martin and Richards 2019; Shi et al. 2021). I propose to confine *evolvability* in macroevolution to phenotypes, with the hypothesis that evolvability is manifested in the behavior of traits and clades in a quantitative morphospace or functional space. An enormous literature exists on factors that promote or damp speciation and taxonomic diversification, but the propensity to achieve reproductive isolation, or to accrue taxonomic richness, probably involves a very different set of organismal and species-level attributes from those promoting the evolvability of form or function (Jablonski 2017b). Thus, expanding evolvability to include taxonomic rates or patterns in terms of evolvability probably is not useful.

One way to operationalize evolvability in macroevolution is as the differential (phenotypic) ability to take advantage of, or respond to, opportunity. This comparative approach is broadly analogous to the measurement of evolvability in terms of differential responses of traits to a unit strength of directional selection (Hansen and Pélabon 2021). Both intrinsic and extrinsic factors can create the opportunities—the acquisition of a novel structure, developmental pathway, or mode of life; entry into a novel ecosystem by surviving a mass extinction, invading a new landmass, or encountering newly evolved or introduced resources—and the analysis entails comparison of how clades performed in response (for useful discussions of evolutionary opportunity, see Losos 2010 and Gillespie et al. 2020). The difficulty for macroevolutionary analysis, of course, is that no two convergent evolutionary novelties are truly identical, and no two clades are likely to experience an environment in identical ways. However, because we can set prior expectations for the consequences of at least some confounding factors, we can frame hypotheses incorporating them that can be tested in a meaningful way. The aim is to frame a macreovolutionary equivalent of a common-garden experiment, analyzing the behavior of clades under shared or similar circumstances.

This phenotypic approach, predicated on *net* phenotypic shifts or gains of disparity in morphology or function, also differs from a view of evolvability as a capacity for a species or clade to realize variation in any direction from a starting phenotype (Brown 2014; i.e., minimal developmental bias; see Uller et al. 2018). The "bias" approach would allow clades to be evaluated in isolation and perhaps may be useful over short timescales, but it is insufficient for macroevolutionary purposes. Many clades traced through multivariate morphospaces ("phylomorphospaces") undergo much movement in morphospace with little net expansion or shift compared to related clades; see, for example, the contrasting echinoid clades in figure 17.1. Similarly, frequent changes in discrete characters, even if apparently isotropic around a given starting point, need not yield extensive net change when homoplasy is common across the phylogeny, so that clades undergo many state changes but capture few of the new states (see Foote 1997; P. Wagner 2000; Oyston et al. 2015). This is one reason for heterogenous results on the correlation between (morpho)speciation rates and overall phenotypic evolution: Much total change can occur while repeatedly traversing a limited range of morphologies. The larger question remains: whether or how often among-clade differences in apparent evolvability can be understood, and predicted, in terms of intrinsic differences rather than simply reflecting the operation of extrinsic pressures. Of course, the intrinsic-extrinsic distinction is not clearly demarcated, and both factors operate in concert to some degree; but the abovementioned macroevolutionary common-garden approach can help tease apart intrinsic among-clade differences.

The two major arenas for macroevolutionary analysis—the fossil record and comparative data on extant taxa—are essentially historical or retrospective, each with strengths and weaknesses; they are most powerful when applied in concert, although integrating them is difficult (among many others, see, e.g., Quental and Marshall 2010; Jablonski 2017b; Mitchell et al. 2019). Neontological approaches (mostly) begin with genetic or developmental data thought to indicate evolvability and attempt to recognize how they have shaped the large-scale dynamics of the clade leading to the present day; paleontological analyses (mostly) begin with the phenotypic dynamics and attempt to exclude confounding factors to recognize differences in intrinsic evolvability among clades. In either domain, the first step is to frame *comparative* analyses, potentially identifying the role of intrinsic biological properties relative to the myriad extrinsic factors that can drive differences in evolutionary tempo and mode among clades in time and space.

17.2.1 Observations on Extant Organisms

As noted, one approach measures attributes in extant populations that might impose or reflect differing degrees of evolvability of traits or clades, and then tests predictions retrospectively (i.e., by analyzing macroevolutionary outcomes or estimated dynamics of those traits or clades). Some intriguing analyses have done just that (e.g., Goswami and Polly 2010 on primates versus carnivores [with important later work incorporating extensive fossil data]; Haber 2016 on ruminants; Houle et al. 2017 on *Drosophila*). Such analyses require some strong or poorly understood assumptions. These include:

• The stability of G-matrices that capture aspects of the genotype-phenotype map, and thus the utility of extrapolating from present-day data (see Hansen and Pélabon 2021; Hansen, chapter 5; Pavličev et al, chapter 8), and their roles in determining properties, such as the distribution of accessible phenotypes around a given starting point, at these scales, with a variety of empirical outcomes; further analyses in a multispecies phylogenetic framework would be valuable, with an urgent need for new genetic and developmental model systems that have robust fossil records (see Love et al. 2022; Voje et al., chapter 14).

• The robustness of taxonomic or morphological dynamics derived from the topology of large molecular phylogenies. Some progress has been made here, but separating speciation and extinction rates from net diversification—potentially important for testing hypotheses of cause and effect in morphospace occupation (as in Huang et al. 2015) remains challenging (e.g., Louca and Pennell 2020; Love et al. 2022), as does the problem of inferring ancestral character states from extant taxa alone (Slater et al. 2012; Betancur-R et al. 2015; Marshall 2017); and more generally, evolutionary modeling is demonstrably improved and results shift when fossils are incorporated (Mongiardino Koch 2021, citing twelve studies).

• The focal clade is today at its maximum morphological breadth; this is a generally unstated assumption required for phylogenies containing only extant species, but it is patently false for many clades having a reasonable fossil record, from oysters to cephalopods to elephants to horses to hominins. The extinct forms are often not simply extensions along existing morphogenetic lines but variations that might seem highly improbable, given today's representatives, for example, giant ground sloths (terrestrial *and* aquatic), rainforest-dwelling carnivorous kangaroos, sharks with coiled tooth arrays, uncoiled or spiny nautiloids, and sea urchins with periscope-like extensions (see Jablonski 2020 for references; even the quintessential static lineage, the horseshoe crabs, has exhibited bursts of phenotypic diversification that pushed beyond their current limited repertoire—see Bicknell et al. 2022).

17.2.2 Observations in the Fossil Record

Paleontological analyses pertaining to evolvability are beset by a different set of strong assumptions. Sampling and preservation can distort or even generate apparent patterns, although increased understanding of such potential biases have reduced their impact. Only post-embryonic, phenotypic data are available for most extinct taxa, and so the developmental and genetic underpinnings of observed contrasts must be inferred. Particularly challenging is the assessment of negative evidence (also an issue for neontological data, of course), and of the role of intrinsic and extrinsic factors in determining vacancies or boundaries of a clade's morphospace. Some vacancies are longstanding and phylogenetically localized, and thus may represent a lack of developmental capacity, at least for the clades presented with these opportunities (Vermeij 2015; Jablonski 2020). Others may reflect extinction and insufficient time to re-occupy vacated morphospace (consider mammalian body sizes in the Americas, although humans have surely now blocked that evolutionary route). Furthermore, morphospace occupation can be limited by preemptive occupation or later, displacive conquest of portions of the space by competing clades. Displacive competition seems to be scarce at macroevolutionary scales, but preemptive, incumbency patterns or priority effects seem relatively common (see Jablonski 2008a, 2017b; Benton 2009; Tilman and Tilman 2020; and Tomiya and Miller 2021 for a study that may find both effects). Other negative interactions, such as predation and parasitism, can promote or impede phenotypic or taxonomic diversification, as can positive interactions such as mutualism, and either type can sometimes increase extinction probabilities (see Vermeij 1987; Jablonski 2008b; Hembry and Weber 2020). Comparative analyses of clades presented with similar opportunities can control for some of these uncertainties, and temporal and spatial paleo-data can be especially valuable, with insight not just into extinct phenotypes demonstrably accessible to a clade but lacking today, but also into potential interactions: clades cannot impede one another if they did not co-occur.

Despite these drawbacks and complications, many analyses do suggest among-clade and temporal differences in evolvability, with macroevolutionary consequences. Some of these are discussed in the following sections.

17.3 Features Enhancing Evolvability of Clades

17.3.1 Modularity

The developmental property most often proposed as associated with evolvability is modularity. The general view has been that greater modularity enhances evolvability (e.g., Wagner and Altenberg 1996; Love et al. 2022; Vermeij 1974, 2015 as "versatility," which he associates with modularity in the later paper). However, many different types of modules are recognized, including functional, developmental, genetic, and evolutionary modules (see the references in Jablonski 2017a), and we lack clarity on how they are related, with mixed results on the positive, negative, or negligible relation between the strength of modularity and macroevolution (Rhoda et al. 2021 and references therein). For modularity to enhance evolvability, the intrinsic structure of modules—that is, genetic or developmental modules—must be configured along viable lines, which may or may not be the case (e.g., Pavličev and Hansen 2011; Pavličev et al., chapter 8), and align with internal selection (the need for body parts to function together) and external selection by the environment. Otherwise the covariation of traits within modules can instead impede evolution. In principle, the covariation structure imposed by morphological integration—not strictly the antithesis of modularity but useful in this context—can enable more rapid and extensive evolutionary change in certain directions than would emerge from strictly isotropic or unbiased variation (Goswami et al. 2014; Felice et al. 2018; Uller et al. 2018; Jablonski 2020; Love et al. 2022). Thus, in the special circumstance when selection (i.e., an opportunity) is aligned with such (viable) lines of genetic least resistance in Schluter's (1996) sense, integration rather than modularity might promote greater evolvability (see also Evans et al. 2021, and Voje et al., chapter 14, on instances where highly integrated traits appear to have been most evolvable). These contingent aspects of modularity would seem to disallow generalizations, and macroevolutionary predictions become difficult, although retrospective understanding of a role for modularity in specific cases is not a trivial insight.

Despite these issues, the ubiquity of mosaic evolution (the evolution of different characters at different rates), and more broadly, of incompatible character transformations across phylogenies (Jablonski 2017a), indirectly supports the view that evolution is more often facilitated by the ability of traits to change independently. Furthermore, among-clade differences may exist: arthropods seem to be masters of modularity, not just in terms of dissociating morphological modules for independent growth and transformation (e.g., Nijhout and McKenna 2017), but perhaps also at the molecular level. For example, arthropods apparently more readily deploy the *Distal-less* pathway in new locations to generate novel structures (e.g., horns, wings: see Shubin et al. 2009; Bruce and Patel 2020) than do tetrapods, with the arthropod pathway largely dedicated to regulating outward growth but the vertebrate homolog Dlx involved not just in the early development of limbs, but in the placenta, forebrain, branchial arches, and other tissues (Panganiban and Rubenstein 2002; Sumiyama and Tanave 2020).

A related view sees evolvability as a positive function of the dimensionality of form (Vermeij's 1974 argument), which need not be directly related to modularity per se: Limpet shells can be described by fewer mathematical parameters than can helically coiled shells with complex apertures, and thus have lower dimensionality, but different snail lineages have not been analyzed from this perspective (for more on the positive associations between dimensionality and the rate or extent of diffusion in morphospace, see Foote 1991, 129; Pie and Weitz 2005, E9; Holzman et al. 2011 on evolvability as a positive function of the number of traits determining organismal performance). In a sense this is a "degrees of freedom" hypothesis: More components mean more avenues to evolve along, or, in Vermeij's (2015) view, for alleviating functional trade-offs.

Central to all these ideas from a macroevolutionary perspective is the still-open question of the long-term stability of genetic and phenotypic modularity (see Urdy et al. 2013), and how to operationally distinguish modules maintained by intrinsic factors resistant to change from those maintained by selection and thus readily altered at these large scales. Here too, retrospective macroevolutionary analyses of clades with demonstrable present-day differences in modularity would be a powerful merger of paleontological and neontological data. Ideally, we could compare two clades differing in modularity but presented with a similar opportunity, such as survival of a mass extinction, or arrival in a relatively unoccupied archipelago or larger landmass (potential examples, still lacking the paleontological dimension, include Galapagos finches versus mockingbirds, and Hawaiian honeycreepers versus thrushes; see Lovette et al. 2002).

Given the array of skeletal types that constitute almost all of the fossil record, we might ask whether developmental and evolutionary modularity—and thus potentially evolvability—differ across body plans involving many-element, articulating skeletons (e.g., vertebrates, echinoderms, and arthropods) and those having just one or two discrete elements and accretionary growth (e.g., corals, mollusks, and brachiopods; see, for example, Edie et al. 2022). The remarkable range of molluscan shell shapes (that is, scaphopods, nautiloid cephalopods, chitons, snails, and bivalves) and ornamentation patterns suggest exquisite local control in the sheet of tissue that generates those shells; but does the extra level of morphogenetic control and interaction afforded by articulating skeletons create a correspondingly enlarged evolvability at macroevolutionary scales?

An even more profound difference between clades that could be viewed from the modularity/evolvability standpoint involves lineages that sequester the germ line early, versus the plants and clonal colonial animals that sequester the germ line late and so can incorporate somatic mutations into gametes (Schoen and Schultz 2019 and references therein; C. Simpson et al. 2020; Yu et al. 2020). With late sequestration, each plant bud or animal zooid is potentially both a developmental and an evolutionary module, so that novel variants can originate within the colony and propagate both sexually and asexually, conceivably increasing clade evolvability relative to early-sequestration clades. This notion might seem to contradict a widely (though not universally) accepted case of higher-level selection for evolvability: the pervasiveness of sexual reproduction across the tree of eukaryotic life. The Red Queen hypothesis for the maintenance of sex (e.g., parasite-mediated selection for the continual production of novel phenotypes) defines a process playing out at the population, species, and/or clade level (Van Valen 1975; Stanley 1979, 213–227; Nunney 1989; Sterelny and Griffiths 1999, 208–210; Hansen 2011) and thus is a decidedly macroevolutionary hypothesis. But there need not be a contradiction here: Species in most eukaryotic clades that reproduce asexually or parthenogenetically are also capable of sexual reproduction. Testing a macroevolutionary hypothesis of the consequences of evolvability as imposed by sex could involve asking whether lineages in which sexual reproduction is rare or involves a limited number of individuals are less prolific phenotypically than lineages in which sex is the norm. Because sexually produced individuals or colonies can be distinguished from asexually produced ones in several well-fossilized groups (foraminiferans, corals, and bryozoans), this question could be addressed empirically.

17.3.2 Ontogenetic Allometry or Multiphase Life Cycles

As already noted, developmental integration might promote long-term evolvability when the resulting trait covariation is aligned with internal and external selection, and a few analyses have provided examples (e.g., Navalón et al. 2020 on bird craniofacial evolution; Hedrick et al. 2020 on bat cranial evolution). Such covariation may reach its richest macroevolutionary potential in clades that undergo strong changes in form during ontogeny, as continuous variation in ontogenetic allometry (e.g., see the origin of sand dollars [Smith 2001] and brittle stars [Thuy et al. 2022]), or discontinuously in multiphase life cycles. As long recognized (e.g., Gould 1977), such clades have often evolved along ontogenetic trajectories via heterochrony (i.e., evolutionary changes in developmental timing), and in at least some cases, they traverse significantly greater volumes of morphospace than do clades with lesser allometries or more direct development. These clades include canids (e.g., Geiger et al. 2017; Machado et al. 2018, 1413; and for a broader overview, see Sánchez-Villagra et al. 2017), dinosaurs (Chapelle et al. 2020), angiosperms (Armbruster, chapter 15), and perhaps most famously, extant and fossil salamanders that retain larval traits, with modularity clearly a critical part of this capability (see Johnson and Voss 2013; Urdy et al. 2013; Fabre et al. 2020).

17.3.3 Novel Traits

Evolutionary novelty in the broad sense often seems to increase evolvability by creating new features for further variation and allowing clades to access new adaptive zones (G. Simpson 1944): the origin of limbs, lungs, the amniote egg, and feathers are certainly associated with an expansion in the morphological disparity (and taxonomic diversity, and functional repertoire) of the clades bearing them. However, we have surprisingly few robust examples of this key-innovation phenomenon, in which a novel feature directly triggers diversification (see Rabosky 2017, Martin and Richards 2019, and Erwin 2021a for catalogs and critiques of the many definitions of "key innovation"). Many putative key innovations have proven to be part of a chain of derived characters, or associated with "key opportunities" (i.e., extrinsic events), prior to phenotypic expansions (Donoghue and Sanderson

2015; Stroud and Losos 2016; Jablonski 2017a). Such contingencies are most clearly seen in macroevolutionary lags, the geologically long interval between the inception of a novelty or clade and its taxonomic or phenotypic diversification (Jablonski and Bottjer 1990), which appears to be widespread or even the general rule (Jablonski 2017a; Halliday et al. 2019; Kröger and Penny 2020; Ramírez-Barahona et al. 2020; Simões et al. 2020; Erwin 2021a). Such lags can provide a novel framework for evaluating intrinsic and extrinsic factors; they are generally tracked using taxonomic diversity, however, and more analyses are needed that treat them in morphospace and incorporate functional variety (as in Slater 2013 and Folk et al. 2019).

We do not know how often evolutionary novelties in the strict sense—that is, a trait lacking a homolog in the ancestor (G. Wagner 2014)—also fail to trigger diversification. As these true novelties often define clades, analyses of lags will need to operate across broad evolutionary trees, but effects seemingly imposed by intrinsic constraints and their removal or absence may also present a useful set of test cases. For example, mammals are highly constrained in the number of cervical vertebrae (Galis, chapter 16), but it is unclear, and worth testing, whether this constraint has impaired mammalian functional or morphological evolution relative to tetrapods that have circumvented it, such as sauropod dinosaurs, plesiosaurs, and long-necked birds (Müller et al. 2010; Taylor and Wedel 2013; Marek et al. 2021).

Another intriguing modification of development, little considered from the standpoint of evolvability, is the breaking of bilateral symmetry, which has occurred throughout plant and animal phylogeny, by a variety of developmental mechanisms (Palmer 2004). Bivalve mollusks are a system that would reward macroevolutionary analysis, as most species are bilaterally symmetrical, aside from small developmental adjustments allowing interlocking, hinged valves (Moulton et al. 2020; recall that the plane of symmetry lies between the two valves, not down the midline of a single valve). Some bivalve clades have strongly diverged from bilaterality, including the extinct, perhaps photosymbiotic, rudists, which evolved a conical-cylindrical right valve and a cap-shaped left valve, among other configurations (Jablonski 2020). Oysters, spiny oysters, scallops, and others have also shed bilateral symmetry in impressive ways (Nicol 1958), with extinct oysters showing a much wider range of shell geometries than do extant species, including planispiral, helical, and conical forms (Seilacher 1984). As many of these lineages are in the Order Pteriomorphia, the question arises whether this clade weakened bilateral patterning early in bivalve history and then could adopt asymmetry according to later opportunities or pressures, and thus had greater evolvability than related bivalve clades.

A shift from radial to bilateral symmetry is associated with a striking contrast in apparent evolvability in sea-urchin history (figure 17.1). The ancestral condition is radial, and the survivors of the end-Paleozoic mass extinction inherited that state, continuing to evolve as the group informally termed "regular" echinoids; they gave rise to many species but remained confined in morphospace. However, one lineage diverged to become the irregular echinoids, a bilaterally symmetrical, burrowing clade that eventually split into two branches typified respectively by heart urchins and sand dollars. The regular and irregular echinoids each contain ~500 extant species, but the irregulars have explored a much broader range of morphospace (Hopkins and Smith 2015). Understanding the developmental basis of this contrast, including a potential change in modularity (López-Sauceda et al. 2014; Saucède et al. 2015),

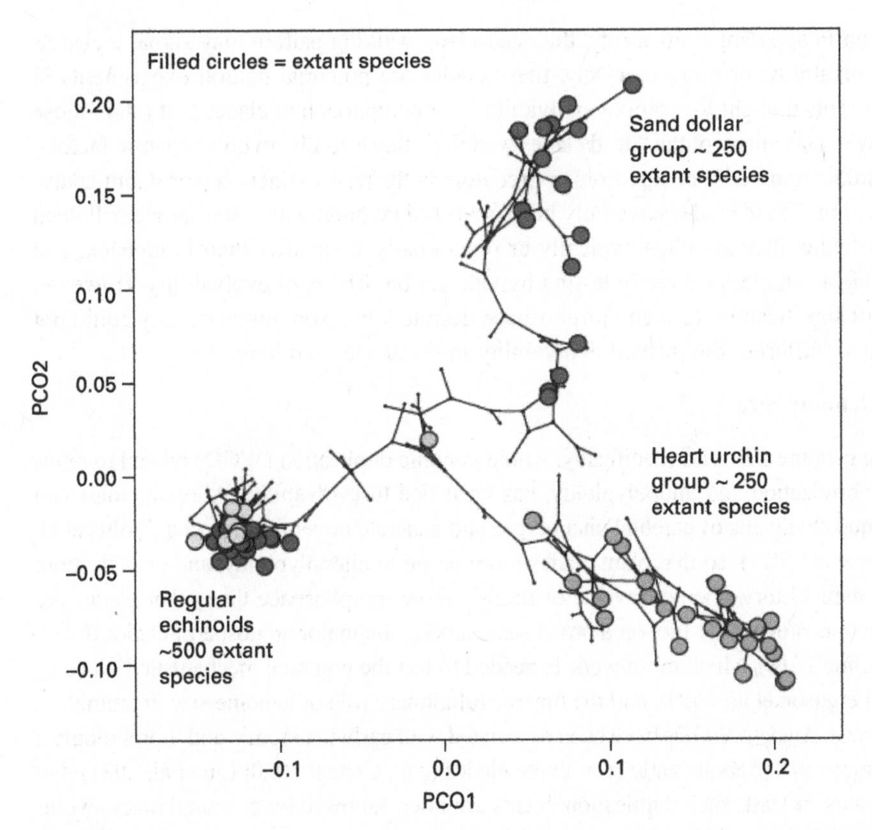

Figure 17.1
Differences in apparent evolvability in the major sea-urchin clades, portrayed in a phylomorphospace based on principal coordinates analysis of a character matrix. Modified after Hopkins and Smith (2015), used by permission.

and then testing an intrinsic evolvability hypothesis against alternatives—for example, ecological opportunities afforded by adoption of the burrowing, deposit-feeding habit—would create an exceptional model system for exploring macroevolutionary issues. One factor may be a profound developmental change near the origin of irregulars (Smith 2005) that allowed their plates to grow predominantly in place throughout ontogeny (as opposed to ontogeny via a combination of plate growth and insertion in regulars), making it easier to differentiate the upper and lower surfaces of the test, and thus to become burrowers, or, as in sand dollars, to use the upper surface as a feeding sieve. Shifts from radial to bilateral symmetry may also promote diversification in angiosperms, separately or in combination with other traits (O'Meara et al. 2016; Armbruster, chapter 15; but see Vamosi et al. 2018), but the effect has only been evaluated in terms of species richness and not phenotypic evolvability. Comparative analysis of floral evolution in morphospace according to floral symmetry would be a valuable next step.

Finally, the converse of a macroevolutionary lag is the dead-clade-walking phenomenon, where a clade suffers a sharp decline (e.g., during a mass extinction) and then persists for some time without rediversifying (Jablonski 2002). Like macroevolutionary lags, such clades appear to be widespread (Barnes et al. 2021), and just as lags appear to signal a belated gain in apparent evolvability, the dead-clade-walking pattern may signal a clade's loss of evolvability, or more precisely, these clades are potential natural experiments in the *loss* of traits thought to promote evolvability, for comparison to clades that retain those traits. As with lags, many of the "dead clades walking" may actually involve extrinsic factors, such as limits imposed by competitors or predators in the post-extinction world, but analyses are lacking. These clades have only been analyzed taxonomically, so that we still need to know whether they are phenotypically or functionally static after their bottleneck, and thus provide a vehicle for directly testing hypotheses on drivers of evolvability. However, if they shift significantly through morphospace despite low taxon numbers, they could not be viewed as suffering diminished evolvability in the sense used here.

17.3.4 Genome Size

For plants, genome size, and specifically, whole-genome duplication (WGD) related to interspecific hybridization and allopolyploidy, has been tied to evolvability. Allopolyploids can create unique amalgams of parental phenotypes and generate novel features (e.g., Soltis et al. 2014; Alix et al. 2017), so that plant clades more prone to allopolyploidy, and/or with more WGDs in their history, should traverse or occupy more morphospace than other clades do. This prediction is evidently met on a broad scale among the major angiosperm clades (Clark and Donoghue 2018). Much more work is needed to test the potential mechanistic link (e.g., see Zenil-Ferguson et al. 2019), and the macroevolutionary role of genome size in animals is even less clear. Ancient WGDs have been associated with early taxonomic and morphological diversifications in vertebrate and invertebrate clades (e.g., Conant 2020; Liu et al. 2021), but for vertebrates, at least, such duplication events are often followed by extended macroevolutionary lags (Glasauer and Neuhauss 2014; Davesne et al. 2021), and these events may even impede diversification (Kraaijeveld 2010), raising questions about a causal role.

What plants and animals do share is the potential to track genome size directly in the fossil record, allowing for more rigorous analysis without reliance on ancestral character-state estimation from extant species (animals: Thomson and Muraszko 1978; Organ et al. 2007, 2011; Hunt and Yasuhara 2010; Davesne et al. 2021; plants: Masterson 1994; Lomax et al. 2014; McElwain and Steinthorsdottir 2017). Of course, genomes can enlarge for reasons other than duplication, and one potential direction for macroevolutionary investigation in this area is the relative impact of transposon proliferation and WGD on clade survivorship and diversification, which might be assessed retrospectively when phylogenetic analysis shows a constant ploidy level, but fossil data indicate shifts in genome size. There are many ideas on the evolutionary role of mobile elements, some of them plausible, including the potential for cross-level conflicts, but the macroevolutionary impact of among-clade differences in transposon content—active or not—remains uncertain.

17.3.5 Elevated Speciation Rates

Over geologic timescales, most species tend to be morphologically static (i.e., oscillate within limits) or nondirectional over their histories, affording speciation a potential role in the extent and direction of morphospace occupation for many clades (e.g., Gould 1982; Hunt 2007; Jablonski 2017b; and from a very different perspective, Gorné and Diaz 2019). Some authors include high speciation rates in their definition of evolvability (e.g., Hedrick et al. 2020) although I argued against such a broad definition in section 17.2. In any case, we can ask whether clades having higher speciation rates for intrinsic reasons—that is, owing to traits that increase the probability of reproductive isolation (see Jablonski 2008a for an inventory)—have higher rates or extents of net morphospace occupation. (Such analyses will not be circular if performed with care, even in the fossil record, where speciation is necessarily recognized phenotypically, because the critical variable is *net* differences in morphospace occupation.) A rough correlation between speciation rate and morphological change is seen for many clades at various points in their histories, albeit with considerable heterogeneity and an array of counterexamples (Stanley 1979; Rabosky et al. 2013; Crouch and Ricklefs 2019; Cooney and Thomas 2021, and many more citing and cited in these publications; see section 17.4.1 for temporal changes, such as early bursts).

The potential association between speciation and morphologic change is relevant to evolvability for at least three reasons.

(1) Speciation may tend to occur preferentially in the direction of intraspecific variation (Hunt 2007; Love et al. 2022), providing a potential link between standing variation and both developmental bias and macroevolutionary evolvability, with high-speciation clades moving more rapidly across morphospace per unit time, and doing so more efficiently in that fewer species go in the opposing direction over the course of the trend—what Gould (1982) called a direction bias in clade dynamics (see also Jablonski 2020). The potential role of speciation rates in evolvability may depend on the shape and stability of the variational envelope around the taxa within a clade (e.g., Haber 2016; Watanabe 2018), including the resistance of that envelope to external pressures, but little is known about among-clade intrinsic differences that determine such features at this scale, or their mechanistic underpinnings.

(2) Traits can hitchhike on high speciation rates, proliferating in the clades that generate more species per unit time (see Jablonski 2017b). Thus, any attribute that tends to confer high speciation rates, such as low dispersal ability (see Jablonski 2008a; and for a recent discussion on birds, Tobias et al. 2020), might promote the proliferation of other traits that happen to covary with it among lineages. This hitchhiking aspect of species selection in the broad sense is likely to be widespread (Jablonski 2017b; Polly et al. 2017), so that the apparent evolvability of a trait, or of a clade, should be analyzed in a framework that takes both direct organismic selection and this indirect, cross-level effect into account.

(3) Directionality aside, clades having high speciation rates potentially generate more phenotypic experiments per unit time than low-rate clades. And if high-speciation clades tend to *accumulate* species, all else being equal, this will tend to reduce the clade's extinction risk and thus extend its duration, giving the clade more time to explore morphospace. However, counterexamples are well documented, particularly situations where high speciation rates lack commensurate expansions in morphospace (see the discussion of "non-adaptive radiations" in Rundell and Price 2009; Czekanski-Moir and Rundell 2019); even clades showing considerable movement through morphospace via speciation may ricochet within a confined portion of the space, as for the "regular" urchins in figure 17.1. Further undermining a simple relation between speciation rates and evolvability, high speciation rates are often accompanied by a "macroevolutionary trade-off" (Jablonski 2008a, 2017b), in which traits that confer high speciation rates also impose high extinction rates (e.g., Gould and Eldredge 1977; Stanley 1979, 1990; Van Valen 1985; Valentine 1990; Marshall 2017). Nonetheless, blanket statements that "diversity and disparity appear to be fundamentally decoupled" (Oyston et al. 2015; Guillerme et al. 2020; and many more) are an

oversimplification. The observation is certainly true for a single moment in geologic time, such as the present day, but the dynamics are more complex. The two currencies can accrue at different rates and even at different times, as implied by macroevolutionary lags, but when disparity increases, it tends to do so via branching events (i.e., via taxonomic diversification). Thus, while the wide range of potential relationships between diversity and disparity is crucial for understanding the evolutionary process, there is an important mechanistic association, albeit an imprecise one, and a more nuanced, quantitative approach is needed.

Given the broad range of potential relationships between speciation and a clade's movement or expansion in morphospace, the clades with the greatest evolvability might be viewed as the ones that disproportionately explore morphospace relative to their speciation rates. Broad morphospace occupation relative to species numbers at a point in time can also be produced by extinction (either random with respect to position in morphospace or against "average" morphologies; see Foote 1993, 1996), inflating the apparent relationship between diversity and disparity, so that time-series using fossil time-slices in diversity-disparity plots is the most informative approach (Jablonski 2017b; Wright 2017; see also P. Wagner 2010). This method has mostly been applied to clades originating under differing conditions (figure 17.2), but comparative analyses of clades responding to the same opportunity, advocated above, would be a valuable extension—for example, revisiting Eble's (2000) work comparing holasteroid and spatangoid echinoids, or the contrasting echinoderm clades in the Cambro-Ordovician interval (Deline et al. 2020). Testing potential factors in evolvability, clades having greater modularity or stronger ontogenetic allometry (for example) might tend to fall well above the diagonal in figure 17.2, while less modular or more isometric clades lie below or closer to it.

This diversity-disparity approach can also shed light on how evolvability changes over the history of a clade, by indicating where phenotypic productivity exceeds the stochastic

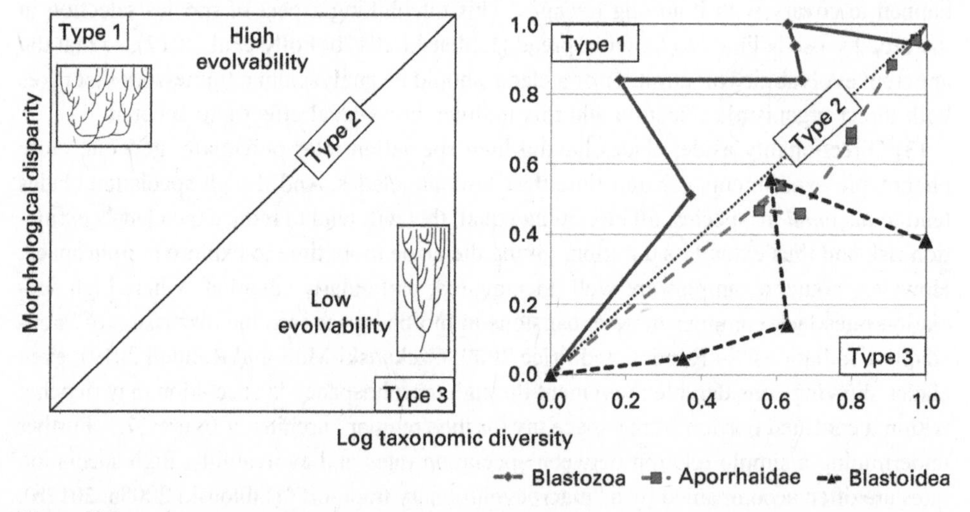

Figure 17.2
Evolution in diversity-disparity space. (Left): Type 1, morphology outstrips taxonomic diversification; Type 2, morphology is concordant with taxonomic diversification; Type 3, morphology trails behind taxonomic diversification. (Right): 3 empirical trajectories, for Cambrian-Ordovician blastozan echinoderms, Jurassic-Cretaceous aporrhaid gastropods, and Ordovician-Carboniferous blastoidean echinoderms. From Jablonski (2017b), which cites sources.

expectation from taxonomic diversification. For example, an early pulse of phenotypic invention might be followed by later confinement in morphospace (i.e., shifting from the upper left to lower right in figure 17.2, as implied in many studies, e.g., Cooney et al. (2017) on bird bill evolution). Alternatively, such a burst might be followed by diffusion in morphospace more proportionate to taxonomic diversification, trending from the upper left in figure 17.2 toward the diagonal. If ecological opportunity is an important factor interacting with intrinsic evolvability, then combining and partitioning clades within functional categories may yield new insights, as in the underappreciated finding that carnivorous mammals as a functional group show a significant burst in form relative to taxonomic richness after the end-Cretaceous extinction, but the constituent clades individually do not (Wesley-Hunt 2005).

17.4 Temporal and Spatial Patterns: Intrinsic or Extrinsic Factors?

Evolvability does not appear to be constant in time and space. The most frequently cited temporal patterns involve greater evolvability in early metazoan history, and at the inception of clades, regardless of their absolute geologic age. Such changes within time series may be best assessed as disparity relative to taxonomic diversity trajectories, as in figure 17.2, with high evolvability taken as a disproportionate occupation of morphospace relative to taxonomic richness in a time bin. Such discordance between diversity and disparity suggests that something unusual is going on, and as discussed below, the challenge is to separate intrinsic evolvability from extrinsic opportunities as the primary factor.

17.4.1 Temporal Patterns

Debates on the driver(s) of the Cambrian explosion of metazoan form, and its slowdown later in the Paleozoic and to the present day, are essentially asking whether evolvability has changed over time, on a grand scale. The evidence largely supports the view that major clades, and Metazoa overall, underwent a spectacular expansion of morphological and functional breadth in a geologically brief episode that significantly outpaced taxonomic diversification relative to later events in the history of life (Erwin and Valentine 2013; Jablonski 2017b; Deline et al. 2020, Erwin 2021b). However, mechanisms are still controversial: first, did intrinsic or extrinsic factors drive the rapid expansion in form and function, and second, what then slowed it down? Phylogenetic and paleontological data suggest that many of the developmental tools for building metazoans evolved well before the Cambrian, with a macroevolutionary lag that ended with an extrinsic trigger or opportunity, still not clearly identified (Erwin and Valentine 2013; Erwin 2021b). Thus the simple dichotomy between developmental (i.e., intrinsic) and ecological (i.e., extrinsic) mechanisms might be replaced by a "perfect storm" model of mutually reinforcing factors that successively fell into place, neither factor being sufficient on its own (Jablonski et al. 2017; Jablonski 2017b; see Love and Lugar 2013 for a tabulation of hypothesized mechanisms). Increases in gene-regulatory capacity certainly were associated with the Cambrian radiation (reviewed by Erwin 2021b), but much of that radiation appears to be associated with the redeployment and differentiation of existing developmental pathways. The failure to duplicate the Cambrian burst after the massive end-Permian extinction had been viewed as an argument for a post-Cambrian decline in intrinsic evolvability (but see Foote 1999), very much in the spirit of the comparative approach suggested here. However, we now know that functional diversity barely dropped after the Permian event despite severe taxonomic losses (Foster and Twitchett 2014; Edie et al. 2018; and a truly pioneering study by Erwin et al. 1987), suggesting that post-Permian ecological opportunity was not comparable to that of the Cambrian. Most authors currently seem to view the slowdown of the Cambrian explosion in terms of ecological filling of marine habitats, but more extensive comparative studies of variation in Cambrian and post-Cambrian are needed.

At lower taxonomic levels, evolvability might decline over a clade's history, regardless of when it originated. This long-standing idea has mixed support: Harmon et al. (2010) detect few early bursts, Hughes et al. (2013) detect many, and Slater and Pennell (2014) attribute Harmon et al.'s result to a lack of statistical power. Integrating the early-disparity findings of Hughes et al. (2013) with the macroevolutionary-lag findings of Kröger and Penny (2020), superficially contradictory but actually dealing in different currencies, should clarify matters, but several non-exclusive mechanisms might be operating. The rate of production of *new* character states does seem to slow in many clades, even when character-state transitions do not (Oyston et al. 2015). However, nearly all clades produce some new character states throughout their history, rather than reiterating old states after maximum disparity is reached (Oyston et al. 2015); if the slowdown is attributable to intrinsic reductions in evolvability rather than ecological crowding, this implies a relatively weak effect. The apparent tendency for taxonomic diversification to slow with clade age (Henao Diaz et al. 2019) is at least as consistent with crowding effects as with regular, among-clade changes in intrinsic factors such as evolvability.

Perhaps the most provocative evidence for declines in intrinsic evolvability during a clade's history comes from P. Wagner's (2018) analysis of character-state correlations in the fossil record. Data from a large set of character matrices support a model breaking up correlations among characters and forming new ones, arguably analogous to reorganizing the structure of phenotypic variances and covariances—the P-matrix (see Love et al. 2022)—and thus presumably the G-matrix. Developmental data are needed to test this "correlated change-breakup-relinkage" model, and a key question is whether these changing linkages unfold across the appropriate timescales and have the limiting effects on overall phenotypic change that appear to typify the clade histories analyzed by Oyston et al. (2015).

Others suggest that evolvability tends to *increase* through a clade's history. Vermeij (2015) argues that younger branches within major animal and plant clades explore a greater portion of morphospace than older ones, explaining this pattern in terms of selection to alleviate energetic tradeoffs. This view implies a ratcheting effect not seen in the analyses by Oyston et al. or P. Wagner cited above, but those data are at a much finer scale than Vermeij's examples. Vermeij argues that "versatility" (which as noted above includes but is not restricted to modularity) has increased overall through time; Goswami et al. (2014) also argue for a net tendency of modularity to increase and integration to decline—implying that the relinkage in P. Wagner's "correlated change-breakup-relinkage" model is more localized within the phenotype than the ancestral state. These are plausible viewpoints that require testing in a common framework. One unexplored possibility is that the increase and later decline of evolvability occurs only at the origin of clades that are founded via an evolutionary novelty sensu G. Wagner (2014), that is a trait lacking homology in the ancestor or that has radically and irreversibly changed from the ancestral state. Testing for declines (or increases!) in apparent evolvability of clades that originated in this way, vs those that more clearly arose in the context of ecological opportunity, may be one way to integrate these rather heterogeneous arguments (Jablonski 2020). Comparative analysis could also use evolutionary accelerations after mass extinctions to differentiate evolvabilities among contemporaneous clades, to test for differences in expansions in form or function among clades of different ages when encountering the same post-extinction opportunity to test for clade-age effects.

17.4.2 Spatial Patterns

Hypotheses for spatial variation in evolvability have long focused on the tropics, stunningly rich in taxonomic diversity and phenotypes, and the fossil record presents an additional, unexpected pattern, with disparity repeatedly emerging in marine invertebrate clades in onshore habitats. Comparing clade dynamics in morphospace across latitudes is challenging in terms of data required and the need to control for the strong latitudinal bias in both paleontological and neontological sampling. Great caution is warranted when maxima in origination rates or standing diversity or disparity are found to lie in the best-sampled regions, usually in the present-day temperate zone, as biases can be so strong that standard methods for factoring them out are ineffective (Valentine et al. 2013). A study that factored out sampling bias in two different ways found a significant tendency for marine invertebrate Orders, as a proxy for significant evolutionary novelty, to originate in the tropics over the past 250 Myr (Jablonski 1993; Martin et al. 2007), although data were lacking to test whether higher taxa originated more frequently in the tropics on a per-species basis.

A far less intuitive pattern occurs along marine depth gradients. Orders of marine invertebrates, again used as proxies for evolutionary novelty, preferentially originated in onshore habitats, that is at depths regularly subject to storms or normal wave disturbance (Jablonski and Bottjer 1990, Jablonski 2005). This pattern is independent of clade-specific bathymetric diversity gradients, turnover rates, or origination frequencies of constituent genera or within-clade traits, with low-level lineages originating offshore in certain clades and therefore sometimes expanding onshore as well as offshore (see also Jablonski et al. 1997; Tomašových et al. 2014; Bribiesca-Contreras 2017; Franeck and Liow 2019). The lone morphospace analysis to date is consistent with this finding: two Orders of irregular echinoids show greater divergences in disparity at their onshore origins than seen within the clades at any depth once established (Eble 2000; it would be interesting to plot the branch lengths in Figure 1 against their bathymetric context). Early vertebrate clades also first appear onshore (Sallan et al. 2018, who unfortunately exaggerate differences with the invertebrate patterns).

As with many other aspects of this overview, we have some provocative patterns, potentially indicating greater evolvability in tropical settings, and in onshore marine environments. New kinds of data and analyses are needed to bring these results more fully into the framework discussed here, and then to address the fundamental question: are they driven by intrinsic factors, as tentatively proposed by Jablonski (2005), or are they promoted by the extrinsic environmental gradients that define them (e.g., Vermeij 2012)? In other words, do organisms, species, or clades that inhabit warm, shallow settings have properties that enhance evolvability, presumably indirectly selected for by those environments, or do those environments directly promote greater phenotypic change?

17.5 Conclusion

Taken together, the data do suggest that intrinsic factors can influence the rate and scope of morphological and functional evolution at large scales. However, major challenges remain in converting these suggestions into a rigorously defined field. Perhaps the central difficulty for macroevolution lies in separating the intrinsic factors from the multitude of potential extrinsic biotic and abiotic drivers in determining vacancies, boundaries, or extents of expansion or transformation in a clade's morphospace or functional repertoire. When extrinsic factors can be excluded or accounted for, the issue becomes how apparent intrinsic evolvability differences map onto the potential causes of evolvability differences explored here and elsewhere in this volume, and the consequences of those different causes for the persistence or evolutionary lability of clades.

The most powerful analyses will be comparative, with the operational approach advocated here involving tests for among-clade (and perhaps across-time) differences in responses to a shared opportunity, in a macroevolutionary analog to a common-garden experiment. New methods for integrating fossil and present-day data are becoming available, and for macroevolutionary purposes this integration will be essential; one aim of this chapter has been to show that there is much raw material and a growing toolkit for moving the field forward. Every among-clade comparison of morphospace occupation or functional diversification is the potential basis for a study of evolvability, particularly when the occupation pattern is informed by phylogeny or explicitly structured over geologic time. We need a more active two-way exchange, predicting macroevolutionary patterns from short-term evolvability estimates, and predicting short-term evolvability and its developmental and genetic underpinnings from macroevolutionary dynamics. Such an exchange should come closer to testing underlying mechanisms and how they play out on the macroevolutionary stage. Evolvability could then become a powerful bridge between micro- and macroevolution. This would not involve simple extrapolation from lower to higher levels, but a way to understand and systematize the many nonlinearities and indirect effects inherent in a multilevel system, as we now understand organic evolution to be.

Acknowledgments

I thank the editors for inviting me to participate in this exciting project; I. Brigandt, K. S. Collins, N. M. A. Crouch, M. Distin, S. M. Edie, F. Galis, T. F. Hansen, S. Huang, M. Pavličev, C. Pélabon, and N. H. Shubin for discussions and/or reviews, though none should be blamed for remaining misconceptions; and NSF, NASA, and the John Simon Guggenheim Foundation for support.

References

Alix, K., P. R. Gérard, T. Schwarzacher, and J. S. Heslop-Harrison. 2017. Polyploidy and interspecific hybridization: Partners for adaptation, speciation and evolution in plants. *Annals of Botany* 120: 183–194.

Barnes, B. D., J. A. Sclafani, and A. Zaffos. 2021. Dead clades walking are a pervasive macroevolutionary pattern. *PNAS* 118: e201920811.

Benton, M. J. 2009. The Red Queen and the Court Jester: Species diversity and the role of biotic and abiotic factors through time. *Science* 323: 728–732.

Betancur-R, R., G. Ortí, and R. A. Pyron. 2015. Fossil-based comparative analyses reveal ancient marine ancestry erased by extinction in ray-finned fishes. *Ecology Letters* 18: 441–450.

Bicknell, R. D., J. Kimmig, G. E. Budd, et al. 2022. Habitat and developmental constraints drove 330 million years of horseshoe crab evolution. *Biological Journal of the Linnean Society* 136: 155–172.

Bribiesca-Contreras, G., H. Verbruggen, A. F. Hugall, and T. D. O'Hara. 2017. The importance of offshore origination revealed through ophiuroid phylogenomics. *Proceedings of the Royal Society B* 284: 20170160.

Brown, R. L. 2014. What evolvability really is. *British Journal for the Philosophy of Science* 65: 549–572.

Bruce, H. S., and N. H. Patel. 2020. Knockout of crustacean leg patterning genes suggests that insect wings and body walls evolved from ancient leg segments. *Nature Ecology & Evolution* 4: 1703–1712.

Chapelle, K. E., R. B. Benson, J. Stiegler, A. Otero, Q. Zhao, and J. N. Choiniere. 2020. A quantitative method for inferring locomotory shifts in amniotes during ontogeny, its application to dinosaurs and its bearing on the evolution of posture. *Palaeontology* 63: 229–242.

Clark, J. W., and P. C. J. Donoghue. 2018. Whole-genome duplication and plant macroevolution. *Trends in Plant Science* 23: 933–945.

Conant, G. C. 2020. The lasting after-effects of an ancient polyploidy on the genomes of teleosts. *PLoS One* 15: e0231356.

Cooney, C. R., J. A. Bright, E. J. R. Capp, et al. 2017. Mega-evolutionary dynamics of the adaptive radiation of birds. *Nature* 542: 344–347.

Cooney, C. R., and G. H. Thomas. 2021. Heterogeneous relationships between rates of speciation and body size evolution across vertebrate clades. *Nature Ecology & Evolution* 5: 101–110.

Crouch, N. M. A., and R. E. Ricklefs. 2019. Speciation rate is independent of the rate of evolution of morphological size, shape, and absolute morphological specialization in a large clade of birds. *American Naturalist* 193: E78–E91.

Czekanski-Moir, J. E., and R. J. Rundell. 2019. The ecology of nonecological speciation and nonadaptive radiations. *TREE* 34: 400–415.

Davesne, D., M. Friedman, A. D. Schmitt, et al. 2021. Fossilized cell structures identify an ancient origin for the teleost whole-genome duplication. *PNAS* 118: e2101780118.

Deline, B., J. R. Thompson, N. S. Smith, et al. 2020. Evolution and development at the origin of a phylum. *Current Biology* 30: 1672–1679.

Donoghue, M. J., and M. J. Sanderson. 2015. Confluence, synnovation, and depauperons in plant diversification. *New Phytologist* 207: 260–274.

Eble, G. J. 2000. Contrasting evolutionary flexibility in sister groups: Disparity and diversity in Mesozoic atelostomate echinoids. *Paleobiology* 26: 56–79.

Edie, S. M., D. Jablonski, and J. W. Valentine. 2018. Contrasting responses of functional diversity to major losses in taxonomic diversity. *PNAS* 115: 732–737.

Edie, S. M., S. C. Khouja, K. S. Collins, N. M. A. Crouch, and D. Jablonski. 2022. Modularity, integration and disparity in an accretionary skeleton: Analysis of venerid Bivalvia. *Proceedings of the Royal Society B*: 20211199

Erwin, D. H. 2021a. A conceptual framework of evolutionary novelty and innovation. *Biological Reviews* 96: 1–15.

Erwin, D. H. 2021b. Developmental capacity and the early evolution of animals. *Journal of the Geological Society, London* 178: jgs2020–245.

Erwin, D. H., and J.W. Valentine. 2013. *The Cambrian Explosion*. Greenwood Village, CO: Ben Roberts.

Erwin, D. H., J. W. Valentine, and J. J. Sepkoski, Jr. 1987. A comparative study of diversification events: The early Paleozoic versus the Mesozoic. *Evolution* 41: 1177–1186.

Evans, K. M., O. Larouche, S. J. Watson, S. Farina, M. L. Habegger, and M. Friedman. 2021. Integration drives rapid phenotypic evolution in flatfishes. *PNAS* 118: e2101330118.

Fabre, A. C., C. Bardua, M. Bon, et al. 2020. Metamorphosis shapes cranial diversity and rate of evolution in salamanders. *Nature Ecology & Evolution* 4: 1129–1140.

Felice, R. N., M. Randau, and A. Goswami. 2018. A fly in a tube: Macroevolutionary expectations for integrated phenotypes. *Evolution* 72: 2580–2594.

Folk, R. A., R. L. Stubbs, M. E. Mort, et al. 2019. Rates of niche and phenotype evolution lag behind diversification in a temperate radiation. *PNAS* 116: 10874–10882.

Foote, M. 1991. Morphological and taxonomic diversity in a clade's history: The blastoid record and stochastic simulations. *Contributions from the Museum of Paleontology, University of Michigan* 28: 101–140.

Foote, M. 1993. Discordance and concordance between morphological and taxonomic diversity. *Paleobiology* 19: 185–204.

Foote, M. 1996. Models of morphological diversification. In *Evolutionary Paleobiology,* edited by D. Jablonski, D. H Erwin, and J. H. Lipps, 62–86. Chicago: University of Chicago Press.

Foote, M. 1997. The evolution of morphological diversity. *AREES* 28: 129–152.

Foote, M. 1999. Morphological diversity in the evolutionary radiation of Paleozoic and post-Paleozoic crinoids. *Paleobiology* 25 (Suppl. to no. 2): 1–115.

Foster, W. J., and R. J. Twitchett. 2014. Functional diversity of marine ecosystems after the Late Permian mass extinction event. *Nature Geoscience* 7: 233–238.

Franeck, F., and L. H. Liow. 2019. Dissecting the paleocontinental and paleoenvironmental dynamics of the great Ordovician biodiversification. *Paleobiology* 45: 221–234.

Geiger, M., A. Evin, M. R. Sánchez-Villagra, D. Gascho, C. Mainini, and C. P. Zollikofer. 2017. Neomorphosis and heterochrony of skull shape in dog domestication. *Scientific Reports* 7: 13443.

Gillespie, R. G., G. M. Bennett, L. De Meester, et al. 2020. Comparing adaptive radiations across space, time, and taxa. *Journal of Heredity* 111: 1–20.

Glasauer, S. M., and S. C. Neuhauss. 2014. Whole-genome duplication in teleost fishes and its evolutionary consequences. *Molecular Genetics and Genomics* 289: 1045–1060.

Gorné, L. D., and S. Díaz. 2019. Meta-analysis shows that rapid phenotypic change in angiosperms in response to environmental change is followed by stasis. *American Naturalist* 194: 840–853.

Goswami, A., and P. D. Polly. 2010. The influence of modularity on cranial morphological disparity in Carnivora and Primates (Mammalia). *PLOS One* 5: e9517.

Goswami, A., J. B. Smaers, C. Soligo, C., and P. D. Polly. 2014. The macroevolutionary consequences of phenotypic integration: From development to deep time. *Philosophical Transactions of the Royal Society B* 369: 20130254.

Gould, S. J. 1977. *Ontogeny and Phylogeny.* Cambridge, MA: Harvard University Press.

Gould, S. J. 1982. The meaning of punctuated equilibrium and its role in validating a hierarchical approach to macroevolution. In *Perspectives on Evolution,* edited by R. Milkman, 183–104. Sunderland, MA: Sinauer.

Gould, S. J. 2002. *The Structure of Evolutionary Theory.* Cambridge, MA: Belknap.

Gould, S. J., and N. Eldredge. 1977. Punctuated equilibria: The tempo and mode of evolution reconsidered. *Paleobiology* 3: 115–151.

Guillerme, T., N. Cooper, S. L. Brusatte, et al. 2020. Disparities in the analysis of morphological disparity. *Biology Letters* 16: 20200199.

Haber, A. 2016. Phenotypic covariation and morphological diversification in the ruminant skull. *American Naturalist* 187: 576–591.

Halliday, T. J., M. Dos Reis, A. U. Tamuri, H. Ferguson-Gow, Z. Yang, and A. Goswami. 2019. Rapid morphological evolution in placental mammals post-dates the origin of the crown group. *Proceedings of the Royal Society B* 286: 20182418.

Hansen, T. F. 2011. Epigenetics: Adaptation or contingency. In *Epigenetics: Linking Genotype and Phenotype in Development and Evolution,* edited by B. Hallgrímsson and B. K. Hall, 357–376. Berkeley: University of California Press.

Hansen, T. F., and C. Pélabon. 2021. Evolvability: A quantitative-genetics perspective. *AREES* 52: 153–175.

Harmon L. J., J. B. Losos, T. J. Davies, et al. 2010. Early bursts of body size and shape evolution are rare in comparative data. *Evolution* 64: 2385–2396.

Hembry, D. H., and M. G. Weber. 2020. Ecological interactions and macroevolution: A new field with old roots. *AREES* 51: 215–243.

Hedrick, B. P., G. L. Mutumi, V. D. Munteanu, et al. 2020. Morphological diversification under high integration in a hyper diverse mammal clade. *Journal of Mammalian Evolution* 27: 563–575.

Henao Diaz, L. F., L. J. Harmon, M. T. Sugawara, E. T. Miller, and M. W. Pennell. 2019. Macroevolutionary diversification rates show time dependency. *PNAS* 116: 7403–7408.

Holzman, R., D. C. Collar, R. S. Mehta, and P. C. Wainwright. 2011. Functional complexity can mitigate performance trade-offs. *American Naturalist* 177: E69–E83.

Hopkins, M. J., and A. B. Smith. 2015. Dynamic evolutionary change in post-Paleozoic echinoids and the importance of scale when interpreting changes in rates of evolution. *PNAS* 112: 3758–3763.

Houle, D., G. H. Bolstad, K. van der Linde, and T. F. Hansen. 2017. Mutation predicts 40 million years of fly wing evolution. *Nature* 548: 447–450.

Huang, S., K. Roy, and D. Jablonski. 2015. Origins, bottlenecks, and present-day diversity: Patterns of morphospace occupation in marine bivalves. *Evolution* 69: 735–746.

Hughes, M., S. Gerber, and M. A. Wills. 2013. Clades reach highest morphological disparity early in their evolution. *PNAS* 110: 13875–13879.

Hunt, G. 2007. Evolutionary divergence in directions of high phenotypic variance in the ostracode genus *Poseidonamicus. Evolution* 61: 1560–1576.

Hunt, G., and M. Yasuhara. 2010. A fossil record of developmental events: Variation and evolution in epidermal cell divisions in ostracodes. *Evolution & Development* 12: 635–646.

Jablonski, D. 2002. Survival without recovery after mass extinctions. *PNAS* 99: 8139–8144.

Jablonski, D. 2005. Evolutionary innovations in the fossil record: The intersection of ecology, development and macroevolution. *Journal of Experimental Zoology Part B: Molecular and Developmental Evolution* 304B: 504–519.

Jablonski, D. 2008a. Species selection: Theory and data. *AREES* 39: 501–524.

Jablonski, D. 2008b. Biotic interactions and macroevolution: Extensions and mismatches across scales and levels. *Evolution* 62: 715–739.

Jablonski, D. 2017a. Approaches to macroevolution: 1. General concepts and origin of variation. *Evolutionary Biology* 44: 427–450.

Jablonski, D. 2017b. Approaches to macroevolution: 2. Sorting of variation, some overarching issues, and general conclusions. *Evolutionary Biology* 44: 451–475.

Jablonski, D. 2020. Developmental bias, macroevolution, and the fossil record. *Evolution & Development* 22: 103–125.

Jablonski, D. 2022. Evolvability and macroevolution: Overview and synthesis. *Evolutionary Biology* 49: 265–291.

Jablonski, D., and D. J. Bottjer. 1990. The origin and diversification of major groups: Environmental patterns and macroevolutionary lags. In *Major Evolutionary Radiations,* edited by P. D. Taylor and G. P. Larwood, 17–57. Oxford: Clarendon Press.

Jablonski, D., S. Lidgard, and P. D. Taylor. 1997. Comparative ecology of bryozoan radiations: Origin of novelties in cyclostomes and cheilostomes. *Palaios* 12: 505–523.

Jablonski, D., S. Huang, K. Roy, and J. W. Valentine. 2017. Shaping the latitudinal diversity gradient: New perspectives from a synthesis of paleobiology and biogeography. *American Naturalist* 189: 1–12.

Johnson, C. K., and S. R. Voss. 2013. Salamander paedomorphosis: Linking thyroid hormone to life history and life cycle evolution. *Current Topics in Developmental Biology* 103: 229–258.

Kraaijeveld, K. 2010. Genome size and species diversification. *Evolutionary Biology* 37: 227–233

Kröger, B., and A. Penny. 2020. Skeletal marine animal biodiversity is built by families with long macroevolutionary lag times. *Nature Ecology & Evolution* 4: 1410–1415.

Liu, C., Y. Ren, Z. Li, et al. 2021. Giant African snail genomes provide insights into molluscan whole-genome duplication and aquatic–terrestrial transition. *Molecular Ecology Resources* 21: 478–494.

Lomax, B. H., J. Hilton, R. M. Bateman, et al. 2014. Reconstructing relative genome size of vascular plants through geological time. *New Phytologist* 201: 636–644.

López-Sauceda, J., J. Malda-Barrera, A. Laguarda-Figueras, F. Solís-Marín, F. and J. L. Aragón. 2014. Influence of modularity and regularity on disparity of Atelostomata Sea Urchins. *Evolutionary Bioinformatics* 10: EBO-S14457.

Losos, J. B. 2010. Adaptive radiation, ecological opportunity, and evolutionary determinism. *American Naturalist* 175: 623–639.

Louca, S., and M. W. Pennell. 2020. Extant timetrees are consistent with a myriad of diversification histories. *Nature* 580: 502–505.

Love, A. C., M. Grabowski, D. Houle, et al. 2022. Evolvability in the fossil record. *Paleobiology* 48: 186–209.

Love, A. C., and G. L. Lugar. 2013. Dimensions of integration in interdisciplinary explanations of the origin of evolutionary novelty. *Studies in History and Philosophy of Science Part C: Studies in History and Philosophy of Biological and Biomedical Sciences* 44: 537–550.

Lovette, I. J., E. Bermingham, and R. E. Ricklefs. 2002. Clade-specific morphological diversification and adaptive radiation in Hawaiian songbirds. *Proceedings of the Royal Society B* 269: 37–42.

Machado, F. A., T. M. G. Zahn, and G. Marroig. 2018. Evolution of morphological integration in the skull of Carnivora (Mammalia): Changes in Canidae lead to increased evolutionary potential of facial traits. *Evolution* 72: 1399–1419.

Marek, R. D., P. L. Falkingham, R. B. Benson, J. D. Gardiner, T. W. Maddox, and K. T. Bates. 2021. Evolutionary versatility of the avian neck *Proceedings of the Royal Society B* 288: 20203150.

Marshall, C. R. 2017. Five palaeobiological laws needed to understand the evolution of the living biota. *Nature Ecology & Evolution* 1:0165.

Martin, C. H., and E. J. Richards. 2019. The paradox behind the pattern of rapid adaptive radiation: How can the speciation process sustain itself through an early burst? *AREES* 50: 569–593.

Martin, P. R., F. Bonier, and J. J. Tewksbury. 2007. Revisiting Jablonski (1993): Cladogenesis and range expansion explain latitudinal variation in taxonomic richness. *Journal of Evolutionary Biology* 20: 930–936.

Masterson, J. 1994. Stomatal size in fossil plants: Evidence for polyploidy in majority of angiosperms. *Science* 264: 421–424.

McElwain, J. C., and M. Steinthorsdottir. 2017. Paleoecology, ploidy, paleoatmospheric composition, and developmental biology: A review of the multiple uses of fossil stomata. *Plant Physiology* 174: 650–664.

Mitchell, J. S., R. S. Etienne, and D. L. Rabosky. 2019. Inferring diversification rate variation from phylogenies with fossils. *Systematic Biology* 68: 1–18.

Mongiardino Koch, N. 2021. Exploring adaptive landscapes across deep time: A case study using echinoid body size. *Evolution* 75: 1567–1581.

Moulton, D. E., A. Goriely, and R. Chirat. 2020. Mechanics unlocks the morphogenetic puzzle of interlocking bivalved shells. *PNAS* 117: 43–51.

Müller, J., T. M. Scheyer, J. J. Head, et al. 2010. Homeotic effects, somitogenesis and the evolution of vertebral numbers in recent and fossil amniotes. *PNAS* 107: 2118–2123.

Navalón, G., J. Marugán-Lobón, J. A. Bright, C. R. Cooney, and E. J. Rayfield. 2020. The consequences of craniofacial integration for the adaptive radiations of Darwin's finches and Hawaiian honeycreepers. *Nature Ecology & Evolution* 4: 270–278.

Nicol, D. 1958. A survey of inequivalve pelecypods. *Journal of the Washington Academy of Sciences* 48: 56–62.

Nijhout, H. F., and K. Z. McKenna. 2017. The origin of novelty through the evolution of scaling relationships. *Integrative and Comparative Biology* 57: 1322–1333.

Nunney, L. 1989. The maintenance of sex by group selection. *Evolution* 43: 245–257.

O'Meara, B. C., S. D. Smith, W. S. Armbruster, et al. 2016. Non-equilibrium dynamics and floral trait interactions shape extant angiosperm diversity. *Proceedings of the Royal Society B* 283: 20152304.

Organ, C. L., A. Canoville, R. R. Reisz, and M. Laurin. 2011. Paleogenomic data suggest mammal-like genome size in the ancestral amniote and derived large genome size in amphibians. *Journal of Evolutionary Biology* 24: 372–380.

Organ, C. L., A. M. Shedlock, A. Meade, M. Pagel, and S. V. Edwards. 2007. Origin of avian genome size and structure in nonavian dinosaurs. *Nature* 446: 180–184.

Oyston, J. W., M. Hughes, P. J. Wagner, S. Gerber, and M. A. Wills. 2015. What limits the morphological disparity of clades? *Interface Focus, Royal Society of London* 5: 20150042.

Palmer, A. R. 2004. Symmetry breaking and the evolution of development. *Science* 306: 828–833.

Panganiban, G., and J. L. R. Rubenstein. 2002. Developmental functions of the *Distal-less*/Dlx homeobox genes. *Development* 129: 4371–4386.

Pavličev, M., and T. F. Hansen. 2011. Genotype-phenotype maps maximizing evolvability: Modularity revisited. *Evolutionary Biology* 38: 371–389.

Pie, M. R., and J. S. Weitz. 2005. A null model of morphospace occupation. *American Naturalist* 166: E1–E13.

Polly, P. D., J. Fuentes-Gonzalez, A. M. Lawing, A. K. Bormet, and R. G. Dundas. 2017. Clade sorting has a greater effect than local adaptation on ecometric patterns in Carnivora. *Evolutionary Ecology Research* 18: 61–95.

Quental, T. B., and C. R. Marshall. 2010. Diversity dynamics: Molecular phylogenies need the fossil record. *TREE* 25: 434–441.

Rabosky, D. L. 2017. Phylogenetic tests for evolutionary innovation: The problematic link between key innovations and exceptional diversification. *Philosophical Transactions of the Royal Society B* 372: 20160417.

Rabosky, D. L., F. Santini, J. Eastman, S. A. Smith, B. Sidlauskas, J. Chang, and M. E. Alfaro. 2013. Rates of speciation and morphological evolution are correlated across the largest vertebrate radiation. *Nature Communications* 4: 1958.

Ramírez-Barahona, S., H. Sauquet, and S. Magallón. 2020. The delayed and geographically heterogeneous diversification of flowering plant families. *Nature Ecology & Evolution* 4: 1232–1238.

Rhoda, D., P. D. Polly, C. Raxworthy, and M. Segall. 2021. Morphological integration and modularity in the hyperkinetic feeding system of aquatic-foraging snakes. *Evolution* 75: 56–72.

Rundell, R. J., and T. D. Price. 2009. Adaptive radiation, nonadaptive radiation, ecological speciation and nonecological speciation. *TREE* 24: 394–399.

Sallan, L., M., Friedman, R. S. Sansom, C. M. Bird, and I. J. Sansom. 2018. The nearshore cradle of early vertebrate diversification. *Science* 362: 460–464.

Sánchez-Villagra, M. R., V. Segura, M. Geiger, L. Heck, K. Veitschegger, and D. Flores. 2017. On the lack of a universal pattern associated with mammalian domestication: Differences in skull growth trajectories across phylogeny. *Royal Society Open Science* 4: 170876.

Saucède, T., R. Laffont, C. Labruère, et al. 2015. Empirical and theoretical study of atelostomate (Echinoidea, Echinodermata) plate architecture: Using graph analysis to reveal structural constraints. *Paleobiology* 41: 436–459.

Schluter, D. 1996. Adaptive radiation along genetic lines of least resistance. *Evolution* 50: 1766–1774.

Schoen, D. J., and S. T. Schultz. 2019. Somatic mutation and evolution in plants. *AREES* 50: 49–73.

Seilacher, A. 1984. Constructional morphology of bivalves: Evolutionary pathways in primary versus secondary soft-bottom dwellers. *Palaeontology* 27: 207–237.

Shi, J. J., E. P. Westeen, and D. L. Rabosky. 2021. A test for rate-coupling of trophic and cranial evolutionary dynamics in New World bats. *Evolution* 75: 861–875.

Shubin, N., C. Tabin, and S. Carroll. 2009. Deep homology and the origins of evolutionary novelty. *Nature* 457: 818–823.

Simões, T. R., O. Vernygora, M. W. Caldwell, and S. E. Pierce. 2020. Megaevolutionary dynamics and the timing of evolutionary innovation in reptiles. *Nature Communications* 11: 3322.

Simpson, C., A. Herrera-Cubilla, and J. B. C. Jackson. 2020. How colonial animals evolve. *Science Advances* 6: eaaw9530.

Simpson, G. G. 1944. *Tempo and Mode in Evolution.* New York: Columbia University Press.

Slater, G. J. 2013. Phylogenetic evidence for a shift in the mode of mammalian body size evolution at the Cretaceous-Palaeogene boundary. *Methods in Ecology and Evolution* 4: 734–744.

Slater, G. J., L. J. Harmon, and M. E. Alfaro. 2012. Integrating fossils with molecular phylogenies improves inference of trait evolution. *Evolution* 66: 3931–3944.

Slater, G. J., and M. W. Pennell. 2014. Robust regression and posterior predictive simulation increase power to detect early bursts of trait evolution. *Systematic Biology* 63: 293–308.

Smith, A. B. 2001. Probing the cassiduloid origins of clypeasteroid echinoids using stratigraphically restricted parsimony analysis. *Paleobiology* 27: 392–404.

Smith, A. B. 2005. Growth and form in echinoids: The evolutionary interplay of plate accretion and plate addition. In *Evolving Form and Function: Fossils and Development,* edited by D. E. G. Briggs, 181–196. New Haven, CT: Yale Peabody Museum.

Soltis, P. S., X. Liu, D. B. Marchant, C. J. Visger, and D. E. Soltis. 2014. Polyploidy and novelty: Gottlieb's legacy. *Philosophical Transactions of the Royal Society B* 369: 20130351.

Stanley, S. M. 1979. *Macroevolution.* San Francisco, CA: W.H. Freeman.

Stanley, S. M. 1990. The general correlation between rate of speciation and rate of extinction: Fortuitous causal linkages. In *Causes of Evolution,* edited by R. M. Ross and W. D. Allmon, 103–127. Chicago: University of Chicago Press.

Sterelny, K., and P. E. Griffiths. 1999. *Sex and Death: An Introduction to Philosophy of Biology.* Chicago: University of Chicago Press.

Stroud, J. T., and J. B. Losos. 2016. Ecological opportunity and adaptive radiation. *AREES* 47: 507–532.

Sumiyama, K., and A. Tanave. 2020. The regulatory landscape of the *Dlx* gene system in branchial arches: Shared characteristics among *Dlx* bigene clusters and evolution. *Development, Growth & Differentiation* 62: 355–362.

Taylor, M. P., and M. J. Wedel. 2013. Why sauropods had long necks; and why giraffes have short necks. *PeerJ* 1: e36.

Thomson, K. S., and K. Muraszko. 1978. Estimation of cell size and DNA content in fossil fishes and amphibians. *Journal of Experimental Zoology B* 205: 315–320.

Thuy, B., M. E. Eriksson, M. Kutscher, J. Lindgren, et al. 2022. Miniaturization during a Silurian environmental crisis generated the modern brittle star body plan. *Communications Biology* 5: 14.

Tilman, A. R., and D. Tilman. 2020. Evolution, speciation, and the persistence paradox. In *Unsolved Problems in Ecology,* edited by A. Dobson, D. Tilman, and R. D. Holt, 160–176. Princeton, NJ: Princeton University Press.

Tobias, J. A., J. Ottenburghs, and A. L. Pigot. 2020. Avian diversity: Speciation, macroevolution, and ecological function. *AREES* 51: 533–560.

Tomašových, A., S. Dominici, M. Zuschin, and D. Merle. 2014. Onshore–offshore gradient in metacommunity turnover emerges only over macroevolutionary time-scales *Proceedings of the Royal Society* B 281: 20141533.

Tomiya, S., and L. K. Miller. 2021. Why aren't rabbits and hares larger? *Evolution* 75: 847–860.

Uller, T., A. P. Moczek, R. A. Watson, P. M. Brakefield, and K. N. Laland. 2018. Developmental bias and evolution: A regulatory network perspective. *Genetics* 209: 949–966.

Urdy, S., L. A. B. Wilson, J. T. Haug, and M. R. Sánchez-Villagra. 2013. On the unique perspective of paleontology in the study of developmental evolution and biases. *Biological Theory* 8: 293–311.

Valentine, J. W. 1990. The macroevolution of clade shape. In *Causes of Evolution,* edited by R. M. Ross and W. D. Allmon, 128–150. Chicago: University of Chicago Press.

Valentine, J. W., D. Jablonski, A. Z Krug, and S. K. Berke. 2013. The sampling and estimation of marine paleodiversity patterns: Implications of a Pliocene model. *Paleobiology* 39: 1–20.

Vamosi, J. C., S. Magallón, L. Mayrose, S. P. Otto, and H. Sauquet. 2018. Macroevolutionary patterns of flowering plant speciation and extinction. *Annual Review of Plant Biology* 69: 685–706.

Van Valen, L. 1975. Group selection, sex, and fossils. *Evolution* 29: 87–94.

Van Valen, L. M. 1985. A theory of origination and extinction. *Evolutionary Theory* 7: 133–142.

Vermeij, G. J. 1974. Adaptation, versatility, and evolution. *Systematic Zoology* 22: 466–477.

Vermeij, G. J. 1987. *Evolution and Escalation*. Princeton, NJ: Princeton University Press.

Vermeij, G. J. 2012. Crucibles of creativity: The geographic origins of tropical molluscan innovations. *Evolutionary Ecology* 26: 357–373.

Vermeij, G.J. 2015. Forbidden phenotypes and the limits of evolution. *Interface Focus, Royal Society of London* 5: 20150028.

Wagner, G. P. 2014. *Homology, Genes, and Evolutionary Innovation*. Princeton, NJ: Princeton University Press.

Wagner, G. P., and L. Altenberg. 1996. Complex adaptations and the evolution of evolvability. *Evolution* 50: 967–976.

Wagner, P. J. 2000 Exhaustion of morphologic character states among fossil taxa. *Evolution* 54: 365–386.

Wagner, P. J. 2010. Paleontological perspectives on morphological evolution. In *Evolution since Darwin: The First 150 Years,* edited by D. Futuyma, J. Levinton, M. Bell, and W. Eanes, 451–478. Sunderland, MA: Sinauer.

Wagner, P. J. 2018. Early bursts of disparity and the reorganization of character integration. *Proceedings of the Royal Society B* 285: 20181604.

Watanabe, J. 2018. Clade-specific evolutionary diversification along ontogenetic major axes in avian limb skeleton. *Evolution* 72: 2632–2652.

Wesley-Hunt, G. D. 2005. The morphological diversification of carnivores in North America. *Paleobiology* 31: 35–55.

Wright, D. F. 2017. Phenotypic innovation and adaptive constraints in the evolutionary radiation of Palaeozoic crinoids. *Scientific Reports* 7: 13745.

Yu, L., C. Boström, S. Franzenburg, T. Bayer, T. Dagan, and T. B. Reusch. 2020. Somatic genetic drift and multilevel selection in a clonal seagrass. *Nature Ecology & Evolution* 4: 952–962.

Zenil-Ferguson, R., J. G. Burleigh, W. A. Freyman, B. Igić, I. Mayrose, and E. E. Goldberg. 2019. Interaction among ploidy, breeding system and lineage diversification. *New Phytologist* 224: 1252–1265.

18 Conclusion: Is Evolvability a New and Unifying Concept?

David Houle, Christophe Pélabon, Mihaela Pavličev, and Thomas F. Hansen

The title of the workshop from which this volume grew asserted that the concept of evolvability is both novel and unifying. In this chapter, we consider the sense in which these assertions are true. Evolvability concepts were clearly, if sparingly, used in the last half of the 20th century, but the term only came into widespread use starting around 1990. What was new at that time was a growing awareness that inheritance should be separated from natural selection and that the properties of the inherited system themselves evolve. This pair of ideas catalyzed productive evolvability research programs that differed substantially from earlier work. The rise of the targeted study of evolvability has proved to be unifying in two senses. It emphasized connections among evo-devo, systems biology, and population biology, resulting in intellectual exchange among these subdisciplines. On a deeper level, the recognition that evolvability is a disposition, an ability only expressed when certain conditions are present, unifies different uses of the term. The diversity of contexts in which we want to quantify the disposition to evolve brings to the fore different aspects of biology that are themselves dispositions: the disposition to mutate, the disposition for mutations to have phenotypic effects, and the disposition of populations to harbor variation. The underlying unity of evolvability as a disposition to evolve is consistent with the fact that the properties that best predict evolvability will differ, depending on the context.

18.1 Introduction

As explained in the Introduction (Hansen et al., chapter 1),[1] this book is the outcome of a yearlong workshop aimed at understanding the use of the term evolvability as a label for intrinsic dispositions of organisms, genetic systems, or populations to evolve. The title of the workshop was "Evolvability: A New and Unifying Concept in Evolutionary Biology?" We want to address the implied claims in that title here: Is evolvability new? Is evolvability unifying?

Nuño de la Rosa's (2017) bibliometric work documents the increased usage of the term evolvability starting in the 1990s (see also Villegas et al., chapter 3). More importantly, usage increased in a wide variety of evolutionary subdisciplines, and these tended to cite work from a variety of the other subdisciplines that also adopted the term. Price (1965) termed such citation networks "research fronts," and we adopt this term as a neutral descriptor of the totality of evolvability research over the past 30 years.

1. References to chapter numbers in the text are to chapters in this volume.

Bibliometry alone, however, leaves unsettled the questions of novelty and coherence of the research itself. As documented by Nuño de la Rosa's interviews (chapter 2), many see evolvability as a fashionable relabeling of older concepts. Similarly, the unity of evolvability research is widely questioned, both by those who see value in their own restricted usage of the concept (Sterelny 2007; Pigliucci 2008; Brown 2014; Riederer et al. 2022), and by those documenting the diversity of usage (Love 2003; Nuño de la Rosa 2017).

Unsurprisingly, those participating in this project share the sense that there is something more than terminology to the rise of evolvability research, and many of us see ways in which evolvability can unify aspects of evolutionary biology. Despite this, there is no clear consensus about the answers to these two questions among us, and the chapters of this volume lay out a range of possible answers. We take up the question of novelty in section 18.2, unificatory potential in section 18.3 and highlight ongoing research in section 18.4.

18.2 Is Evolvability New?

The question of novelty is foremost one about recent history: What caused the use of the term to increase dramatically between 1990 and 2010? Nuño de la Rosa's interviews (chapter 2) probe the degree to which participants and observers in evolvability research fronts perceive it as a novel topic. With few exceptions, the interviewees identify earlier work that considered the same or related issues as current evolvability studies. These precursors are clear in the foundational population genetic work of Fisher (1930) and Wright (1932). Their conflicts over the shape of adaptive landscapes; the importance of drift; and whether genetic features, such as dominance (Mayo and Bürger 1997), are adaptations; are still actively debated by those working on evolvability today (Frank 2012). Similarly, the idea that clades (Simpson 1953; Vermeij 1987; Jablonski, chapter 17) or traits (Armbruster, chapter 15) differ in their ability to evolve is of longstanding importance in macroevolutionary studies. In evo-devo research, an important precursor to evolvability research is the European structuralist tradition exemplified by the work of Waddington (1957) and Riedl (1977, 1978). In addition, interviewees point to precursors of the effort to integrate genetics, development, and macroevolution, such as Lewontin's (1974, 12–16) conceptualization of the genotype-phenotype (GP) map as the hole in evolutionary genetics that needed to be filled by incorporation of such processes as development (e.g., Raff 1996) and biochemistry (e.g., Kacser and Burns 1981) into evolutionary thinking.

We will make a case that, despite these precursors, the emergence of the evolvability research front was catalyzed by two major conceptual advances that clarified how to think about evolvability. The first is the separation of the concept of natural selection from that of inherited variation, which enabled the recognition of evolvability as a disposition to evolve should the right stimuli occur. The second is the articulation of the idea that the ability to evolve can itself evolve. Both conceptual shifts set the stage for investigation of the processes that shape evolvability. This change in perspective catalyzed new research programs, in which novel concepts were brought into play, while drawing on intellectual precursors.

Natural selection has long been confounded with inheritance. For Darwin and most of his followers, natural selection was not separated from inheritance. Endler's (1986) influential book on natural selection made the inclusion of inheritance explicit. His favored definition of natural selection included three elements: phenotypic variation, fitness differences among phenotypes, and inheritance, and has been widely used by others (e.g., Lewontin 1970; Mayr 1982; Ridley 1998, 2002). A few previous authors, albeit influential ones (Fisher 1930; Haldane 1954; Van Valen 1965), had instead adopted definitions involving only phenotypic variation and fitness differences. Endler noted this viewpoint but explicitly rejected it by definition: "If there is no inheritance then the process of natural selection cannot occur" (Endler 1986, 13).

Two important developments, one theoretical and the other empirical, tipped the balance of thinking toward separation of selection and inheritance. The theoretical development was the Price theorem (Price 1970; 1972; Hansen, chapter 5), which expresses the change in phenotype as the sum of the effects of selection and of transmission bias that includes inheritance. The Price theorem was not widely understood or applied before the 1990s (Frank 1995), but it then became a staple of conceptual analyses of selection that allowed clarification of previously confusing debates about levels of selection and the like, which are indeed complicated when inheritance or genetics is mixed into the picture but are quite simple from the perspective of selection alone (but see Okasha and Otsuka 2020).

The empirical development was the emergence of evolutionary quantitative genetics with its operational tools for measurements of evolution, genetic variance, inheritance, and selection as separate entities. Lande (1979) first showed that the multivariate response to selection could be represented as

$$\Delta\bar{\mathbf{z}} = \mathbf{G}\boldsymbol{\beta},$$

where $\Delta\bar{\mathbf{z}}$ is the vector of predicted changes in trait mean values, $\mathbf{G}$ is the additive genetic variance matrix, and $\boldsymbol{\beta}$ is the selection gradient vector. The Lande equation neatly separates selection from genetic variation and inheritance. Lande and Arnold (1983) went on to demonstrate that $\boldsymbol{\beta}$ could be estimated as the multivariate regression of relative fitness on the trait vector. These papers popularized the representation of selection separately from inheritance, and they demonstrated how to estimate selection on multiple traits from data obtained using the standard observational and experimental methods of the ecologist. This started an industry of investigations that has provided thousands of field estimates of the strength and mode of natural selection. Similarly, inheritance and genetic variation in the form of the G-matrix, which describes the heritable component of genetic variation (Hansen, chapter 5), could be studied in the lab or in the field with classical or modern genetic methods without worrying about the connection to selection.

With these advances, the modern evolutionary biologist is well primed to think about evolution by natural selection as a two-step process. The first step is the appearance of heritable variation, and the second is the action of natural selection on this variation. The term "evolvability" filled a newly created need to talk about the variational preconditions for natural selection separately from discussion of natural selection itself. Terms such as facilitated variation, evolutionary drivers, adaptability (Anpassungsfähigkeit), and adaptive versatility were used for this purpose in the 1970s and 1980s, but they never came into general use. This gain in conceptual and empirical separation of selection and inheritance also precipitated a change in emphasis from constraints as forbidden or discouraged directions of evolution to quantification of how evolvable the phenotype is in each direction using properly justified measures of evolvability (Gould 1989; Schluter 1996; Hansen and Houle 2008).

While applied quantitative geneticists were already primed to regard selection as something different from inheritance due to the fact that selection is under the control of experimenter (e.g., Falconer 1981), the recognition of this critical separation was ironically hampered by the univariate breeder's equation familiar to quantitative geneticists,

$$\Delta\bar{z} = h^2 S,$$

where h^2 is the heritability, the proportion of variation that is additive genetic, and S is the covariance between relative fitness and the trait value, also known as the selection differential. Although the breeder's equation is a correct formulation of the response to selection, it invites the user to interpret h^2 as a measure of evolvability and S as a measure of selection. This interpretation is incorrect, as explained in chapters 5 and 6. Heritability is a dimensionless quantity, thus displacing all the scale information into S. Both the magnitude of genetic variance and the strength of selection affect S, thus confounding selection and inheritance. The assumption that h^2 represents what we want to know about inheritance precludes measurement of the disposition to respond to selection. Indeed, the first use of the term evolvability in the quantitative genetic literature was to make this point (Houle 1992). This misconception about the nature of h^2 persists (Hansen et al. 2011; Hansen and Pélabon 2021).

The second conceptual advance that led to the emergence of an evolvability research front was the recognition that the disposition to evolve itself has the capacity to evolve (Conrad 1983; Dawkins 1989; Pigliucci 2007; Hansen and Wagner, chapter 7). The recognition of the evolution of evolvability as a productive research area has origins in the fields of evo-devo and computer modeling of evolutionary processes. The earliest use of the phrase "evolution of evolvability" was by Dawkins (1989) in an essay on creating computer models of "artificial life." Dawkins argued that implementing open-ended evolution of computer programs under selection for increased performance required that the variation introduced fulfill special conditions, and that the existence of these conditions in biological organisms was in itself an interesting and understudied problem. Other theoreticians had previously identified the problem of what organismal features enable evolution without using the term evolvability (e.g., Lewontin 1978; Riedl 1978; Conrad 1983; Wagner 1984).

The rise of thinking about the evolution of evolvability is closely tied to the concept of the genotype-phenotype map, Lewontin's (1974) term for the set of processes by which genetic effects result in the phenotype. Alberch (1991) was among the first to explicitly associate evolvability with properties of the GP map, but the key paper that merged the study of evolvability with the GP map is Wagner and Altenberg (1996), who, like Dawkins, drew on evolutionary computer science concepts. Following ideas of Riedl (1978), they argued that evolvability could only be achieved if the effects of genes on traits could be parceled out into modules that can be changed in a quasi-independent fashion. The idea that evolvability requires modularity was paradigmatic in the emerging field of evolutionary developmental biology, evo-devo (e.g., Raff 1996). This is, for example, manifest in the emphasis on changes in cis-regulatory modules as the source of morphological evolution (e.g., Stern 2000; Carroll 2008).

A second important distinction introduced by Wagner and Altenberg (1996) is between variability and variation. While mutation was always recognized as the ultimate source of genetic variation, Wagner and Altenberg identified variability, the disposition of mutations to produce phenotypic effects, as the property of the GP map that affects evolvability. Subsequently many have argued that the defining feature of evo-devo is the study of how evolvability is determined by the structure and evolution of the GP map (von Dassow and Munro 1999; Hendrikse et al. 2007; Brigandt 2015; Love 2015; Minelli 2017). In this view, evo-devo is a field devoted to the study of the first of the two steps in evolution by natural selection; that is, to the variational preconditions for natural selection.

Identification of variability as distinct from both mutation and the maintenance of genetic variation mirrors the identification of evolvability as the features that enable a response to selection. It is important to realize that these two related conceptions of evolvability arose essentially independently from different intellectual precursors—one from incorporating organismal processes such as development into mainstream evolutionary biology, and the other from a combination of evolutionary theory and quantitative genetics. We argue that the adoption of the term evolvability independently by these two intellectual traditions is what catalyzed the initial synthetic power of the evolvability research front (Nuño de la Rosa 2017, chapter 2; Villegas et al., chapter 3).

The degree of novelty in these advances is a matter of debate. Pigliucci (2007, 2008) featured the idea that the concept of the evolution of evolvability had no intellectual precursors in the modern synthesis prior to 1990 and used this novelty as a key argument that an "extended synthesis" of 20th-century evolutionary biology is occurring. In contrast, we prefer to treat the current work on the evolution of evolvability as a case of "endogenization" (Okasha 2021), in which the abstract principles of Darwinian mechanisms are applied to explain previously recognized but less well understood phenomena.

Adopting the idea that the process of endogenization is a key form of novelty, suggests a mechanism for separating "mere historical precedence" from novel intellectual traditions of the kind implied by Nuño de la Rosa's bibliometric work. We point to three additional areas of endogenization in evolvability research currently being explored: the direct studies of the evolution of GP maps (Pavličev et al., chapter 8; Hallgrímsson et al., chapter 9), the role of robustness in evolution (A. Wagner, chapter 11), and the roles of plasticity and environmental interactions in determining evolutionary direction and rate (West-Eberhard 2003; Laland and Sterelny 2006; Paenke et al. 2007; Scott-Phillips et al. 2014), a topic regrettably underrepresented in this volume.

A rather different kind of novelty in the evolvability research front is the explicit use of measurement theory (Houle et al. 2011, Houle and Pélabon, chapter 6) to justify particular choices of both empirical measures of evolvability and the theoretical constructs that they represent. These arguments feature in the justification of the separation of natural selection and evolvability (Hansen, chapter 5; Houle and Pélabon, chapter 6). A second example of the use of measurement theory is in linking conceptions of genetic variation of the GP map to the measurement of epistasis, the interactions of alleles at different genetic loci in determining phenotypes. Hansen (chapter 5) lays out the conceptual basis for this change in viewpoint. Pélabon et al. (chapter 13) and G. Wagner (chapter 10) discuss empirical situations in which directional epistasis is expected to lead to a correlation between evolvability and trait mean.

A final novel element to evolvability studies is the incorporation of the concept of dispositions into the consciousness of biologists. Some of the earliest proponents of evolvability recognized that it was a disposition (Wagner and Altenberg 1996; Hansen 2006), but perhaps more important to the introduction of dispositional concepts was that biological dispositions like evolvability attracted attention from philosophers (Love 2003; Sterelny 2007; Brown 2014; Brigandt et al., chapter 4). For biologists, this engagement clarifies the separation between the causal basis for the capability of evolution, and the stimuli that may actually convert capability into change (Prior et al. 1982). This clarification helps generalize the concept of evolvability by incorporating stimuli other than natural selection, including exceptionally rare changes, such as those leading to evolutionary novelties such as new body plans (Galis, chapter 16).

18.3 Is Evolvability Unifying?

This volume addresses a wide variety of phenomena related to evolvability from different perspectives. Observers of the evolvability research front have frequently expressed frustration at the diversity of phenomena to which the term is applied, and they have suggested that it should be restricted to some subset of current usage (Pigliucci 2008) or expanded to be more comprehensive (Brown 2014). Consequently, evolvability researchers are primed to consider the question of whether there is a unified basis to evolvability studies. We see several kinds of unity at work in the evolvability research front.

Nuño de la Rosa (2017) has suggested that evolvability studies can be seen as an intellectual "trading zone" (chapter 9 in Galison 1997; Winther 2015). This idea draws an analogy between locations where individuals from different human cultures exchange goods and intellectual arenas where individuals from multiple scientific subcultures find it worthwhile to engage. The key idea is that the meaning and value of the "goods" exchanged can vary from culture to culture and yet still contribute value on both sides of the exchange.

Villegas et al. (chapter 3) foreground the diversity of roles that concepts such as evolvability can play in scientific activities, including setting a research agenda, characterization, explanation, prediction, and control. For example, the idea that evolvability is correlated with mutational robustness (A. Wagner, chapter 11) can be used as a tool for prediction of evolvability, or as a target in studies that assess the strength or cause of the correlation. Villegas et al.'s catalog of roles that evolvability can play in research expands the variety of goods potentially exchanged in the evolvability trading zone, where one scientist's explanation is used for prediction by another researcher.

This evolvability project was assembled as an instantiation of a trading zone. Its attractiveness to a diverse array of biologists exemplifies the value of the evolvability trading zone to practicing scientists. It attracted effort at least partly because it promised intellectual interchange among scientists working in different specialties. This accepting attitude to diverse work on evolvability was fostered by the philosophers and historians of science in this project, who have explicitly valued their roles as observers and documenters of, rather than judges of, scientific practice (Love 2003; Brigandt and Love 2012; Nuño de la Rosa 2017). This inclusive approach to evolvability is to be expected in a trading zone, and it represents a kind of unification of evolvability studies.

We believe that there is also a deep conceptual unification in the concept of evolvability. In particular, the recognition of the idea of evolvability as a disposition (Wagner and Altenberg 1996) provides a basis for recognizing common elements in different usages of the term. What unites the different uses of the term evolvability is, first, the recognition that each describes a disposition to evolve intrinsic to the organism or population (e.g., Brown 2014). Second, these organismal and populational uses of evolvability are themselves linked to other dispositions. Wagner and Altenberg's (1996) made the distinction between variability, which is the propensity of a genetic system to yield variants that affect the phenotype; and evolvability, the propensity to evolve should the relevant cause, such as natural selection, occur. They assumed that the supply of adaptive variants limits the rate of evolution, a situation under which variability of a typical individual is the direct cause of evolvability. The quantitative genetic conception of evolvability focuses on the potentially adaptive variation in populations (Houle 1992; Hansen and Houle 2008). These two dispositions are directly linked, because variation at the population level would not exist without the ability of individuals to generate that variation. Similarly, we can identify the disposition of genomes to undergo mutation as an even more fundamental disposition that underlies both variability and variation. Consequently, the study of variability is intimately linked to the study of variation (Lewontin 1974; Houle et al. 1996; Houle 1998; Houle and Fierst 2013). Thus, there are several dispositional steps in a causal chain, and different notions of evolvability focus on different links in this chain.

Houle and Pélabon (chapter 6) use this logic to argue for a unified definition of evolvability as a disposition to evolve that can be applied in specific instances by drawing on different aspect of this linked set of dispositions. They note that there are many different aspects of organisms that could evolve, different stimuli that might trigger evolution, and different time scales of interest to biologists, for which Houle and Pélabon adopt the phrase *Of, Under,* and *Over.* Villegas et al. (chapter 3) point out that this unifying proposal carries over to two separate roles. First, it points to definitional unification. As long as research addresses a disposition to evolve, we have a study of evolvability. More important in their view is that this provides a basis to identify a research agenda based on the concept of evolvability. Houle and Pélabon's figure 6.1 (chapter 6) and Villegas et al.'s figure 3.3 (chapter 3) diagram different versions of this agenda.

18.4 What Is Ahead

In section 18.3, we emphasized those novel features of evolvability studies that originally catalyzed the evolvability research front. Current research focuses on important unsolved issues that we are optimistic will yield further advances in our understanding of evolvability. We highlight four specific areas in which progress is foreseeable: Modeling genotype-phenotype maps, phenomics, measurement of evolvability, and comparative evolvability.

18.4.1 Increasingly Realistic Genotype-Phenotype Maps

The genotype-phenotype (GP) map was arguably proposed to highlight the lack of attention by evolutionary biologists to how genes make phenotypes (Lewontin 1974). Since then, evolutionary biologists have enthusiastically joined in the effort to incorporate the GP map into the field, as reviewed by Pavličev et al. (chapter 8) and Hallgrímsson et al. (chapter 9). These trends are particularly apparent in the rise of studies of the role of GP maps in the evolution of development, featured in this volume and in evolutionary systems biology (Soyer and O'Malley 2013). With the metaphor of the GP map now firmly entrenched as a target of research, we can look forward to progress on several fronts.

The first of these is that the actual description of the pathways between genotype and phenotype is rapidly becoming more complete due to basic research in every area of biology, from molecular genetics and developmental biology, through physiology and behavior. Although this progress is most apparent for model organisms and humans, it also enables research on the GP map in a wider variety of systems. Basic research emphasizes the use of manipulations to test hypotheses about causal links between genotype and various functions. These data are increasingly being used to build detailed models of the genotype to phenotype relationships. Particularly promising is the trend toward causally-cohesive genotype-phenotype (cGP) models (Rajasingh et al. 2008; Houle et al. 2010; Omholt 2013; Pavličev et al., chapter 8) that relate the effects of variation through a biologically motivated and explicit network of processes that extend, with varying degrees of realism, from genotypes to phenotypes. It is important to recognize that even the most sophisticated of such models is, and is likely to remain, limited to a small portion of the total GP map. Although we cannot hope to build detailed GP maps of all phenotypes, the development and validation of a modest number of such maps, instantiated as cGP models, should be sufficient to reveal whether we can expect to discover generalizations about the relationship between GP maps and evolvability.

The second trend is the increasingly sophisticated use of genome-wide association studies (GWAS) that generate hypotheses about the potential causes of genetic variation in the phenotype. The many inferential challenges of naïve GWA studies are gradually being overcome by increasingly sophisticated statistical approaches, and by the existence of larger and larger data sets, driven by both more complete and cheaper genotyping, and by increased phenotyping capacity.

The development of GP map concepts reveals how important it is to estimate pleiotropic and epistatic effects in GWAS. Pleiotropy was largely unaddressed in the initial phases of mapping, and then addressed only indirectly (Pitchers et al. 2019). Similarly, the statistical challenges of detecting particular epistatic interactions are formidable, calling into question the validity of findings based on P-values. By transferring our attention to higher level aggregate properties, such as directional epistasis or modularity of genetic effects, we can expect more robust inferences about those aspects of pleiotropy and epistasis that shape evolvability.

Merging the basic information about organismal function with detailed GWAS data has the potential to transform our understanding of GP maps. When a cGP model is paired to data on the effects of both experimentally-induced and natural variation, we can anticipate a virtuous cycle, where researchers can predict the effects of genetic variation using systems models and then test those predictions using experimentally validated information. Where the predictions fail, the model can be improved.

18.4.2 Phenomics, Natural Selection, and Fitness

Phenomics is the laudable aspiration to comprehensively study the phenotype as a whole (Houle et al. 2010). Adaptation by natural selection depends on both the fitness consequences of an unknown number of traits subject to selection, and on the pleiotropic effects of the variants that cause variation in all those unknown traits. To be sure, there are striking examples where the genotype-phenotype-fitness relationships seem satisfyingly simple (e.g., Linnen et al. 2013), but these may be unusual, rather than typical cases. Only by broadening our attention to include a more comprehensive view of phenotypes can we investigate this possibility.

Current genetic studies of evolution are by and large carried out on a handful of phenotypes; thus they can only address a few phenotypic dimensions. Even studies that characterize organisms with highly multivariate data are almost always limited to one sort of phenotype, such as gene expression (Aguet et al. 2017), morphological shape (McGlothlin et al. 2018), or abundance of biomolecules (Chenoweth and Blows 2008).

Limited phenotyping hampers the study of pleiotropy. Although there have been some fairly large-scale attempts to assess pleiotropy (Wagner and Zhang 2011), these are hard to generalize due to their reliance on statistical testing to infer pleiotropic effects and on the study of gene knockouts that are not representative of natural variation (Paaby and Rockman 2013). As a result, contradictory views about pleiotropy remain viable. Is pleiotropy a by-product of evolution, incapable of responding to natural selection (Wagner et al. 2007), or a key target of selection that has been shaped to maximize evolvability or robustness (Wagner and Altenberg 1996)? Is pleiotropy a source of evolutionary constraint (Orr 2000; Hansen and Houle 2004), or is it so variable that organisms can respond to any selective pressure (Pavličev and Hansen 2011; Pavličev and Wagner 2012)? Is pleiotropy "universal," so that most mutations affect all traits to some extent (Paaby and Rockman 2013; Boyle et al. 2017), or modular, restricted to a few related traits (Wagner et al. 2007; Wagner and Zhang 2011)? The continued viability of these alternatives reveals profound ignorance about pleiotropy and its role in evolution.

The significance of properties that are clearly important to evolvability, such as pleiotropy, modularity (Pavličev et al., chapter 8), and robustness (A. Wagner, chapter 11), depend on the full range of phenotypic effects that a variant has. Equally important is that the evolutionary impact of properties such as modularity and robustness on evolvability depend on how selection affects multiple traits simultaneously (Houle and Rossoni 2022). The same modular structure that promotes the response to selection aligned with the modules will hamper responses to selection in other directions (Hansen 2003; Welch and Waxman 2003; Houle and Pélabon, chapter 6). Similarly, the potential advantages to evolvability of nearly neutral networks depends on the variety and accessibility of phenotypes at the edges of that network, and whether those particularly accessible phenotypes enhance the response to actual selection pressures (Mayer and Hansen 2017).

Technical advances in several areas are expanding our ability to phenotype individuals. Chief among these is the ever-expanding ability to measure gene expression, the causal foundation of much phenotypic variation (Aguet et al. 2017). Coupled with better knowledge of GP maps (section 18.4.1), this ability could allow more sophistiated predictions about phenotypic consequences, guiding further phenotyping efforts toward variants and traits with consequential effects on fitness. The range of high-throughput phenotyping platforms is increasing. Image processing approaches can now rapidly extract a variety of measurements of morphological features from images of any taxon (Martins et al. 2015; Porto and Voje 2020). Specialized high-throughput phenotyping has been implemented for model and economically important species, including crop plants (Yang et al. 2020) and *Drosophila* (Medici et al. 2015).

Some of the uncertainties that pleiotropy poses for inferences about evolvability could be resolved with estimates of fitness, rather than comprehensive phenotyping. For example, if fitness is included in the set of phenotypes in a GWAS, one could compare how much variation in fitness is explained by a specific variant, and how much is explained by the effects of the variant on the measured phenotypes. Similarly, with replicated genotypes, one could measure the proportion of fitness variation explained by the measured phenotypes relative to the variation in genotypic fitnesses. Close correspondence would suggest that the traits important to fitness have been measured. Unfortunately, measuring fitness in a manner relevant to evolution of natural populations is itself a challenging task.

Until this problem of unmeasured traits receives attention from experimentalists, theorizing about aspects of evolvability that depend on pleiotropy will remain speculative.

18.4.3 Development and Measurement of Evolvability Parameters

One of the important products of the evolvability research front is the expansion of our roster of measurable features that we can relate to evolvability. Thirty years ago, these included mutation (Kimura 1967), genetic variance (Falconer 1981), integration (Olson and Miller 1958), and modularity (Raff 1996; Wagner 1996; Wagner and Altenberg 1996). In the intervening years, the relevance of new concepts has been developed, including plasticity, niche construction, and regulatory evolution. We focus here on three for which the relationship to evolvability is quantifiable: robustness, directional epistasis, and conditional evolvability.

Robustness is the tendency for DNA mutations to have no phenotypic consequences (A. Wagner, chapter 11). Andreas Wagner and others (Wagner 2005; Masel and Trotter 2010) note evidence that effect sizes of particular mutations change, depending on the genotype in which they occur. Genotypes that are more robust then give rise to mutants with different spectra of descendant genotypes than less robust ones. Some models predict that selection will often push genotypes to regions of genotype space that are more robust. The consequences of this can either enhance or suppress evolvability, depending on the genotype-fitness relationships (Wagner 2008; Mayer and Hansen 2017). Clever experimental work has revealed the causes and evolutionary consequences of robustness in viruses and bacteria, in which spectra of mutational effects can be rapidly screened (A. Wagner, chapter 11). This work suggests that robustness of individual biomolecules to random events, such as misfolding, reliably predicts their ability to produce adaptive variation when their function is challenged in a novel way. The challenge is to generalize these results to robustness of more complex systems and more complex organisms.

Hansen and colleagues have focused attention on changes in the average properties in the GP map as a function of the position of a genotype in phenotype space. This is reflected in *directional epistasis,* where variants that change the phenotype in one direction have systematic epistatic effects that increase or decrease the average effect sizes (Carter et al. 2005; Hansen et al. 2006). There has been relatively little experimental work on directional epistasis, although methods to estimate it are available (Álvarez-Castro and Carlborg 2007; Le Rouzic 2014). This is beginning to change (Pélabon et al., chapter 13), and we expect a great deal more experimental work on the relationship between phenotypic means and genetic effects in the near future.

Conditional evolvability measures evolvability of one trait while holding other traits constant (Hansen 2003; Hansen et al. 2003; Hansen and Houle 2008; Hansen, chapter 5). The ratio of conditional to unconditional evolvability provides a dimensionless measure of integration and modularity in the context of a particular set of selection pressures.

Although conditional evolvability is readily calculated given a **G** matrix (Hansen and Houle 2008), even when we have an empirical estimate of the direction of selection, we often lack information on the traits that may be subject to stabilizing selection, and always lack a complete inventory of traits potentially correlated with the focal trait under selection. Useful empirical work on conditional evolvabilities awaits advances in phenotyping and the measurement of fitness landscapes.

The usefulness of these new measures of evolvability, as well as more familiar ones, will continue to increase as biologists consider the scale on which parameters are measured, i.e., interpret the meaning of measurements with explicit reference to their units. The case for doing so has been laid out on numerous occasions (e.g., Houle et al. 2011; Hansen, chapter 5), and yet many studies and reviews attempt to address quantitative questions about evolvablity with quantitatively uninterpretable summary measures, such as heritabilities and correlation matrices, or P-values as substitutes for effect sizes.

We hope that the relatively new measures of evolvability that we have mentioned here are not the last to be developed. For example, some believe that organisms vary in their ability to generate novel phenotypes, a feature not captured by the measures of evolvability we have in hand.

18.4.4 Comparative Evolvability

A final important research area is characterization of the variation in evolvability in a wider array of taxa. We currently have few well-estimated **M** matrices and a slightly larger variety of **G** matrices, supplemented by a relatively large number of **P** matrices. Similarly, the GP map properties that underly variability and variation have only been studied in a handful of model organisms. Broadening these studies to include more populations and in particular a wider taxonomic diversity would allow us to generalize what we know about evolvability.

Another productive direction to expand evolvability studies would be to integrate studies of GP maps and contemporary evolvability statistics with paleontological data. Paleontological data provide unique information on evolutionary rates and directions over very long periods. The key to making use of such data is to find opportunities to distinguish between natural selection and evolvability as a cause of variation in evolutionary rate (Jablonski 2017, chapter 17; Jackson 2020). Love et al. (2021) outline three different research programs that could potentially uncover an evolvability signal in paleontological data by drawing on neontological research in development, quantitative genetics, and comparative biology.

The combination of comparative and quantitative genetic data has already suggested that such signals exist. Voje et al. (chapter 14) outline the strong and consistent evidence that the variation within single populations predicts the rates of evolution in the clade in which it resides. This observation leads to the striking prediction that the variational properties of diverging populations must be conservative. Armbruster (chapter 15) makes the case that such conservative patterns of variability in vegetative and floral traits shape the evolution of flowering plants. This correspondence between evolvability and evolutionary rates is surprising, because we know that the population processes discussed by Sztepanacz et al. (chapter 12) and Pélabon et al. (chapter 13) can lead to large changes in within-population genetic variation over fairly short time scales. Furthermore, a correspondence between variation and the rate of evolution is not expected under any of the simple models that are currently available (Bolstad et al. 2014; Houle et al. 2017; Voje et al., chapter 14).

Performing such analyses in taxa with an informative fossil record would deepen our understanding of the strength and longevity of these patterns.

Hansen and Wagner (chapter 7) point out that a wide variety of non-exclusive hypotheses exist for what drives the evolution of evolvability. These broadly fall into adaptive explanations (in which evolvability is the direct target of selection) and non-adaptive explanations, in which evolvability emerges as a by-product of other evolutionary forces. Although there has been little empirical work aimed at distinguishing between adaptive and non-adaptive hypotheses, we see opportunities to do so by investigating how evolvability changes under natural or artificial selection or following experimental alterations of mean phenotype. For example, evolve and resequence experiments could be used to infer the identity of haplotypes that increase in frequency under particular selective regimes. Then the variability of variants on favored and non-favored backgrounds could be compared by engineering specific variants into both backgrounds. An adaptive hypothesis is that persistent directional selection would increase variability regardless of the direction of selection. A non-adaptive hypothesis predicts that variability would be a monotonic function of the trait mean, regardless of the form of selection.

18.5 Conclusion

We have made the case that evolvability research is both novel and unifying, albeit with some qualifications.

The events that triggered an evolvability research front and rendered it interdisciplinary were the independent arrival of similar concepts of evolvability in different fields. This convergence generated connections between different intellectual traditions, resulting in the research documented in this volume. As we have argued in section 18.4, evolvability research continues to generate groundbreaking research. There is reason to hope that the research set in motion by the individuation of the evolvability concept will precipitate answers to the many questions about evolvability that this book identifies.

We believe one of the reasons that evolvability research has been and will continue to be productive is that it has provided a theme that unites disparate fields. The strong version of this unifying influence lies in the realization that evolvability depends on a set of linked dispositions—the disposition to mutate, the disposition of mutations to produce phenotypic variation, and the disposition of the populations carrying that variation to evolve under a variety of stimuli. This unity serves to explain the disposition of geneticists, developmental and systems biologists, and paleontologists to exchange ideas about evolvability.

The ultimate goal of research into evolvability and variability is to develop a theory to explain and predict the linked dispositions that is on par with our well-developed theory of how natural selection acts on the manifest variation. We are aware of formidable challenges that stand in the way of a full realization of the promise of evolvability theory. Until we can better characterize phenotypes and their inheritance, the form of natural selection on them, and improve our ability to measure fitness, it will be difficult to apply concepts such as robustness, conditional evolvability, and directional epistasis in a grounded way. In addition, the study of longer-term differences in evolvability as they apply to rare but critical events in the history of life, such as the origin of novelty, remains speculative.

References

Aguet, F., K. G. Ardlie, B. B. Cummings, et al. 2017. Genetic effects on gene expression across human tissues. *Nature* 550: 204–213.

Alberch, P. 1991. From genes to phenotype: Dynamical systems and evolvability. *Genetica* 84: 5–11.

Álvarez-Castro, J. M., and O. Carlborg. 2007. A unified model for functional and statistical epistasis and its application in quantitative trait loci analysis. *Genetics* 176: 1151–1167.

Bolstad, G. H., T. F. Hansen, C. Pélabon, M. Falahati-Anbaran, R. Pérez-Barrales, and W. S. Armbruster. 2014. Genetic constraints predict evolutionary divergence in *Dalechampia* blossoms. *Philosophical Transactions of the Royal Society B* 369: 20130255.

Boyle, E. A., Y. I. Li, and J. K. Pritchard. 2017. An expanded view of complex traits: From polygenic to omnigenic. *Cell* 169: 1177–1186.

Brigandt, I. 2015. From developmental constraint to evolvability: How concepts figure in explanation and disciplinary identity. In *Conceptual Change in Biology: Scientific and Philosophical Perspectives on Evolution and Development,* edited by A. C. Love, 305–325. Dordrecht: Springer.

Brigandt, I., and A. C. Love. 2012. Conceptualizing evolutionary novelty: moving beyond definitional debates. *Journal of Experimental Zoology B* 318: 417–427.

Brown, R. L. 2014. What evolvability really is. *British Journal for the Philosophy of Science* 65: 549–572.

Carroll, S. B. 2008. Evo-devo and an expanding evolutionary synthesis: A genetic theory of morphological evolution. *Cell* 134: 25–36.

Carter, A. J. R., J. Hermisson, and T. F. Hansen. 2005. The role of epistatic gene interactions in the response to selection and the evolution of evolvability. *Theoretical Population Biology* 68:179–196.

Chenoweth, S. F., and M. W. Blows. 2008. Q_{ST} meets the **G** matrix: The dimensionality of adaptive divergence in multiple correlated quantitative traits. *Evolution* 62: 1437–1449.

Conrad, M. 1983. *Adaptability: The Significance of Variability from Molecule to Ecosystem.* New York: Springer Science & Business Media.

Dawkins, R. 1989. The evolution of evolvability. In *Artificial Life 6, SFI Studies in the Sciences of Complexity,* edited by C. Langton, 201–220. Redwood City, CA: Addison-Wesley.

Endler, J. A. 1986. *Natural Selection in the Wild.* Princeton, NJ: Princeton University Press.

Falconer, D. S. 1981. *Introduction to Quantitative Genetics.* London: Longman.

Fisher, R. A. 1930. *The Genetical Theory of Natural Selection.* Oxford: Clarendon Press.

Frank, S. A. 1995. George Price's contributions to evolutionary genetics. *Journal of Theoretical Biology* 175: 373–388.

Frank, S. A. 2012. Wright's adaptive landscape versus Fisher's Fundamental Theorem. In *The Adaptive Landscape in Evolutionary Biology,* edited by E. I. Svensson and R. Calsbeek, 41–57. Oxford: Oxford University Press.

Galison, P. 1997. *Image and Logic: A Material Culture of Microphysics.* Chicago: University of Chicago Press.

Gould, S. J. 1989. A developmental constraint in *Cerion* with comments on the definition and interpretation of constraint in evolution. *Evolution* 43: 516–539.

Haldane, J. 1954. The measurement of natural selection. *Proceedings of the IXth International Congress of Genetics* (*Caryologica* supplement) 1: 480–487.

Hansen, T. F. 2003. Is modularity necessary for evolvability? Remarks on the relationship between pleiotropy and evolvability. *Biosystems* 69: 83–94.

Hansen, T. F. 2006. The evolution of genetic architecture. *AREES* 37: 123–157.

Hansen, T. F., and D. Houle. 2004. Evolvability, stabilizing selection, and the problem of stasis. In *The Evolutionary Biology of Complex Phenotypes,* edited by M. Pigliucci, and K. Preston, 130–150. Oxford: Oxford University Press.

Hansen, T. F., and D. Houle. 2008. Measuring and comparing evolvability and constraint in multivariate characters. *Journal of Evolutionary Biology* 21: 1201–1219.

Hansen, T. F., and C. Pélabon. 2021. Evolvability: A quantitative-genetics perspective. *AREES* 52: 153–175.

Hansen, T. F., W. S. Armbruster, M. L. Carlson, and C. Pélabon. 2003. Evolvability and genetic constraint in *Dalechampia* blossoms: Genetic correlations and conditional evolvability. *Journal of Experimental Zoology B* 296: 23–39.

Hansen, T. F., J. M. Álvarez-Castro, A. J. R. Carter, J. Hermisson, and G. P. Wagner. 2006. Evolution of genetic architecture under directional selection. *Evolution* 60: 1523–1536.

Hansen, T. F., C. Pélabon, and D. Houle. 2011. Heritability is not evolvability. *Evolutionary Biology* 38: 258–277.

Hendrikse, J. L., T. E. Parsons, and B. Hallgrímsson. 2007. Evolvability as the proper focus of evolutionary developmental biology. *Evolution & Development* 9: 393–401.

Houle, D. 1992. Comparing evolvability and variability of quantitative traits. *Genetics* 130: 195–204.

Houle, D. 1998. How should we explain variation in the genetic variance of traits? *Genetica* 102–103: 241–253.

Houle, D., and J. Fierst. 2013. Properties of spontaneous mutational variance and covariance for wing size and shape in *Drosophila melanogaster. Evolution* 67: 1116–1130.

Houle, D., B. Morikawa, and M. Lynch. 1996. Comparing mutational variabilities. *Genetics* 143: 1467–1483.

Houle, D., D. R. Govindaraju, and S. W. Omholt. 2010. Phenomics: The next challenge. *Nature Reviews Genetics* 11: 855–866.

Houle, D., C. Pélabon, G. P. Wagner, and T. F. Hansen. 2011. Measurement and meaning in biology. *Quarterly Review of Biology* 86: 3–34.

Houle, D., G. H. Bolstad, K. Van Der Linde, and T. F. Hansen. 2017. Mutation predicts 40 million years of fly wing evolution. *Nature* 548: 447–450.

Houle, D., and D. M. Rossoni. 2022. Complexity, evolvability, and the process of adaptation. *AREES* 53: 137–159.

Jablonski, D. 2017. Approaches to macroevolution: 1. General concepts and origin of variation. *Evolutionary Biology* 44: 427–450.

Jackson, I. S. C. 2020. Developmental bias in the fossil record. *Evolution & Development* 22: 88–102.

Kacser, H., and J. A. Burns. 1981. The molecular basis of dominance. *Genetics* 97: 639–666.

Kimura, M. 1967. On the evolutionary adjustment of spontaneous mutation rates. *Genetical Research* 9:23–34.

Laland, K. N., and K. Sterelny. 2006. Perspective: Seven reasons (not) to neglect niche construction. *Evolution* 60: 1751–1762.

Lande, R. 1979. Quantitative genetic analysis of multivariate evolution applied to brain:body size allometry. *Evolution* 33: 402–416.

Lande, R., and S. J. Arnold. 1983. The measurement of selection on correlated characters. *Evolution* 37: 1210–1226.

Le Rouzic, A. 2014. Estimating directional epistasis. *Frontiers in Genetics* 5: 198.

Lewontin, R. C. 1970. The units of selection. *ARES* 1: 1–18.

Lewontin, R. C. 1974. *The Genetic Basis of Evolutionary Change*. New York: Columbia University Press.

Lewontin, R. C. 1978. Adaptation. *Scientific American* 239: 212–231.

Linnen, C. R., Y. P. Poh, B. K. Peterson, et al. 2013. Adaptive evolution of multiple traits through multiple mutations at a single gene. *Science* 339: 1312–1316.

Love, A. C. 2003. Evolvability, dispositions, and intrinsicality. *Philosophy of Science* 70: 1015–1027.

Love, A. C. 2015. Conceptual change and evolutionary developmental biology. In *Conceptual Change in Biology: Scientific and Philosophical Perspectives on Evolution and Development,* edited by A. C. Love, 1–54. Dordrecht: Springer.

Love, A. C., M. Grabowski, D. Houle, et al. 2022. Evolvability in the fossil record. *Paleobiology* 48: 186–209.

Martins, A. F., M. Bessant, L. Manukyan, and M. C. Milinkovitch. 2015. R²OBBIE-3D, a fast robotic high-resolution system for quantitative phenotyping of surface geometry and colour-texture. *PLOS One* 10: e0126740.

Masel, J., and M. V. Trotter. 2010. Robustness and evolvability. *Trends in Genetics* 26: 406–414.

Mayer, C., and T. F. Hansen. 2017. Evolvability and robustness: A paradox restored. *Journal of Theoretical Biology* 430: 78–85.

Mayo, O., and R. Bürger. 1997. The evolution of dominance: A theory whose time has passed? *Biological Reviews* 72: 97–110.

Mayr, E. 1982. *The Growth of Biological Thought: Diversity, Evolution, and Inheritance*. Cambridge, MA: Harvard University Press.

McGlothlin, J. W., M. E. Kobiela, H. V. Wright, et al. 2018. Adaptive radiation along a deeply conserved genetic line of least resistance in Anolis lizards. *Evolution Letters* 2: 310–322.

Medici, V., S. C. Vonesch, S. N. Fry, and E. Hafen. 2015. The FlyCatwalk: A high-throughput feature-based sorting system for artificial selection in *Drosophila. G3: Genes | Genomes | Genetics* 5: 317–327.

Minelli, A. 2017. Evolvability and its evolvability. In *Challenges to Evolutionary Theory: Development, Inheritance and Adaptation,* edited by P. Huneman and D. M. Walsh, 211–238. New York: Oxford University Press.

Nuño de la Rosa, L. 2017. Computing the extended synthesis: Mapping the dynamics and conceptual structure of the evolvability research front. *Journal of Experimental Zoology B* 328: 395–411.

Okasha, S. 2021. The strategy of endogenization in evolutionary biology. *Synthese* 198: 3413–3435.

Okasha, S., and J. Otsuka. 2020. The Price equation and the causal analysis of evolutionary change. *Philosophical Transactions of the Royal Society B* 375: 20190365.

Olson, E. C., and R. L. Miller. 1958. *Morphological Integration.* Chicago: University of Chicago Press.

Omholt, S. W. 2013. From sequence to consequence and back. *Progress in Biophysics and Molecular Biology* 111: 75–82.

Orr, H. A. 2000. Adaptation and the cost of complexity. *Evolution* 54: 13–20.

Paaby, A. B., and M. V. Rockman. 2013. The many faces of pleiotropy. *Trends in Genetics* 29: 66–73.

Paenke, I., B. Sendhoff, and T. J. Kawecki. 2007. Influence of plasticity and learning on evolution under directional selection. *American Naturalist* 170: E47–E58.

Pavličev, M., and T. F. Hansen. 2011. Genotype-phenotype maps maximizing evolvability: Modularity revisited. *Evolutionary Biology* 38: 371–389.

Pavličev, M., and G. P. Wagner. 2012. A model of developmental evolution: Selection, pleiotropy and compensation. *TREE* 27: 316–322.

Pigliucci, M. 2007. Do we need an extended evolutionary synthesis? *Evolution* 61: 2743–2749.

Pigliucci, M. 2008. Opinion—Is evolvability evolvable? *Nature Reviews Genetics* 9: 75–82.

Pitchers, W., J. Nye, E. J. Márquez, A. Kowalski, I. Dworkin, and D. Houle. 2019. A multivariate genome-wide association study of wing shape in *Drosophila melanogaster. Genetics* 211: 1429–1447.

Porto, A., and K. L. Voje. 2020. ML-morph: A fast, accurate and general approach for automated detection and landmarking of biological structures in images. *Methods in Ecology and Evolution* 11: 500–512.

Price, D. J. D. S. 1965. Networks of scientific papers. *Science* 149: 510–515.

Price, G. R. 1970. Selection and covariance. *Nature* 227: 520–521.

Price, G. R. 1972. Fisher's "fundamental theorem" made clear. *Annals of Human Genetics* 36: 129–140.

Prior, E. W., R. Pargetter, and F. Jackson. 1982. Three theses about dispositions. *American Philosophical Quarterly* 19: 251–257.

Raff, R. A. 1996. *The Shape of Life: Genes, Development, and the Evolution of Animal Form.* Chicago: University of Chicago Press.

Rajasingh, H., A. B. Gjuvsland, D. I. Våge, and S. W. Omholt. 2008. When parameters in dynamic models become phenotypes: A case study on flesh pigmentation in the Chinook Salmon (*Oncorhynchus tshawytscha*). *Genetics* 179: 1113–1118.

Ridley, M. 1998. *Evolution.* Oxford: Oxford University Press.

Ridley, M. 2002. Natural selection: An overview. In *Encyclopedia of Evolution,* edited by M. Pagel, 797–804. Oxford: Oxford University Press.

Riederer, J. M., S. Tiso, T. J. van Eldijk, and F. J. Weissing. 2022. Capturing the facets of evolvability in a mechanistic framework. *TREE* 37: 430–439.

Riedl, R. 1977. A systems analytical approach to macro-evolutionary phenomena. *Quarterly Review of Biology* 52: 351–370.

Riedl, R. 1978. *Order in Living Organisms.* Brisbane: Wiley.

Schluter, D. 1996. Adaptive radiation along genetic lines of least resistance. *Evolution* 50: 1766–1774.

Scott-Phillips, T. C., K. N. Laland, D. M. Shuker, T. E. Dickins, and S. A. West. 2014. The niche construction perspective: A critical appraisal. *Evolution* 68: 1231–1243.

Simpson, G. G. 1953. *The Major Features of Evolution.* New York: Columbia University Press.

Soyer, O. S., and M. A. O'Malley. 2013. Evolutionary systems biology: What it is and why it matters. *BioEssays* 35: 696–705.

Sterelny, K. 2007. What is evolvability? In *Philosophy of Biology,* edited by M. Matthen and C. Stephens, 163–178. Amsterdam: Elsevier.

Stern, D. L. 2000. Perspective: Evolutionary developmental biology and the problem of variation. *Evolution* 54: 1079–1091.

Van Valen, L. 1965. Selection in Natural Populations. III. Measurement and Estimation. *Evolution* 19: 514–528.

Vermeij, G. J. 1987. *Evolution and Escalation: An Ecological History of Life.* Princeton, NJ: Princeton University Press.

von Dassow, G., and E. Munro. 1999. Modularity in animal development and evolution: Elements of a conceptual framework for EvoDevo. *Journal of Experimental Zoology B* 285: 307–325.

Waddington, C. H. 1957. *The Strategy of the Genes*. New York: Macmillan.

Wagner, A. 2005. *Robustness and Evolvability in Living Systems*. Princeton, NJ: Princeton University Press.

Wagner, A. 2008. Robustness and evolvability: A paradox resolved. *Proceedings of the Royal Society B* 275: 91–100.

Wagner, G. P. 1984. Coevolution of functionally constrained characters: Prerequisites for adaptive versatility. *Biosystems* 17: 51–55.

Wagner, G. P. 1996. Homologues, natural kinds and the evolution of modularity. *American Zoologist* 36: 36–43.

Wagner, G. P., and L. Altenberg. 1996. Perspective: Complex adaptations and the evolution of evolvability. *Evolution* 50: 967–976.

Wagner, G. P., M. Pavličev, and J. M. Cheverud. 2007. The road to modularity. *Nature Reviews Genetics* 8: 921–931.

Wagner, G. P., and J. Z. Zhang. 2011. The pleiotropic structure of the genotype-phenotype map: The evolvability of complex organisms. *Nature Reviews Genetics* 12: 204–213.

Welch, J. J., and D. Waxman. 2003. Modularity and the cost of complexity. *Evolution* 57: 1723–1734.

West-Eberhard, M.-J. 2003. *Developmental Plasticity and Evolution*. Oxford: Oxford University Press.

Winther, R. G. 2015. Evo-devo as a trading zone. In *Conceptual Change in Biology*, edited by A. Love, 459–482. Dordrecht: Springer.

Wright, S. 1932. The roles of mutation, inbreeding, crossbreeding, and selection in evolution. *Proceedings of the 6th International Congress of Genetics* 1: 356–366.

Yang, W., H. Feng, X. Xiang, et al. 2020. Crop phenomics and high-throughput phenotyping: Past decades, current challenges, and future perspectives. *Molecular Plant* 13: 187–214.

Contributors

J. David Aponte, Department of Cell Biology and Anatomy, University of Calgary, AB, Canada

W. Scott Armbruster, School of Biological Sciences, University of Portsmouth, UK

Geir H. Bolstad, Norwegian Institute for Nature Research, Trondheim, Norway

Salomé Bourg, Department of Biology, Centre for Biodiversity Dynamics, Norwegian University of Science and Technology, Trondheim, Norway

Ingo Brigandt, Department of Philosophy, University of Alberta, Edmonton, AB, Canada

Anne L. Calof, Department of Developmental and Cell Biology, University of California Irvine, CA, USA

James M. Cheverud, Department of Biology, Loyola University, Chicago, IL, USA

Josselin Clo, Department of Botany, Charles University, Prague, Czechia

Frietson Galis, Naturalis Biodiversity Center, Leiden, The Netherlands

Mark Grabowski, Research Centre in Evolutionary Anthropology and Palaeoecology, Liverpool John Moores University, UK

Rebecca Green, School of Dental Medicine, University of Pittsburgh, PA, USA

Benedikt Hallgrímsson, Department of Cell Biology and Anatomy, University of Calgary, AB, Canada

Thomas F. Hansen, Department of Biosciences, CEES and Evogene, University of Oslo, Norway

Agnes Holstad, Department of Biology, Centre for Biodiversity Dynamics, Norwegian University of Science and Technology, Trondheim, Norway

David Houle, Department of Biological Science, Florida State University, Tallahassee, FL, USA

David Jablonski, Department of Geophysical Sciences, University of Chicago, IL, USA

Arthur D. Lander, Department of Developmental and Cell Biology, University of California Irvine, CA, USA

Arnaud Le Rouzic, Université Paris-Saclay, CNRS, IRD, Évolution, Génomes, Comportement, Écologie, Gif-sur-Yvette, France

Alan C. Love, Department of Philosophy, College of Liberal Arts, Twin Cities, MN, USA

Ralph S. Marcucio, Department of Orthopaedical Surgery, School of Medicine, University of California, San Francisco, CA, USA

Michael B. Morrissey, School of Biology, University of St. Andrews, UK

Laura Nuño de la Rosa, Complutense University of Madrid, Spain

Øystein H. Opedal, Department of Biology, Lund University, Sweden

Mihaela Pavličev, Department of Evolutionary Biology, University of Vienna, Austria

Christophe Pélabon, Department of Biology, Centre for Biodiversity Dynamics, Norwegian University of Science and Technology, Trondheim, Norway

Arthur Porto, Department of Biological Sciences, Louisiana State University, Baton Rouge, LA, USA

Jane M. Reid, School of Biological Sciences, University of Aberdeen, UK

Heather Richbourg, Department of Orthopaedical Surgery, School of Medicine, University of California, San Francisco, CA, USA

Jacqueline L. Sztepanacz, Department of Ecology and Evolutionary Biology, University of Toronto, ON, Canada

Masahito Tsuboi, Department of Biology, Lund University, Sweden

Marta Vidal-García, Department of Cell Biology and Anatomy, University of Calgary, AB, Canada

Cristina Villegas, Konrad Lorenz Institute for Evolution and Cognition Research, Klosterneuburg, Austria

Kjetil L. Voje, Natural History Museum, University of Oslo, Norway

Andreas Wagner, Department of Evolutionary Biology and Environmental Studies, University of Zurich, Switzerland

Günter P. Wagner, Systems Biology Institute and Department of Ecology and Evolutionary Biology, Yale University, CT, USA

Nathan M. Young, Department of Orthopaedical Surgery, School of Medicine, University of California San Francisco, CA, USA

Index